Bill Braunworth

TEXTBOOK OF CYTOGENETICS

TEXTBOOK OF CYTOGENETICS

WALTER V. BROWN, Ph.D.

Professor of Botany, University of Texas,
Austin, Texas

with 301 illustrations

THE C. V. MOSBY COMPANY SAINT LOUIS 1972

Copyright © 1972 by The C. V. Mosby Company

All rights reserved. No part of this book may be reproduced in any manner without written permission of the publisher.

Printed in the United States of America

International Standard Book Number 0-8016-0834-1
Library of Congress Catalog Card Number 79-185523

Distributed in Great Britain by Henry Kimpton, London

PREFACE

Cytogenetics is specifically the hybrid study of those aspects of genetics that can be studied by use of microscopes. Put another way, cytogenetics and certain methods of molecular biology study the physical basis of the heredity of viruses and of organisms, from bacteria to bandicoots. At the ultimate level cytogenetics can visualize the gene and even the gene in action and can see some of the results of the molecular biologist's studies of nucleic acid hybridization. The interrelationships of cytology and genetics with molecular biology are obvious since it is possible to "see" a nucleic acid polymer almost as well from density gradient centrifugation techniques as with the electron microscope.

In a broad sense cytotaxonomy and chromosomal evolution are part of the field of cytogenetics, for the studies of genetic systems and chromosomal changes are the bases of much of the modern concepts in classification, phylogeny, and evolution. Thus cytogenetics is mostly the study of the chromosomes of viruses, Prokaryota and Eukaryota—what they are, how many there are, what they do and have done, what can be done with them, and how what they can do and have done relates to heredity and evolution. There is also a small amount of cytogenetics of extranuclear genetic material, of the genophores of mitochondria and plastids.

The process of meiosis, both typical and atypical, is a chief concern of the cytogeneticist. But to understand the essentials of reduction and recombination, the fine structure of chromosomes and their behavior during simple division (mitotic or otherwise) is essential knowledge, although recent findings indicate that mitosis is not as simple as is generally described. But meiosis is more than mitosis and a model of its evolution is presented. Pairing as distinct from synapsis is proposed and the unique although variable cytological structure, called the synaptinemal complex, must be somehow involved in genetic recombination. The classical scheme of meiosis is modified by cytogenetic evidence that pairing of homologous chromosomes is distinct from synapsis, which suggests that somatic crossing-over may be more common than generally believed. The contributions of salivary gland chromosome studies to cytogenetics, evolution, and development are discussed and are supplemented by evidence of gene amplification, as earlier reported in oocytes.

Thus the cytogeneticist is usually more of a cytologist than a geneticist and, therefore, this book leans heavily toward cytology. Nevertheless, an effort has been made to include only those cytological matters that are fundamental to or clearly part of cytogenetics. And at all times the purpose and function of a textbook, to present clearly and concisely the fundamentals of the subject as well as inadequacies of old and new knowledge, and objective presentation of alternative models, hypotheses, concepts, and evidence, have directed and controlled the content. Some previously unpublished models and ideas are presented, such as the concept of diplopolyploidy, the model of division in the higher fungi, and the mechanism of crossing-over with the synaptinemal complex.

Two of the chapters in this book have been reviewed by experts—Chapter 3 (Polytene Chromosomes) by my colleague at the University of Texas at Austin, Dr. C. Pavan; Chapter 19 (Human Cytogenetics) by Dr. Matti S. Al-Aish, of the National Institutes of Health, Washington, D. C. Although their advice, criticisms, and illustrations are gratefully acknowledged, any errors of omission, commission, or evaluation are mine, not theirs. I am also greatly in debt to the many generous contributors of illustrative material, to all scientists who have reported the facts and ideas that have made this book possible, and especially to my wife, Helen, who not only endured the preparation of the manuscript but actually took an active part in it by learning to type and then typing most of it.

Walter V. Brown

CONTENTS

1 Historical survey, 1
2 The nondividing nucleus, 7
3 Polytene chromosomes, 23
4 Mitosis, 38
5 Somatic pairing and recombination, 57
6 Meiosis, 72
7 Sex determination, 103
8 Mitotic metaphase chromosomes, 113
9 Supernumerary chromosomes, 128
10 Chromosome numbers, 135
11 Haploidy and aneuploidy, 148
12 Polyploidy, 157
13 Translocation, 174
14 Inversions, 196
15 Karyotypes, 207
16 Cytogenetics of hybrids, 215
17 Permanent heterozygosity, 233
18 Apomixis, 241
19 Human cytogenetics, 252
20 Cytogenetics of prokaryotic types, 276

Literature cited, 295

CHAPTER 1

Historical survey

> The most fundamental contribution of cell-research to the theory of heredity is the law of genetic continuity by cell division.
>
> E. B. Wilson, 1909
>
> The chief interest of cytology at the present time probably lies in the relation which it bears to the subject of heredity.
>
> L. W. Sharp, 1921

Cytogenetics has been described as a hybrid science having cytology as one parent and genetics as the other. That being so, cytogenetics was born rather abruptly in 1902. In order to understand that vigorous child and how it was to grow, the lives of its parents up to the birth date are meaningful. Actually, there was little in common between cytology and genetics until a few years before 1900. Of the two parents cytology was by far the more vigorous during the nineteenth century, when nearly all the minimum biological facts necessary for comprehension of the chromosomes, mitosis, meiosis, and reproduction of organisms were acquired. Genetics, as distinct from empirical breeding programs, made very little progress as a discrete science. Even Mendel's work stood outside the field of heredity until 1900.

Previous to the nineteenth century a few preliminary genetic ideas had been derived from observation or experimentation and had been expressed about inheritance. Generally it was believed, from most breeding work, that a hybrid is intermediate between the parents and subsequent crosses or backcrosses continued to produce mostly intermediates. Nevertheless, a few observations might be considered genetic. Tabernaemontanus in 1588 recorded that grains on a cob of *Zea mays* may be of different color. In 1592 Gerard observed that tulips (hybrids) do not breed true from seed, and in 1716 Cotton Mather determined that plants can be hybridized by cross-pollination.

True genetic experiments with unit characters began about 1800 when, in 1799, Knight crossed gray peas to white peas and recovered all gray hybrids. The backcross to the white parental strain produced a variety of colors, including white.

About 20 years later both Seton and Goss reported, in 1822, that green crossed to white peas gave all green hybrids, but hybrid of those F_1 hybrids gave both green and white progeny but not peas of intermediate color. Goss also produced progeny from both the homozygous and heterozygous dominant phenotypes. Knight also, in 1823, reported dominance in the F_1 and both dominant and recessive phenotypes when the heterozygous F_1 were backcrossed to the homozygous recessive parent. Of course, the terminology used here was not employed 150 years ago. In 1826 Sageret reported that, following a cross between a muskmelon and a cantaloupe, there was an independent assortment of unit characters. About the middle of the century Dzieron, in 1845, confirmed that in honey bees the drones (males), in contrast to females (queens and workers), are produced from unfertilized eggs. Later, in 1856, he recorded that hybrid queens produced equal numbers of parental types (German and Italian) of drones. The same year that Mendel published his first and most significant paper, Verlot (1865) reported that certain individuals of the F_1 breed true whereas others do not.

Even with this background of contributions, which were probably known to him, Mendel's genius is obvious. He made many preliminary tests by which he reduced about 30 characters to 7 and, remarkably, to 7 non-linked characters, one on each of the 7 chromosomes of the pea. He must have incidentally accrued considerable data on linkage. His analysis was statistically much more definite than his predecessors', and he combined all of the previously isolated data into a unified theory. It was a theory, simple enough to us now, but until 1900 (and even after for some biologists) it was incomprehensible. Apparently there continued to be hybridization experiments between 1865 and 1900, and at least 11 references to Mendel are known (Olby and Gantrey, 1968); but they contributed little to progress of the discipline except for concepts of unit hereditary factors such as Darwin's (1868) "gemmules," De Vries' (1889) "pangenes," and Weismann's "biophores" and the concept of the continuity of the germ plasm in animals (Nussbaum, 1880; Weismann, 1885).

In contrast to genetics cytology enjoyed great progress during the nineteenth century, ramifying into embryology, cell physiology, and general biological concepts such as epigenesis. Pre-nineteenth century cytology started in about 1665 when Hooke, Grew, Swammerdam, Leeuwenhoek, and Malpighi started observing organic matter under magnifying lens systems. Only random observations were added during the eighteenth century. Almost with the opening of the nineteenth century, however, numerous cytologists began studies of plants and animals more intensively, especially in and around Germany.

By 1833 the concept of the cell had nearly completely changed from the cell wall of Hooke to protoplasm, the living stuff, the physical basis of life (von Mohl, 1833, 1846; Cohn, 1850). Brown in 1831 first generalized that a nucleus is a characteristic structure of cells. The cell theory, that large organisms are constructed of cells and that cells arise only from pre-existing cells by division, developed during 30 years but was eventually implied or stated adequately by Purkinje (1837); Schleiden, a botanist (1838), and Schwann, a zoologist (1839); and finally Virchow (1855). Gametes were proved to be cells (Schwann, 1839; Schweigger-Seidel, 1865; LaVallette St. George, 1865); and fertilization is the union of gametes (Thuret, 1857; Newport, 1854; Pringsheim, 1856; Strasburger, 1877) derived from two parents (Hertwig, 1875) rather than the ancient concept of preformation; and the central feature of fertilization is nuclear fusion (Hertwig, 1875; Strasburger, 1877, 1884). Thus the concept of epigenesis—that each new organism is a new creation that forms by development from a zygote—was formally established. Cells increase in number by division rather than otherwise (de novo or from a "cytoblastoma"); and nuclei divide (Nägeli, 1844; Kolliker, 1845; Remak, 1853), the division including chromosomes (Hofmeister, 1848; Schneider, 1873; Strasburger, 1875) and spindles (Kovalevsky, 1871; Fol, 1873; Bütschli, 1875). These chromosomes separate longitudinally into two daughter chromosomes (Flemming, 1880, 1882; van Beneden, 1883) that go to opposite poles of the spindle (van Beneden, 1884; Heuser, 1884). Each species has a definite chromosome number (haploid or diploid) Rabl, 1885; Th. Boveri, 1885), each chromosome of the set is qualitatively different from the others (Boveri, 1903), and so cell division is preceded by nuclear division (Kolliker, 1845; Remak, 1853); and the permanence from generation to generation of chromosomes of recognizable form was assumed (van Beneden, 1883; Rabl, 1895; Boveri, 1887) and proved (Sutton, 1902; Richards, 1917).

Haeckel in 1866 proposed that the nucleus is the principal agent of inheritance, the cytoplasm the principal agent of adaptation. Cytologists between 1883 and 1885 (Roux, Hertwig, Kolliker, Weismann, Waldeyer, and Strasburger) accepted Haeckel's concept but implicated the chromosomes as hereditary bodies that segregate independently at anaphase.

If fertilization occurs each generation with the consequent doubling of the chromosome number, there must be a compensating reduction. Meiosis was established as occurring during gametogenesis in animals and sporogenesis in plants (Strasburger, 1888; Overton, 1893; Henking, 1891; Rückert, 1891; Farmer, 1894). Henking and Rückert observed that each chromosome during the late first meiotic division is a bivalent, each half derived from a different parent, and exchange may perhaps occur between these maternal and paternal synapsed chromosomes (Rückert). Winiwarter (1900), McClung (1900), Montgomery (1901), Sutton (1902), Boveri (1904), and others finally demonstrated that homologous chromosomes are paired side-by-side to form the meiotic diplotene and metaphase bivalents. But ever since and even today cytologists, cytogeneticists, and geneticists have continued to work on the details of early meiotic prophase, the structural changes that actually occur previous to diplotene during what are now called leptotene, zygotene, and pachytene.

Thus, cytology had provided enough of the minimum necessary knowledge of the physical basis of heredity at almost the same time (1900) that Correns, De Vries, and Tscharmak rediscovered the laws of Mendel, dormant since 1865, of independent assortment of unit characters, of dominance and recessiveness, of homozygosity and heterozygosity, of purity of the genes, and of segregation and recombination. Cytogenetics was "born" when in 1901 Montgomery, in 1902 Sutton, Wilson, Cannon, Boveri, and Greyer, and in 1903 Sutton again demonstrated that the hereditary laws of Mendel could probably be explained completely by chromosomal activities at meiosis and fertilization. This cytological basis for genetic theory is often called the Sutton-Boveri theory of chromosomal inheritance.

Late in the nineteenth century it was generally believed that the sex of an individual was determined by one or more environmental factors. Many researches were carried out to prove that light, temperature, moisture, nutrition, and other factors were capable of determining the sexes of the progeny. Cuenot (1899) finally proposed that sex is determined by something in the nucleus, and Strasburger supported that concept in 1900 for plants. The chromosomal mechanism of sex determination was established during the next few years by McClung (1902) and Wilson (1905) in XO male grasshoppers, the genetic XY condition in "male" plants of *Lychnis dioica* by Shull (1910), the genetic XX male in a moth by Doncaster and Raynor (1906), and the haploid male in Hymenoptera by Newell in 1915 and 1925.

Early in the twentieth century the cytological basis of linkage was proposed by Boveri, and Bateson and Punnett in 1906 reported a case of linkage of flower color and pollen grain shape in sweet peas. Morgan in 1910 demonstrated a case of eye color and sex linkage in *Drosophila melanogaster*. Crossing-over, predicated by Rückert in 1891, was again proposed by Boveri in 1904 to explain the small amount of genetic recombination reported by geneticists and was demonstrated genetically by Morgan in 1911. By 1915 about 100 genes were known in *D. melanogaster* and were known to be in four linear linkage groups on the four known chromosomes (Morgan, Sturtevant, Muller, and Bridges, 1915).

The independent assortment of linkage groups was proved for chromosomes by Carothers (1913, 1917, 1921) by showing independent assortment of visibly unlike homologues in relation to one another and to the single X chromosome in males of a number of grasshopper species.

In 1912 Wilson "proved" that homologous chromosomes are separate until zygotene of early meiosis when they come together in intimate side-by-side association. Thus, he (and others) established the dogma of pairing only at zygotene in contrast to evidence by other equally competent cytologists that the pairing of homologues may occur long before meiosis. Earlier Janssens (1909) had proposed the chiasmatype theory, that genetic crossing-over is probably represented by visible cytological chiasmata between homologues during meiotic prophase and that there is a breakage and reunion of chromosomal strands of the two homologues. Bridges (1916) proposed on genetic evidence that the breakage and reunion occurs at the four-strand stage of chromosome bivalents and involves the breakage and reunion of only two strands at any one locus.

The nonmendelian (maternal or cytoplasmic) inheritance of chlorophyll was known as early as 1909 (Correns).

About 1920 new hypothetical "chromosomes" were added by geneticists to the cytological concepts of metaphase and interphase chromosomes. One of the genetic "chromosome" models was the linkage map; such a chromosome was a linear series of genes. Another model was the meiotic bivalent chromosome that must consist of four strands of genes when crossing-over occurs (Bridges, 1916). Also, about 1930, the cytologist Darlington proposed his precosity theory of meiosis. He claimed that at early prophase of meiosis each chromosome, instead of being double as at mitotic prophase, is single, and so there is a tendency for homologues to pair at zygotene, doubleness being the natural, the "satisfied" condition. Longitudinal splitting then followed to produce the four-stranded condition required by Bridges.

During the first 20 years of the twentieth century two sorts of mutation were characterized. The point mutation is a strictly genetic mutation, a change in a single gene. The second sort of mutation was found to be chromosomal and is a cytogenetic mutation, one that can often be studied under the microscope. At first cytogenetic mutations were detected in the plant genus *Oenothera* originally made famous by De Vries. It was shown that the *gigas* forms were usually tetraploid (Tupper and Bartlett, 1916), and tetraploids, as in tetraploid *Primula sinensis* (Gregory, 1914), have twice the number of genetic allelomorphs as do diploids. Other forms of mutant *Oenothera* "species" had only one extra chromosome (Gates, 1909). Furthermore, polyploid series of species within genera were found, such as the wheats 7, 14, 21 (Sakamura, 1919) and *Chrysanthemum* 9, 18, 27, 36, 45. In other genera intraspecific aneuploidy was noted, as in *Crepis* with 3, 4, 5, 8, 9, and 21 pairs of chromosomes (Rosenberg, 1918), and variation in chromosome sizes was noted.

Polyploidy was characterized as the instantaneous evolution of species, in contrast to the very slow evolution by gene mutations.

Comparisons of chromosome numbers and forms among related species and genera permitted estimates of relationships, as in the Drosophilidae (Metz, 1916) and in grasshoppers (Robertson, 1916).

Geneticists were detecting aberrations in linkage groups and ratios, which were given a physical basis by cytogenetic techniques, such as the complex heterozygotes of *Oenothera* and trisomic "mutants" of *Datura* (Blakeslee and Belling, 1924).

Cytogenetics had a burst of activity about 1930. Muller (1927) and Stadler (1928) demonstrated that X rays can produce mutations that are similar to natural mutations but are far more common. Actually there had been at least 10 papers on X-ray effects (such as Mavor, 1922). McClintock and a number of graduate students at Cornell University established the cytogenetics of pachytene (meiotic) chromosomes of *Zea mays* by the acetocarmine squash technique. They were able to demonstrate inversions and deletions, to correlate genetic crossing-over with cytological chiasmata, to relate linkage groups to specific chromosomes, etc.

Similar types of analysis became easy in *Drosophila* and other flies when Painter in 1934 demonstrated that such chromosomal aberrations could be almost diagrammatically observed in the "synapsed" giant salivary chromosomes far more easily and with hundreds of times more resolution than in normal metaphase chromosomes (Dobzhansky, 1929; Painter and Muller, 1929). Subsequently thousands of cytogenetic studies have been made of plant pachytene and fly salivary gland chromosomes by the squash technique.

Cytotaxonomy, often considered a subdivision of cytogenetics, originated in the late nineteenth century when it was discovered that species often differed in the number and/or form of chromosomes of the set at meiotic or mitotic metaphase. Slowly pertinent data accumulated, especially for plants, until it was possible to compare species, genera, and higher taxa. Avdulow's 1929 study of the grass family is an outstanding early example in which chromosome numbers, sizes, nucleoli, starch grains, etc. were compared and a new classification of tribes resulted that has remained with only minor changes ever since.

Genome analysis, as a method of determining the diploid ancestors of polyploid species of plants, was first employed to determine the progenitors of the diploid, tetraploid, and hexaploid cultivated wheats (Sax, 1922; Kihara, 1930; and others). This method consisted of the analysis of meiotic pairing in the hybrids on the assumption that if both parents contributed at least one genome (a set of homologous chromosomes) in common, pairing should occur among them. The method has subsequently contributed to knowledge of evolution of many cultivated and wild polyploid species such as cotton, tobacco, strawberries.

New techniques and discoveries for cytological or genetic study were added from time to time from the 1920's to the 1960's, such as: a qualitative and quantitative measurement of DNA content of nucleic acid (Feulgen and Rossenbeck, 1924); ultraviolet quantitative microscopic measurements of protein and nucleic acids (Caspersson and Schultz, 1940); the fundamental nature of DNA in heredity (Avery, MacLeod, and McCarty, 1944) as earlier suggested by Schultz (1941) and Mirsky (1943); a method of detecting biochemical mutations (Beadle and Tatum, 1941); "sexuality" and recombination in bacteria, at least in *Escherichia coli* (Lederberg and Tatum, 1946), and by transduction using viral carriers (Zinder and Lederberg, 1952); electron microscopes (Ruska and Knoll, 1931) and study of biological materials (Marton, 1934; Krause, 1938); reproduction and recombination (experimental at least) in viruses (Luria and Dulbecco, 1949); various contributions leading to the 1953 Watson and Crick proposal for DNA double helical structure and significances of that structure; "chromosomes" in mitochondria (Nass and Nass 1963; Nass, 1969) and plastids (Gibor and Izawa, 1963; Tewari and Wildman, 1966), which "explained" in part certain aspects of maternal and cytoplasmic inheritance. A technique for removing chromosomes from accumulated synchronized metaphases and spreading them on slides (Hsu, 1952; Tjio and Levan, 1956; Moorhead et al., 1960) has permitted the recent development of the active field of human (Priest, 1969) and other vertebrate (Hsu and Benirschke, 1967) cytogenetics; the relationship of cytochemistry with gene action is contributing advances in that field; and techniques of spreading interphase and metaphase chromosomes for electron microscopy (Du Praw, 1968; Wolfe, 1968) are contributing new ideas of chromosome ultrastructure at last.

During the last 8 years techniques for separating the two strands of DNA and keeping them separate long enough to combine (anneal) with other DNA or RNA (hybridization) have been developed. Such methods have indicated by the amount of hybridization recovered such information as phylogenetic relationships (McCarthy and Bolton, 1963) and genetic redundancy of magnitudes to about one million (Britten and Kohne, 1968). Thus in higher organisms much DNA seems to be highly redundant, but the genes usually studied by geneticists are unique.

A still more recent technique (Harris and Watkins, 1965) uses appropriate viruses to cause cells of very different origins, such as human and chicken, to fuse into one binucleate cell. Often the two very different nuclei also will fuse to create a very wide hybrid of unusual genetic possibilities (Watkins and Grace, 1967).

The electron microscopy of isolated genophores (chromosomes) of viruses (Westmoreland, Szybalski, and Ris, 1969) and bacteria annealed (by hybridization) to a genophore of a mutant form opens a field of cytogenetics reminiscent of the study of pairing of homologues at pachytene or of synapsed salivary gland cell giant polytene chromosomes.

Construction of cytogenetic ultrastructural or molecular models of chromosomes, genes, cistrons, replicons, other structures, and DNA replication and transcription has been popular since the late 1950's (Taylor, 1957; Schwartz, 1955; Freese, 1958; Ris, 1961; Stahl, 1961; Taylor, 1963; Peacock, 1963; Uhl, 1965; Whitehouse, 1963, 1965; Callan, 1967; and others). Such models are made to conform to the basic double helix structure of DNA, hypothetical schemes of structural activity of the helix during replication, amount of DNA in the chromosome relative to the time of its replication, replication only in the direction from 5' to 3', possible lateral redundancy of helices, circularity of prokaryotic genophores, spatial relationship to histones and other proteins such as necessary enzymes, the semiconservative method of structural replication, etc. These are all admittedly merely hypothetical models and should not be considered any more reliable than good guesses.

Thus, cytogenetics is no longer an entirely microscopic science. For bacteria and viruses it includes many molecular biological techniques, and "seeing" the genophores results from genetic and biochemical methods as well as the electron microscope.

CHAPTER **2**

The nondividing nucleus

After many years of research, comparatively little is known about the finer constitution of the nuclear reticulum.

L. W. Sharp, 1926

The nondividing cell, whether interphase or postmitotic, is much more a subject of biochemical genetics than of cytogenetics. It, especially the nucleus, is not characterized as much by observable structures as by physiological activities. The cytogenetics of the cytoplasm, at least of plastids and mitochondria, is discussed in Chapter 20. The cytogenetics of the nondividing nucleus is the subject of this chapter.

At the end of mitotic telophase the chromosomes "disappear" into the nondividing, the energic, nucleus. The designation *energic* emphasizes its biochemical activity and at the same time does not imply that it is either interphase or postmitotic. Nondividing nuclei and cells are of two sorts, (1) those in interphase, which are preparing for another division, and (2) postmitotic cells, which will normally never divide again. One common difference between them is that postmitotic nuclei usually do not replicate their DNA during a synthetic (S) period but usually differentiate in a permanent G_1 condition. For example, postmitotic nuclei of human lymphocytes are characterized by considerably more heterochromatin than interphase nuclei (Tokuyasu et al., 1968). Phytohemagglutinin (and other natural or experimental mitotic stimulators) converts certain postmitotic nuclei to interphase nuclei, which involves conversion of heterochromatin to euchromatin, phosphorylation and dephosphorylation of nuclear proteins (Kleinsmith et al., 1966) thereby stimulating RNA and protein synthesis and enzymatic activity, DNA replication, and eventually mitosis.

One of the first energic activities of a telophase nucleus is the formation of one or more nucleoli. The cytogenetics of nucleoli is only now being developed, mostly by biochemical techniques such as nucleic acid hybridization but also by electron microscopy.

THE NUCLEOLUS

In most eukaryotic cells one nucleolus or more forms in each telophase nucleus (see Vincent and Miller, 1966, for a recent detailed discussion). Thus nucleoli are not permanent cellular structures in higher plant and animal cells. It has now been well established (McClintock, 1934; McLeish, 1954; Lin, 1954; Crosby, 1957; Longwell and Svihla, 1960; Ritossa and Spiegelman, 1965) that nucleoli are formed by specific regions of specific chromosomes, called "nucleolus organizers" (McClintock, 1934). These short segments of chromosomes are usually located in short arms of certain chromosomes (Longwell and Svihla, 1960). When they lie near the ends of chromosome arms, the tip of the arm beyond the nucleolus organizer constriction is called a *satellite*.

Nucleolus organizers can be broken into two parts, as in translocation in maize (McClintock, 1934), and still function (produce small nucleoli) on new chromosomes. Beerman (1960) found that the fly *Chironomus tentans*, which has two pairs of nucleolus organizers, when crossed to *C. pallidivittatus*, which has one pair that, interestingly, does not correspond with those of *tentans* in position, can produce progeny having from six (three pairs) to none. All are functionally equivalent, as is not always true.

Nucleolus organizers are usually detectable in metaphase chromosomes in higher plants, in which they are usually few, but in animals they are usually revealed with difficulty and are often rather numerous. The electron microscope can show them, and in some animal species they lie near or at the ends of chromosome arms (Fig. 3-5). In a cell of a diploid plant species there are at least two, one in each chromosome set, but there are diploid species known with four. In most species of *Drosophila* there are usually two, one on the X and one on the Y chromosomes (Ritossa and Spiegelman, 1966). In man there are as many as 10.

Nucleoli (Brown and Bertke, 1969) are large nuclear organelles that are not membrane bounded. They usually lie in the nucleus and are usually rather spherical except often during early prophase of the first meiotic division when they are flattened against the inner surface of the nuclear envelope. They are very dense protoplasm consisting of thin 100 Å strands of chromatin (DNA-protein) and a great bulk of 150 Å granules of ribonucleoprotein. They often contain canals, vacuoles, or crystals. Often there are two regions of different fine structure, an inner dense "core" containing the nucleolus organizer chromatin and an outer granular "cortex" of ribonuclear particles (Miller et al., 1970a). There is no doubt that each such nucleolus is associated with one or more chromosomes at their nucleus organizer regions since the nucleolar organizer chromatin forms loops within the nucleolus. Extrachromosomal nucleoli are produced during prophase of the first meiotic division as many free nucleoli in amphibian oocytes (Gall, 1968; MacGregor, 1965).

The nucleolus (of eukaryotic but not prokaryotic cells) represents the first phases of ribosomal formation. Evidently DNA genes (or cistrons) of the nucleolus organizer region (r-DNA) aided by RNA polymerase transcribe nucleolar RNA (n-RNA). Protein seems to be synthesized immediately in the nucleolus or elsewhere to form the 150 Å ribonucleoprotein granules that constitute the bulk of the nucleolus. The present evidence is that each n-RNA polymer is about 0.33μ long and has a sedimentation coefficient of 40s (Svedberg units). The RNA strand, perhaps after it leaves the nucleolus, is "cut" into (1) an 18s piece, (2) a 28s portion, and (3) a portion that is apparently degraded (Brown and Weber, 1968; Weinberg et al., 1967). The 18s RNA fragment ends up in the smaller, 50s, subunit of the eukaryotic ribosome. Also present in ribosomes is 5s RNA, but it seems to be produced independently of the 18s and 28s. In *Xenopus* there are from 40,000 to 56,000 copies of the 5s sequence per diploid cell compared to about 900 copies of r-DNA genes in the nucleolus organizers (Brown and Weber, 1968). In *Drosophila melanogaster*, however, there are between 130 and 260 copies of each of the genes coding for the 18s and 28s RNA but only about 200 5s genes (Tartof and Perry, 1970). Wimber and Steffensen (1970) have located the locus of the 5s genes at the 56E-F region of the right arm of chromosome No. 2 of *D. melanogaster*. This was accomplished by feeding young larvae tritiated uridine so they could make radioactive 5s RNA, which was extracted and purified. Squashes of third instar larval salivary gland chromosomes prepared by a series of treatments were hybridized with the radioactive 5s RNA. After 2 months' exposure to photographic emulsion the radioactivity was always concentrated at the 56E-F locus on the right arm of the second chromosome. The authors had in the salivary gland chromosome a redundancy of about 2×10^5 5s genes. They speculated that the known redundancy of about 13 or more per haploid chromosome set for transfer RNA genes could also be detected by this method since there would be 1.3×10^4 cistrons per locus in a polytene salivary chromosome.

Redundancy

Cytogenetic study of the nucleolus is concerned with the r-DNA of the nucleolus organizer, but much knowledge derives from nucleic acid hybridization studies (Ritossa et al., 1966). Only recently have true cytogenetic study of those genes been accomplished (Miller and Beatty, 1969; Miller et al., 1970a; Miller et al., 1970b). The amount of functional r-DNA in the nucleolus organizer coded for only one type of RNA message (n-RNA) is very high, often about 1 to 2% of the total nuclear DNA (Brown and Weber, 1968; Ritossa and Scala, 1969). This nucleolar r-DNA in all species of eukaryotic organisms, and probably prokaryotes as well, consists of a large number of repeated genes (cistrons), all of which make the same n-RNA messages. For example there are about 100 such redundant r-DNA cistrons in the chick, perhaps 1,600 in *Xenopus* (Wallace and Birnstiel, 1966), 250 in *D. melanogaster,* and 5 or more even in the bacterium *E. coli.*

This redundancy of nucleolus organizer DNA was first detected by nucleic acid hybridization. It is possible to separate nucleolus organizer DNA from the remainder of the nuclear DNA because it is heavier, having a greater proportion of guanylic and cytidylic acids, about 60% in *Xenopus* for example. Thus, by equilibrium centrifugation in CsCl the r-DNA can be isolated (Wallace and Birnstiel, 1966; Brown and Dawid, 1968). Hybridization of the isolated r-DNA with ribosomal RNA (which the r-DNA had already transcribed) indicates by complicated techniques and computations that there are many repeated r-DNA genes for the production of n-RNA in each nucleolus organizer of all of the many species that have been studied (Brown and Dawid, 1968).

In the oocytes of amphibians during meiotic prophase each whole nucleolus organizer DNA replicates itself many times although the rest of the chromosomes do not. For discussion of the chromosomes at that time see lampbrush chromosomes in Chapter 6. These between 500 and 1,000 chromatin strands of as many as 200μ in length separate individually from the chromosomes and become circular (Miller, 1964), and each produces an extrachromosomal nucleolus, in the large nucleus, called the germinal vesicle (Brown and Dawid, 1968). Each of these separate DNA strands is many microns long, circular, and about one-third covered with RNA replicase, which transcribes the many polymers of n-RNA (Miller and Beatty, 1969). Such an unfertilized oocyte, because of its approximately 1,000 nucleolus organizers and as many as 1,000 extrachromosomal nucleoli, can produce about 1,000 times as many ribosomes per unit of time as an ordinary cell with only two organizers. Therefore, the oocyte becomes loaded with both functioning polysomes and nonfunctioning ribosomes. At fertilization all of this stops, and the many nucleoli break down when the nuclear envelope breaks down before metaphase of the first meiotic division. The ribosomes and chromosomal nucleolus organizers remain nonfunctional throughout embryogenesis until the early gastrula stage (Karasaki, 1965) of about 300,000 cells. At that stage all of the extrachromosomal DNA is gone, two chromosomal nucleoli form, and ribosomes are organized into polysomes and, therefore, protein synthesis begins.

Generally among animal embryos during cleavage the DNA is synthesized proportionately to division rate, protein synthesis varies among forms little to much, but RNA transcription is nil until the blastula in insects (Lockshin, 1966; Harris and Forrest, 1967) or gastrula in sea urchins, fish, and amphibians (Brown, 1966) but starts at the four-celled stage in *Ascaris* (Kaulenas et al., 1969). Emerson and Humphreys (1971), however, reported that r-RNA is produced in typical amounts during cleavage, but because mitotic divisions are so rapid, nucleoli cannot form.

True cytogenetics of nucleolus organizers of amphibian oocytes consists of seeing by electron microscopy these genes in action (Miller, 1964; Peacock, 1965; but especially Miller and Beatty, 1969, Miller et al., 1970a). The extrachromosomal nucleolus organizer DNA strands can be prepared by bursting the germinal vesicles and spreading the DNA strands of the

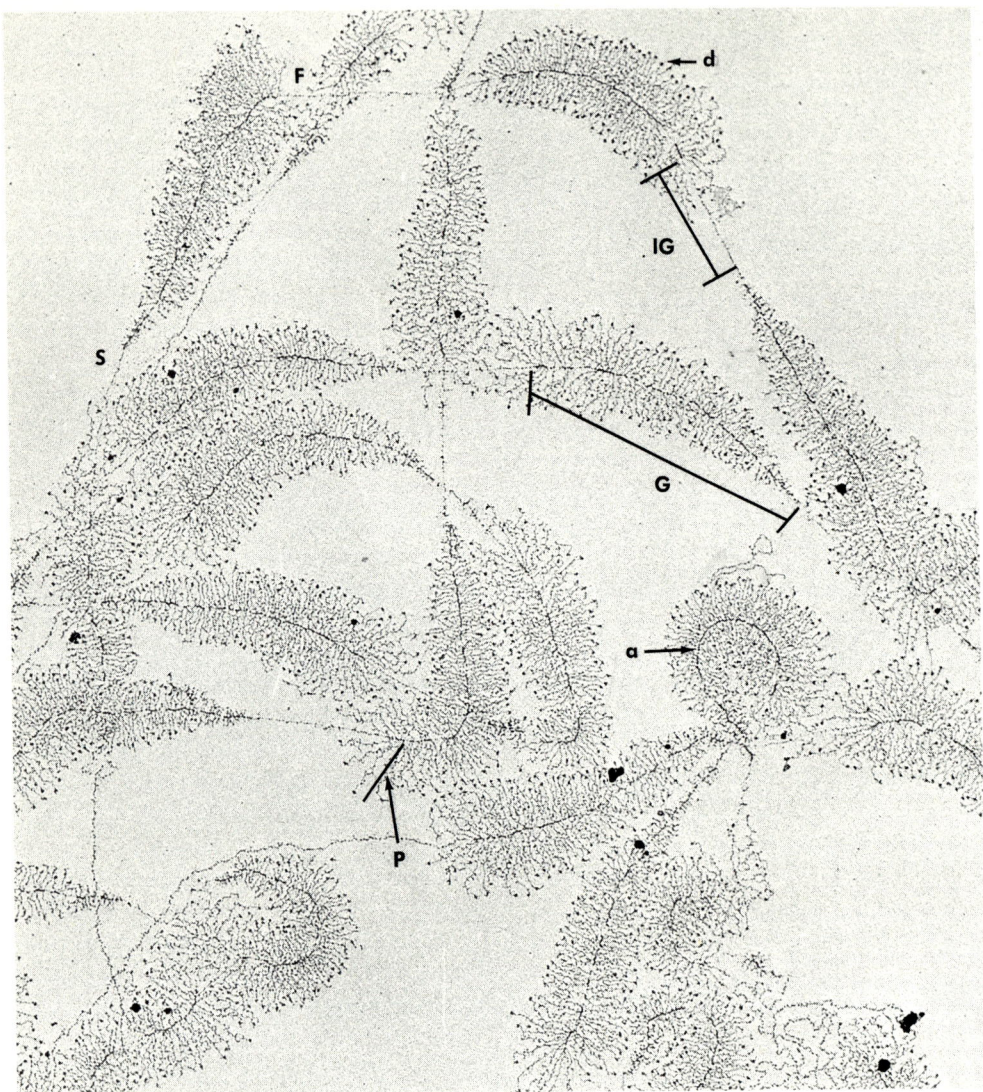

Fig. 2-1. Genes in action. Nucleolar genes, about 25 of them, from an extrachromosomal nucleolus of an amphibian (the spotted newt, *Triturus viridescens*) oocyte. From each such nucleolus a continuous and circular chromatin strand (*a*) of hundreds of such genes is extracted. Along this strand are many alternate nucleolar RNA-producing genes, delimited by *G*, or *S* to *F*, and intergene segments (*IG*). The axial fiber of a gene *a* is about 100 to 300 Å thick but after trypsin treatment, with protein removed, has a thickness of about 30 Å, the dimension of a DNA double helix. The axial filament of one gene (*G*, or *S* to *F*) is about 2.0 to 2.5μ long. Apparently a molecule of RNA polymerase starts to transcribe n-RNA at about *S* and "moves" along the axial filament of the gene. As it "progresses," the n-RNA grows longer and longer until at *F* the RNA strand is complete (*P*) and about 0.7μ long, long enough to contain the 28 and the 18 ribosomal RNA pieces plus some n-RNA that is degraded. Thus the length of the produced n-RNA polymer is about one fourth the length of the r-DNA gene that produces it. At any time there are about 100 RNA polymerase molecules producing about 100 n-RNA strands on each gene in various stages (lengths) of completion. The intergene segments do not seem to function in n-RNA production. At the outer (first-formed) ends of the RNA strands there are granules (*d*) of unknown function. Approximately 25,000×. (From Miller, O. L., and B. R. Beatty. 1969. Science **164**:955-957. Copyright 1969 by the American Association for the Advancement of Science.)

1,000 nucleoli on a water surface (Fig. 2-1). The extrachromosomal DNA strands are many microns long and are coated with protein. Each consists of a great many r-DNA cistrons each from 2.0 to 2.5μ long, but adjacent cistrons are separated by, in these preparations, "inert" DNA. Each cistron is made evident by its transcriptional activity. That is, along *each* cistron about 100 n-RNA polymers in a graded sequence of lengths and stages of synthesis extend laterally. Apparently 100 RNA polymerase molecules "move" along the cistron synthesizing as they go with the completed part of each such 40s RNA polymer extending outward from the cistron. At one end of each cistron, the beginning of synthesis, each n-RNA polymer is very short; at the terminal end of each cistron the completed n-RNA polymers are about 0.7μ long. That is, each cistron is transcribing about 100 RNA polymers simultaneously. If there are 100 cistrons in each of 1,000 nucleoli, each producing 100 n-RNA polymers at one time, the minimum number of ribosomal monomers being produced during the time necessary to form one such polymer is 10 million. But it is certain that each cistron must function repeatedly and for a long time. Thus are produced enough polysomes for all of the synthesis going on in the growing oocyte as well as nonfunctional ribosomal monomers to adequately stock the 300,000 gastrula cells before new nucleoli are finally formed. Karasaki (1965) and Brown and Dawid (1968) proposed that a single developing oocyte synthesizes r-RNA equal to about 200,000 liver cells, and each cell has tens of thousands of ribosomes.

Related to the above cytogenetics of amphibian oocytes is the famous "mutation" in the African clawed toad, *Xenopus laevis*, that prevents nucleolus formation and function, probably as a deletion, with the symbol *nu*. The wild type has two nucleolus organizers and two nucleoli and is designated 2-*nu*. The homozygous mutant is O-*nu*, having no nucleoli, and the heterozygote is 1-*nu* with one nucleolus per nucleus.

Perkowska et al. (1968) reported that oocytes of both 1-*nu* and 2-*nu* females contain similar numbers of extrachromosomal nucleoli per oocyte and have similar amounts of DNA in spite of the 2-*nu* oocyte having two nucleolus organizers to one in the 1-*nu* oocyte. Thus this mutation is recessive, but some equalizing mechanism makes the two quantitatively different forms equal. The O-*nu* individuals die as young larvae since during the gastrula stage their cells, lacking r-DNA, cannot form the necessary nucleoli.

In *Drosophila* the cytogenetics of nucleolus organizers is related to the morphological phenotype called the *bobbed* locus (Ritossa and Spiegelman, 1965; Ritossa et al., 1966a; Gersh, 1968). Apparently, there are varying intensities of this "mutation" depending on the percentage of the nucleolus organizer deleted, the percentage deleted of the 130-260 repeated cistrons in each X-chromosomal nucleolus organizer (Ritossa et al., 1966b). That is, the amount deleted produces a specific *bobbed* bristle phenotype. The nucleolus organizers in the genus *Drosophila* are located on the X and Y (sex) chromosomes (as they are also in the mouse *Mus musculus*, Ohno et al., 1957). In fact sexual species and strains of some species, such as *D. affinis*, can persist following the complete loss of Y chromosome (Patterson and Stone, 1952; Miller and Stone, 1962; Voelker, 1967).

The r-DNA of the genophore of *E. coli* naturally is not in a nucleolar organizer, although the bacterial r-DNA cistrons may be contiguous in some bacteria (Colli and Oishi, 1969), although probably not in *E. coli* (Cutler and Evans, 1967). Miller et al. (1970a) have been able to photograph what they consider are r-DNA cistrons of *E. coli* (which are not contiguous) transcribing fibrils of r-RNA and which are immediately coated with what is apt to be ribosomal protein (see Chapter 20). The ribosomal cistron is about 1.5μ long, as it should be according to the molecular weight of the 16s plus 23s r-RNA, and each is producing from about 60 to 70 short fibrils of ribosomal materials.

Somewhat similar electron microscopic figures of m-RNA formation on viral reaction cores in vitro have been published by Gillies et al. (1971). The reaction cores contain the viral RNA and transcriptase enzymes after the capsid has been removed. Apparently the densely compacted RNA is able to be transcribed, a very interesting fact indeed.

There is now a considerable body of knowledge that nucleolus organizers are subject to genetic control just as are chromosomes, cells, and organisms. Navashin in 1934 observed in species hybrids of the low chromosome number plant genus *Crepis* that some nucleolus organizers that should have been evident at mitotic metaphase were often undetectable. As he explained it, the satellite, instead of being separated by the unstained organizer region from the rest of the chromosome, was attached directly to the chromosome. Which of the two nucleolus organizers was formed and which was not evident depended upon the species involved in the cross, but was the same in reciprocal crosses, so that the cytoplasm had no apparent influence.

Although Navashin supposed that there are "strong" and "weak" organizers competing for available materials, this case may rather represent a form of genetic regulation of the r-DNA by the genes of the "stronger" genome present in the hybrid cell. *Crepis capillaris* was his standard species, and its genome suppresses the nucleolus organizers in hybrid cells of such species as *alpina, dioscorides, neglecta,* and *tectorum.* The organizer of *capillaris* is in turn repressed by the genome of *parvula*. Evidently *capillaris* and *setosa* genomes repress each others' organizers equally, the result being both nucleoli of reduced sizes in the hybrid.

Hexaploid wheat is an allohexaploid derived by hybridization and chromosome doubling from the three diploid species *aegilopoides, speltoides,* and *squarrosa*. The *aegilopoides* and *speltoides* ancestors must have introduced two nucleolus organizer chromosomes each and the *squarrosa* ancestor, one. Cytological study, however, reveals only four rather than five pairs of nucleolus organizer chromosomes in hexaploid wheat, on chromosomes No. 1, No. 10, No. 14, and No. 18. Of these, No. 1 and No. 10 are "strong" whereas No. 14 and No. 18 are "weak" organizers; the latter two normally function less (produce less nucleolar material) in the presence of the former two (Crosby, 1957; Longwell and Svihla, 1960).

Levenbook et al. (1958) have indicated in unfertilized (diploid oocytes) eggs of *Drosophila* that the nucleolus organizer of the Y chromosome is repressed in the presence of XX but that the presence of the Y affects the qualitative activity of the two nucleolus organizers on the two X chromosomes in the same cell. In man there are five pairs of satellited chromosomes (the three D and two G pairs); and yet it has been repeatedly observed that satellites are never found on all of these in the same cell. Thus, whether nucleolus organizers function or not seems to be influenced by the genome or genomes present within the cell.

The proof that nucleolus organizers, and those chromosomal regions only, produce the nucleolar material and the nucleoli themselves now seems adequate. In *Pisum* (Koller, 1938), hyacinth (La Cour, 1952), rye (Rees, 1955), *Vicia faba* (McLeish, 1954), and wheat (Crosby, 1957) when chromosomes form micronuclei atypically, it is only in those nuclei containing nucleolar chromosomes that nucleoli are formed. Furthermore, increase in nucleolar volume and RNA content is produced only when nucleolus organizer chromosomes or arms of chromosomes are increased in number; other chromosome increase has no effect, as in maize (Lin, 1954) and wheat (Longwell and Svihla, 1960). Nevertheless, the often repeated observation of a correlation between nucleolar volume increase with an increase in cellular synthetic activity and vice versa implies genetic regulation of organizer (r-DNA) activity, since increased ribosomal production must precede protein and all other synthetic activity in the cell.

CHROMATIN

The chromatin within the nondividing nucleus is mostly invisible to the optical microscope although it can be examined with the electron microscope either in sections (Ris, 1956; Hay and Revel, 1963) or as burst and spread nuclei (Du Praw, 1968; Wolfe, 1968; Comings and Okada, 1970a). The biological significance of the nondividing nucleus is its physiological activity, which has been worked out in considerable detail during the past 25 years. Although it is not strictly cytogenetics, comprehension of nuclear function is essential to the cytogeneticist since it often is related directly to his findings and is necessary to his development of sound concepts.

It is probable that every college undergraduate biology major comprehends the overall scheme of DNA code, the various kinds of transcription RNA's including the coded messenger, and the polymerization or translation of proteins by messenger RNA polysomes, transfer RNA's, enzymes, etc. There are, of course, many hypothetical schemes of the double activity of the classic Watson-Crick double DNA helix of (1) replication into two double helices and (2) transcription of the various RNA's by replicases. Also of common knowledge are models of gene regulation such as the original by Jacob and Monod of 1961 and the more recent one of Britten and Davidson (1969).

The biochemical techniques of nucleic acid hybridization, either DNA-DNA or RNA-DNA (McCarthy and Bolton, 1963), have recently been providing new and unexpected understanding about the DNA sequences in the nucleus such as repeated sequences (Britten and Kohne, 1968) in addition to the nucleolus organizer. These investigators reported hundreds, thousands, hundreds of thousands, and even one million (in mouse) of duplicate sequences (genes?), which may account for as much as one half of the DNA of the nucleus (40% of calf DNA). Thus, only about two thirds of the DNA in a nucleus may be composed of the "unique" unduplicated genes of classical genetics. Such repeated sequences have been found in eukaryotic organisms from all groups of plants and animals investigated (Brown and Dawid, 1968). The function or functions of all of this duplicate sequence DNA are presently unknown.

It is also common knowledge that during an interphase between two mitotic divisions the DNA and histones (Robbins and Borun, 1967) replicate during the synthetic (S) period. The time between the preceding telophase and the beginning of the S period is called the G_1 period, and the time between the end of the S period and the beginning of prophase is the G_2 period. Mitosis itself is called the M period. In diploid organisms the relative amount of DNA in a (haploid) gamete is described as the 1C amount. At telophase and during the G_1 period of diploid cells there is, therefore, the 2C amount; and from G_2 to telophase a diploid cell has the 4C amount of DNA. These facts can be determined by various quantitative techniques of cytochemistry or analytical cytology.

It is also evident that cells can function with one set of chromosomes (haploid), two sets (the common diploid), or more sets (triploid, tetraploid, etc.) in one nucleus, at least if such conditions are normal for the species or do not produce inviability. Examples of haploid cells are common in plant gametophytes and male Hymenoptera (bees, wasps, etc.).

Some cytologists have asked the question, Does the nondividing nucleus contain anything except "chromosomes" and nucleolus? That is, is there any region of the interphase nucleus in which there is only nucleoplasm? In considering this question another question arises, namely, What is the nature of an interphase "chromosome"? If one defines it as only the chromatin, then there is nucleoplasm. But if one assumes that the material around the chromatin strands, often called matrix, is part of the chromosome, then there probably is little or no matrix during interphase. The loose loops and gyres of the chromatin of one chromosome must often lie very close to the chromatin loops of others to permit aberrations such as translocations and somatic crossing-over.

The current concept of the interphase "chromosome" (Fig. 2-2) is that the 250 Å thick chromatin strands (Ris, 1966; Gall, 1966; Du Praw, 1965b; Wolfe, 1968) are loosely dispersed throughout the nucleus and are separated by matrix or nucleoplasm (Du Praw, 1968). One should imagine that the condensed telophase chromosomes expand just as they lie at late anaphase (Fig. 4-5). Thus the telophase positional interrelationships among chromosomes and their internal structures remain unchanged during interphase but in an uncondensed, expanded, loosened form (Comings, 1968). The telophase chromosomes expand, during the S period they replicate their DNA, at some presently unspecified time they double structurally (whatever that means), and at prophase the chromosomes condense (perhaps doubling structurally at that time). The expanded interphase chromosomes have some of their chro-

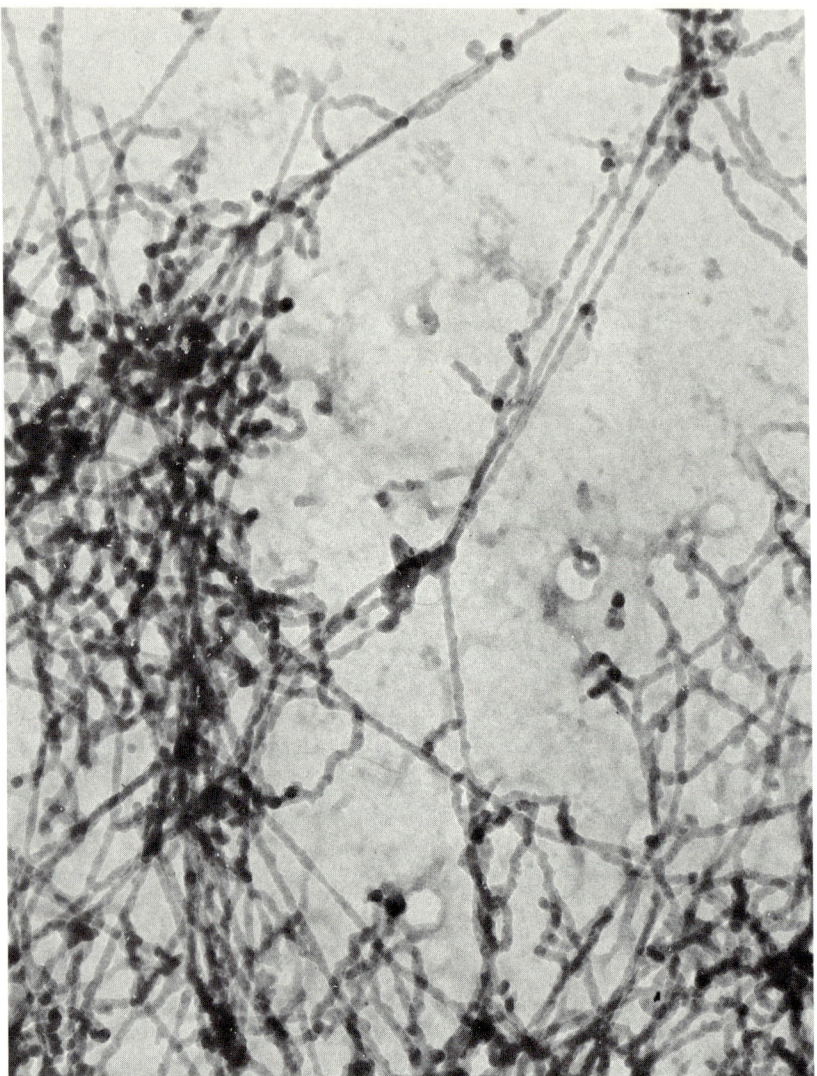

Fig. 2-2. Interphase chromatin strands isolated by bursting cells and nuclei, photographed by electron microscopy. By such methods chromatin strands have a diameter of about 250 Å and presumably consist of a coiled double helix of DNA plus proteins. It is in this uncondensed condition that DNA can replicate and transcribe. The treatment completely disrupts the in vivo order and arrangement of the strands. (From Wolfe, S. L. 1968. J. Cell Biol. **37:**610-620.)

matin strands attached to the nuclear envelope at the annuli of the pores especially at their "ends" but at other loci as well (Du Praw, 1968; Comings and Okada, 1970a).

This looseness of the chromatin in energic nuclei (Fig. 2-2) seems necessary for the replication of DNA and transcription of RNA during G_1 and possibly G_2 in higher organisms, but lower forms seem to have somewhat condensed chromosomes during interphase, as in the symbiotic flagellates (Cleveland, 1953), dinoflagellates (Kubai and Ris, 1969), etc. Condensed chromosomes of higher organisms are physiologically inactive (Brown, 1966) during mitosis; and, presumably, condensed chromatin (heterochromatin, prochromosomes, etc.) within nondividing nuclei is also inactive in replication and transcription.

The amount of DNA in nuclei varies very widely among organisms, from about 0.05 to about 300μμg (Vendrely, 1955; Sunderland and McLeish, 1961). Expressed another way, the total length of DNA double helix within a nucleus is centimeters or decimeters long. The human X chromosome alone contains about 4.4 cm of DNA (McKusick, 1964), DNA molecules from sea urchin sperm were measured at from 50 to 93μ (Solari, 1965), and DNA fibers 2.2 cm long have been recovered from human lymphocytes (Sasaki and Norman, 1966). Du Praw (1968) considers further that the 200 to 300 Å chromatin fiber itself (Fig. 2-2) contains about 60 times its length of DNA. Therefore, a 6-cm DNA strand would be contained within a 1-mm length of the commonly observed 250 Å chromatin fiber, a fiber still 100 times in length the diameter of a 10μ nucleus. In the human nucleus containing 46 chromosomes, therefore, 174 cm of DNA would constitute a total of 250 Å chromatin fiber about 3,000 times the diameter of the nucleus. Obviously, the nucleus has to be full of convoluted 250 Å chromatin strands.

As stated above, the DNA and histone double during the S period of interphase. That is easily demonstrated by quantitative studies. The problem of the how and when of structural doubling, at all levels, on the other hand, is still unresolved. Somehow the telophase double helix replicates itself quickly—about 0.5μ/minute in HeLa cells (Cairns, 1966) and from 1 to 2μ/minute in hamster cells (Taylor, 1968)—and neatly without tangling during the S period. Then at prophase there appear two separate chromatids composing each chromosome. And that is about all that is certain.

For 50 years or more there has been an unresolved question as to how many strands large enough or almost large enough to be seen in the light microscope there are in the telophase chromosome: one, two, four, or more. Electron microscopy has so far failed to determine the correct answer. It is known with considerable certainty that the DNA of each chromosome stays together as a unit. This is evident from the constancy of chromosome form and from the permanence of genetic linkage groups. But what goes on *structurally* within the interphase and/or the prophase nucleus is still unknown.

Postmitotic cells and nuclei usually remain in the G_1 condition of 2C; they do not usually enter the S period. Thus, most differentiated cells contain the 2C amount of DNA. Perhaps that condition is best for RNA transcription.

DNA INCONSTANCY

Since before the turn of the century it has been assumed that within a species the chromosome number is constant. Subsequently it was determined as likely that the individuals of a species also have a common idiogram. During the 1950's, when amounts of DNA were determined for species, it was concluded that species within major classification groups, such as mammals, birds, reptiles, etc., had rather constant although distinctive amounts. Furthermore, the amounts of DNA from various tissues and organs of an individual are the same. That is, there is DNA constancy at many levels.

Nevertheless, biologists considering the problem of differentiation often proposed (Huskins and students, 1947-1955) that perhaps chromosome number doubling (endoploidy) or reduction might be a mechanism of cellular differentiation. Mather (1948), for example, discussed the "significance of nuclear changes in differentiation," as including (1) ontogenetic chromosomal elimination, (2) replication of whole chromosome sets by polyteny or endoploidy, and (3) manifestation. The last category includes such observable chromosomal changes as the formation of lampbrush chromosomes, the heterochromatization of one or more but not all chromosomes of a cell, Balbiani rings, nuclear size difference following a division, etc.

It has now been well established that there is considerable variation in DNA per nucleus among species of a genus, among subspecies of a species, rarely among populations of a species, and among cells of an organism. Hughes-Schrader (1953) reported that two species of the mantid genus *Liturgousa* have 1.5 times as much DNA per nucleus as two other species. Ullerich (1966) reported of toads that *Bufo viridis* has 50% more DNA per cell than *B.*

calamita or *B. bufo*. Miksche (1967) reported DNA amounts (in 10^{-12} g) of 75, 84, and 139 among three species of pine trees that have the same chromosome number. Martin (1968) reported arbitrary amounts of DNA in 12 diploid species of *Vicia*: 17, 19, 22, 23, 25, 28, 51, 57, 59, 83, and 100; the last, *Vicia faba*, was taken as the standard. Martin discussed this matter in detail in flowering plants as an aspect of evolution and favors lateral redundancy as presently the best "explanation." In *Betula*, Grant (1969) found a direct correlation between chromosome number and relative amount of DNA in birches with chromosome numbers of 2n=28, 42, 56, and 70; but the 84-chromosome types, relative to the 28-chromosome species, had only 2.25 times as much DNA per nucleus rather than the expected 3 times as much. There are many other examples in the literature.

An example of DNA variation among populations of one species was published by McLaren et al. (1966) in the marine copepod *Pseudocalanus minutus*. Large forms had much larger cells, nuclei, and chromosomes than the small form and seven times as much DNA per nucleus. The authors assumed that the large chromosomes with more DNA are more highly polytene. They described the two forms as "cryptic species" and conclude from the literature that other cryptic polytenic species of copepods exist.

The usual model of such increase in DNA amount is that of lateral polyteny (Hughes-Schrader, 1956; McLaren et al., 1966; Martin, 1968). If this is the correct explanation, such polyteny must be very easily achieved.

Variation in DNA amount within an organism can also exist by increase of chromosome number, by endopolyploidy, or by some postmitotic nuclei being in the permanent G_1 and C_2 condition whereas others may differentiate in the G_2 and C_4 stage. The reports of Herman and Lapham (1968) and others of certain neuron cells of mammals having twice as much DNA as other cells of the nervous system may be either cases of polyploidy or differentiation in the C_4 condition.

Blood chimeras can even be produced. Volpe and Gebhardt (1966) joined in parabiosis during tailbud stage of embryonic development 17 diploid with triploid leopard frog pairs, which remained as parabionts until late metamorphosis. Cultured blood of the adults 1 and 6 months after metamorphosis indicated that each of a pair was chimeric since each had both diploid and triploid leukocytes by chromosome counts. About 15% of cells were of the type normal to the other member of the pair (donor) at both 1 and 6 months. Thus no change was indicated. Furthermore, each of a pair tolerated skin graft from the other but not from any other individuals. It was assumed "that the donor-type leucocytes are descendants of primordial blood cells that had been conveyed through vascular anastomoses into the circulation of the host in embryonic life and had settled in the host's hematopoietic tissues. Each exparabiont thus has blood-forming tissues capable of producing two kinds of blood cells, its own kind and that of its former partner." They did propose, however, that perhaps certain types of leukocytes have long lives and were always only circulating cells. This has not been demonstrated for frogs, but in human beings it is known that leukocytes continue to circulate for many years.

Thus it seems that DNA constancy is not as constant as used to be believed, and greater accuracy may well demonstrate more subtle variations.

DNA AMPLIFICATION

At the present time it is well known that certain cell types of plants and animals are regularly polytene and/or endoploid, but whether this is cause or effect of differentiation is often difficult to determine. But both of these are certainly genomic amplification of DNA and are correlated with differentiation. For example, among flies the extent of polyteny of chromosomes among tissues varies remarkably, and even within the salivary gland it varies in time and among the cells of the gland (Pavan, 1965). Endoploidy, which increases DNA by replicating chromosomes without nuclear division, is *very* common as in the liver of mammals and in the differentiated cells of the anther tapetum of flowering plants.

Partial genome, whole chromosome, amplification is not rare for sex chromosomes as in the mammal *Mecrotus oregoni* in which the XO female zygote achieves the XX condition in the germ line cells. In the nurse cells of *Rhynchosciara angelae* all of the chromosomes become cryptopolytene and later polytene, except that the "L" chromosome remains completely unamplified (Basile, 1969).

Gene amplification (Pavan and da Cunha, 1969), as a specific term, however, refers generally to DNA replication of one or a few specific loci of particular chromosomes while most of the DNA replicates at a lower rate. The example given above of nucleolus organizer DNA amplification is the outstanding example. The above characterization of "gene amplification" may be too restrictive, however. Perhaps it should include, as an antithesis, DNA replication of most genes of the chromosomes although certain chromosomal regions do not amplify themselves at the same rate, as in salivary gland polytene chromosomes of *Drosophila*. It now seems proved that the centric regions of the otherwise polytene chromosomes in salivary glands (and probably cells of other tissues of flies that form polytene chromosomes) replicate at a lower rate than the arms (Rudkin and Schultz, 1961; Rudkin, 1969). That is, in such cells most genes divide many times by polyteny, but a small percentage do so at a lower rate.

Other cases (see Pavan and da Cunha, 1969, and Swift, 1969, for discussions, and Chapter 3) are the formation of DNA-containing "micronucleoli" or "nucleoloids" within the nuclei of salivary gland cells of the flies *Hybosciara fragilis* and *Sciara ocellaris* (Perondini and Dessen, 1969). These evidently are not derived from nucleolus organizers because the different "regions of the chromosomes which release micronucleoli with a DNA core behave differently and are active with different intensity as well as at different times during larval development." DNA *puffs* of salivary gland cell polytene chromosomes of numerous Diptera are similar to RNA puffs except that additional DNA is produced; the amplified genes do not leave the chromosomes but function within the chromosome puff (Breuer and Pavan, 1955; Henderson, 1967; Pavan and da Cunha, 1969; da Cunha et al., 1969).

Other probable examples of amplified genes are the extrachromosomal DNA body reported by Lima-de-Faria and Moses (1966) in the oocytes of the flies *Tipula* and *Pales* and beetle *Dytiscus*, of the cricket *Acheta* (Lima-de-Faria et al., 1969), and of the beetles *Dytiscus* and *Colymbetes* (Gall et al., 1969). Roberts et al. (1969) found DNA granules from polytene chromosomes in cells of the foot pad of the fly *Sarcophaga*, and Piko et al. (1967) detected cytoplasmic DNA in sea urchin eggs. Ammermann (1969) reported that DNA is released from the macronucleus of the protozoan *Stylonychia* and remains as such in the cytoplasm. A possible example of limited amplification has been reported in human beings having the chromosomal mutation "fragile site" on chromosome 16 (Magenis et al., 1970). They observed occasional cells with from 2 to 12 copies of the separate and detached terminal segments and concluded that these fragments had been produced by selective endoreduplication. That is, while the whole genome did not replicate, these specific terminal segments did.

There are two possible ways a gene or segment of genes might replicate independently of adjacent genes or segments. One is that the two or more redundant segments arrange themselves in tandem, in series. The other is that they lie parallel and side by side, thereby increasing the thickness of the strand of DNA. The latter is more popular. Such redundancy may be temporary or permanent. If permanent, it can account for the markedly different amounts of DNA in related species.

Since detectable gene amplification is now known to occur on a grand and measurable scale in many actively metabolizing cells such as oocytes and salivaries, perhaps undetectably small segments of DNA, such as single genes or operons, may be similarly amplified in ordinary cells. Such duplicated genes might remain associated with the parent gene to produce local polyneme conditions or separate from the parent chromatin strand to float freely in the nucleus or even cytoplasm. As Pavan (1965) stated succinctly, "As this thing occurs at the level of chromosomes, why should we not have it also at the level of genes?"

Perhaps the "information cytoplasmic DNA" (I-DNA) which is nonmitochondrial, 7S, and of nuclear origin, recently reported by Bell (1969, 1971) and suggested to represent small copies of the nuclear DNA of embryonic chick and other tissue, may represent such amplification of single genes or short segments of chromosomal DNA. Bell proposed the model that chromosomal (structural) genes produce I-DNA which becomes associated with protein in the cytoplasm as I-somes. This I-nucleoprotein in the cytoplasm then replicates the coded m-RNA for protein translation. Some researchers (Bloch, unpublished) actually are thinking of a model such as nuclear DNA→nuclear RNA→cytoplasmic I-DNA→m-RNA or r-RNA.

Gene magnification

In *Drosophila melanogaster* a type of adaptation of the nucleolus organizer has been reported that is called "magnification" (Ritossa, 1968). He found that *bobbed* (bb) strains (that are deficient for part of the nucleolus organizer and have a low amount of redundancy for the r-DNA cistron) revert to normal after a few generations both for the phenotype and the amount of nucleolar redundancy. Ritossa and Scala (1969) also reported that different stocks of *D. melanogaster* have widely different optima for r-RNA gene redundancy.

More recently Tartof (1971) has demonstrated in four different chromosomal strains that when a fly has only one X nucleolus organizer (NO), as in XO males or XX NO-deficient females, the increase in redundancy per X nucleolus organizer increased during ontogeny from (1) 255 to 430; (2) 256 to 386; (3) 103 to 248; (4) 255 to 395. That is, there were increases in redundancy of 175, 130, 145 and 140 cistrons. Apparently the organism "senses" the deficiency and is able to compensate for the deficiency by increased redundancy. Probably this occurs also in heterozygous mutant *Xenopus*. Perhaps there is normally an ontogenetic change in the amount of redundancy from zygote to later stages.

This condition of increased redundancy either persisted or developed anew in those *Drosophila* progeny that had only one nucleolus organizer. When such increased nucleolus organizers in the progeny were accompanied by a wild type Y chromosome with its NO, the nucleolus of X was normal again (Tartof, 1971).

DNA DIMINUTION

Diminution of DNA, like amplification, varies greatly. The most obvious and extreme is the elimination of the whole nucleus. Well-known examples are the ejection of the whole nuclei from the differentiating mammalian erythrocyte and eye lens cells, and the sieve tube element of the phloem of vascular plants. These are examples of DNA reduction to zero.

Examples of somatic reduction, from the two genomes of the diploid to the one of the haploid condition as discussed in Chapter 6, represent less extreme diminution. Coe (1954), for example, found occasional root tip cells containing 24 chromosomes in plants that were 2n = 48 of *Zephyranthes* and *Cooperia*. Such diminution presumably, but not necessarily, by nondisjunction may occur all at once, in one division, or gradually, one or two chromosomes at a time, over a number of divisions. Extreme examples of this sort occur in a specific cell layer of the roots of cycads where reduction may end with differentiated cells having 1 or 2 chromosomes from an original 22 (Storey, 1968) and the "polymitotic" mutant of maize, which causes similar repeated chromosome reduction divisions in the microspore (Beadle, 1931). Such genome reduction is part of all parasexual cycles. Nur (1966), however, reported the nonreplication of the whole male set of heterochromatic chromosomes in a mealy bug, *Planococcus citri*.

Other examples of diminution by one or a few chromosomes are the elimination of L chromosomes of *Sciara* and the E chromosomes of the Cecidomyiidae or certain sex chromosomes of some species.

Diminution of part of chromosomes is a less well established phenomenon. One often-cited case is the irreversible loss of the ends of the long germ line chromosomes of *Ascaris megalocephala* in the primordial somatic cells during cleavage. Reversible loss of between 5 and 10% of DNA during spermiogenesis has been reported in two species of crabs (Vaughn, personal communication). Walker (1968) might consider the crab satellite DNA as somewhat specific because he concluded that certain nuclear satellite DNA's, of presently unknown function, show marked intergeneric variation in base ratios and total amount. Certainly amount of DNA varies among related species. Henigardner (1968), for example, reported that among more than 200 species of teleost fish the DNA amount varied from 0.4 to 4.4 picograms per cell. Variation among species of *Crepis* is 20 times. Since DNA variation among species can be increased or decreased, it is possible to think also of evolutionary amplification and diminution.

HETEROCHROMATIN

At telophase in many nuclei all of the chromatin does not completely decondense into euchromatin but remains somewhat less condensed than at anaphase (Fig. 5-1). Thus in a nonmitotic nucleus there is decondensed and invisible "chromosome," called *euchromatin*, and condensed visible chromatin, called *heterochromatin*. These heterochromatic masses vary greatly in size, number, and position in the chromosomes, among chromosomes and among species (Bianchi and Bianchi, 1969); they are often called *prochromosomes*. The commonest position of heterochromatin within a chromosome is on each side of the centromere, but it may occur locally anywhere along the chromosome arms, such as the knobs of maize. Whole chromosomes may be heterochromatic such as the sex chromatin body in cells of female mammals or the sex chromosomes during premeiotic interphase. The genetic evidence is that heterochromatin during interphase is inert and any genes located within it are repressed (Brown, 1966). Electron microscopy indicates that the heterochromatic masses are structurally continuous with the euchromatin (Frenster, 1965). Biochemically it seems that histones that are the same in all Eukaryota (de Lange et al., 1968) are formed in the cytoplasm by small polysomes (Robbins and Borum, 1967) and are involved in the difference between euchromatin and heterochromatin (Comings, 1967b). For example, the interconversion of one to the other may be caused by amounts of lysine-rich histone present (Littau et al., 1965). Although histones are always present in chromatin, there are various sorts such as, at least, lysine-rich and arginine-rich. Furthermore, they are alterable by phosphorylation, as by cyclic AMP (Langan, 1968) or acetylation (Pogo et al., 1967) or methylation (Pogo et al., 1969). Such chemical alterations doubtless are correlated with histone function. They have long been implicated in gene repression of a general rather than a specific nature (Huang and Bonner, 1962; Bonner et al., 1968) such as in heterochromatin. It seems likely that heterochromatin becomes diffuse, becomes euchromatin, when its DNA replicates (Milner, 1969) but can then immediately change back to heterochromatin still during interphase.

The time of replication of heterochromatin in a particular S period is later than the time of replication of the euchromatin in the same chromosomes or cell; it is "late replicating." This is determined by autoradiography (Fig. 19-5). When interphase cells are fed thymidine ^{3}H and then examined at various times afterward, the first metaphases showing label were in the late S period (La Cour and Pelc, 1958) and the locations of label correspond to loci of heterochromatin (Lima-de-Faria, 1959); Bianchi and Bianchi, 1965; Abraham et al., 1968; and many others). Braun et al. (1965) working with a slime mold and Mueller and Kajiwara (1966) working with human cells determined that particular DNA that replicates early in one division replicates early in the next division also. Another concept is that replication tends to occur early at the ends of chromosome arms and late near the centromere (Schmid, 1963; Moorhead and Defendi, 1963; Stubblefield and Mueller, 1962; Hsu, 1964; Taylor, 1958a). This correlates with the generalization that there are usually regions of heterochromatin on each side of the centromere. Bianchi and Bianchi (1969) have found that similar chromosomes and similar loci on chromosomes are late replicating (are heterochromatic) among species of a genus and sometimes among genera of a family. The same authors in 1965 reported that in human beings the smallest chromosomes start replication later but finish earlier than the larger ones, but that various chromosomes start and finish at different times.

Such late labeling of heterochromatin is often noted for sex chromosomes, especially the Y, the sex bivalent during meiosis, and the region of the centromeres at diakinesis and metaphase I (Odartchenko and Pavillard, 1970). These authors also found that late-replicating regions of paired chromosomes at pachytene or metaphase I lie exactly side by side, indicating the pairing of corresponding parts of homologous chromosomes.

Recently (Brown, 1966) two sorts of heterochromatin have been tentatively specified: (1) "facultative heterochromatin," and (2) "constitutive heterochromatin." These are very different as to the form of the contained DNA and why they are heterochromatic. Constitutive heterochromatin is such because of its nature and is always such; facultative heterochromatin is euchromatin that has been heterochromatinized. It has been determined that constitutive heterochromatin is redundant DNA and is also satellite DNA (Yunis and Yasmineh, 1970, 1971). *Constitutive heterochromatin* is the usual form, it is redundant, satellite, and somehow redundancy is related to the heterochromatic condition during interphase. It is the heterochromatin associated with the centromere, often near telomeres, often as specific regions of chromosome arms, and often at the nucleolus organizer region (at least at metaphase). *Faculative heterochromatin* is not related to redundancy, it is chromatin that may or may not be condensed at interphase. It is most evident as one or more of the Barr body X chromosomes in interphase cells of female mammals, or as the set of chromosomes in males derived from the male parent in certain sexual coccid insects, or spontaneous production of one set of heterochromatic chromosomes in 5% of genetically identical progeny of certain parthenogenetic soft scale insects, or other chromosomal heterochromatinizations in other coccids, some of which are reversible at meiosis (Brown and Nur, 1964).

There are three new methods for revealing the specific locations on metaphase chromosomes of constitutive heterochromatin (Pardue and Gall, 1970). One method involves the preparation of radioactive DNA, separation of the satellite (redundant) DNA, hybridization of it to intact metaphase or salivary gland chromosomes, and preparation of autoradiograms. The radioactive redundant satellite DNA hybridizes only with those regions of chromosomes known to be heterochromatic or indicates those regions that are regions of redundant DNA. A second method (Borgaonkar and Hollander, 1971) employs quinacrine fluorescent dyes and again, the dye is restricted to known regions of heterochromatin or indicates regions of the chromosomes that are heterochromatic. The third and simplest technique (Arrighi and Hsu, 1971; Yunis et al., 1971) consists of denaturing the DNA of metaphase chromosomes by heat or alkali followed by reassociation of *redundant DNA only* under controlled conditions. The unique (euchromatic) material does not combine with the stain. The chromosomes are subsequently treated with the Giemsa stain. The result is that the regions of chromosomes that are stained are regions of heterochromatin. All of these techniques, as well as late labeling, agree in specifying the constitutive heterochromatin, and since the pattern of eu- and heterochromatin is often different among nonhomologous chromosomes, otherwise identical chromosomes may be clearly distinguished, as in human cytogenetics (Yunis et al., 1971) and can be utilized in evolutionary studies (Barr and Ellison, 1971).

Brown and Nur (1964) presented strong evidence in support of the general conclusion that genes in heterochromatin are inactive or that there are no "unique" genes within heterochromatin. They did show, however, that the presence of some minimum amount of heterochromatin is necessary for fertility in a coccid and that an adequate amount of heterochromatin from another species caused death of the nymph.

REPLICATION

Another cytogenetic study that employs a pulse of thymidine ^{3}H during interphase followed by study of subsequent metaphase chromosomes by optical microscopy and autoradiography is the attempt to determine how the new nucleotides are structurally related in strands to the old strand when DNA replicates during the S period. There are three possibilities starting with double Watson-Crick helices, the dispersive, the conservative, and the semiconservative schemes.

The "dispersive" scheme, which is not supported by theory or observation, would occur if new ("hot" or radioactive) nucleotides are inserted among the old "cold" nucleotides of both polymers of the double helix. Therefore, at each subsequent metaphase, both chromosomes must be radioactive, but the radioactivity of the chromosomes would become weaker and weaker after each division.

The "conservative" scheme proposes that each cold polymer of the G_1 double helix would replicate a hot polymer but then the two cold strands would reassociate to form a cold double helix as would also the two hot strands. Thus the two chromatids of each chromosome at the immediately following metaphase would be different, one hot and one cold, and at the next metaphase again one daughter cell only would have one hot chromatid. This scheme is occasionally reported, but most evidence supports the next scheme.

The "semiconservative" scheme assumes that after replication of two hot polymers each hot strand remains associated with the cold polymer that produced it. At the immediately following metaphase, each chromatid of each chromosome is labeled. At the next (second) metaphase, one chromatid of each chromosome should be labeled in both daughter cells. Nearly all of such studies find this semiconservative scheme to be generally true (Delbruck and Stent, 1957; Djordjivic and Szybolski, 1960; Filner, 1965; Walen, 1965) even in the premeiotic interphase (Taylor, 1965); but it is often somewhat modified in places on some chromosomes (Fig. 2-3).

Often it has been observed at metaphase of the second division after the interphase pulse labeling that instead of only one chromatid of each chromosome showing label, there is some label in both chromatids. However, in such cases, at second metaphase, each chromatid has usually one segment labeled and the remainder unlabeled, and corresponding regions of the two chromatids are different; where one is labeled the other is not and vice versa. These are considered to be "sister chromatid ex-

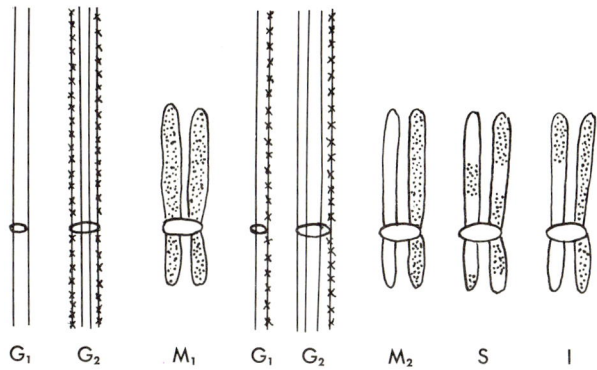

Fig. 2-3. Chromatid labeling and replication as observed (M_1, M_2, S, and I) and in theory (G_1's and G_2's); x's and dots indicate radioactive label. According to theory, a G_1 chromosome (far left) consists of a single double helix, indicated diagrammatically by two parallel lines indicating unlabeled (cold) polymers. During an S period *in* radioactive thymidine the double helix replicates semiconservatively so that each new double helix (G_2) has one radioactive (hot) polymer and one cold polymer. At first metaphase (M_1) it is observed generally that both chromatids are labeled throughout. At the next G_1 each daughter chromosome consists of one hot and one cold polymer. During a second S period, *not* in radioactive thymidine, DNA replication occurs to produce one cold double helix and one hot and cold (G_2). At second metaphase (M_2) it is generally observed that one chromatid is labeled and one is not. Occasionally (S) both chromatids are partially labeled due, presumably, to earlier sister strand crossing-over. Labeling of comparable portions of both chromatids (I) is described as isolabeling.

changes" (Taylor, 1958a). Some of these apparently occur between the pulse label and the first metaphase and are called "twin exchanges" because they show up as identical patterns at the second metaphase in both daughter cells of, if kept together, a tetraploid cell. Exchange occurring between the first and second metaphases' "second generation exchange" produces the condition in only one chromosome of one cell.

A second and more difficult problem has also been noted (Taylor, 1958a; Peacock, 1963; Zweidler, 1964), called "isolabeling." This condition is seen also at the second metaphase after labeling and is evident as both chromatids of a chromosome labeled side by side in corresponding regions. Isolabeling is generally "explained" as resulting from each telophase being double stranded (an old idea in cytology) at least in certain regions. Thus at DNA replication the two genetically identical strands are labeled, but at second metaphase isolabeled regions of sister chromatids still contain label in one or other of the parallel strands of each chromatid.

Du Praw (1968, p. 571) has presented an alternative hypothesis based upon his interpretation of chromosomal fine structure, that the continuous chromatin fiber of a chromosome does not end at the end of a chromosomal arm (a telomere) but turns around and continues back along the chromosome arm toward the centromere. He assumes that a sister chromatid exchange at the corresponding telomeres of two sister chromatids would result in isolabeling.

The cytogenetically significant result of these studies is the strong evidence that exchange does occur during mitosis between sister strands of chromatin. It relates, therefore, to somatic crossing-over and meiotic recombination, which are discussed later.

It is evident that the cytogenetics of energic nuclei involves a great deal of biochemical methodology and indicates the coming together of structure and function at the molecular level.

The giant salivary gland chromosomes of flies represent a special nonmitotic interphase condition of nuclei. But since this study is so extensive, it constitutes the subject of Chapter 3.

CHAPTER 3

Polytene chromosomes

Ever since the formulation of the chromosome theory of heredity cytologists and geneticists alike have dreamed of the day when some one would find somewhere an organism in which the chromosomes were so large that it would be possible to see qualitative differences along their lengths corresponding to the different genes which we know must reside there.

T. S. Painter, 1934

CYTOLOGY

The term "polytene" is applied to chromosomes that have, or may have, more than one closely associated longitudinal strand of chromatin. In some hypothetical schemes of the fine structure of typical chromosomes it is proposed that the normal chromosome may have two or more such parallel and genetically identical strands, in which case the chromosome is often designated as "polyneme." Polyteny is also often invoked as an hypothetical explanation of the wide range of DNA content among some related species. For example, the lungfish have about 100 $\mu\mu$g of DNA per nucleus, but carp only 3.3. The difference may be that the lungfish chromosomes are much more polytene than those of the carp.

The polytene chromosomes, however, sometimes called "giant interphase chromosomes" or "salivary gland chromosomes" or just "salivaries," are spectacular cytological structures. They have been known at least since 1881 when Balbiani figured them (see Painter, 1934c, for review of early literature). They are very long and very wide and are contained within very large nuclei of large cells (Fig. 3-1). They are at least 100 times as long as ordinary metaphase chromosomes and are up to 10,000 times the cross-sectional area of univalent mitotic metaphase chromosomes (Beermann, 1952). Typical of all flies (the Diptera), they occur in many larval tissues, but they are largest in the salivary gland cells. Among flies, however, salivary gland chromosomes vary greatly in clarity of bands and interbands. In many genera or families they are useless for cytogenetic studies.

Polytene chromosomes of massive size, such as those of the Diptera, are rare in other organisms, although they seem to be present in the macronuclei of ciliate protozoa (Perez-Silva and Alonso, 1966; Ammermann, 1969) and have been reported in the salivary gland cells of one Yugoslavian collembolan (insect), *Bilobella massoudi*, in which they were similar to fly salivaries for bands and interbands but were short and unpaired but had puffs and "solid" swellings (Cassagnau, 1968). Polytene chromosomes of sorts have also been reported in the suspensor cells of some leguminous plant embryos (Nagl, 1969). For the rest of this chapter and book, however, the giant polytene chromosomes of the salivary gland cells of various Diptera only will be discussed unless otherwise specified.

As has been known for more than 50 years, the homologous chromosomes of all or most

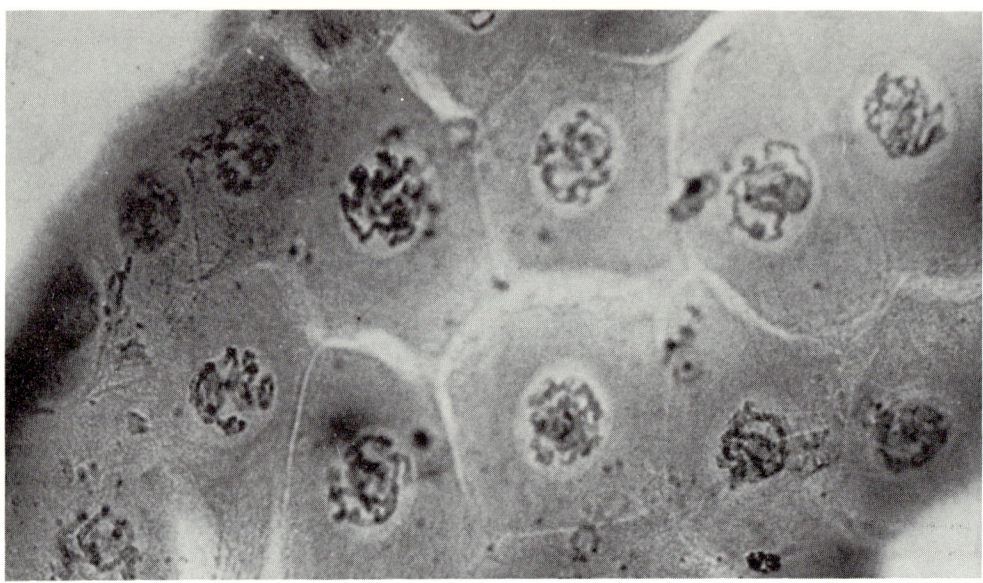

Fig. 3-1. Salivary gland cells of *Diptera* are large and have large nuclei, and the polytene chromosomes are distinctive.

Diptera are always paired (they lie closely side by side) in all or nearly all cells (Metz, 1916a) (Fig. 13-7). Therefore, it is not surprising that the interphase homologues are paired also in salivary gland cells. Alverdes (1912) and Painter and Griffin (1937) studied years ago the development of salivary gland chromosomes from early embryos to pupae in *Drosophila*. In the very young embryos the homologues as they first become evident in the interphase nuclei are loosely paired, as is probably generally true in all interphase fly cells. They elongate, show chromomeres, and divide longitudinally. They then accomplish what Painter described as "somatic synapsis." The four strands may visibly divide again. As development continues, the "synapsed" chromosomes continue to elongate and show longitudinally more and more chromomeres. Evidently, at a submicroscopic level, the chromatin strands continue to replicate DNA and divide structurally but do not separate. Growth in length and diameter continues throughout larval life; therefore comparisons of banding patterns should be made between larvae of similar developmental stages.

In the light of modern knowledge of polytene chromosome structure, it is doubtful if elongation accompanies the development of polytene chromosomes. And rather than the statement that more and more chromomeres appear, the concept of chromomere separation and of their becoming more clearly defined as polyteny increases seems more correct (Pavan, personal communication).

The homologous pairs of polytene chromosomes vary in the extent of "somatic synapsis" among genera. In *Simulium* (Pasternak, 1964; Kunze, 1953), *Bibio* (Heitz and Baur, 1933), etc. the homologues are "synapsed" in only a few short regions along their length. In *Drosophila*, *Chironomus*, etc. the two homologues are usually paired so intimately throughout their lengths that each pair appears as "one apparently single structure" (Painter, 1934c). In some Cecidomyiidae, in some particular tissues of other species, and in certain infected cells (Pavan et al., 1969), the polytene chromosomes split into a number of their polytene chromosomes to produce a combination of polyteny and polyploidy in a single nucleus (White, 1948a; Matuszewski, 1965). Furthermore, pairing may occur in some, but certainly not all, hybrids (Bock, 1971). Buzzati-Traverso (1950)

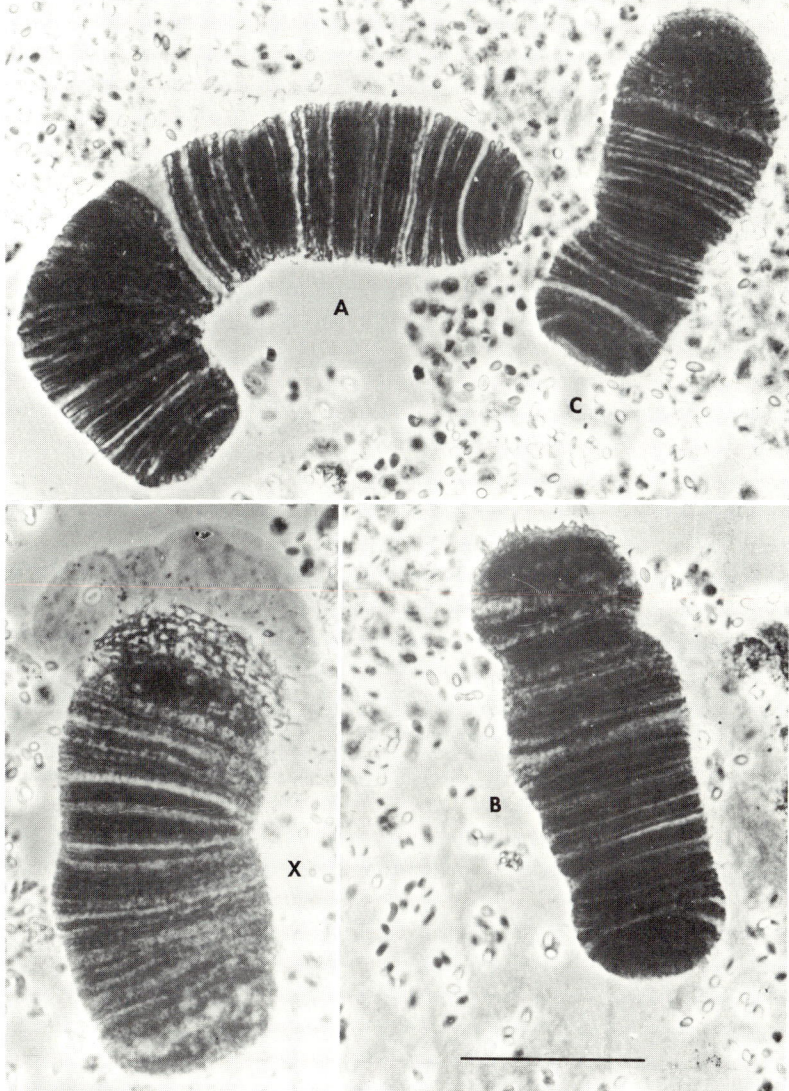

Fig. 3-2. The four super giant polytene chromosomes of an infected salivary gland cell of *Rhynchosciara angelae*. The dark granules in the background represent spores and other developmental phases of the microsporidia (*Protozoa-Sporozoa*) that are the infecting agents in this cell. Infected salivary gland cells of *R. angelae* may become more than 50 times the volume of uninfected cells of the same tissue. In uninfected cells the salivary chromosomes may have from 8,000 to 16,000 times the amount of DNA present in a gamete, whereas these super giant chromosomes have from 250,000 to 1 million times the DNA of a gamete and, when stained, are visible to the naked eye. Compare for shape and size (note difference in magnification) these chromosomes from infected cells with normal chromosomes in Figs. 3-3 and 3-5. Scale, 50μ. (Courtesy Dr. C. Pavan, University of Texas at Austin.)

reported that in hybrids of *D. pseudoobscura* or *persimilis* × *ambigua* (female) the hybrid larvae develop but salivaries were not paired.

Apparently each chromatin strand replicates its DNA during a slow period (about 30 hours compared to about 8 hours in mitosis of *Drosophila*), and structural doubling of the chromonemata follows. But no mitosis occurs. Rather, the G_2 period is followed directly by another G_1 period, another S period, G_2, etc. until each chromosome consists of about 1,000 times as much DNA as a typical nonpolytene chromosome. At full development it consists of about 1,000 to 2,000 (Kurnick and Herskowitz, 1952; Swift and Rasch, 1954; Beermann and Bahr, 1954) to 20,000 (Beermann and Clever, 1964) fine, continuous chromatin strands (Beermann and Pelling, 1965). These 200 to 250 Å strands (Gay, 1956; Rae, 1966) and chromomeres of tightly packed and folded strands (Rae, 1966) along their lengths seem to be embedded in a copious proteinaceous matrix (Painter, 1941).

The excellent review of polytene chromosomes in the dipterous *Rhynchosciara* and other Sciaridae by Pavan and da Cunha (1969a), the symposium contribution by Pavan (1965), and the earlier paper of Breuer and Pavan (1955) present extensive data that contrast to some extent with the findings reported for *Drosophila* salivaries. They report significant differences of extent of polyteny and "synapsis" among tissues and even in different regions of the salivary glands of the same individual as well as during larval development. *Rhynchosciara angelae* salivary gland cells and each chromosome were found to have about 8,000 times as much DNA as would be present in a haploid gamete (the 1C amount of DNA) or chromosome. Such chromosomes have about 8,000C and 8,000 chromonemata. But certain natural infections by microsporidian protozoa cause some salivary gland cells to become very large and to contain very, *very* large polytene chromosomes (Fig. 3-2). These super chromosomes when stained are visible to the naked eye and each may have from 250,000 to 1,000,000C and, therefore, that many chromonemata. The presence of the parasites affects polytene chromosomes differently in different tissues of the same larva but inhibition of puffs, formation of "partial puffs," formation of constrictions, polyploidy, extreme shortening, and turning one or more chromosomes each into one big puff (pomponlike) were observed.

Similar varied responses occur also in *Sciara* (Pavan et al., 1969). Some of such abnormally large cells and chromosomes in *Rhynchosciara* and *Sciara*, instead of breaking down in the pupa, have been observed in the adult insect. It has also been shown by uridine ^{3}H incorporation that these super chromosomes are very active in RNA synthesis. Since from 16 to 18 repeated doublings of the original 2C chromosomes of the embryonic presalivary gland cells occur, there is also considerable DNA synthesis in these huge chromosomes.

Ever since Balbiani first figured salivary gland chromosomes, it has been known that each chromosome may have few (in *Drosophila*) to numerous (*Sciara* and *Rhynchosciara*) small to large swellings on it (Fig. 3-3). Such swellings are called Balbiani rings and puffs; but rings and puffs have certain structural differences, and puffs can be as large as rings. Balbiani rings seem to be limited to Chironomidae, as far as known, whereas puffs occur in all Diptera (Pavan, 1965).

Painter (1934c) has claimed that "somatic synapsis" is different from the somatic pairing typical of nonpolytene chromosomes; but it is likely that there is a much greater difference between "somatic synapsis" and meiotic synapsis. It is now known that synapsis as it occurs during meiosis is always associated with the special and unique structure, the synaptinemal complex which is not found in salivaries. Furthermore, in a triploid *Drosophila* the three homologous polytene chromosomes were paired very intimately throughout their lengths (Cooper, 1938; Schultz and Hungerford, 1953), but meiotic synapsis never occurs between more than two chromosomes at any particular region. Perhaps the interesting asynapsis of salivaries in certain species hybrids (see later) may also indicate a difference from meiotic synapsis. In the plant *Haplopappus gracilis* during premeiotic mitotic divisions the homologues are also very intimately paired throughout their lengths (Fig. 5-3), but it is not synapsis

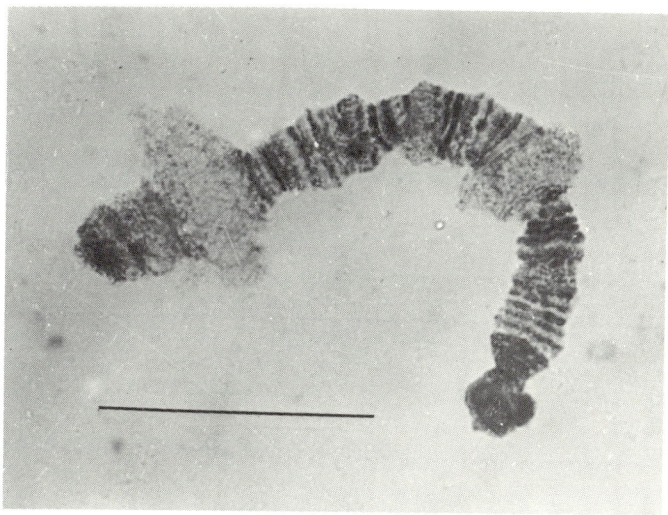

Fig. 3-3. *Rhynchosciara angelae*, chromosome C of a prepupal salivary gland cell. This chromosome, as do the others also, shows puffs that were not present at earlier stages of larval development (see chromosome C in Fig. 3-6). These two puffs show an increased amount of DNA, by gene amplification, and so are called DNA puffs. Scale, 50μ. (Courtesy Dr. C. Pavan, University of Texas at Austin.)

(Brown and Stack, 1968). It seems best to apply the term "paired" rather than "synapsed" to the homologues in salivary gland chromosomes.

As stated above, these chromosomes become polytene by passing through repeated S periods and structural doubling without ever entering mitosis. The S period is not very unusual. Replication in all salivary chromosomes of a cell is coordinated, when DNA and the protein histone is doubled (Cave, 1968). Similar coordinated DNA and histone synthesis occurs (Prescott, 1966) in the polytene chromosomes of the macronuclei of ciliates (Ammermann, 1969). There is also typical early and late replication in salivary chromosomes. There is disagreement as to whether all replicons begin DNA synthesis at the same time, are synchronized (Keyl and Pelling, 1963), or whether they are asynchronous but patterned, the S period beginning and ending in a very few loci at any time (Howard and Plaut, 1968). Plaut (1968) proposed a pattern of replication higher than mere early and late replication. Because salivary gland cells of some Diptera replicate DNA only during S periods, only about 1 cell in 25 takes up tritiated thymidine at any one time in *Chironomus* (Beermann, 1965). The frequency of cells that show incorporation of tritiated thymidine varies from species to species as well as from time to time within the same species or individual. Pavan (1965) reported incorporation following one injection into larvae of *Rhynchosciara* in from 30 to 100% of salivary gland cells (depending upon stage of development) but in only 5% or less of nuclei of Malpighian tubule or median intestine cells.

At the salivary chromosome ends there seem to be specific optical (Warters and Griffen, 1950) and electron microscopically visible elements, called "telomeres" by Berendes and Meyer (1968), and the ends of the chromosomes seem to be associated (obviously nonhomologously) with one another (Bauer, 1936; Hinton and Atwood, 1941; Warters and Griffen, 1950; Berendes and Meyer, 1968). In many fly species, especially of *Drosophila*, there is a lightly staining, irregular mass associated with the chromosomes, called the "chromocenter" (Prokofyeva-Belgovskaya, 1935) (Fig. 3-4). The chromocenter when present contains the heterochromatic centromere regions of the chromosomes and, in some species, a whole small heterochromatic chromosome (Rudkin, 1965). These heterochromatic regions replicate dif-

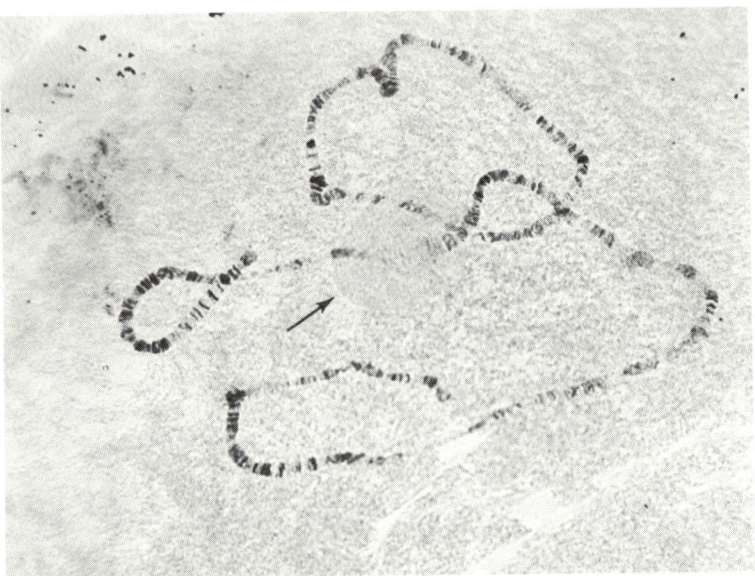

Fig. 3-4. A squash of a salivary gland cell of an unidentified species of *Chironomus* showing the chromocenter (arrow) as a lightly stained mass in which parts of the chromosomes are embedded.

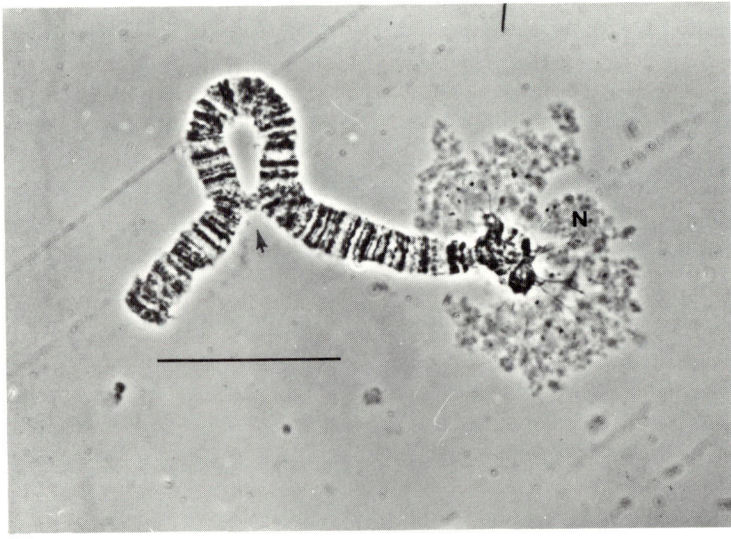

Fig. 3-5. *Rhynchosciara hollaenderi*, the salivary gland X chromosome of an early fourth instar larva. The nucleolus *(N)* is well developed, and fibers of DNA (dark spots and fibers) radiate into the nucleolus from the terminal nucleolus organizer. The X chromosome of various species of *Rhynchosciara* shows in the median region repeated groups of bands that are rather distant from each other but that pair (arrow) to each other typically. Thus, at the paired region the two X's are doubly paired homologously. Scale, 50μ. (Courtesy Dr. C. Pavan, University of Texas at Austin.)

ferently in amount compared to the euchromatic arms (Rudkin and Schultz, 1961; Rudkin, 1965, 1969), like the reverse of gene amplification. Poulson and Metz (1938) described the nucleolus-forming region of the fourth salivary gland chromosome of *Chironomus* as terminal and modified from the banded condition into a heavy network in which solid disks are replaced by interconnected spheres and granules extending outward into the nucleolus. Characteristics of the polytene chromosome that are noteworthy to cytogenetics are the alternating, transverse, darkly staining "bands" and the much lighter "interbands" along their lengths (Fig. 3-6).

It is generally accepted that each band consists of a transverse row, actually a disk, of closely associated chromomeres. Bands vary greatly in thickness and visibility, from near invisibility of a few small dots to a continuous thick line which itself may be double, triple, etc. Their spacing also varies from essentially no interband to rather long ones that then separate widely the two consecutive bands. Nevertheless, the pattern along the same chromosome is, in two individuals of a species, "constant to a most extraordinary degree" (Painter, 1934c) and is so distinctive that any particular segment of a salivary can be recognized even when translocated to some other chromosome. Segments in two different but closely related species are often identical in banding pattern.

Although a band usually runs continuously across both paired polytene chromosomes, it is obvious that an individual heterozygous for a missing band or a few bands in one homologue would show this heterozygosity in its salivaries by half a band. Also in an individual heterozygous for a band that may or may not puff in the salivary gland (Fig. 3-7), the puff will form on only one side of the chromosome as a "heterozygous puff" (Hsu and Lin, 1948; Pavan and Perondini, 1967; Perondini and Dessen, 1969).

Bands in polytene chromosomes of *Drosophila* are often double (Morgan et al., 1935; Demerec and Hoover, 1939). That is, many are duplicate, as many as one third of all bands, according to Green (1955). Either such dupli-

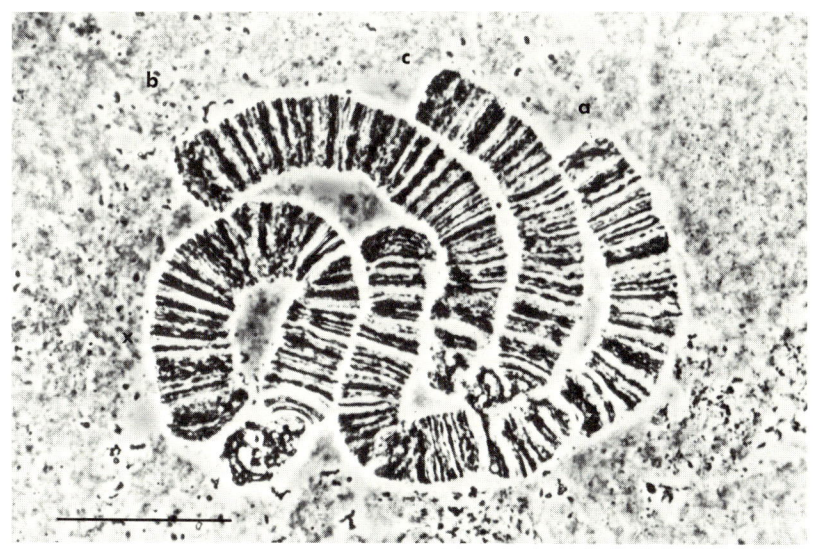

Fig. 3-6. *Rhynchosciara angelae*, the four polytene chromosomes (*a, b, c,* and *x*) of a salivary gland cell of a late fourth instar female larva. Each chromosome is a pair of homologues, each homologue of the pair consisting of 4,000 or 8,000 parallel chromonemata, bands, interbands, etc. In contrast to *Drosophila*, there is no chromocenter in salivary gland cells but in intestine and malpighian tubule cells of *R. angelae* it can be found. Compare this chromosome C from a fourth instar larva that has no puffs with the prepupal chromosome C in Fig. 3-3, which has two DNA puffs. Scale, 50μ. (Courtesy Dr. C. Pavan, University of Texas at Austin.)

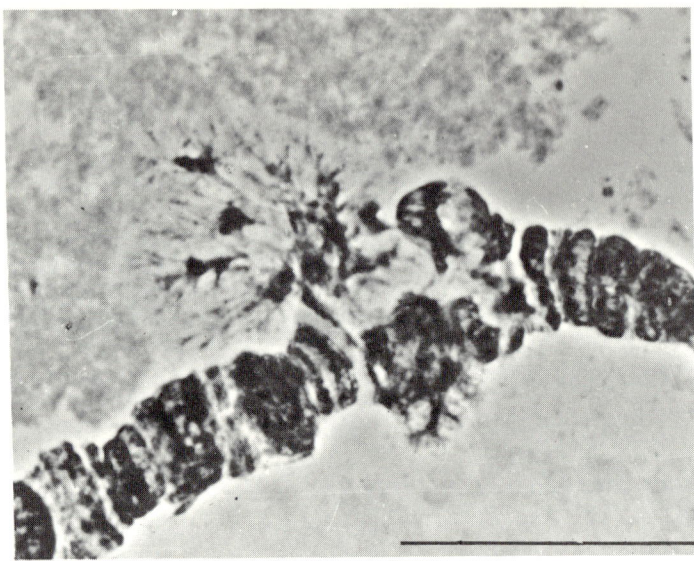

Fig. 3-7. Heterozygous puff of salivary chromosome A of *Sciara ocellaris*. The bands in the two homologues have different amounts of DNA before the puffing. The dark fibers radiating from the central part of the puff—half to the periphery—are of DNA. The locus of the other homologue is just starting to puff. Scale, 25μ. (Courtesy Drs. C. Pavan and A. L. P. Perondini, University of Texas at Austin.)

cated "genes" are not the sort that geneticists study, which are usually present singly as "unique" genes; or, if the duplicated bands represent duplicated unique genes, then one or other must have mutated into another distinct unique gene. Otherwise simple mendelian segregation would not occur as it usually does if a band represents a gene.

Beermann (1952) in *Chironomus tentans* and Pavan and Breuer (1952) in *Rhynchosciara angelae* have determined that the banding pattern is the same in polytene chromosomes from different tissues of the same individual or species (Beermann and Clever, 1964); but in the same tissue or cell at different stages of development they may differ considerably in thickness (Breuer and Pavan, 1955).

A particular band usually extends across both paired polytene chromosomes, and each band has been claimed to be the site of one gene, the one band–one gene concept of Bridges (1938), Berger (1940), Welshons (1965), and others. Kosswig and Shengun (1947), however, reported different banding patterns of polytene chromosomes from different tissues of *Chironomus* and concluded that bands (disks), as contrasted to interbands, do not specifically represent genes.

Fujita and Takamoto (1963) considered that bands represent inactive heterochromatic loci but interbands contain active, uridine-incorporating, m-RNA-producing, euchromatic genes, as had Kostoff (1938a); although Steffenson (1963) reported that DNA is essentially confined to bands. On this concept of heterochromatic bands and euchromatic interbands, genes are continuous along all of the hundreds of longitudinal strands; but some are inactive within the heterochromatic chromomeres of the bands, whereas others within the euchromatic interbands are active. This concept seems sound in the light of present cytogenetic knowledge of euchromatin and heterochromatin. Nevertheless, regardless of presently uncertain location of active genes, in bands or interbands (Pavan and da Cunha, 1969b), the genes of the paired polytene chromosomes are paired gene-for-gene as exactly as in meiosis.

CYTOGENETICS

All of this detailed study of polytene chromosomes of flies began about 1933, shortly after McClintock and others had established the great cytogenetic value of maize pachytene chromosomes from 1929 to 1932. Kostoff (1930) indicated the apparent similarity of the banded salivary chromosomes and genetic maps. In the fall of 1932 Painter (1934c), using the same acetocarmine squash technique employed so successfully in the study of maize pachytene chromosomes, started a cytogenetic study of salivary gland chromosomes of *Drosophila melanogaster*, drawing on the excellent collection of genetically known chromosomal mutant strains already established by his geneticist colleagues Muller, Patterson, and Stone at the University of Texas. Painter quickly established the great cytogenetic value of salivary chromosomes (Painter, 1933, 1934a, 1934b, 1934c, 1935). Almost immediately geneticists and cytogeneticists in many countries adopted the technique and applied it to many cytogenetic and evolutionary problems, for example Bridges (1935), Fujii (1936), and Bauer (1936). More recently studies of the puffs and Balbiani rings that occur on salivaries have yielded valuable knowledge of gene action and development.

During the 1920's geneticists and cytogeneticists of *D. melanogaster* had worked out genetic maps of the four mitotic chromosomes and had related gene loci to positions on the mitotic chromosomes (Fig. 3-8). Much of this cytogenetic work followed Muller's discovery in 1927 that X rays cause chromosomal aberrations as well as gene mutations. Thus there were two sorts of maps by the early 1930's, genetic and mitotic (Muller and Painter, 1932). One of the first cytogenetic efforts of the study of salivaries was to map the bands of the various chromosomes and to relate the bands of chromomeres to known linkage maps of *D. melanogaster* (Painter, 1934c; Bridges, 1935; Mackensen, 1935) and occasionally to refine and revise them (Bridges, 1938; Bridges and Bridges, 1939; and others). Banding patterns were determined for other species such as *D. virilis* (Fujii, 1936; Hughes, 1936) *D. azteca* and *D. athabasca* (Bauer and Dobzhansky, 1937), *D. pseudoobscura* (Tan, 1936), and others.

The cytogenetic study of salivary gland chromosomes of *Drosophila* individuals heterozygous for genetically known chromosomal

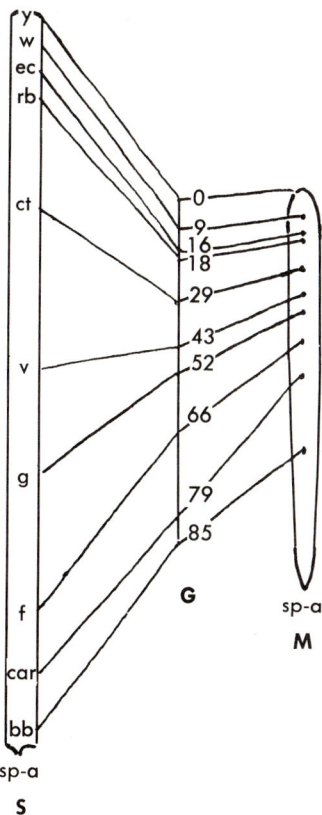

Fig. 3-8. Comparison of salivary *(S)*, genetic *(G)*, and mitotic *(M)* linkage of genes on the X chromosome of *Drosphila melanogaster*. *sp-a* is an accepted indication of the location of the centromere. The order of genes is the same in all three; but the absolute positions and distances between genes vary among the three models. (Data derived from Muller, H. J., and T. S. Painter. 1932. Z. Ind. Abst. Vererb. **62**:316-365, and Bridges, C. B. 1938. J. Hered. **29**:11-13.)

alterations or other observable markers such as ends and centromeres permitted the assignment of the markers to physical locations within the chromosome. They could be located exactly with respect to specific bands. Since the locations of such chromosomal markers are also known genetically with respect to particular genes in the linkage map derived from crossover percentages among genes, genes can be assigned to specific bands. Thus there are three different maps for *D. melanogaster*, genetic, mitotic, and salivary. The three resulting types of maps, the genetic and the two cytological (Fig. 3-8), can be compared for gene loci, map units, and bands (or interbands).

When such comparison was made (Painter, 1934c; Bridges, 1938; Bridges and Bridges, 1939) it was found that the three types of maps are identical for *order of genes* (map units and bands) but they often differ with respect to distances between certain genes and they differ as to general location within the chromosomal arms as to where most known genes are located. For an example of a comparison of the genetic and salivary maps, near the end of the X chromosome the regions from *fa* to *w* and from *w* to *y* each have about 1.5% of crossover frequency; but *fa* to *w* on the chromosome map is a short segment of the chromosome with about 5 bands whereas the *w* to *y* region is a long stretch, 16 times as long, consisting of about 75 bands. Crossing-over must occur very frequently in the short region between *w* and *fa* but quite rarely, about one sixteenth as often, in any comparable length between *y* and *w*. It can be concluded that crossover frequency is not uniform along a chromosome.

Comparison of genetic with chromosomal maps has also yielded the conclusion that, aside from local differences, the maps of the X chromosome of *D. melanogaster* agree completely in the gene sequences (Bridges, 1938). Along the arms of chromosomes No. 2 (Fig. 3-9) and No. 3, however, on the chromosomal map, most of the known genes lie in the distal half whereas on the genetic map they are spaced right up to the centromere. The gene *pr* lies 0.4 crossover units from the centromere in the left arm of chromosome No. 2, yet on the chromosomal map it lies about 25% of the distance from the centromere toward the end of the arm. Almost without exception the known genes lie further toward the ends of the arms on the cytological than on the genetic map. One reason for this is that the proximal regions of the arms are heterochromatic and contain few known genes, and crossing-over is much less common than in the distal halves of the arms.

The cytological study of the salivary gland and mitotic chromosomes of *Drosophila* has added greatly to our knowledge of the distribution of genes on the real chromosomes that the theoretical linkage map "chromosomes" could not reveal. On the other hand mere cytological mapping of the bands of salivaries without correlation with genetic maps would contribute little. It is the union of cytology with genetics that makes the cytogenetic contribution.

Within a few years after 1933 a number of significant cytogenetic contributions had been recorded in the genus *Drosophila* from salivary gland chromosome studies. Among them were cytological confirmation and location of genetically known inversions (very common), deletions (rare according to Bridges et al., 1935; Demerec and Hoover, 1936), duplications (uncommon), translocations (few according to Morgan, 1939, although rare according to Dobzhansky and Socolov, 1939), and linkage groups. Inversions, especially paracentrics (not including the centromere), have been found to be common, for example more than 20 in *D. pseudoobscura*, and 17 in 19 strains of *D. athabasca* (Novitski, 1946), and 115 among 69 Hawaiian species (Carson, 1970; Bock, 1971). Inversions have been widely utilized as indicating relationships among species (Hughes, 1939; Patterson et al., 1940), varieties, or strains (Dobzhansky and Sturtevant, 1938; Dobzhansky and Socolov, 1939) and within species (Epling et al., 1953) and are still being used in that manner (Carson, 1970; Bock, 1971). Use of inversions is justified because "There is no evidence that the same rearrangement has occurred even twice spontaneously" (Patterson and Stone, 1952). Therefore, an inversion becomes a permanent marker of evolutionary change. When the same inversion is found on chromosome segments of similar banding arrangements, those particular taxa are related.

The nature, cause, and effects of inversions will be discussed in Chapter 14. At this point a few examples of their use in evolutionary studies of salivary gland chromosomes are given.

In recent years Frizzi, Kitzmiller, Baker, their associates, and others have been making detailed cytogenetic studies of mosquitos, including detailed studies of salivary gland chromosomes and their banding patterns (Kitzmiller, 1963; Kitzmiller et al., 1967). It is possible to make interspecific hybrids, and the

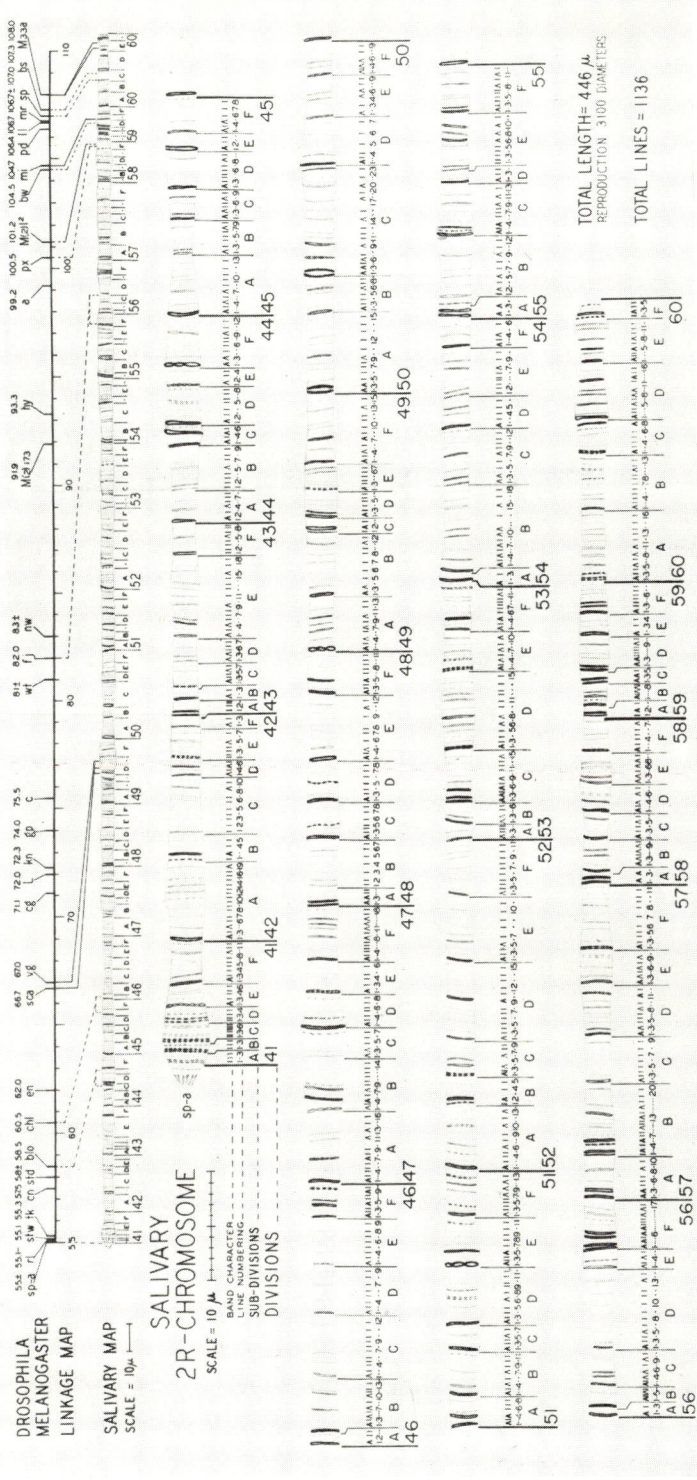

Fig. 3-9. Linkage map (top) compared to the salivary map, next below. *sp-a* at left end of the genetic map and elsewhere designates the location of the centromere, and all of this is for the right arm only of the second chromosome. It is evident that the salivary map places the gene loci closer to the end of the arm (toward the right) than does the genetic map. The remainder of the figure is a plot of bands and their characters as well as the accepted divisions and subdivisions of the arm. (From Bridges, C. B., and P. N. Bridges. 1939. J. Hered. **30**:475-476, plus map.)

stage of development reached is assumed to indicate closeness of relationship, as in freshwater fish. A second indication of relationship is a distinctive sequence in banding differences and aberrations in salivary gland chromosomes. As Baker, et al. (1968) proposed, "It is suggested that in the genus *Anopheles* certain evolutionary chromosome patterns occur in a sequence. First, there are X chromosome differences in closely related species. Second, as species become more divergent, changes also occur in the autosomes. Third, the ends of the arms remain fairly uniform within subgenera, but differ among subgenera." Complex aberrations occur in the centromeric regions.

An interesting condition of X chromosomes in *Anopheles* species hybrids is complete asynapsis of the X chromosomes in salivary gland cells in spite of *identical* banding pattern. This asynapsis is reflected in failure of somatic pairing, in germ line cells at least. The above papers refer to this salivary gland asynapsis as "genetic differences which are independent of the banding pattern." The significance of this concept for the argument of location of genes in bands or interbands is potentially great. Perhaps bands and interbands are unrelated directly to genes. "*The banding pattern of the chromosomes is an expression of the phenotype, and not of the genotype*" (Kitzmiller et al., 1967). Similar asynapsis of nearly homologous salivaries in species hybrids of *Drosophila pseudoobscura* × *ambigua* has been observed (Buzzati-Traverso, 1950). It is also interesting that the ends of autosome arms of species show least and last changes. Complete or almost complete asynapsis of homologous autosomal salivaries, even autosome arms with *identical* band patterns, was reported in such F_1 hybrids as *A. freeborni* × *punctipennis* and *earlei* × *freeborni*. In other hybrids there were different degrees of asynapsis (Kitzmiller et al., 1967) but never very much.

As in other dipterous salivaries, puffs are reported on mosquito salivary gland chromosomes but they have not been studied in relation to RNA puffs, DNA puffs, or development.

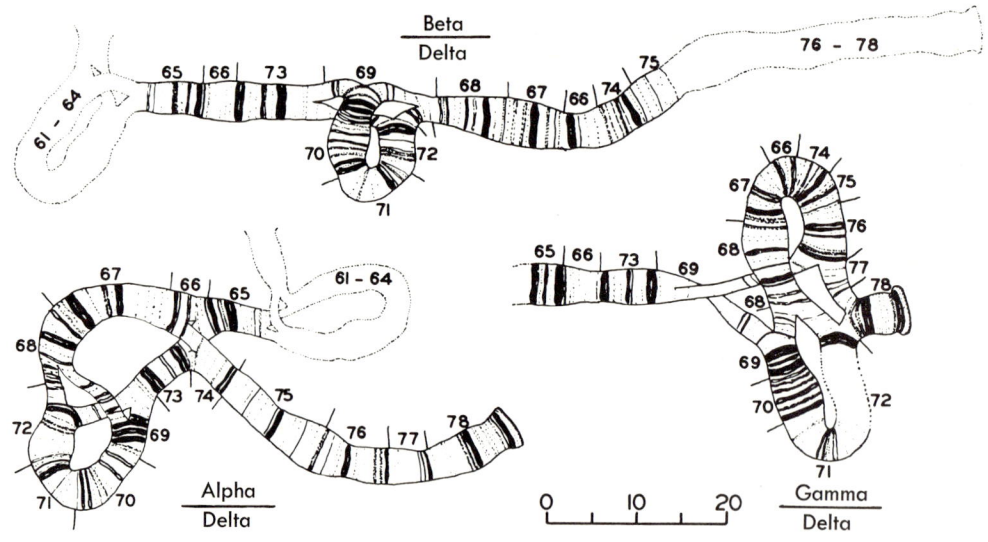

Fig. 3-10. Inversion loops of varying complexities in the B^1 arm of salivary gland chromosomes of F_1 hybrids within the species *Drosophila azteca*, the parents being designated as alpha, beta, gamma, and delta. In the beta/delta hybrid there are two simple inversions, in the 61-64 region and in the 69-72 region (No. 2, Fig. 3-11). The latter inversion is present also in the alpha/delta and in the gamma/delta hybrids, but other inversions also are present. In the alpha/delta hybrid there are two included inversions (No. 3, Fig. 3-11), one from 66 to 74 including the smaller one from 69 to 72. In the gamma/delta hybrid there are two overlapping inversions (No. 4, Fig. 3-11.) (See legend for Fig. 3-11.) (From Dobzhansky, T., and D. Socolov. 1939. J. Hered. **30**:3-19.)

For example, Baker et al. (1968) report 30 puffs and one Balbiani ring typically on the three salivary gland chromosomes of *A. pulcherrimus*.

EVOLUTIONARY STUDIES

Drosophila pseudoobscura is rather unique because many individuals are heterozygous for inversions, especially on the third chromosome. Dobzhansky, in a series of publications, reported that the proportions of different inversions in specific populations vary seasonally or change progressively during a number of years (1943, 1947, 1948), with altitude or weather (1952). Epling et al. (1953) continued this 14-year study by attempting to determine whether or not heterozygotes for inversions were superior to homozygotes in two contrasting environments. The heterozygous conditions were found to be not obviously superior to the homozygous but "by their presence inversion may assist in canalizing the fitness of a local population to its particular milieu by reason of their restriction on recombination and that they may thus conserve the adaptability of the species," probably because inversions, at least short ones, reduce crossing-over in the region of the inversion (Sturtevant and Beadle, 1936).

In their study of *D. azteca*, Dobzhansky and Socolov (1939) crossed homozygous strains so that the inversion loops were evident in the salivaries of the heterozygous hybrids (Figs.

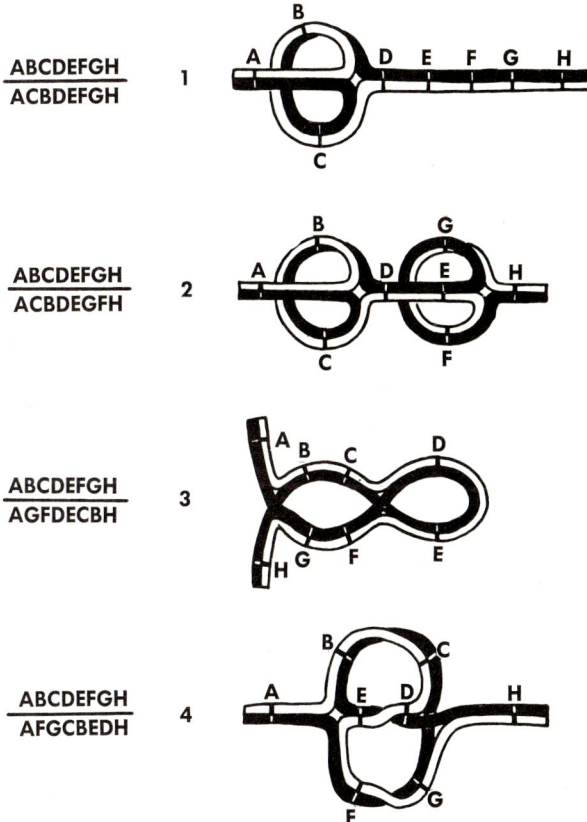

Fig. 3-11. Types of inversions and how they pair gene-for-gene in the heterozygous salivary gland cell. **1,** A simple inversion loop with reverse pairing. **2,** Two independent inversions. **3,** Two included inversions, a small one *(DE)* within a larger one *(BCDEFG)*. **4,** Two overlapping inversions, which requires first that *ABCDEFGH* form *ABCGFEDH;* later a second inversion occurred, including some of the same genes, *ABCGFEDH* forming *AFGCBEDH.* (From Dobzhansky, T. and D. Socolov. 1939. J. Hered. **30:**3-19.)

3-10 and 14-3). One strain was taken as the "standard," and other strains were crossed with it. They observed single loops, two and three independent loops, included loops (when one shorter inversion occurs within a longer one), and overlapping complex loops, the last type forming when two inversions overlap one another so that they both include one section in common. Diagrammatic representations of such loops are discussed as a restatement of a more detailed discussion of the theory of multiple inversions in Dobzhansky and Sturtevant (1938) (Fig. 3-11). Since each inversion originated only once and probably at a different time from all other inversions, it is possible to arrange strains in a phylogenetic sequence relative to the "standard arrangement"; but the "standard" may actually have the inversions and not the strain mated to it. As a result the direction of the phylogenetic changes, the chronological or evolutionary sequence of formation of inversions, is unknown. Either the standard is arbitrarily chosen or it is the strain most likely to represent the original type in the species (Fig. 14-11 as an example) (Bock, 1971).

A recent attempt to determine evolution of some Hawaiian species possessed a unique advantage to such studies, a known point in time, the age of the island of Hawaii, about 700,000 years (Carson, 1970). Thus, the endemic species of "picture-winged" *Drosophila* on that island must have evolved within that time from one or very few "founders" from the adjacent and older island of Maui. Because hybridization proved impossible among the species, as is true of most species in the genus, salivary gland chromosomes were compared for inverted segments to a photograph of the standard. As an example, it was determined that two of the endemic species of Hawaii, *D. silvestris* and *D. heteroneura*, have exactly the same inversions as *D. planitibia*, endemic to Maui. Therefore it is assumed that the probably younger species of the younger island of Hawaii are derived from a common ancestor of the older Maui species, and that speciation, including reproductive isolation (they are all intersterile), took less than 700,000 years. Other species groups from Maui and Hawaii when compared also have similar formulas of inversions but differ somewhat, indicating acquisition of some new inversions on both Maui and Hawaii since the "founders" crossed the 30 miles to Hawaii. Among these Hawaiian species of *Drosophila* morphological and reproductive evolution seems to be so rapid that similarity of inversion formulas is often unexpected.

DEVELOPMENTAL STUDIES

During the past 20 years the puffs of salivary gland chromosomes have been used in cytogenetic studies of development, as simply reviewed by Beermann and Clever (1964) and Beermann (1965). A puff arises from a single band, as though all of the heterochromatic chromomeres of the particular band loosen and extend outward as a ring of chromatin loops embedded in copious nonhistone protein (Mechelke, 1953). The chromatin of the loops is active in m-RNA transcription, as indicated by incorporation of radioactive uridine. In fact, RNA transcription seems to be the almost exclusive activity of puffs except in some sciarid and a few other flies, in which some puffs previously or simultaneously produce large amounts of DNA (Breuer and Pavan, 1955; Rudkin and Corlette, 1957; Gabrusewycz-Garcia, 1964; Amabis and Cabral, 1970).

Puffs vary in size from little more than somewhat swollen bands to the largest extreme form. The experts have concluded that in *Chironomus* approximately 10% of bands may be more or less puffed in any nucleus (Beermann, 1965), although many hardly appear so. But there are other cases in which higher percentages of bands are active (Pavan, 1965). Evidence indicates the probability that a band of a polytene chromosome is inactive but that if it forms a puff it has become somewhat active in RNA transcription and if it becomes a huge, ball-like, diffuse puff or a Balbiani ring it has become very active. Therefore the particular bands that have puffed determine the particular activities of the cell. It is not surprising, then, that cells of different tissues or the same cells at different periods of development have different arrays of puffs. Evidence for this concept has been found in *Chironomus* by Beermann (1961) and in *Rhynchosciara* (Mechelke, 1953; Breuer and Pavan, 1955).

Chironomus tentans and *C. pallidivittatus* are two closely related species that can be hybridized and produce fertile hybrids, although a number of inversions prevent much crossing-over in the hybrid. The salivary *glands* of these species differ, especially in one particular region that consists of four cells. In *pallidivittatus* these four cells secrete a particular material not secreted by *tentans*; otherwise they are the same. It has been shown that in these cells of *pallidivittatus* there is among others a particular Balbiani ring on the fourth chromosome. In *tentans* there is no such lobe-specific ring in that position. Genetically the production of the particular material product is a simple mendelian dominant phenotype, and cytologically the same band is present in both species and has not been deleted in *tentans*. In the hybrid the ring forms only on the *pallidivittatus* half of the fourth polytene chromosome and some of the specific material is produced. It seems, therefore, that *tentans* has the gene (the band), and so the genetic information seems to be there. But in the hybrid the *pallidivittatus* gene only is operationally able to respond to some "operator" gene. But if bands are phenotypes and not genes, this interpretation may fail.

In *Chironomus* specific bands become puffs in response to the hormone ecdysone, which is produced by the larva especially at shedding. The response of those bands to ecdysone is followed by that of some other specific bands and still later others, in a chain reaction. It has also been shown that bands that respond to ecdysone also respond to changes in ionic concentration (increase in K^+ ions within the cell), as though ecdysone is the "first messenger" that effects the enzymatic synthesis of a "second messenger," such as 3', 5' adenosine monophosphate (Rasmussen, 1970). The resulting altered permeability of the plasma membrane permits an increase in K^+ concentration within the cell which activates one or more genes, and certain puffs form.

It is now known that there are several chemical and physical agents capable of changing the pattern of puffs in different species of Diptera (Kroeger and Lezzi, 1966), and recently Pavan et al. (1969) have shown that the puffing pattern can be changed by biological agents.

The contributions of polytene chromosomes in the new field of "gene amplification" were discussed in Chapter 2.

CHAPTER 4

Mitosis

The biological domain of the physical world has conservation principles of its own, operating within the larger ambit of the conservation principles of physics. The most general principle is that biological systems must reproduce themselves in order to sustain their very existence over long periods of time.

D. Mazia, 1961

Mitosis, next to DNA replication, is the fundamental form of reproduction in all organisms. Life can exist at no level lower than the complete cell with its hereditary and informational DNA, its organelles, its enzymes, and other materials required for all of the myriad processes necessary for life. Mitosis generally divides this mass into more or less equal and self-sufficient daughter cells after the DNA and other essentials have been exactly or approximately doubled. During the interphase preceding mitosis, the DNA is exactly doubled quantitatively and qualitatively by the semiconservative process, followed immediately or later by structural doubling into chromosomes consisting of two genetically and cytologically identical chromatids. Subsequent events divide the chromatids into two identical groups, and the cytoplasm is divided usually into two very similar portions (Fig. 4-1). Thus the daughter cells almost always are genetically identical to each other, and each is genetically identical to the cell from which it came. The division of the cytoplasm does not need to be exact, and often it is not; in some cases it does not occur at all.

Aside from the chromosomes at metaphase and their anaphase separation, there is rather little cytogenetics in the process of mitosis. Like all biological processes, all aspects of mitosis must be gene controlled, although during mitosis itself nearly all genes are inactive. It is assumed, therefore, that the RNA transcription and other preparations for all aspects of the process have been completed no later than about mid-prophase when the chromosomes have become quite condensed and the genes inactive (Mazia, 1961; Van't Hoff, 1966; Webster and Van't Hoff, 1969, 1970). There is an unlikely possibility that the genes located in the region of the centromere may still be active, thereby producing the uncondensed region called the primary constriction. The same may be true of the uncondensed region called the nucleolus organizer. Nevertheless, autoradiographic study of chromosomes during mitosis (probably during late prophase, metaphase, and anaphase) indicates that the centromeres do not replicate DNA during that time (Comings, 1966).

"Nonmitotic" nuclei are of two sorts: (1) *interphase* nuclei, or those that will normally enter mitosis as soon as ready, and (2) *postmitotic* nuclei (or cells), or those that will normally never again divide. In general the interphase nucleus replicates its DNA whereas the postmitotic nucleus usually remains in the G_1

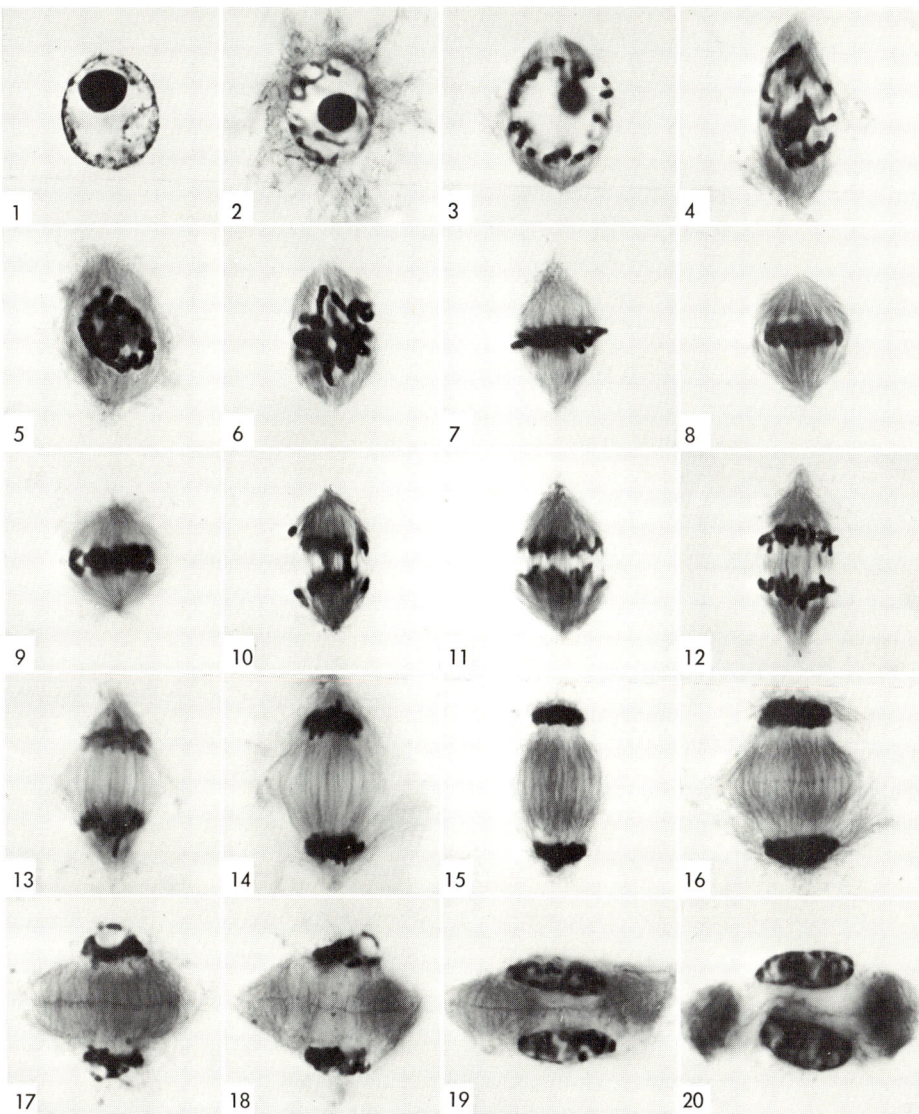

Fig. 4-1. Mitosis in maize endosperm. These divisions occurred in the late "plasmodial" stage of endosperm development and were isolated as whole spindles. *1*, Interphase. *2*, Early prophase with adhering, radially oriented fibrous material that is probably the beginning of extranuclear spindle fiber formation. *3*, Prophase with "polar caps," early spindle formation. In *1* to *3*, notice that chromatin is mostly just inside or against the nuclear envelope. *4*, Prometaphase shortly after nuclear membrane breakdown. *5* and *6*, Prometaphase chromosome movement. *7*, Metaphase. *8* to *14*, Progress of anaphase showing progressive development and buildup of the phragmoplast (interzonal fibers). *15* to *20*, Progressive stages of telophase, especially of the phragmoplast and cell plate. *18* to *20*, Centrifugal formation of a ring of new fibers moving outward and extending the cell plate. These illustrations seem to support the contention that the phragmoplast is as much a distinct cell structure as the spindle and is distinct from the spindle. (From Duncan, R. E., and M. D. Persidsky. 1958. Amer. J. Bot. **45**:719-729.)

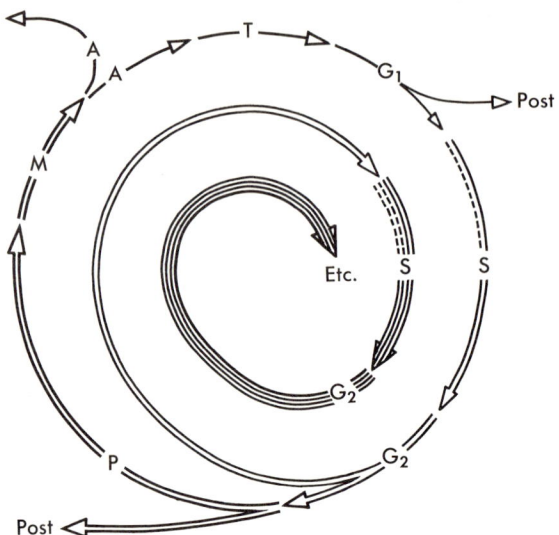

Fig. 4-2. Diploid mitotic cycle. Single arrows = 2C amount of DNA, double arrows = 4C, quadruple arrows = 8C, etc. G_1 extends from telophase, T, to the beginning of the synthetic phase, S, the nucleus and cell having the 2C amount of DNA. It is likely that most differentiated cells (in the 2C condition) become postmitotic from G_1. Synthetic phase is characterized by the gradual doubling of DNA to the 4C amount. Histone protein also doubles during that phase. G_2 is characterized by no DNA or histone synthesis, but other proteins and RNA increase. In the G_2 phase a cell *may* (outer track) become postmitotic and differentiate at the 4C level (although some investigators believe differentiation *always* occurs at the 2C level). It may (inner track) undergo one or many further DNA doublings without nuclear division, as occurs in polytene, endomitotic, and other cells. G_2 may (middle track) terminate at the beginning of the prophase, P, of another division and pass through metaphase, M, and anaphase, A, to another telophase. Haploid mitosis, as occurs in plant gametophytes, has 1C telophase and 2C prophase; otherwise, it is comparable to diploid mitosis. (From Brown, W. V., and E. M. Bertke. 1969. Textbook of Cytology. The C. V. Mosby Co., St. Louis.)

phase without doubling its DNA (Fig. 4-2). Cells may enter mitosis from interphase rather inevitably, or cells in the postmitotic condition may be stimulated naturally or by treatment to do so. In the latter case some sort of inhibition must be removed to stimulate DNA replication (Mazia, 1961). In plants stimulation may be produced by wounding, by application of auxin; in animals by phytohemagglutinin, etc. It has been demonstrated that a response of white blood cells to phytohemagglutinin is almost immediate, followed by DNA replication, increase in RNA transcription, and enzyme activity (Nowell, 1960; Kleinsmith et al., 1966). Since most postmitotic cells are arrested in the G_1 phase, stimulation to mitosis must restore the chromosomes to the DNA replication condition of the S phase by removal or neutralization of inhibition.

Mazia (1961) considers the interphase as preparation for and continuous with mitosis, mitosis being the normal condition. Therefore, the beginning of mitosis is some arbitrarily chosen condition called the beginning of prophase. Mitosis can occur at any level of whole chromosome sets (haploid, diploid, polyploid) as well as in aneuploid cells having extra chromosomes or lacking some, just so long as the genetic constitution is viable.

Within a cell population, such as mammalian cells in culture, between 90 and 100% are viable (MacDonald and Bruce, 1968), and Luykx (1970) concluded from analysis of pertinent data that the average rate of spontaneous mitotic nondisjunction for a particular chromosome is once in from 1,000 to 10,000 divisions. Thus, mitotic divisions are not quite perfectly regular. Luykx (1970) has recently discussed cellular mechanisms of chromosome distribution in detail.

PROPHASE

Prophase is characterized cytologically (and not necessarily biochemically) as the first appearance of chromosomal condensation. As prophase progresses, the chromosomes become shorter, thicker, and more clearcut (Fig. 4-3) and tend to condense on the nuclear envelope where they are attached to it, leaving the center of the nucleus rather free of chromosomes (Comings and Okada, 1970c) (Fig. 4-4). It is obvious as prophase advances in many species, not in all, that the chromosomes are not randomly arranged within the nucleus (Vanderlyn, 1948; Comings, 1968). Apparently, they have in prophase the same arrangement as at the previous telophase with the centromeres all grouped together at one region, the *pole*, close to the nuclear envelope and with the arms extending around the inside of the nuclear envelope. The ends of some or all of the chromosome arms, depending upon whether the chromosomes are acrocentric or metacentric, are attached at the *antipodal* region of the nucleus to the envelope (Fig. 5-4). Whether this polarized arrangement of the prophase chromosomes is related to the orientation of the future spindle axis is unknown. If it is, then the nuclear rotation, emphasized by Vanderlyn

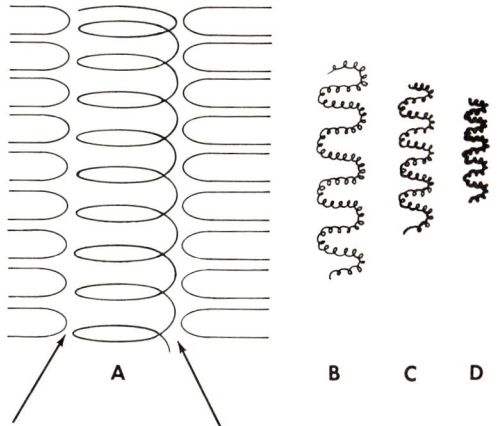

Fig. 4-3. Highly diagrammatic representation of possible chromosomal changes. *A-D,* Condensation, interphase to metaphase. *D-A,* Expansion, telophase to interphase. The chromosome stays in the same place but expands or contracts. Heterochromatin might be a portion like B or C inserted in A. At interphase *(A)* the chromonemata of two homologous or nonhomologous chromosomes or different parts of the same chromosome will lie close enough (arrow) for reassociation following two breaks to produce somatic crossing-over, translocation, or inversion. Although the form of a chromosome is here represented as coiled, the true course of a chromonema from one end to the other is unknown.

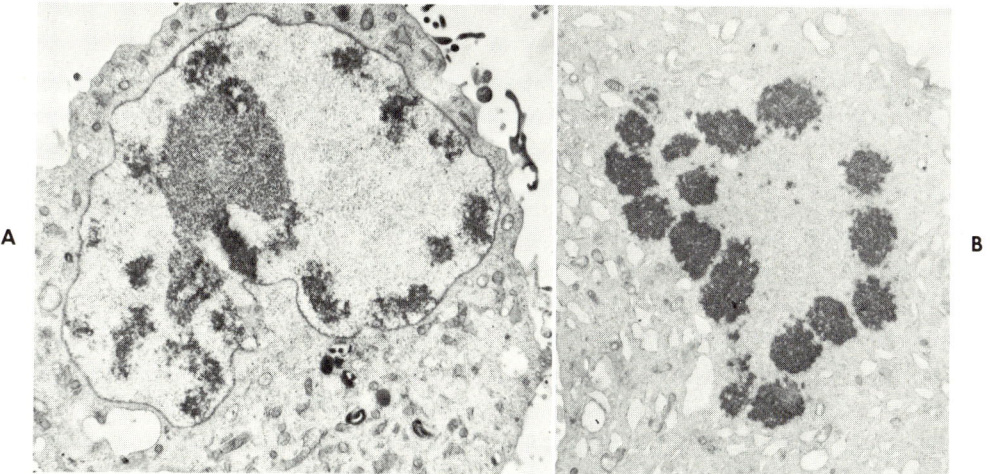

Fig. 4-4. Prophase, **A,** and prometaphase, **B,** of mitosis in cultured cells of the male Indian muntjac (2n = 7). In **A** the chromosomes lie against the nuclear envelope or close to it. The same arrangement persists in early prometaphase after disappearance of the nuclear envelope. (From Comings, D. E., and T. A. Okada. 1970. Exp. Cell Res. **63**:471-473.)

(1948), is necessary if the spindle axis of the new division is to be different from that of the previous mitosis. Pickett-Heaps and Northcote (1966) have noticed prophase microtubules near the plasma membrane where cytokinesis will occur at telophase. Obviously the cell "knows" at or before prophase how it is going to divide even if the division is very asymmetrical (Stebbins and Shah, 1960). Further evidence that the prophase chromosomes are not randomly arranged derives from metaphase and anaphase somatic pairing of homologues, as in Diptera and other animals (Smith, 1942) and various plants (Watkins, 1935; Brown and Stack, 1968; Stack and Brown, 1969a), as well as the claimed end-to-end pairing of prophase chromosomes (Wagenaar, 1969) (see Chapter 5).

Wagenaar (1969) and Wagenaar and Sadasivaiah (1969) have demonstrated in a few low chromosome number plants from squashed preparations that such prophase chromosomes may be associated nonrandomly end-to-end, both homologously and nonhomologously. Nonhomologous nonrandom end-to-end association at metaphase and anaphase has been observed during the otherwise atypical mitotic divisions called *premeiotic* in the $2n = 4$ species of plant *Haplopappus gracilis* (Brown and Stack, 1968) (Fig. 5-3). Electron microscopy of burst metaphase chromosomes does indicate chromatin connections between nonhomologous chromosomes (DuPraw, 1968; Wolfe and John, 1965) that may tie them together. Much more study is necessary before it can be assumed whether such end-to-end association is a general condition or not.

Later in prophase in higher plants and animals, the nuclear envelope seems to become fragmented, probably enzymatically. Either the fragments move out into the cytoplasm and remain intact, or they are destroyed, or they are converted into endoplasmic reticulum, no one knows which. There have been numerous reports from electron microscopists, for example Comings and Okada (1970b), that fragments of nuclear envelope of some mammals remain attached to chromosomes through metaphase and anaphase. These could function as templates for the formation of the telophase nuclear envelopes.

Evidence from burst cells and nuclei examined by electron microscopy has shown that chromatin filaments of the interphase chromosomes seem to be attached to the nuclear envelope (Chapter 2). During prophase such connections must be removed thereby freeing the nuclear envelope fragments so they can move away, if they really do. In many lower forms the nuclear envelope is permanent, division occurs within it, and chromosomes are free.

Also starting in mid-prophase, rather typically in higher organisms, is the dissolution of the nucleolus or nucleoli: it usually fades away to nothingness before metaphase. In many lower forms, such as algae, fungi, and protozoa, the nucleolus is a permanent organelle of the nucleus and cell. At metaphase or anaphase it pinches in half, and the two small daughter nucleoli move to the poles and function in the two daughter cells.

Mrs. Jane Robb (personal communication) has shown in the fungus *Ustilago hordei* that the nucleolus is apparently permanently attached to the end of a chromosome. When that chromosome divides at anaphase, the nucleolus is "pulled apart" into two daughter nucleoli that accompany the two anaphase chromosomes into the daughter nuclei.

In some beetles the two or more sex chromosomes remain attached to the persistent nucleolus through meiotic metaphase I, and the nucleolus is considered to have a function in sex chromosome segregation by Wahrman and Friedländer (1966).

In some higher forms, such as certain flowering plants (Brown and Emery, 1957) and some mammals, as well as some fungi (Robinow and Caten, 1969), the nucleolus only partially fades away during prophase (it probably loses some of its molecular constituents). A faint "ghost" of it remains, which may divide at metaphase or anaphase but is usually excluded from the daughter nuclei (Hsu et al., 1965).

In those organisms having centrioles, all animals and some algae, the centrosome divides during prophase, and the two daughter centrosomes move apart just outside the nucleus and eventually stop at the opposite positions that will become the poles of the spindle.

By some unknown mechanism the prophase cell "tells" the centrioles where to stop so that each will be at a spindle pole. It is not likely that the positions occupied by the centrioles determine the axis of division; that matter must be "decided" earlier by the nucleus or the cell as a whole. Essentially nothing is known of how the axis of division is determined, but it is certainly not random. The interphase centrosome lies just outside the nucleus, often within an invagination of the nucleus, and contains the two centrioles from a previous telophase division. As the daughter centrosomes move apart during prophase, the contained centrioles somehow produce many radiating and elongating microtubules that form the asters. Centrioles, centrosomes (also called centers), and asters face each other across the nucleus by late prophase and may be involved in formation of at least some of the microtubules of the spindle.

Friedländer and Wahrman (1970) have concluded that the positions at the poles of centrioles are a result, not a cause, of spindle formation, and that by late prophase centriolar microtubules are gone and there is no connection of centrioles to spindle microtubules. They agree with Dietz (1966) that the spindle is, in part, a basal body (centriole) distributor.

At some time, possibly during prophase or during late interphase—there is little knowledge of time or mechanism of this structural doubling—each chromosome becomes structurally longitudinally double, consisting of two parallel chromatids held together at or near the centromere (Chapter 8 and Figs. 19-1, 19-2, etc.). This double arrangement is finally broken at the beginning of anaphase. However, the doubleness is not always evident during prophase nor always during metaphase. It is probable that the centromeres, whatever they may be, are formed during late prophase, although no study has followed their telophase fate and prophase formation. They may be permanent structures.

PROMETAPHASE

By the end of prophase in typical mitosis the nucleolus and nuclear envelope are gone, the axis of division is evident, the spindle is forming, and the chromosomes are nearly as short as they ever will be. Prometaphase, as the term implies, is the transition from prophase to metaphase, called "metakinesis" by Mazia (1961). That is, it is a dynamic period during which the centromeres of the chromosomes (or chromatids) acquire spindle fibers to opposite poles and during which the chromosomes move or are moved to their metaphase positions, usually equidistant from the two poles of the spindle. It is inappropriate to discuss chromosome movement in a cytogenetics text, so suffice it to say that there is no generally acceptable model of chromosome (or any protoplasmic) movement. It is evident, however, that during prometaphase the chromosomes do move about (Mazia, 1961) and usually end up on the metaphase plate. It is likely that spindle fibers may be necessary for this organized and directional movement since chromosomes seem unable to achieve metaphase plate arrangement in the presence of colchicine (O'Mara, 1939; Eigsti, 1942; Berger and Witkus, 1943; Allen et al., 1950; Hadder and Wilson, 1958), a drug that is known to destroy spindle fibers.

Part of the prometaphase movement is the orientation of the chromatids, at least in the region of the "centromere," such that one of them lies toward each of the spindle poles. This essential orientation can be expressed in a different way. According to the new terminology, the specific region or structure of the chromosome is called the centromere which at metaphase has two specialized substructures, one on each side and exactly related to the orientation of the two chromatids, called *kinetochores* (Chapter 8). So at metaphase the chromosome at the centromere is so oriented that the two kinetochores of that centromere are oriented toward opposite poles. This arrangement is attained by metaphase and is apparently absolutely necessary for orderly separation at anaphase of unicentric chromosomes.

Centromeres are important not only in mitosis and meiosis but in evolution of chromosomes and chromosome sets. The general assumption is that a centromere is a "thing" (see Chapter 8), a structure of the chromosome that cannot form de novo and that can be broken or lost. Most chromosomes have only one centromere and are therefore described as being *unicentric*. McClintock (1932) found evidence that a centromere can be broken crosswise and both halves can function fairly well as separate centromeres. Some plants (sedges and rushes) and insects (Lepidoptera, Hemiptera, and Homoptera) that have smaller than average chromosomes seem to have either many centromeres along each chromosome or one centromere the length of the chromosome, called *diffuse* centromeres. And lastly, Rhoades and Vilkomerson (1942) have described *neocentromeres* in maize and in rye. These form in mutants at the ends of normally unicentric meiotic chromosomes under the influence of gene mutations. Luykx (1970) has recently reviewed the current knowledge of centromeres.

Electron microscopy of sectioned centromeres at metaphase indicates that the chromosomal material is continuous with the arms but that there are two proteinaceous pads, called kinetochores, one on each side of the centromere (Brinkley and Stubblefield, 1966, etc.) (Fig. 8-8). Metaphase chromosomes that have been "exploded" out of cells, spread on a water surface, and examined in the electron microscope show centromere constrictions as being composed of the same sort of chromatin strands as in the chromosome arms but fewer of them; the kinetochores are usually not present with this technique. More details of likely centromere structure are given in Chapter 8. Centromeres are somewhat similar in appearance to nucleolus organizing regions of chromosomes and, like them, are specialized regions.

It is likely that a centromere, like a nucleolus organizer, is a region of typical chromatin strands but contains also special genes and is equally subject to gene action, such as being repressed or stimulated. Therefore, one might think of a chromosome with a diffuse centromere as being a long centromere containing other genes as well. Any isolated portion of such a centromere (or chromosome) can function perfectly well (Nordenskiöld, 1956). It is impossible to explain neocentromeres at this time.

In plants it seems likely to some investigators, such as McClintock (1932), that centromeres cannot usually function at the very end of a chromosome, although others (Marks, 1957) believe that truly telocentric plant chromosomes do occur. In animals, however, it seems evident that there are many truly telocentric chromosomes and that not only can they break crosswise to make two telocentric chromosomes from one metacentric but two telocentrics can unite with one another (true centric fusion) to make one metacentric. It is well known that broken chromosome arms are able to recombine with other broken ends, so broken terminal centromeres may remain unfused for generations or heal and cease to be able to recombine. Broken centromeres would be brought every close together by the convergent spindle at telophase; and if somatic pairing is very common (Brown and Stack, 1968), then reunion of the same broken centromere could often occur.

It is also difficult to "explain" the centromere condition in *Parascaris (Ascaris) equorum,* which has one or two pairs of large chromosomes in its germ line although related roundworms have many small ones. Furthermore, in *P. equorum*, as cleavage cells become somatic, these few large germ line chromosomes regularly break up into many small ones as part of differentiation. It is not known whether the compound chromosomes of *P. equorum* are primitive or derived or what sort of centromeres exist on the fragmented chromosomes. Certainly the long germ line chromosomes seem to have diffuse centromeres. Perhaps each small fragment chromosome is essentially a centromere. The nature of centromeres of *polycentric* or *holokinetic* chromosomes, the so-called diffuse centromeres as occur in some plants and animals, is unknown (see Bauer, 1967, for general review). The whole chromosome seems to be a centromere, any fraction of which can function adequately during division. Centromeres, therefore, are of cytogenetic significance. The agmatoploidy of *Luzula* (Chapter 12) is cytogenetically very distinct from diploidy and polyploidy of species with unicentric chromosomes. The chromosome number is controlled by unicentric chromosomes; but occasional misdivision may play a part in chromosome number changes in evolution or in the cytogenetics of a species (Chapter 10).

METAPHASE

Metaphase is characterized in most organisms by the chromosomes being fully condensed and being arranged in a plane equidistant from the poles of the spindle (Fig. 4-1). The so-called metaphase produced by colchicine, however, is not so characterized. Rather, the chromosomes still occupy their prophase or early prometaphase positions within the nucleus (O'Mara 1939; Eigsti, 1942; Berger and Witkus, 1943); and so, in squashed preparations, they spread better than if they had congressed upon the metaphase plate (Chapter 19). That is the reason for such wide use of colchicine in certain cytological techniques.

During metaphase the chromosomes of the set or sets are not always arranged at random. The metaphase plate may be devoid of chromosomes at the center; the small chromosomes may be aggregated at the center; very often, if not usually, the centromeres lie toward the center of the plate with the arms directed outward; homologous chromosomes may be closely associated in somatic pairing; sex chromosomes may be regularly positioned at the center or outside, etc. Such arrangements are discussed in Chapter 8.

During metaphase the spindle must continue to develop. The convergence of the spindle fibers at the poles is essential to crowd the chromosomes at late anaphase so that all those at one pole will be included in one nucleus.

Somehow all the chromosomes "know" when to separate at the centromeres and start anaphase. They all do this at the same time even if most have to wait for one laggard to arrive late on the metaphase plate.

ANAPHASE

Anaphase begins when sister kinetochores separate from one another and move toward opposite poles. Again the cause or causes of chromosomal movement will not be discussed since it is not known (see Luykx, 1970, for recent detailed discussion of proposed mechanisms of chromosome distribution). That there is movement is obvious, and the integrity of the spindle or spindle fibers is essential to the achievement of that movement. The centromeres of the anaphase chromosomes (that were metaphase chromatids) lead toward the poles; the arms *seem* to follow passively or to be dragged. Often the ends of the arms of the sister chromatids seem to stick together so that they must eventually be pulled apart (Fig. 5-3, *D*). Interchromatid strands of chromatin are often revealed by electron microscopy (Trosko and Wolff, 1964; DuPraw, 1968).

There is one anaphase error that is of cytogenetic and biological importance, namely *nondisjunction*. This occurs whenever both sister kinetochores move to the same pole or at least when both sister chromomatids of a metaphase chromosome end up in the same telophase nucleus. Nondisjunction may be a regular, gene-determined process, as in some insects that eliminate certain chromosomes during cleavage divisions, or it may be a mistake. The latter condition produces cells that are aneuploid if one or a few of the chromosomes only of the set undergo nondisjunction. That is supposed to be the cause of extra (XXX) or deficient (XO) sex chromosomes, or trisomy 21, or monosomy 21 in human beings (Chapter 19) or other organisms, or mosaics, organisms having within them two different chromosomal conditions.

Anaphase of mitosis is usually very regular; but if there have been any chromosomal breaks and reunions so that an anaphase chromatid is formed with two centromeres (and another with none) and if the two centromeres go to opposite poles, a bridge is formed along with an acentric fragment. Other chromosomal aberrations have been seen and studied during mitotic divisions, such as ring chromosomes. Pieces of unicentric chromosomes that do not have centromeres are called *acentric fragments*, and they are incapable of anaphase movement to the poles so that they and their contained genes are not included in either daughter nucleus.

As was remarked earlier, the convergence of the spindle fibers at the poles is very important to the exact accomplishment of mitosis. Because of the convergence, the chromosomes are jammed together as anaphase comes to an end so that they are all included within one nuclear envelope and one genetically complete nucleus at each pole rather than in two or more incomplete nuclei. In some species normally, however, divergent spindles form and a number of small nuclei form; but those formed at each pole stick together as one compound nucleus (Sharp, 1926, p. 172). Divergent spindles resulting from mutation in a species that normally has convergent spindles produce many micronuclei at each pole and the condition is lethal.

Also normal but unusual are the unipolar spindles of meiosis in the fly *Sciara* (Metz, 1933) or the segregation of univalents in *Leucopogon juniperinus* (Smith-White, 1948, 1955) and *Rosa canina* (Gustafsson, 1944) (Chapter 17). Abnormal multipolar nuclei are often seen in meiosis of hybrids (Fig. 4-4, *A*) but also at mitosis occasionally (Chapter 5). In *Sciara*, not only is the spindle unipolar but even more remarkable is the fact that some of the chromosomes back away from the pole, and still more remarkably only paternal chromosomes back away. This inexplicable ability of maternal and paternal chromosomes, which may be genetically identical, to behave differently at anaphase is evident in orderly and genetically controlled nondisjunction during cleavage divisions in some flies such as *Miastor* and *Sciara* (White, 1954).

The metaphase and anaphase of the first meiotic division is remarkable in what selection it can achieve, such as in the example just given or in the controlled alternate disjunction in *Oenothera* or alternate disjunction of multiple sex chromosome aggregates. But here mitosis is being considered, and meiosis must not be thought of as the same as mitosis nor should the two be discussed interchangeably.

When the mitotic anaphase chromosome groups stop moving apart, anaphase is complete and telophase begins.

TELOPHASE

Telophase is the termination of mitosis, characterized by the conversion of anaphase chromosomes into interphase "chromosomes," the reformation of the nuclear envelope, formation of a second centriole, the formation of one new nucleolus or more, disappearance of spindle and astral fibers, and usually the division of the cell into two complete and distinct cells. Cytologically it is the most difficult phase of mitosis to study because the chromosomes are so jammed together that it is difficult to see details clearly.

It is now well established by electron microscopy (Fig. 4-5) that the new nuclear envelope forms at telophase in contact with exposed surfaces of the telophase mass of chromosomes such as the regions of centromeres, ends of long arms, and the arms of those chromosomes at the periphery of the mass of chromosomes (Brown and Bertke, 1969). It is generally assumed that pieces of endoplasmic reticulum or fragments derived from the earlier prophase nuclear envelope "move" (or are still attached) to the telophase chromosomes and make contact with them. Subsequently, the pieces grow laterally and make contact with other fragments until a complete but irregular envelope is formed. If it is derived from fragments of endoplasmic reticulum, the latter must be modified, at least to the extent that pores are formed.

Meanwhile, one or more new nucleoli are being formed by the ribosomal DNA of the two or more nucleolus organizers. Apparently the DNA of those regions (composed of a great many redundant genes for ribosomal RNA transcription) extends as loops and forms ribosomal RNA (Chapter 2 and Fig. 3-5). Protein is also synthesized, and the resulting mass of nucleoprotein is the nucleolus. Often, two or more small nucleoli will fuse to form a larger one. Subsequently, the nucleoli continue to produce r-RNA and protein as long as they persist.

About the only cells lacking nucleoli (and their function) are the cells of some animals during cleavage (Karasaki, 1968). Such rapidly dividing cells of many animals get along on the r-RNA produced earlier by the ovum (Gilcrist, 1968, p. 71). It is not until the blastula that nucleolar structures and functions are resumed. In other animals nucleoli appear in early or later cleavage cells where r-RNA is synthesized

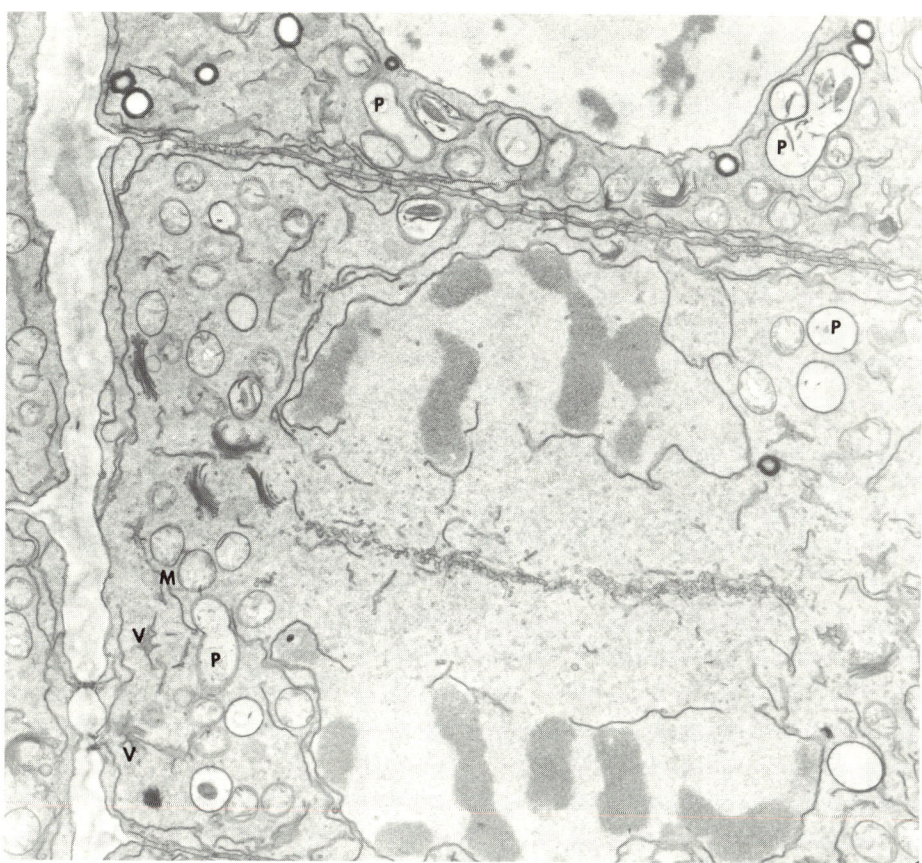

Fig. 4-5. Telophase of mitosis showing the formation of the nuclear envelopes, still incomplete. The nuclear envelope fragments lie very close to the exposed surfaces of the chromosomes. 7,000×. (Courtesy Dr. M. Dauwalder, University of Texas Cell Research Institute. From Brown, W. V., and E. M. Bertke. 1969. Textbook of Cytology. The C. V. Mosby Co., St. Louis.)

(Kaulenas et al., 1969). Emerson and Humphreys (1971), however, believe that ribosomal RNA is produced during cleavage of a sea urchin in quantity equal to that of other stages, and there are no typical nucleoli because interphases are too short for them to form.

The highly condensed anaphase chromosomes become uncondensed during telophase. Recent concepts of this process imply merely the loosening of the folds and loops of the condensed chromosome arms so that the chromosomes occupy more and more volume as the telophase nucleus grows in volume (Fig. 4-3 in reverse). The statement of Darlington that the nucleus *is* the chromosomes (except for the nucleoli and membrane) does express the concept quite well. The question of how much of the interphase nucleus is *not* chromosomal has not yet been answered, but there is probably very little.

In those cells that have centrioles, the doubling of those organelles occurs at telophase. Centriole reproduction is not by division; rather, a "baby" centriole forms perpendicular to and close to the proximal region of the "mother" centriole. Subsequently, the astral rays disappear, and the centrosome continues to lie adjacent to the new nucleus.

Cytokinesis, which usually follows anaphase as part of telophase, is the final division of the original cell into two daughter cells. Through the end of anaphase, the genes and chromosomes have been doubled and divided equally so that each daughter nucleus is genetically equal to the other and each is genetically equal to the original cell. Cytokinesis divides the cytoplasm with all its ribosomes, organelles, enzymes, etc. into sufficiently adequate portions; but unlike nuclear division, cytokinesis not only is not always equal but often does not occur. Unequal division is called *asymmetrical*, and when it occurs two cells of unequal sizes are produced. The result of no cytokinesis following nuclear division is a binucleate cell. If repeated, a plasmodium or coenocyte of numerous nuclei in one common mass of cytoplasm is formed. Three well-known examples are the young endosperm of many flowering plants, the early embryogeny of arthropods, and the plasmodium of slime molds. Partial cytokinesis *before* mitosis occurs in yeasts by budding.

Asymmetrical division often produces cells of different sizes and/or shapes as well as of different contents. Often the smaller cell may contain more than half of the mitochondria, or the larger cell may contain most of the vacuole. As a result, they are physiologically different and give rise to different sorts of differentiated cells (Avers, 1958; Avers and Grimm, 1959; Lowary and Avers, 1965; Cutter and Feldman, 1970a, 1970b).

Restitution nuclei are of considerable cytogenetic significance whether formed fortuitously, artificially, or normally under gene control. If at anaphase and telophase the chromosomes are spread from pole to pole or if metaphase chromosomes cannot move because the spindle has been prevented from forming by colchicine, the nuclear envelope as it forms against exposed regions of chromosomes will include all of the chromosomes. That is, a restitution nucleus will form and contain twice as many chromosomes as were present in the original prophase nucleus.

Fortuitous and gene-controlled restitution nuclei form most commonly at telophase of the first meiotic division as in hybrids and in many apomictic plants. Chromosome number doubling in somatic tissues is usually accomplished by endoreproduction of some sort (see Chapter 12).

Artificially and occasionally fortuitously, however, restitution nuclei do form in somatic cells and give rise to polyploid cells. This is accomplished artificially by application of certain chemicals such as colchicine or calcemid, chloral hydrate, etc. Many polyploid plants have been produced in this way. In the preparation of lymphocytes for human or other vertebrate karyotype analysis, colchicine or calcemid treatment is routinely employed to prevent spindle formation and, thereby, the accumulation of many cells in the metaphase condition. It was early determined that too long exposure to colchicine produced polyploid cells by restitution nuclei, sometimes after a second division (Fig. 19-17).

Radiations, virus infections, tissue culture, and malignancy also seem to upset the orderly anaphase distribution of chromosomes so that restitution nuclei and polyploid cells are occasionally formed. Tissue in culture can also produce somatic pairing and even pseudochiasmata in animal (Boss, 1954) and plant (Mitra and Steward, 1961) cells.

ATYPICAL FORMS OF MITOSIS
Prokaryote division

The mechanism of division of bacteria, especially division of the prokaryon or "nucleus," is still something of a mystery; the electron microscope has enabled some indecisive observations to be made, permitting, however, proposal of schemes impossible by light microscopy alone.

A small number of light microscopists, such as DeLamater (1952) DeLamater and Hunter (1951), did report observations of "mitosis" in some of the largest bacteria, which were not taken seriously (Bisset, 1952) by most cytologists. Electron microscopy has subsequently provided observations of the actual separation of one "tangle" of probable DNA filaments gradually into two separate "tangles" or prokaryons. Of course there has to be some such separation, but the details are still somewhat unclear. A number of observations and interpretations seem to indicate a contact of at least a point of the single circular genophore or bacterial chromosome with the plasma membrane (Chen, 1966; Jacob et al., 1963; Lark, 1966; Grula et al., 1968; Pontefract et al., 1969) or more specifically via a mesosome (Ryter and Jacob, 1963, 1964), a mesosome being a local, convoluted, invaginated, spherical mass of modified plasma membrane (Fig. 4-6). That the membrane may well be involved derives also from the results of Grula et al. (1968) that penicillin (which only affects cell wall formation) also, probably secondarily, prevents prokaryon division in *Erwinia*. They concluded "that deposition of cell wall mucopeptide, membrane function, and division of the nuclear body are somehow related." Schwarz et al. (1969) reported similar results with *Escherichia coli* in a very narrow equatorial band of the wall. It is also proposed that the particular point of the circular genophore (the anchor) that attaches to the plasma membrane or mesosome (the anchorage) is that specialized region of the genophore which is also the origin point (O) of initial DNA replication, called the "replication origin" (Wolf et al., 1968) or the "replicator" (Jacob et al.,

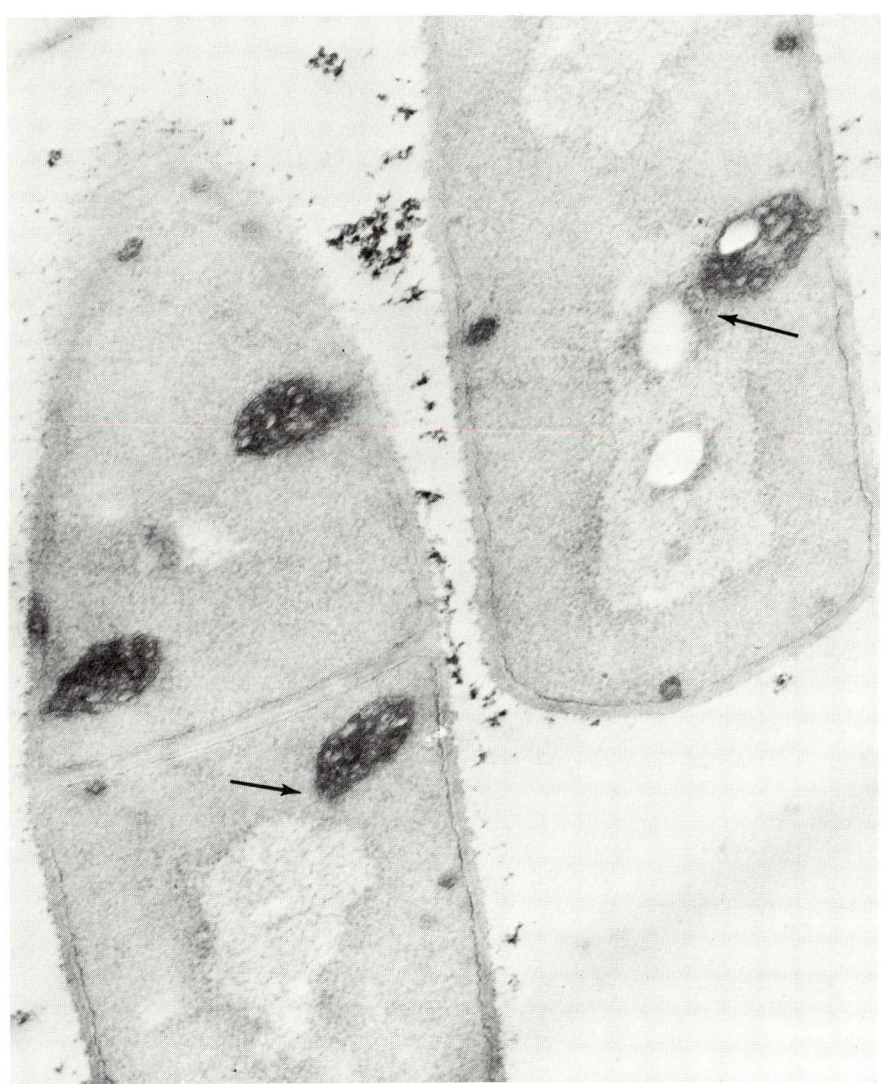

Fig. 4-6. Mesosomes in a prokaryote. The darkly stained structures attached to the plasma membrane are mesosomes, which often make contact (arrows) with prokaryons, which may be of evolved function in division or merely inevitable by their sizes and positions. (Courtesy Dr. P. Fitz-James, University of Western Ontario. From Brown, W. V., and E. M. Bertke. 1969. Textbook of Cytology. The C. V. Mosby Co., St. Louis.)

1963). Presumably the two daughter anchors, one at the origin and one at the terminus of the circular genophore (Snyder and Young, 1969), early or late in DNA replication attach (or are attached at all times) to the two anchorages such as the mesosome (Ryter, 1968) which may be morphologically modified in the region of contact with the genophore (Ellar et al., 1967). These then move apart according to the localized unilateral replication concept of wall growth described by Cole and Hahn (1962) thereby separating the anchors, but at most about 1 or 2μ. Furthermore, Highton (1969) reported in a species of *Bacillus* that is from 1.5 to 6μ long and with one to four prokaryons that there is usually only one mesosome and it is associated with cross-wall formation rather than genophore separation. Luykx (1970) presented three models of genophore separation but drew the circular genophore about 1 or 2μ long. Such models become meaningless with the 1,000 to $2,000\mu$ long genophores that actually exist. The whole genophore in most bacteria is from 1,000 to $2,000\mu$ long, so that something more than 1μ separation is demanded to really separate the two long interconvoluted genophores at all meaningfully. Sueoka (1968) has proposed, theoretically, some sort of genophore "condensation," somewhat similar to that of eukaryotic prophase chromosomes, to achieve the final separation. That hypothesis is certainly testable but at present has no observational validity.

Sueoka also proposed the analogy of the anchor of the genophore to the centromere of the chromosome, and the plasma membrane anchorage to the "centriole" of, perhaps, the nuclear envelope of yeast and other fungi. However, he reports nothing comparable to the eukaryotic spindle. The last word has not been said on the mechanism of prokaryon division, but a reasonably acceptable scheme is in the making.

Cytokinesis in Prokaryota is the function of the plasma membrane (Chapman, 1959; Pankratz and Bowen, 1963), but in some species the mesosome (Ellar et al., 1967; Highton, 1969) is involved. When the mesosome seems involved, it positions itself at the center of the long cell and the cytokinesis grows through it, thereby cutting it into two portions. Perhaps this is more a mechanism for dividing the mesosome so that each daughter cell receives a mesosome than that the mesosome is responsible for cell division. Cytokinesis may commence at any time from long before prokaryon division to long after prokaryon division. In some species the membrane and wall material together grow inward to achieve separation. If cell separation follows, the wall material itself is subsequently split centripetally. Membrane and wall growth are important in division.

Gonomery

In many eukaryotic animal species and some plants the first (and sometimes subsequent) mitotic division in the fertilized egg and/or early cleavage is unusual. That is, the two haploid pronuclei (one from sperm and one from egg) do not completely unite at karyogamy in the interphase condition, nor mix the two sets of chromosomes, nor then divide by typical mitosis. Rather, the two haploid nuclei lie side by side as they pass through prophase. They form adjacent and parallel spindles, and the metaphase plates lie in the same plane. Then, in some species such as *Drosophila* (Huettner, 1924) the two spindles fuse, but the two sets of chromosomes remain separate during anaphase. At telephase the two sets at each pole are finally enclosed in one diploid nucleus. In some species the two sets of chromosomes appear segregated during metaphases at one or more subsequent divisions. That is the simplest form of gonomery.

The other extreme occurs in the water flea *Cyclops* (Rückert, 1895) in which the two haploid nuclei remain close to each other but completely separate throughout a number of mitotic divisions of cleavage, finally fusing gradually as embryonic structures are forming.

A somewhat similar condition occurs in the basidiomycete fungi and to some extent in ascomycete fungi during the dikaryophase.

Somatic pairing

The pairing or close parallel association of homologous chromosomes in somatic cells from prophase through anaphase of mitosis is well known. It seems to occur in all tissues from cleavage on in all tissues of all Diptera (Metz, 1916a) as well as in numerous plants (Watkins, 1935), such as *Yucca*. It is apparently common in lower organisms as well as occasionally in man (Schneiderman and Smith, 1962; German, 1964). Somatic pairing seems to be a necessary prerequisite to somatic crossing-over that is detected in genetic studies of asexual fungi (Pontecorvo, 1958) and has been detected in *Drosophila* (LeClerc, 1946), especially in the male (Stern, 1936) which has no crossing-over during meiosis. Thus it seems that a small amount of crossing-over (genetic recombination) inevitably accompanies somatic pairing, perhaps at a rate of one five-hundredth that of meiotic recombination (Pontecorvo, 1958) (see Chapter 5 for extended discussion of this and the next two topics).

Premeiotic pairing

A special case of somatic pairing occurs in gonial cells of animals (S. G. Smith, 1942) and in premeiotic sporogenous cells of plants (Brown and Stack, 1968; Chauhan and Abel, 1968; Stack and Brown, 1969a). In Diptera and some plants a loose somatic pairing may occur throughout the organism and throughout ontogeny. In others it may occur throughout the flower or only in sporogenous or gonial tissues. Thus, premeiotic pairing is that somatic pairing which immediately precedes meiosis; otherwise it and somatic pairing are probably the same. Numerous cytogeneticists (Maguire, 1965, 1967; Feldman, 1966) and cytologists now believe that premeiotic pairing is the essential first step in "meiosis" of plants, animals, and fungi, but it occurs one or more divisions before the interphase preceding the first meiotic division (see Brown and Stack, 1968). If this generalization proves to be correct, it means that a small amount of nonmeiotic recombination may occur in most species of sexual and asexual organisms.

Stack (1971) has found in the liliaceous plant *Ornithogalum virens* ($n = 3$) that there is progressive change during ontogeny from "typical" mitosis without pairing, large effective metaphase plate, long chromosomes, and loosely associated sister chromatids in the seedling root to pairing of homologous chromosomes, small effective metaphase plate, shorter chromosomes, and tightly associated sister chromatids in the flower parts. In this species, then, there is a gradual change from "typical" mitosis to the premeiotic mitosis. These developmental changes include changes, in addition to the pairing of homologues, that can be considered to be differentiation changes necessary for accomplishment of the premeiotic pairing, which seems to be the last change accomplished.

Thus, premeiotic pairing may be essential for and the beginning of "meiosis."

Nonrandom segregation

It is generally assumed that the chromatids moving to either pole at anaphase are a random mixture that probably differs at each succeeding mitosis. This may not be entirely true. Lark et al. (1966); Lark (1967); and Rosenberger and Kessel (1968) using autoradiographic techniques have shown that all or most of the chromatids formed by DNA replication during one particular S period during one particular interphase move nonrandomly to one, the same, pole at anaphase. That is, following the semiconservative replication the daughter nuclei formed at the second division are different for the interphase during which their DNA was replicated.

This conclusion is not generally accepted yet, and some similar experiments (Heddle et al., 1967; Wolff and Heddle, 1968) do indicate that segregation of chromatids formed during a particular interphase is *random*. If nonrandom anaphase segregation should prove to be generally true, a likely cause is the possible end-to-end (continuous spireme) association of chromosomes during prophase and interphase (Wagenaar, 1969), unilateral chromatid replication (Fig. 5-5), somatic pairing, or a combination of these (see Chapter 5). If true, this condition might have genetic implications as producing interchromosomal linkage of some mutations in somatic tissues.

The topics just discussed, somatic pairing, premeiotic pairing, and nonrandom segregation, will be related to the problem of fungal mitosis.

Mitosis in fungi (hypothetical scheme)

Mitotic divisions in many fungi seem to be very different from divisions in higher plants and animals (Robinow and Marak, 1966; Namboodiri and Lowry, 1967; Robinow and Caten, 1969). One of the most obvious differences is that in these fungi there seems to be no congression of chromosomes on a metaphase plate (Bakerspigel, 1959). The contracted chromosomes seem to move directly from late prophase to an atypical "anaphase." Apparently, although it has not been proved, the chromosomes by "anaphase" are associated end to end in one or more series. Also, there seems to be a great reduction in chromosome volume, that is, chromatin, between prophase and anaphase.

Fungal "anaphase" is characterized by two parallel strands of chromatin lying on the spindle and parallel with the spindle axis (Ramirez and Miller, 1962). Whether these apparent strands represent a number of independent chromosomes or two strands of interconnected chromosomes is not known, although the latter is the more likely (Fig. 4-7). Furthermore, it is not known whether these two strands of chromosomes represent strands of sister chromatids or represent groups of nonhomologous chromosomes; the latter is somewhat more

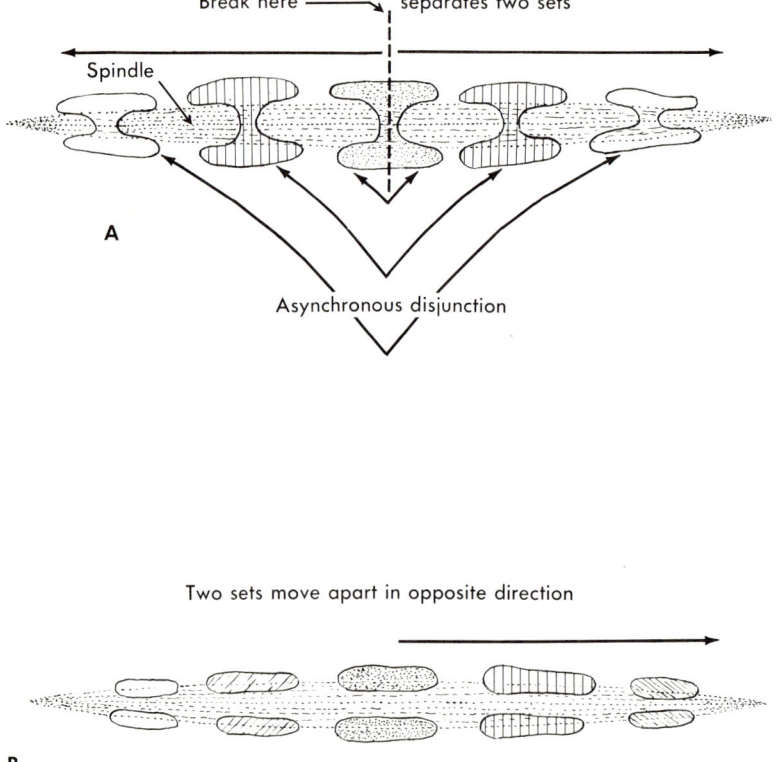

Fig. 4-7. Two tentative schemes for "metaphase-anaphase" separation of sister chromatids on the narrow spindle of at least some fungi—the "railway track" anaphase problem. **A,** The *asynchronous disjunction* scheme, according to which the chromosomes *in turn* and sequentially get onto the spindle at the equator and divide, and the two sisters move toward opposite poles, perhaps attached together as two trains moving in opposite directions from the equator. **B,** The *synchronous disjunction* scheme, by which the double "metaphase" chromosomes get onto the spindle along its length, separate into sisters, perhaps to form two trains, and these trains move in opposite directions along the spindle railway track. (Based on sketches and personal correspondence from Dr. C. F. Robinow, University of Western Ontario. From Brown, W. V., and E. M. Bertke. 1969. Textbook of Cytology. The C. V. Mosby Co., St. Louis.)

likely. If these two parallel "tracks" of chromatin represent strands of nonhomologous chromosomes, then each strand must be unobservably double, of two chromatid strands, by anaphase.

Some fungal divisions do have an intranuclear spindle; the nuclear envelope is permanent throughout division, perhaps necessarily so. Apparently at telophase the greatly elongated spindle dissolves, at least between the two forming daughter nuclei. The portion of the spindle within each daughter nucleus may remain or it may not, it has not been determined. It is known, however, that the ends of spindles, during interphase as in yeast (Robinow and Marak, 1966) or in all cells during division, are attached at both ends to *spindle plaques*. These definite structures (Robinow and Marak, 1966; Robinow and Caten, 1969; Mrs. J. Robb, personal communication) lie in and adjacent to the nuclear envelope. Structurally they are nothing like centrioles and kinetochores, although they have spindle fibers (microtubles) attached to them and probably function in organization and/or formation of the spindles.

Robinow and Caten (1969) and Robinow (personal communication) have figures showing apparent spindle plaque division in *Aspergillus*. Also, it seems that as the two daughter plaques separate during prophase in *Aspergillus*, but not in yeast, a spindle forms between them, curved at first. As the plaques move apart along the permanent nuclear envelope either the spindle elongates to "keep up" with the separating plaques or the elongating spindle pushes the spindle plaques apart (which is not known); the latter is slightly more likely hypothetically. Eventually the spindle extends straight across the nucleus with a spindle plaque at each end.

Figure 35A and B in Robinow and Caten (1969) seems to indicate two convoluted, late prophase chromatin strands, each attached at its ends to the two plaques (or at least near the ends of the short spindle). Presumably, the spindle has by then elongated enough to extend across the whole nucleus, and the two chromatin strands, whether single or double, have straightened out and lie parallel along the spindle.

Following the establishment of the two parallel tracks, the next evident condition seems to be the accumulation of chromatin at the two poles and attenuation of the tracks, as though there is movement of chromatin to the poles at the expense of the intermediate region. Before complete attenuation is achieved, the two tracks are unevenly chromatic, as though there are heavy regions and thin regions of chromatin. The number of heavy regions along each track varies but rarely if ever equals eight (Robinow and Caten, 1969), the haploid chromosome number in *Aspergillus nidulans*. Usually there appear to be about four, but it is very uncertain that these dark blobs actually represent chromosomes.

Late anaphase can be considered to start when there is a gap in one or both tracks about halfway between the poles. The remaining portions of the chromatin strands then "contract" into daughter nuclei to which they are attached, and telophase has been achieved.

The evidence that the two tracks are not homologous compound chromosomal strands, each consisting of a complete set of chromatids (eight in *Aspergillus*), is that the two strands do not seem to move in opposite directions, move past each other to opposite poles of the spindle (Ramirez and Miller, 1962; and others). Rather, both tracks "break" in the middle, and two approximately equal remnants are pulled into each daughter nucleus at late anaphase.

A hypothetical scheme to fit all of these observations (Fig. 4-8) is that each early anaphase track *(E)* contains about one half of the chromosome set, that is, in *Aspergillus* each would contain four of the eight chromosomes. At that time each track would be (unobservably) double, of two chromatid strands. One chromatid strand of each track would be attached to one spindle plaque and the others to the opposite plaque. During anaphase the chromatid strands of each track would move in opposite directions *(F, G, H)*.

If each chromatid strand is submicroscopically unevenly thick, thicker at the chromatids and thinner where they are attached to each other within the strand, then each track would be uneven along the length when the oppositely moving chromatids happen to be side by

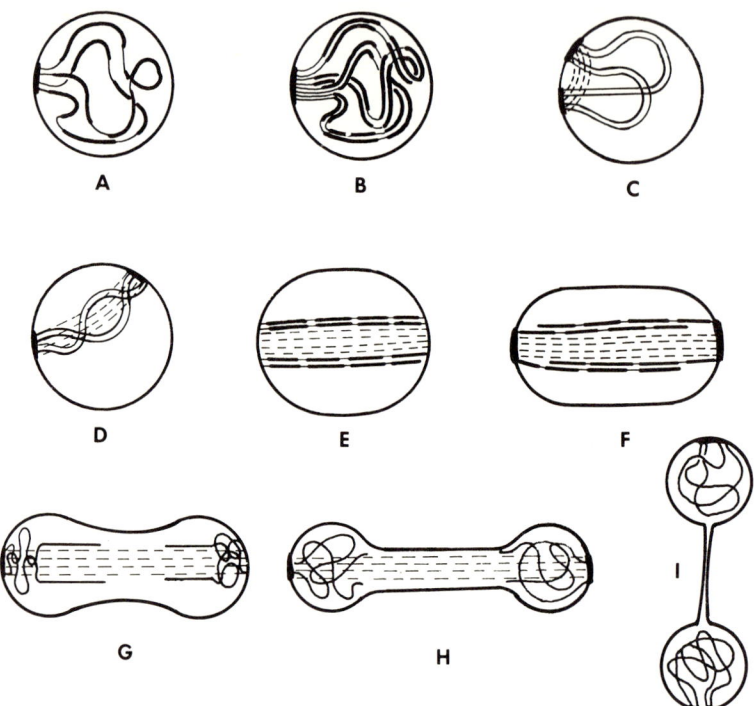

Fig. 4-8. A model of mitosis in some fungi, the *Basidiomycetes* and *Ascomycetes*, including in part the "railway track" metaphase-anaphase (Fig. 4-7). **A,** Interphase with one "centriolar plaque" to which the ends of the one or more "compound chromosomes" are attached. **B,** Chromosome structural replication. **C,** Beginning of "mitosis." The plaque divides with one end of each compound chromosome attached to each. The spindle forms between the plaques, either pushing them apart or, more likely, guiding them as separation is achieved by nuclear envelope differential growth between them. **D,** A later stage as the plaques continue to separate. **E,** Separation can go no farther; "metaphase." **F,** Beginning of "anaphase"; sister chromatids are attached to opposite plaques. Relative opposite movement is achieved in part by elongation of nucleus and by contraction of compound chromatids. **G,** Mid-anaphase. **H,** Late anaphase, continued contraction of chromosomes and elongation of nucleus. **I,** Telophase as daughter nuclei round up. The spindle has broken down and the envelope connection will break. How the loose ends of the chromosomes become attached to the plaques is not specified. This scheme seems applicable to mitosis in *Aspergillus* and meiosis in *Ustilago hordei*. (See text.)

side in a track *(E)* and a darkly stained blob would appear in contrast to the thinness of adjacent side-by-side attachment regions. When the ends of the chromatids of each track have passed each other, more or less simultaneously in both tracks *(G)*, the clear equatorial region would appear and increase as the two chromatid ends approach the daughter nucleus into which each is eventually included *(H)*.

There is cytological justification for end attachment of nonhomologous chromosomes and also for nonrandom segregation of chromatids, as was just proposed (see Chapter 5). There is also observational evidence for attachment of chromatid ends to the nuclear envelope, separation of sister chromatid attachments, and chromosome segregation around the nuclear envelope, at least in a flagellate (Cleveland, 1938). There is also precedence for the proposal that the two compound chromatids of each track of chromosomes in fungi might be separately attached and move as groups to opposite poles (see Chapter 5).

There are, of course, reports of "typical" mitosis in fungi (Somers et al., 1960; Ichida and Fuller, 1968; and others), but the "track" mitosis seems to be the best characterized. Nevertheless, other forms may occur. At present there are not enough unequivocal sequences of stages or adequately clear resolution of "chromosomes" during fungal mitosis to generalize that all fungi have the same type of division or that there may be two or more distinctive types; but this hypothesis has already, in unpublished form, proved useful in interpreting the meiotic divisions of *Ustilago hordei* (Mrs. Jane Robb, personal communication).

Mitosis in dinoflagellates

Another form of atypical mitosis in Eukaryota occurs in the dinoflagellate algae and may be similar in the Euglenineae and the radiolarian protozoa (K. G. Grell, 1964). Not only is mitosis different from the "typical" but the chromosomes are unusual (1) in possibly lacking histones (Dodge, 1964); (2) in remaining condensed all of the time (Dodge, 1963a; Mendiola et al., 1966); (3) in having chromatin strands only about 25 Å thick instead of the usual 250 Å; and (4) in having a peculiar whorled arrangement of the chromatin strands in a repeated pattern along the length of the chromosome (Leadbetter and Dodge, 1967; Kubai and Ris, 1969). The chromatin strands may be naked DNA without permanently associated protein, somewhat comparable to the genophores of prokaryota (Kubai and Ris, 1969) and not coiled and supercoiled (DuPraw, 1968) into 250 Å strands as in higher organisms.

The mitotic process has now been adequately characterized in the dinoflagellates (Chatton, 1920; Dodge, 1963b; Leadbetter and Dodge, 1967; Kubai and Ris, 1969). The division takes place without centrioles and within an intact nuclear envelope, the permanence of which throughout division may well be necessary. As division begins, the condensed interphase chromosomes, which lack centromeres, may even elongate. Eventually the numerous chromosomes arrange themselves parallel to the spindle axis as they do also in the Euglenineae (Leedale, 1958); there is no typical metaphase arrangement and not much chromosomal movement, although in some species the chromosomes at metaphase may be arranged like a picket fence, parallel, with ends toward the poles.

The unique aspect of dinoflagellate mitosis is the formation and form of the "spindle" (Chatton, 1920; Leadbetter and Dodge, 1967; Kubai and Ris, 1969). Apparently a bundle of parallel microtubules forms just external and close to the nucleus, the orientation of which either is determined by or determines the axis of division just as does a typical spindle. This bundle of microtubules moves sideways *into* the nucleus, carrying nuclear envelope and cytoplasm with it but without breaking the nuclear envelope. The bundle breaks up into a number of small parallel groups of tubules and accompanying cytoplasm that push into various parts of the nucleus. The nuclear envelope then closes around the cytoplasm and bundles of tubules to produce a variable number of parallel cytoplasmic tubes or channels extending straight through the nucleus. They and the contained microtubules are all parallel. Thus the "spindle" of microtubules, although extending "through the nucleus," is still in the cytoplasm and separated from the chromosomes by the nuclear envelope that lines the channels. Presumably each chromosome forms two chromatids, which separate from each other starting at one end. Such separation may not be simultaneous, and the chromosomes seem to be scattered randomly adjacent to the cytoplasmic channels, although Kubai and Ris reported that most of the V-shaped chromosomes (presumably at the end of separation) are attached to the nuclear envelope of the channels.

Dodge (1963b) indicated a rather similar anaphase, all chromosomes separating together in two groups and moving endwise toward the two poles. But just how the movement is achieved is unknown. Kubai and Ris proposed the nuclear envelope of the channels as being possibly involved. There certainly are no spindle fibers connected to the chromosomes. In some species the nucleolus divides at anaphase into two portions that move with the chromosomes to the poles. Later the nuclear envelope constricts, leaving two daughter nuclei. The simplest hypothesis is that the anaphase chromosomes move autonomously to the poles, that the channels are merely determined and maintained by the microtubules but the numerous parallel channels act as guides that direct the autonomous chromosomal movement.

SUMMARY

It is obvious now that among organisms there are many methods of achieving chromosomal segregation into two daughter nuclei, all of which can be described as "mitosis." Presumably in all forms the daughter nuclei are genetically identical, although the possibility exists in the dinoflagellates, which have high chromosome numbers, that random distribution of approximately equal numbers of many duplicate chromosomes would suffice, as in the radiolarian *Aulacantha* with its from 1,000 to 2,700 chromosomes that may all be cytologically alike and genetically homologous and equivalent, and just some go to one pole and some to the other (Grell and Ruthmann, 1964).

Mitosis is a basic form of reproduction, and cellular division is the reproduction of unicellular organisms. It is the process by which the genetic constancy is maintained, although occasional errors provide some of the variability necessary for evolutionary change.

CHAPTER 5

Somatic pairing and recombination

All the evidence indicates that the preparation for these two [meiotic] divisions is made at one time. How far back in the spermatogonial period does this preparation extend?

C. E. McClung, 1927

OLD EVIDENCE

Reports of paired homogous chromosomes during mitotic divisions go back to the turn of the century for both some animals (Montgomery, 1906; Stevens, 1908) and some plants (Strasburger, 1904; 1905; Sykes, 1908; Digby, 1910; Němec, 1910). It soon became apparent that homologous chromosomes are paired at all stages of mitosis (but least at metaphase, Moffett, 1936) in all Diptera (Metz, 1916b) in all mitotic division (Huettner, 1924) (Fig. 13-7). Such pairing in salivary glands of many Diptera (Chapter 3), especially *Drosophila*, has provided excellent cytogenetic material since their value as such was discovered by Painter in 1933. By 1935 Watkins could list reports of somatic pairing in 33 species of plants including the most famous, *Yucca*.

Except in the Diptera most so-called somatic pairing in animals derives from the examination of gonial cells of the germ line, the divisions of which precede the meiotic divisions. Therefore, they may not be necessarily comparable to mitotic divisions in the root tips or other somatic tissues of plants. It was the many reports of somatic pairing in gonial cells of animals of many phyla that led S. G. Smith in 1942 to propose that these particular divisions should not be considered as demonstrating so much somatic pairing as a preparation for meiosis. In this he followed Stevens (1908), who stated, concerning such premeiotic pairing in Diptera that there is "in the spermatogonium only a foreshadowing of reduction"; Wilson (1912), "It seems quite possible that the way for synapsis may be prepared already in a very early pre-synaptic stage, by definite regrouping of the chromosomes that may take place before the leptotene loops are formed as such"; and Metz and Nonidez (1923), "it may be possible to bring all types [of Diptera] within a common general scheme in which synapsis occurs in the anaphase or telophase of the final spermatogonial division." The subsequent history of cytology indicates that this concept of meiosis was not accepted.

Some of the older evidence for this concept of premeiotic pairing as preliminary to meiosis from other than the Diptera is quite convincing. In human lice, *Pediculus corporis* and *P. capitis*, somatic (premeiotic) pairing occurs in the primordial spermatogonium six divisions before the spermatocyte (Doncaster and Cannon, 1919; Cannon, 1922; Hindle and Pontecorvo, 1942). That is, the homologous chromosomes are paired during five divisions and probably during six interphases and enter meiosis already paired. Similar somatic reduction was reported in the insects *Icerya purchasi* (Schrader and Hughes-Schrader, 1926) and *Goniodes stylifer* (Perrot, 1934). Oksala (1944), from a study of spermatogonial and meiotic divisions in the dragonfly *Aeschna*, concluded that the pairing of homologous chromosomes occurred gradually during the spermatogonial divisions. Svardson (1941) reported somatic pairing in the fish genera *Salmo* and *Coregonus*, the "attraction between chromosomes being most pronounced during anaphase and telophase." S. G. Smith (1942) listed many other reported cases from most animal phyla of pairing in spermatogonial divisions. In 1927 McClung asked how far back in the spermatogonial period this preparation for the two meiotic divisions extends.

Premeiotic pairing is indicated in many ani-

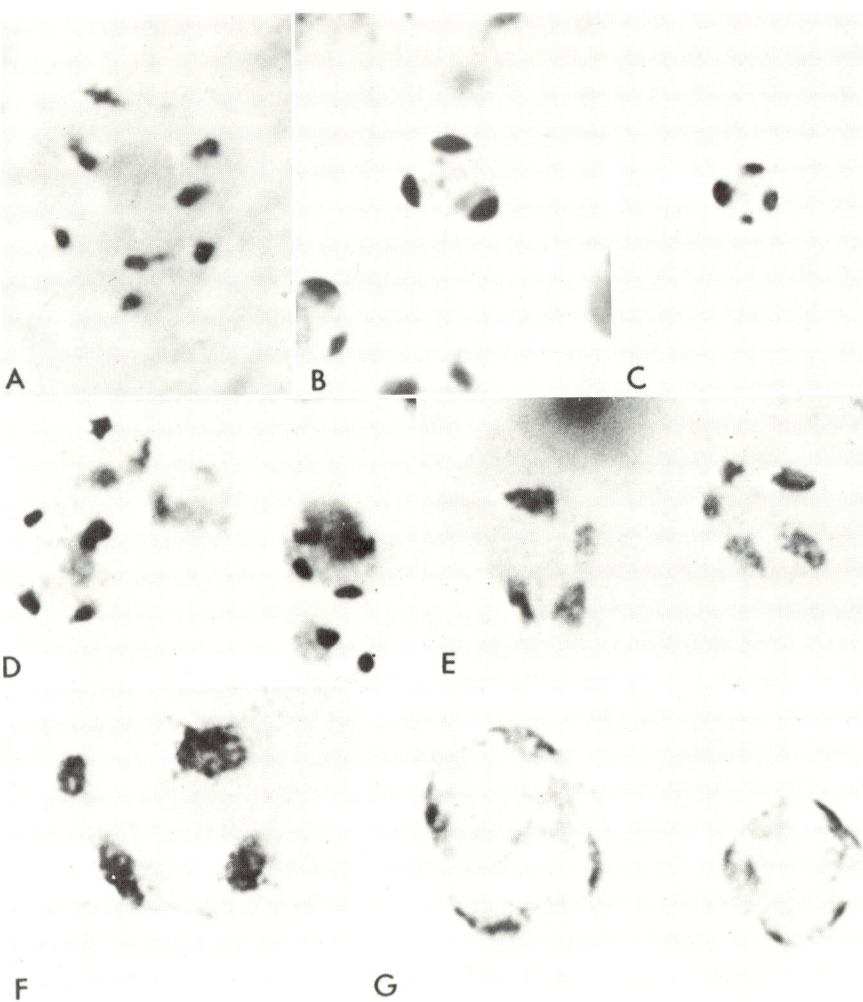

Fig. 5-1. Prochromosomes in interphase nuclei of *Plantago ovata* (2n = 8). **A,** The eight prochromosomes in each cell correspond to the eight distinct chromosomes in root tip nuclei. **B,** In somatic cells of the flower bud, homologous prochromosomes have paired so intimately that there are only four in each nucleus, three of which are visible in the optical plane of the photomicrograph. **C** to **G,** During the interphase just before meiosis in the sporogenous cells of anthers, the four dense, paired prochromosomes, **C,** become more, **D** and **E,** and still more, **F** and **G,** diffuse and they increase in size as the nuclei, **C** to **G,** and cells increase in size preparatory to meiosis. (From Stack, S. M., and W. V. Brown. 1969. Bull. Torr. Bot. Cl. **96:**143.)

mals by the paired condition of the heterochromatic sex chromosomes (the sex bivalent) during the premeiotic interphase, as in grasshoppers (McClung, 1927), reptiles (Nakamura, 1928, 1931), mammals (Tobias, 1956), and man (Painter, 1923, 1924; Evans and Swezy, 1929). It can be assumed that if the sex chromosomes (or any heterochromatic chromosomes or parts of chromosomes) are clearly paired during the premeiotic interphase then the euchromatin and other chromosomes *may* be paired also.

A most significant case of somatic pairing and chromosome number and genome reduction has been reported in mosquito gut lining cells (Berger 1938; Grell, 1946). This is a, perhaps, unique case of somatic meiosis (*mei* meaning "fewer"). For details see endopolyploidy in Chapter 12.

Such pairing during somatic or premeiotic interphases of blocks of heterochromatin (Fig. 5-1) has been reported in the older literature (Strasburger, 1904, 1905, 1909; Overton, 1922; Janaki-Ammal, 1932); but Watkins (1935) listed examples of paired metaphase chromosomes also, such as many studies of *Yucca*, and of spinach (Stomps, 1910). He listed reports of somatic pairing from 33 species in 29 genera of 15 families of angiospermous plants. Peto (1935) and Heilborn (1936) also reported somatic pairing in plants.

It should be remembered that these observations were made by many cytologists examining many species of animals and plants when they were freer to interpret, since it was before the concept of pairing only at meiotic prophase (zygotene) had hardened into dogma and before the precocity theory (Darlington, 1937) had been generally accepted.

During this same early period genetic evidence of somatic crossing-over and recombination was reported. Stern (1936, 1939) reported somatic recombination in male *Drosophila* (which has no meiotic recombination) and also that the so-called *Minutes* (recessive lethal deficiencies) in the heterozygous condition increase the frequency of mitotic but not meiotic crossing-over. Fig. 13-7 shows homologous pairing at mitotic prophase that is similar to pachytene pairing in appearance and indicates that somatic crossing-over could occur during interphase.

In 1939 Stern demonstrated that somatic and meiotic crossing-over are genetically identical. Recombination in both is exclusively between homologous chromosomes at the four-strand stage, but only one strand of each sister pair exchanges at any site, recombination is reciprocal, both give rise to complete recombinants and at precisely the same level, and both yield the same map order. LeClerc in 1946 detected somatic recombination in female *Drosophila* that supported the report of Dobzhansky (1949) of cytological pairing of four chromosomes in somatic cells of a known translocation heterozygote. At the same time Jones (1936, 1937) reported genetic evidence of segregation of color- and growth-regulating genes in somatic tissue of maize.

In 1931 Lawrence reported pairing of homologous and homeologous bivalents at metaphase I of meiosis in polyploid plants which he called "secondary association" (Fig. 5-2). It has subsequently been reported in a number of other plant species (Meurman, 1933; Darlington, 1937; Therman, 1951). These authors assume that this secondary pairing *during meiosis* reflects the homology of the chromosomes and is of the same nature as somatic and typical meiotic pairing according to the precocity theory.

Fig. 5-2. Secondary association at metaphase I of meiosis. Fourteen bivalents are represented, and with a minimum of imagination the bivalents can be grouped into seven pairs of associated bivalents. The assumption is that pairs of bivalents (or larger associations) indicate some homology between the associated bivalents. The published illustrations are not always as convincing as this.

PRECOCITY THEORY

Darlington's precocity theory of 1937 proposed that there is some sort of inherent attraction between two homologous genes and so, by extension, between two homologous chromonemata. He assumed (1) an intrachromosomal mechanism that is represented by the cycle of division undergone by the chromosomes and (2) and extrachromosomal mechanism represented by the spindle. In typical mitosis showing no somatic pairing the two mechanisms act strictly in step: the chromosomes double at the proper time with respect to the external changes so that the genes and chromosomes are double when pairing might act. Therefore, there is no pairing of homologues, the inherent tendency to doubleness being satisfied by the presence of two sister chromatids in each chromosome.

In meiosis, however, the external mechanism is precocious: it starts before the chromosomes double. The result is that homologous chromosomes pair in the single-stranded condition at zygotene. The subsequent doubling that occurs at pachytene again satisfies the inherent tendency to doubleness, and as a result diplotene repulsion occurs.

Somatic pairing and secondary association are considered to be "residual attraction." That is, the doubleness of mitotic chromosomes almost, but not quite, saturates the tendency to doubleness; but there is some left over, and it is this residual attraction that causes the loose somatic or secondary pairing.

This precocity theory of Darlington has dominated the thinking of those cytologists concerned with meiosis and somatic pairing (Svardson, 1941; Oksala, 1944; Therman, 1951; Sturtevant, 1951; and others) until the present time but seems to be fading away now. As Sturtevant stated, it was a beautifully simple and satisfying scheme and there was no better scheme to displace it.

RECENT DEVELOPMENTS

The literature on somatic and premeiotic pairing and somatic recombination since about 1945 is extensive, especially since 1960.

In 1947 Battaglia reported an observation of a meiotic (heterotypic) division in the somatic tissue of the pistil of the plant *Sambucus ebulus*. Sonnenblick in 1950 reported somatic pairing in the early embryogenesis of *Drosophila melanogaster*. Leblond and Clermont in 1952 as well as Tobias (1956) have reported animal gonial mitoses which may be premeiotic with pairing. In 1953 Kaplan found that the *Minutes* of *Drosophila* produce an increased recombination rate in somatic cells but do not affect the meiotic rate. LeClerc had reported in 1946 that the gene C(3)G eliminates meiotic recombination without affecting the somatic rate.

Revell (1953) observed somatic pairing in root tip cells of the plant *Vicia* following X-irradiation and application of radiomimetic compounds. Rhoades (1955) noted that the prophase chromosomes of maize during premeiotic divisions of sporogenous cells are more elongate than in typical mitosis, as if the chromosomes are being made ready for the onset of meiosis, the extremely long and slender leptotene condition. Boss (1954, 1955) observed somatic pairing and chromosome number "reduction" at anaphase in tissue culture cells of the newt. In the latter paper he also reviewed the subject of somatic pairing.

Hiraoka (1958) in a study of the plant *Daphne odora* reported interphase paired prochromosomes; late prophase separation; at metaphase some "bivalents," some loosely paired, and some unpaired chromosomes; and progressive pairing during anaphase in mitotic divisions from root tips, young leaves, young vascular cells, and shoot apices.

Pontecorvo (1953), Roman (1956), Käfer (1961), Hastie (1967), and other geneticists working with imperfect (asexual) fungi (see Pontecorvo, 1958) reported that somatic pairing of homologous chromosomes must occur to provide the recombination detected. Apparently in some of these fungi, such as *Aspergillus*, (Pontecorvo, 1953; Kafer, 1961), *Penicillium, Saccharomyces* (James and Lee-Whiting, 1955), and *Ustilago* (Holliday, 1961), *Verticillium* (Hastie, 1967), although asexual they still produce recombination and must have somatic pairing. The frequency of mitotic crossing-over for each nuclear division is 2% in *Aspergillus* (Käfer, 1961) but 40% in *Verticillium* (Hastie, 1967). Pontecorvo proposed that rate of somatic crossing-over is much higher in asexual than in sexual fungi.

Parasexuality

Pontecorvo (1958) and Roper (1966) have discussed the capabilities of such asexual fungi to mimic sexual and meiotic processes by some apparently random processes. The occasional fusion of cells and nuclei mimics sexuality. Some unknown process, perhaps involving a series of nondisjunctional divisions, produces a reduction in chromosome number to the haploid condition in at least some cells. Detectable recombination implies somatic pairing as a necessary prerequisite. Holliday (1965) has determined in such fungi that the somatic recombination, and therefore exact somatic pairing of homologues, occurs during interphase at the same time as DNA replication.

Such parasexual cycles may occur to a limited extent in asexual plants and animals and in tissue cultures. In all such systems, somatic recombination is estimated to be about 500 times less frequent than in meiosis.

FINAL PROOFS

During the 1960's the study of somatic and premeiotic pairing and somatic recombination accelerated among fungi, plants, and animals. Holliday (1961, 1965) discovered and studied genetically somatic recombination in the fungus *Ustilago maydis*. Studying synchronized cells of *Ustilago* and *Saccharomyces* and using ultraviolet treatment at different stages of the mitotic cycle, he determined that somatic crossing-over occurs during the DNA-replicating S stages of interphase and that genes near the ends of the chromosomes replicate early and those near the centromere replicate late. Early S-stage irradiation produces most crossovers toward the ends of chromosome arms whereas late S-stage irradiation produces crossovers near the centromeres. Thus DNA replication and crossover production seem to occur simultaneously in these species.

In 1962 Robinow observed paired chromosomes during mitotic prophase in the fungus *Allomyces*. Pittenger and Coyle (1963) detected recombination in somatic disomic nuclei of *Neurospora*, confirming the earlier observation of McClintock (1945) and Singleton (1953) that *Neurospora* chromosomes are probably somatically paired and synapse in a condensed condition immediately after nuclear fusion and just before meiosis. Westergaard (1964) in a discussion of the theoretical considerations of the mechanism of crossing-over pointed out the obvious, that somatic crossing-over can occur only in cells having somatic pairing. Stern (1969) invoked somatic pairing to explain gene expression in genetic mosaics.

The strongest argument against homologous chromosomal pairing not later than the last premeiotic division has been that in those forms having no mitotic division between karyogamy and meiosis (zygotic meiosis) such a scheme is impossible. However, numerous recent studies of ascomycetes (Singleton, 1953; Rossen and Westergaard, 1966; Barry, 1969; Furtado, 1969a, 1970) and basidiomycetes (Lu, 1964; Furtado, 1968, 1969b) seem to indicate that as the prefusion nuclei unite in karyogamy, the chromosomes are as contracted as in mitosis and very quickly the two prefusion nucleoli unite as though brought together by pairing of the condensed homologous nucleolar chromosomes. Thus, pairing seems to be achieved by condensed chromosomes as in a premeiotic anaphase. Interestingly, Rossen and Westergaard concluded that the DNA replication, which usually precedes meiosis in the diploid interphase, occurs in the fungus *Neottiella* in the haploid (gamete) nuclei before cellular union and karyogamy.

Among plants Steinitz-Sears (1963) reported probable somatic pairing in various parts of the plant *Arabidopsis thaliana*. Hirono and Rødei (1965) believed they had produced somatic recombination in that species by X-irradiation which was detected by somatic sectoring. They did not detect natural somatic recombination; but since X-irradiation does increase somatic recombination in *Drosophila* (Becker, 1957) and *Aspergillus* (Käfer, 1963), they assumed that natural somatic crossing-over may occur in *Arabidopsis* also. Feldman et al. (1966) reported statistically significant association of recognizable telocentric homologues (but not of nonhomologous telocentrics) in primary root tips of wheat. Since they also found statistically significant association of telocentric nonhomologous arms of the 3B chromosome, they proposed some involvement of the centromere in somatic pairing, a moot point.

Maguire has hypothecated premeiotic pairing in maize to explain a number of cytogenetic observations. In 1965 she proposed premeiotic anaphase pairing as one possible interpretation of the similarity of trivalent frequency at metaphase I and pachytene. In 1966 she again proposed premeiotic pairing and crossing-over (perhaps during interphase) as perhaps being necessary to account for the correspondence of crossover frequencies at metaphase I and pachytene in a heterozygous short paracentric inversion. Still later (1967) a statistical analysis of whether the visible homologous heterochromatic regions of maize chromosomes during premeiotic interphase, in tapetal cells, and in root tip cells in squashed preparations are significantly "paired" revealed that they are.

That method of determining statistically significant interphase pairing of homologous regions of heterochromatin has been used in human metaphase figures (Schneiderman and Smith, 1962) and in oats (Dubuc and McGinnis, 1970). In all cases pairing was indicated.

More convincing evidence of somatic pair-

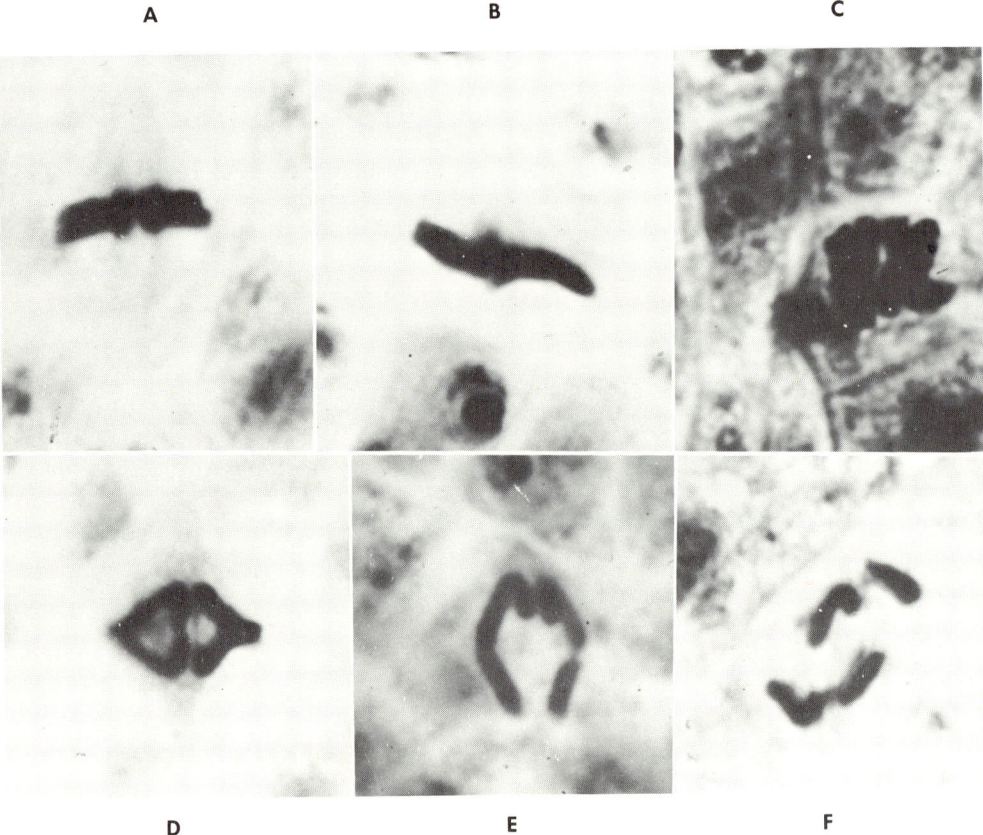

Fig. 5-3. Premeiotic mitosis in somatic cells of flower parts of *Haplopappus gracilis*. **A** to **C**, Metaphase showing the haploid (n = 2) chromosome number. **C**, the four-parted character of such chromosomes. **D** to **F**, Anaphase. In **D** the sticking of chromosome ends to form the "double triangle" appearance of early anaphase is evident. Spindles in **A** to **F** (except **C**) extend toward the top and bottom. In **E** the two paired "bivalents" have nearly separated. Although it looks as though the two chromosomes moving "up" are homologous, as are those moving "down," actually two nonhomologous "bivalents" are moving to each pole, as is evident at later anaphase in **F**, where doubleness of each anaphase chromosome is evident. (Modified from Brown, W. V., and S. M. Stack. 1968. Bull. Torr. Bot. Cl. **95**:369-378.)

ing is the observation of closely paired or fused prochromosomes during the last premeiotic interphase, as Chauhan and Abel (1968) found in the plants *Impatiens* and *Salvia* and as illustrated by earlier workers they cite on the same species.

Most convincing, however, are observations of actually paired chromosomes during mitotic metaphase and anaphase (Kitani, 1963), such as in beets (Butterfass, 1967), *Haplopappus gracilis* (n = 2) by Brown and Stack (1968) (Fig. 5-3), *Plantago ovata* and *P. insularis* (n = 4) by Stack and Brown (1969a) (Fig. 5-1), and *Ornithogalum virens* (n = 3) by Stack (1971). Mitra and Steward (1961) had earlier seen paired chromosomes in tissue culture cells of *Haplopappus gracilis*, supporting the observation of Boss (1954) that tissue culture may induce the condition.

Feldman et al. (1966), using telocentric chromosomes as markers, found that such homologous telocentrics were statistically associated in seedling root tip cells. Sears (1969) concluded from this study that in wheat "it seems likely that such association exists throughout the life of the plant in all somatic cells."

German (1964) supported the conclusion of Schneiderman and Smith (1962) by observing actual, although uncommon, pairing of human homologous chromosomes. Since the preparation of such material of lymphoid cells for observation requires some short-term culture, these observations by German may reflect the action of the treatment. Furthermore, the figures indicate occasional pairing and seem to show that crossing-over had occurred.

During somatic prophase of a mosquito (Diptera) the pairing of the 3 pairs of homologues appears to be almost as intimate as at meiotic pachytene (Breland and Gassner, 1961), but they separate by metaphase.

Further indirect cytogenetic evidence for premeiotic pairing of homologous chromosomes of wheat and supporting the earlier report of Feldman et al. (1966) is that of Feldman (1966). He examined meiotic metaphase pairing of homologous and homeologous chromosomes in the presence of different doses of the bivalent-pairing gene contained on the 5BL chromosome arm. 5BL is the abbreviation for the long arm (L) of the fifth chromosome (5) of the B genome. 5A, 5B, and 5C are homeologous chromosomes derived from the three species that contributed chromosome sets in the evolution of the hexaploid wheat. Feldman proposed the following:

The apparently paradoxical effect of extra dosage of 5BL can be explained in the following way: in plants nullisomic for chromosome 5B, the chromosomes are not randomly scattered in the premeiotic nucleus, but homologues as well as homeologues lie near each other and may perhaps even associate throughout most of their length. The meiotic pairing (synapsis) follows as a second step and brings the closely oriented or loosely paired homologous segments to a full and more intimate contact. It may be assumed that the 5B gene suppresses the premeiotic association and thus tends to cause random distribution of chromosomes in the premeiotic nucleus. Two doses of this gene, while scarcely affecting the homologous chromosomes, keeps the homeologues apart.

The 5BL gene is a "meiotic gene" (Lindsley et al., 1968), as are those reported by Weaver (1960) for genes in *Drosophila* on all chromosomes of the complement that control the frequency of mitotic recombination.

An interesting effect of colchicine on wheat meiosis that parallels the effect of excessive doses of the 5B gene has recently been documented (Driscoll et al., 1967; Driscoll and Darvey, 1970). Colchicine applied *after* the last premeiotic mitosis but continuously until metaphase I produced asynapsis more or less as excessive doses (six) of 5B did but did not affect synapsis between the two homologous arms of isochromosomes. It was assumed that colchicine prevented premeiotic pairing or at least maintenance of such pairing until synapsis. Sears (1969) in commenting on this phenomenon stated, "Since a major effect of colchicine is to destroy the spindle, it appears that somatic pairing may depend on a spindle-like substance, such as could well be attaching the centromeres to the nuclear membrane." Sears also stated that there are weak promoters of somatic pairing in wheat in arms 5BS, 5DS, and possibly 5AS (these are all homeologous short arms of homeologous chromosomes and so may be triplication of the same gene). Sears further states that all diploid species of wheat seem to have a genotype that promotes somatic pairing; in two species, *T. speltoides* and *T.*

tripsacoides, the promoters are strong enough to override the effects of 5BL and bring about homeologous pairing.

Indirect evidence for premeiotic pairing comes from Moens (1964), who reported that the chromosomes of tomato are paired before leptotene in a preleptotene condensation that looks something like a diffuse pachytene. Preleptotene condensation was studied by Stack (1969) in *Haplopappus gracilis* by both optical and electron microscopy. Mericle and Mericle (1967) proposed somatic pairing in the flower of *Tradescantia* to explain their somatic mutations for flower color.

Somatic pairing has long been studied in *Drosophila* both cytologically and genetically. Recently Grell and Chandley (1965) have proposed that DNA replication and genetic recombination are concurrent. Since it is known that DNA replication occurs before the beginning of meiosis in the premeiotic meiocyte, then, according to the authors, crossing-over occurs then also. If true, and most researchers find that crossing-over occurs during pachytene (Henderson, 1966; Hotta et al., 1967; Lindsley et al., 1968; Peacock, 1968), then homologous chromosomes must be already intimately paired long before meiosis begins. Grell (1967) has also postulated that there are two sorts of pairing in *Drosophila* oocytes, (1) "exchange pairing" between homologues that can later result in crossing-over and (2) what she calls "distributive pairing" that can occur between homologues or nonhomologues at some time after crossing-over has occurred between chromosomes that did not engage in or produce crossovers. There is considerable disagreement that nonhomologous distributive pairing has been proved in *Drosophila* oocytes.

SUMMARY

The evidence as presented indicates that somatic pairing is the same as premeiotic pairing except that the latter is by definition restricted to the germ line of animals and sporogenous tissue of plants and immediately precedes meiosis. In some animals (Diptera) and some plants *(Yucca)* the premeiotic or somatic pairing first occurs shortly after the division of the zygote and so occurs throughout the body. In other plants *(Haplopappus)* and some animals (lice) it occurs a number of divisions before meiosis. And finally, in some animals, plants, and fungi *(Neottiella)* the premeiotic pairing occurs or may occur in the last premeiotic division or, in the fungi, between condensed chromosomes at karyogamy.

So far, somatic recombination has been proved in a few fungi and *Drosophila*, and then at a low level, 0.02 in *Aspergillus* (Käfer, 1961) although Hastie (1967) reported 0.4 as the mitotic crossover frequency per nuclear division; but it may be more common than is known, especially in plants (that do not have a germ line). In *Drosophila*, furthermore, somatic pairing seems to permit more subtle genetic effects that are possible only because of very intimate association of genes in a *trans* arrangement (Lewis, 1963).

On the assumption that pairing of homologous chromosomes is the "natural" and evolutionary primitive condition and that the nonpaired condition is derived (Westergaard, 1964), Stack and Brown (1969b) have also proposed that the "natural" condition is for homologous chromosomes to be paired, as in Diptera. The unpaired condition, therefore, is a derived condition produced by genes that inhibit pairing, as in wheat. In any event it is now possible to study the cytogenetics of premeiotic pairing (Feldman, 1966).

This scheme, which proposes premeiotic pairing as a necessary prelude to, and makes possible, synapsis (involving the synaptinemal complex) at zygotene, is in direct conflict with the equally old dogma of the initial pairing of homologues only during zygotene. At this time it seems likely that one scheme is a general condition among meiotic organisms so that the other scheme is doomed to be discarded. At this time both schemes have adherents, and there is evidence for both. The dogma will be discussed in Chapter 6.

CHROMOSOME END ASSOCIATIONS

In 1969 Wagenaar reported end-to-end associations of all chromosomes during mitotic prophase in eight species of plants (*Allium cepa*, $2n = 16$; *Secale cereale*, $2n = 14$; *Nigella arvensis,* 2n = 12; *Callitriche hermaphroditica,* 2n = 6; and four species of *Crepis,* 2n = 6, 8, 8, and 10). Later the same year Wagenaar and Sadasivaiah reported end-to-end associations of homologous and nonhomologous chromosomes during pachytene of the first meiotic division in *Crepis capillaris* (2n = 6).

At first sight these observations are startling; but later one recalls that until about 1925 such end-to-end association of homologous and nonhomologous chromosomes was accepted by some cytologists as proved in at least some species (Mottier, 1905, 1914; Mottier and Nothnagel, 1913; Chipman, 1925). Sharp (1926) stated (p. 143) about mitotic prophase chromosomes in some species, "they seem to be arranged end-to-end in a more or less continuous spireme, which later breaks up into independent chromosomes." The continuous spireme was supposed by some cytologists to occur also during pachytene of meiotic prophase (Sharp, 1926, p. 267; authors cited above). Thus, although Wagenaar's reports are striking at this time, they are, with respect to prophase end-to-end association, by no means novel observations. What cytologists wish to know now, however, is, Were his observations correct specifically or generally, or did he report artifacts or abnormal or unique conditions? His quantitative data were very scanty, but he was perspicacious in selecting a number of plant species with low chromosome numbers for study. Presumably, the fewer chromosomes in a prophase nucleus, the clearer and more reliable the observations. Fig. 5-4, and a similar figure of a brain cell prophase of a mosquito (2n = 6) by Breland and Gassner (1961), indicate that prophase chromosomes may not always be associated end to end; and the concept, that quite generally nonhomologous chromosome ends are associated from telophase into prophase, should not be accepted as proved.

End associations of giant polytene salivary gland chromosomes of *Drosophila* also occur typically (Bauer, 1936; Hinton and Atwood, 1941; Berendes and Meyer, 1968) and are nonhomologous. Ramirez and Miller (1962) and Namboodiri and Lowry (1967) have assumed that the chromosomes of certain fungi are associated end to end during division, in two parallel "tracks." Schrader (1940) described prophase end associations in an insect as "touch-and-go pairing." Zang and Back (1968) from orcein squashed preparations and Hoskins (1969) from electron microscopy of micrurgically isolated metaphase spindles have described interconnections between human chromosomes. Brown and Stack (1968) regularly observed association of the same two nonhomologous ends of premeiotically paired chromosomes at metaphase and early anaphase in normal floral tissues of *Haplopappus gracilis* (n = 2) (Fig. 5-3, *A, B,* and *D*). Therefore, the observations in cultured cells of stem, but not of root, of the same species were not necessarily effects of culture (Mitra and Steward, 1961).

Aside from end association of nonhomologous chromosomes, there is indirect (the "distributive pairing" of Grell, 1964, 1967) and direct (McClintock, 1933; Maguire, 1966; Michel and Burnham, 1969) evidence for nonhomologous side-by-side pairing of whole or parts of chromosomes during meiosis. Thus, nonhomologous pairing is no novelty.

Another possible example of nonhomologous pairing has been reported for maize chromosomes at meiosis in certain *Zea mays* (n = 10) × tetraploid *Tripsacum dactyloides* (n = 36) hybrids (Harlan et al., 1970). These workers found that the 10 nonhomologous (but possibly homeologous) chromosomes of the hap-

Fig. 5-4. Mitotic mid-prophase chromosomes of living root tip cells squashed directly in aceto-orcein of *Bradburia hirtella,* **A** (2n = 6), and *Krigia gracilis,* **B** (2n = 8), both of the family *Compositae.* In neither species were the ends of mid-prophase chromosomes associated. Five such cells of each species were examined. In both species the chromosomes were polarized with the more or less median centromeres near the nuclear pole and the ends of arms near the antipole.

loid maize set, in some of these 46-chromosome hybrids, formed from zero to five pairs frequently. Also, in 50-chromosome hybrids (36 *Tripsacum* and 14 maize) 25 pairs are formed. The maize chromosomes consist of one haploid set of 10 from the pollen parent and 4 only from the irregular female parent. Of these, the four pair with four of the haploid set, and the remaining six nonhomologous maize chromosomes form three pairs. Presumably, meiotic bivalent-pairing genes of the bivalent-forming autotetraploid *Tripsacum* may cause all of this nonhomologous pairing. Disjunction of these nonhomologous pairs at anaphase I seemed normal (except for early separation of the *Zea* chromosomes in the 46-chromosome hybrids); but whether there were chiasmata or not was not determined. It is possibly similar to the "distributive pairing" in *Drosophila* oocytes (Grell, 1964, 1967).

These recent data support the old concept of the continuous spireme during prophase of mitosis. Wagenaar proposed that the end associations are formed at telophase, some are homologous, they persist through interphase and into prophase, and this form of association is related to somatic and premeiotic side-by-side pairing of homologues (S. G. Smith, 1942; Brown and Stack, 1968) in preparation for meiotic synapsis.

The old and the new cytological evidence indicates directly that end associations during prophase and probably during interphase are seemingly common, especially in flowering plants. There is also direct and indirect evidence that the chromosomes are usually not, or never, arranged spatially at random within interphase nuclei (Vanderlyn, 1948 and Fig. 5-4), and this may be related to end associations.

It seems as though there are two sorts of pairing, homologous and nonhomologous, of which Grell's "exchange" and "distributive" are examples. It is likely that these two sorts are at least somewhat different, the nonhomologous not being dependent on genetic similarity and not producing the synaptinemal complex. Certainly, association of ends of nonhomologous chromosomes is very different from the synapsis of homologues.

SOMATIC REDUCTION

Another orderly phenomenon of occasional mitotic divisions is somatic reduction. Although there are numerous cases of irregular disjunction or reduction via aneuploid intermediates, there are too many cases of euploid reduction directly to haploidy to be ignored as "accidents." Some such reductions result from meiosis, but the interesting ones occur during a more or less typical mitosis.

That the chromosome number in a species can be reduced is well known. Meiosis achieves that, and in many types of plants the reduced number persists for a short or long time as a gametophyte. But such reduction is merely from one phase of the life cycle, the 2n or diploid form, to a gamete, spore, zoospore, etc. Among higher plants and animals the haploid cells are short-lived gametes or gametophytes or, if they develop parthenogenetically, produce inviable or sterile "haploids." Exceptions occur in such forms as male Hymenoptera, which have evolved such a system as viable, but, again, as only part of the life cycle.

Among plants, however, viable haploids can occur and persist. This possible evolutionary countercurrent to polyploidy (Raven and Thompson, 1964; DeWet, 1968) requires previous polyploidy. A haploid derived by meiosis and parthenogenesis from a polyploid, called a *polyhaploid*, can be viable and fertile, but only because it is still at a diploid or polyploid level. The process has been recorded, but its frequency in nature and evolution is unknown. Kimber and Riley (1963) listed 94 recorded polyhaploids, of which nearly all had variable chromosome numbers, but at meiosis usually few bivalents or none at all were evident. Only seven species, one in each genus of *Bromus, Dactylis, Sorghum, Medicago, Parthenium, Solanum,* and *Valeriana,* had polyhaploids with regular bivalent pairing, and only three were self-fertile and could represent the beginning of a line of "haploids."

Another well-documented type of successful evolutionary chromosome reduction is aneuploid reduction. This is achieved by translocation of most or all of the gene-containing material of one chromosome onto another and subsequent loss of the geneless centric fragment. That is, 2n = 6 (AA BB CC) becomes DD CC (2n = 4) where D = A + B. This is the type of reduction documented in three genera of Compositae: *Crepis* (Tobgy, 1943), *Haplopappus* (Jackson, 1962), *Chaenactis* (Khyos, 1965), and suspected in many genera of plants and animals *(Drosophila),* especially most higher plant species having n = 3 or 4 (see Chapter 13).

Probably the most controversial form of chromosome reduction is "somatic reduction" or "genome segregation." Ever since the early cytogeneticists proved random anaphase II segregation of chromosomes, it has been assumed that random segregation of chromosomes occurs also at mitotic anaphase. That assumption may not be correct.

Genome segregation has been recorded in fungi, animals, and plants. In fungi that are asexual and basically haploid, such as *Aspergillus, Penicillium,* and *Verticillium,* diploid forms can be induced. These diploids can produce haploids (Käfer, 1961; Hastie, 1967). Since the diploids have somatic crossing-over and genetic recombination, a haploid-diploid–haploid-diploid pseudosexual-pseudomeiotic system exists, called *parasexuality* by Pontecorvo (1958). In *Aspergillus* it seems that reduction from diploidy or 2n + 2 to haploidy is achieved by gradual aneuploid decrease one or, more likely, two chromosomes at a time. It may be, on the other hand, that diploidy to haploidy may occur in one step fairly often.

Genome segregation has been reported in polyploid cells of rat liver (Glass, 1957), in parathenogenetic weevils (Suomalainen, 1969), in chimeric frogs (Everett and Borcherding, 1970), in early segmentation by unequal mitosis in the snail *Aplysia limacina* (2n = 32) into daughter cells with 8 and 24 chromosomes, which is possibly a case of whole haploid set segregation in a tetraploid (Fallieri, et al., 1969).

Chromosome reduction directly from diploidy or polyploidy to haploidy occurs in culture or following colchicine treatment. Buffaloe (1959) produced highly polyploid *Chlamydomonas* cells, but when colchicine was withdrawn the cells quickly achieved the haploid, probably the 2C DNA level (Sueoka et al., 1967), apparently by a sequence of whole genome reductions. Mitra and Steward (1961) and Mitra et al. (1960) recorded haploid cells in cultures of carrot and *Haplopappus gracilis,* respectively, but no aneuploids. Straus (1954) also recorded haploid cells of maize endosperm in culture. Since no aneuploids were recorded in these reports, it seems that there was whole genome reduction directly to the haploid condition.

Somatic reduction in vivo is more controversial but is documented sufficiently to imply reality, such as by Huskins (1948), Huskins and Steinitz (1950), Wilson and Cheng (1949), Huskins and Cheng (1950), Britten and Hull (1948), Sharma and Bhallachayie (1953), Chen and Ross (1963), Storey (1968), Fallieri et al., (1969), and others. Huskins and his associates reported that the prophase chromosome mass separates into balanced haploid parts, at metaphase the two haploid groups remain separate, and four haploid daughter nuclei result. As Patan (1950) concluded, the prophase "group separation is closely determined by what may be called, without physical implications, 'homologous repulsion.' It seems that 'reductional groupings' originate at prophase as random irregularities and that later, presumably during formation of the mitotic spindle(s), only those are kept apart which happen to have a sufficient degree of homologous segregation. Reduced nuclei derived from 'reductional groupings' may, therefore, be expected to have a more or less complete genome." The result should be haploid cells at later divisions, but few of such are recorded. Coe (1954) did report cells having 24 chromosomes in root tips of *Zephyranthes* and *Cooperia* polyploid plants that were 2n = 48.

Brown (1947), Menzel and Brown (1952), and Menzel (1952) have reported that somatic reduction in polyploid hybrid cotton produced branches that were exactly or almost exactly polyhaploid. Franzke and Ross (1957), Chen and Ross (1963), and Simantel and Ross (1964) found exact haploids and polyhaploids among progeny of colchicine-treated *Sorghum.* One such haploid, derived from a heterozygous diploid, was then doubled to produce a homozygous diploid. Knott (1956), on the other hand, reported somatic reduction just before meiosis in hexaploid wheat (2n = 42). The products had 20 and 22 chromosomes, both just off haploid. But meiosis in the derived meiocytes showed only four and five closed bivalents, respectively, as well as many univalents, indicating random assortment at the anaphase when somatic reduction occurred. Genetically controlled elimination of certain chromosomes during embryogenesis in *Sciara* and *Miastor* is well known.

Another form of chromosome "reduction" is the somatic pairing of homologues where they are so intimately paired that they appear in the haploid number. This is true of most polytene chromosomes of the Diptera, but the pairing and reduction (somatic meiosis) in mosquito gut cells (see endopolyploidy in Chapter 12 for details) is very significant. Brown and Stack (1968) recorded and illustrated ordinary mitotic homologous chromosomes of *Haplopappus gracilis* (2n = 4) so closely paired at metaphase and anaphase as to appear as n = 2 (Fig. 5-3). It is probable that Battaglia (1947) saw the same configuration in another plant and called it "heterotypic." Such chromosomes are as reduced in number as chromosomes are at metaphase I and have about the same composition.

Recently there have been reports of somatic reduction from in vitro (Sinha, 1967; Teplitz et al., 1968) and in vivo (Volpe and Earley, 1970) studies of cell hybridization. (But Freed, 1970, "explained" Volpe and Earley's observation as two partially superimposed cells). The polyploid hybrid cells and nuclei produced are able to give rise to daughter cells each of which contains one whole diploid chromosome set (genome) of each parent. Teplitz et al. observed and proposed that such a polyploid hybrid cell containing two genomes of mink and two genomes of cattle chromosomes produces four daughter cells simultaneously by two spindles on one tetrapolar "spindle," each of the cells contains one mink and one cattle genome. It is generally assumed to be achieved by repeated loss of one or more chromosomes at a time during a series of mitotic divisions (Käfer, 1961), but it may occur in one abnormal division. The chromosome number reduction in cycad roots (Storey, 1968) and the polymitotic mutant of maize (Beadle, 1931) seems to be of a different sort. Evidently there is no DNA replication or structural doubling, so that each metaphase chromosome is single. Nevertheless, in the cycad root cells there seems to be equal distribution to the haploid daughter nuclei in the first of such reduction divisions, and distribution is as equal as numerically possible in subsequent divisions. This remarkable example seems to indicate a control that causes equal segregation whether whole or partial genomes are present.

The question of why the observed complete genomes segregate, rather than a miscellaneous random assortment of chromosomes, is not explained, although Teplitz et al. (1968) stated "that in normal tetraploid cells a mechanism (distribution control) strictly regulates movement of a haploid set of mitotic chromosomes into daughter cells upon cell division."

NONRANDOM SEGREGATION

Recently another type of unusual nonrandom mitotic segregation has been reported. In 1965 Walen reported autoradiographic evidence that DNA replication may be nonrandom at the level of chromatids. Lark et al. (1966) demonstrated from autoradiographic studies of tissue culture fibroblast nuclei of mouse that all of the chromatids produced at one particular S period separated *as a group* from sister chromatids formed at an earlier or later S period. Lark (1967) has reported the same nonrandom anaphase segregation of chromatids in plant root tips of *Vicia faba* and diploid *Triticum boeoticum*. Rosenberger and Kessel (1968) have also demonstrated in the fungus *Aspergillus nidulans* that "chromatids containing DNA strands of identical age segregate as a unit during mitosis."

Of course there are other, somewhat similar experiments that indicate the expected *random* anaphase segregation of chromatids (or centromere regions) of the same age (Heddle et al., 1967; Wolff and Heddle, 1968; Goldstein and Lark, (1967). But the technique employed in these later experiments, treatment with 5-amino uracil to synchronize the cells, might have modified the physiology of the cells and might have upset the mechanism that produces nonrandom segregation in animals, plants, and fungi, if it does indeed occur.

HYPOTHESES

An hypothesis to account for the various sorts of nonrandom anaphase segregation at mitosis is that the chromosomes of a genome establish end-to-end associations at telophase. These associations are maintained throughout interphase and at least in prophase, perhaps maintained in part by end attachment to the nuclear envelope. This end association of chromosomes is passed on as end associations of sister chromatids, which are synthesized during the same S period and maintained during structural chromosome doubling. The association must continue to persist to prometaphase and early metaphase to produce a nonrandom arrangement of genomes and of sister chromatids. Therefore, at anaphase, whole genomes can pass to one of four alternative poles in a tetrapolar division to produce the somatic reduction reported, or whole sets of chromatids formed at a particular S period can segregate as a group.

Of course, an alternative hypothetical scheme for somatic genome reduction is for homologous chromosomes to pair at anaphase and maintain the paired condition to the following metaphase, which seems to be a common occurrence (Brown and Stack, 1968). If during interphase there is no DNA replication or splitting of each chromosome into two sister chromatids, much as occurs during the interkinesis between the two meiotic divisions, then the paired homologues could separate at anaphase to form two haploid daughter nuclei; a sort of meiotic mitosis. Such a division has been proposed as an early stage in the evolution of meiosis (Stack and Brown, 1969b), but a complete division of that sort is known to occur in only one organism. The intimate somatic pairing of homologues does occur in *Haplopappus gracilis* (Brown and Stack, 1968), and interphase without DNA replication or chromosome doubling occurs in cycad root tips (Storey, 1968) and maize pollen grains (Beadle, 1931); but there is only one case known of intimate somatic pairing of homologues *and* an interphase without DNA or chromatid replication, in mosquito gut cells during metamorphosis (Berger, 1938; Grell, 1946), although the condition is certainly more widely possible.

This discussion is at best tentative. It relates a number of types of reported nonrandom anaphase segregations, which are themselves tentative, to a condition of homologous and nonhomologous end association of prophase and, by implication, interphase chromosomes. Such end association has not yet been proved to be a general condition, but as an hypothetical scheme it may prove useful. All of these observations do indicate that the classical concept of mitosis may be too simple and that change to a modified concept is in the making at this time.

Luykx (1970) has proposed a different scheme to explain how nonrandom segregation of radioactive chromatids might be achieved. This scheme requires that an anaphase or telophase chromosome be already double, a concept that has been controversial for a long time. His concept is in part based upon observations of Walen (1965), Schwarzacher and Schnedl (1966), and Herreros and Gianelli (1967) that in newly duplicated chromatids the radioactivity is located along the outer lateral surface (see later). That implies an orderly and specific control of chromatid replication and orientation. Luykx proposes that a G_1 chromosome consists of two parallel nonradioactive chromatids but that each G_2 chromosome, after synthesis in radioactive thymidine, consists of four strands, the *two outer* only being radioactive. At first anaphase one radioactive and one cold subchromatid of each anaphase chromosome move to the pole, *radioactive strand leading*. This is supposedly true for all of the chromosomes of the set. During the next synthetic stage, not in radioactive thymidine, the radioactive strand is one of the two interior strands, and in all chromosomes of the nucleus it is in the same relative position to the spindle axis and all are closer to one and the same pole of the nucleus than to the other. This position is maintained throughout interphase by the centromere regions being attached to the nuclear envelope. At the second metaphase all of the radioactive strands of all the chromosomes are still closer to the same pole, and at anaphase all move to that pole.

He emphasizes that this scheme is most or only applicable to cells that in repeated divisions have all of the spindle axes parallel. This is especially true in plant root tips where a series of cell divisions usually forms a linear series of cells.

In addition to the weaknesses of this model just mentioned is the important but highly questionable assumption that at anaphase one chromatid of a double anaphase chromosome leads the other to the pole.

The hypotheses discussed here, which try to "explain" the still controversial matter of nonrandom segregation of chromatids, are very speculative and are themselves based upon questionable "facts." The only facts they use in common are that at G_2 there is some order of new and old chromatids shared by all of the chromosomes of the nucleus and that the telophase arrangement of chromosomes is rigidly maintained to at least prophase and probably to the beginning of anaphase. These are not explanations of the phenomenon of nonrandom segregation. They are no more than speculative hypotheses.

UNILATERAL CHROMATID REPLICATION

Another phenomenon of DNA and structural relationships of chromatids has been observed following a pulse of radioactive thymidine and examination at the *second* mitotic metaphase following the pulse (Walen, 1965; Schwarzacher and Schnedl, 1966; Herreros and Gianelli, 1967). At the first metaphase both chromatids of each chromosome are labeled as expected according to the semiconservative scheme of replication. If the two chromatids are prevented from anaphase separation so that at second metaphase after a second S period in cold thymidine the two pairs of chromatids lie parallel, as diplochromosomes of endoreplication (Fig. 19-17), it has been observed that the two inner chromatids are *not* labeled but the two outer ones *are* labeled, very regularly (Fig. 5-5). It is difficult to interpret the reason for this spatial relationship of the labeled and the later-formed unlabeled chromatids. Of course the chromosomes enter the second S period in a somewhat abnormal condition, more or less double rather than single, depending upon just when the chromatids actually do separate. But it does seem that the two radioactive (old) chromatids form the two unlabeled (new) chromatids between them, symmetrically very predictably, and nonrandomly, but one to the right and one to the left. Walen considered the possibility of the endoreplication not being due to colchicine since the replication cycle of the cells used is about 30 hours but the colchicine treatment lasted only 8 or 9 hours. It is, however, quite probable that the endoreplication (between the first and second metaphases) was due to the colchicine or mercapto-ethanol (Schwarzacher and Schnedl) treatment, according to conclusions by cytogeneticists of human mitotic chromosomes. Herreros and Gianelli examined

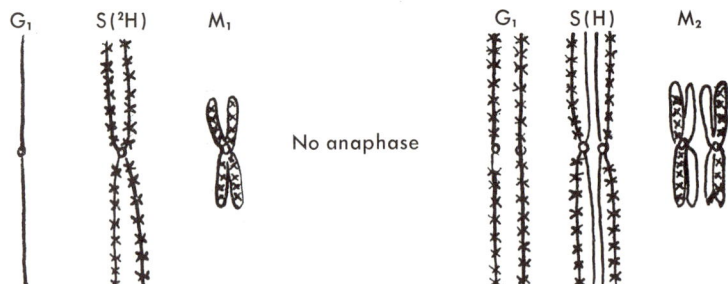

Fig. 5-5. Unilateral chromatid replication. M_1 represents the metaphase$_1$ chromosome as seen with both chromatids labeled after an S period in thymidine ^{3}H. Colchicine or other chemical prevents anaphase so the two chromatids separate but remain closely parallel at the second G_1. At the second S period, in nonradioactive thymidine, replication occurs. At metaphase$_2$ the two adjacent double chromosomes (a diplochromosome) show the outer chromatids only to be radioactive. X's indicate radioactivity. Simple diagrams of chromonemata are represented under G_1 and S periods.

1,273 diplochromosomes after from 2 to 3 days in colchicine and reported 825 outer-outer arrangements of radioactive chromatids but only 23 outer-inner and 2 inner-inner arrangements, and 423 had chromatid exchanges. Schwarzacher and Schnedl and Walen reported the outer-outer arrangement in nearly all diplochromosomes. The significance of this phenomenon may be little or very much in the eventual comprehension of chromosome structural doubling.

CONCLUSION

This chapter discusses and documents a number of observations, concepts, and tentative generalizations that are to some extent at odds with the general concepts of mitosis. Some of them may be the sort of atypical conditions that usually occur among organisms. Some eventually may be demonstrated to be errors of observation or interpretation. Others may eventually become incorporated into the generally accepted dogma. But at this time they do demonstrate that all is not now known about the process of mitosis, even of "typical" mitosis.

I shall end this chapter of debatable topics with two thought-provoking quotations from East (1933):

I should like to suggest that there is a fundamental mechanism for bringing complete genomes into gametes at meiosis and into daughter cells at mitosis. This mechanism, I assume, is a more generalized and less obvious scheme than that usually visualized. Gene-by-gene pairing is thus considered to be a part of a complicated process, occurring wherever possible.

. . .

It appears to me to be unwise, in our current state of knowledge, to become wedded to the conviction that gene-by-gene pairing occurs and is the fundamental mechanism in the distribution of the hereditary units; but perhaps there is another force or mode of action that brings the chromosomes into association through general similarities as a prerequisite to the essential type of pairing which then occurs if possible.

CHAPTER **6**

Meiosis

> Because of the profound alterations in the chromosomal mechanism effected by meiosis and syngamy, these processes represent the two most important crises in the life cycle, so far as nuclear constitution is concerned.
>
> Sharp, 1926

As Sharp indicated in that excerpt, meiosis and sexuality, which compensate for one another chromosomally, are both necessary within any continuing typical life cycle. Sexuality doubles the chromosome number; meiosis halves it. <u>Sexuality evolved as a method for increasing intraspecific variability, and a reduction division, as a means of reducing the chromosome number, had to evolve simultaneously.</u> Meiosis as it now occurs, however, is much more than a means of reducing the chromosome number; it is also supplementary to sexuality in increasing intraspecific variability. Sexuality mixes up whole chromosomes; meiosis mixes up segments of chromosomes by crossing-over, which produces genetic recombination. And the study of recombination is the heart of genetics. Later in this chapter a hypothetical scheme of the evolution of meiosis is presented. Furthermore, recent evidence for a modification of the classical scheme is here supported.

An interesting example of what happens when sexuality occurs but meiotic reduction is not achieved was reported in *Parthenium* by Powers (1945). He found a strain (2n = 72) that produced unreduced eggs by meiotic failure (apomeiosis) but required fertilization. The progeny of aberrants had 2n = 108. These plants also had apomeiosis, diploid eggs, fertilization, and a progeny with 2n = 140 (called hyperaberrants). That seemed to be as far as the process could go. The production of haploids (haploids of polyploids) is the opposite condition, in which meiosis does occur but fertilization is not accomplished. This condition usually requires genes for parthenogenesis in order to be achieved and so, as a regular process, is largely limited to apomictic species such as in *Parthenium, Poa, Dichanthium,* etc. (see Chapter 18).

It is likely that the meiotic process evolved from two sequential mitotic divisions. The first of these is the more unusual and used to be called the *heterotypic mitosis*. The second meiotic division, meiosis II, is less different from a typical mitosis and used to be called the *homeotypic mitosis*. But, in addition to modified mitotic divisions, the interphases preceding each are also differently atypical. The interphase before the first meiotic division is preparatory to the unusual meiosis I. The interphase between them, if it occurs, is called interkinesis and is very unusual; it is unique in that there is no DNA replication or structural doubling of the chromosomes. I know of only three reported examples of such an interphase clearly not related to meiosis, and they occur: in late differentiation and probable death of the specialized cycad root cells (Storey, 1968), in the gut lining of mosquitos during pupal metamorphosis (Berger, 1938), and in pollen grains of maize carrying the "polymitotic" gene (Beadle, 1931). The result in the cycad root cells is a decrease in chromosome number from $2n = 22$ (for example) to $n = 11$, to $n/2 = 5$ or 6, to $n/4 = 2$ or 3, and even to one chromosome. It is possible that the interphase following meiosis in *Chlamydomonas reinhardi* also may be characterized by no DNA replication because it is reported to have, in addition to replication in the gametes, a second DNA replication after crossing-over (pachytene?) (Sueoka et al., 1967) which is supposed to provide for a mitotic division of each of the four cells produced by meiosis. The polymitotic mutant reported by Beadle (1931) is of this sort, also producing after regular meiosis a series of divisions with no DNA replication. As a result many cells are formed, each with one or few chromosomes. The distribution at the irregular anaphase is random. Haploid cells in diploid tissue, which are occasionally reported, may arise in this manner.

The whole meiotic process of two interphases and two divisions is unusual. And new evidence suggests that the pairing of the homologous chromosomes may occur, in some species at least, as long before meiosis as in the zygote or early cleavage divisions.

THE CLASSICAL SCHEME
Premeiotic interphase

Meiosis occurs in certain cells of animals, fungi, and plants called *meiocytes*. Regardless of other names, a cell in which meiosis occurs is called a meiocyte. Spermatocytes and oocytes of animals; sporocytes, including microsporocytes and megasporocytes of land plants; the ascus, basidium, etc. of fungi; zoospores of some fungi and algae are all examples of meiocytes. Many meiocytes (Beasley, 1938), but especially oocytes of animals, may grow considerably before meiosis begins. In some insects (Lima-de-Faria and Moses, 1966) extrachromosomal DNA is formed during the premeiotic interphase whereas such gene amplification occurs in amphibians during meiosis, at diplotene, with lampbrush chromosomes.

The replication of DNA typically occurs during the premeiotic interphase as has been repeatedly demonstrated (Taylor, 1959). It is probable (Stack and Brown, 1969a) that the heterochromatin of premeiotic interphase (Fig. 5-1) must decondense before its DNA and histone replicate, as Milner and Hayhoe (1968) and Milner (1969) reported in human blood cell mitotic interphase. Antropova and Bogdanov (1970) found that histone doubling starts in the DNA S period but is not complete until pachytene. Sheridan and Stern (1967) reported a unique histone synthesized before meiotic prophase begins. Lamb (1969) concluded that there is the synthesis of two heat-sensitive compounds during the premeiotic interphase that affect recombination at pachytene. Lawrence (1961, 1965) concluded that gamma radiation just before or during the S period interferes with the specific synthesis of one or a few compounds concerned directly or indirectly with crossing-over later at pachytene.

An exception to premeiotic interphase replication of DNA occurs in those organisms that do not have a diploid premeiotic interphase, such as some fungi and algae as exemplified by the fungus *Neottiella* (Rossen and Westergaard, 1966) or *Gelasinospora* (Lu, 1967, 1969) and the alga *Chlamydomonas* (Sueoka et al., 1967; Chiang and Sueoka, 1967). In the fungi the DNA does replicate in the haploid gametes before karyogamy, but in *Chlamydomonas* the doubling occurs in the zygospores by two consecutive DNA replications for one cell division.

An unsolved problem is that of the number of strands per homologue during the premeiotic interphase. Some cytologists think that the strands during G_1 are single, some think them double (Smith, 1932; Atwood, 1937), and some (Nebel, 1936; Nebel and Ruttle, 1936) believe each homologue is four-partite. Furthermore, there is very little knowledge of the relationship between DNA replication and chromosome structural doubling. Certainly DNA replicates during interphase; but when does structural doubling first occur? Doubleness of each homologue is not visible until pachytene. Since we do not comprehend the fine structure of chromosomes or their intermediate structure, we do not even know what strands are, let alone how many there may be.

Needless to say, meiosis is genetically controlled, probably by many genes (Rees, 1961). Therefore, mutations of such genes will produce anomalous meiosis. There are, of course, unusual meiotic divisions that have evolved and are characteristic of the species, such as the odd unipolar meiosis in the fly *Sciara*

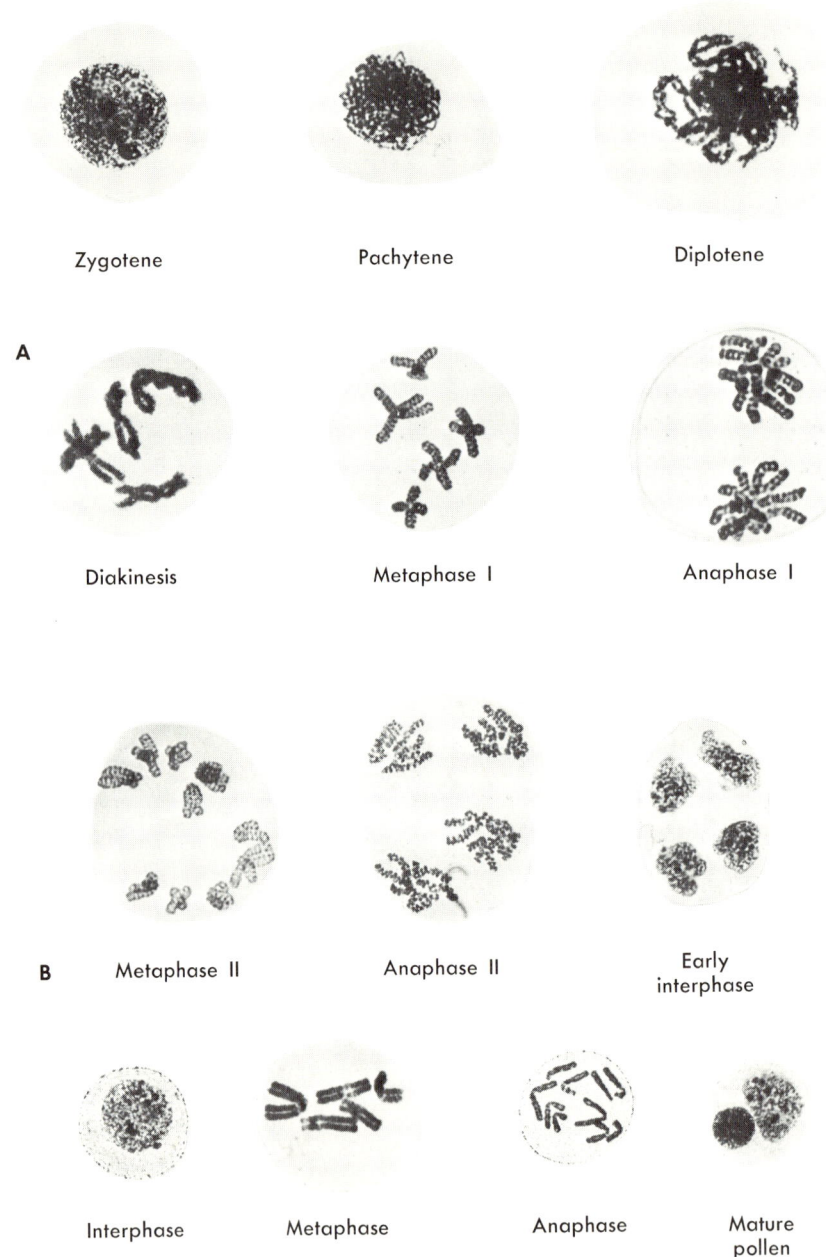

Fig. 6-1. Stages of meiosis in *Trillium* (n = 5) of rather normal, unflattened cells. (See text for details.) (Courtesy Dr. A. H. Sparrow, Biology Department, Brookhaven National Laboratory.)

(Metz, 1938) or the achiasmatic (which follows from asynapsis) in some males such as *Drosophila*. Random single gene mutations, on the other hand, produce the common phenomenon of asynapsis (Soost, 1951) or other upsets of meiosis (Rees, 1961), as in *Drosophila* (Sandler et al., 1968). Such meiotic genes may act before meiosis in the strict sense begins, as during premeiotic interphase or earlier if premeiotic pairing is part of meiosis (Feldman, 1966). Asynapsis, for example, might result from prevention of premeiotic pairing or failure of development of the synaptinemal complex. A discussion of meiotic genes and asynapsis is included at the end of this chapter.

Most significant to cytogenetics, however, are the characteristics and activities of the chromosome during the prophase and anaphase of the first meiotic division and anaphase of the second. It is essential to the geneticist, cytologist, and cytogeneticist to have a clear and detailed knowledge of meiosis (Fig. 6-1).

Meiotic prophase

The end of the premeiotic interphase, or the fusion of the two 2C gamete nuclei of many fungi and algae having zygotic meiosis, leads gradually into meiotic prophase. This period is the heart and soul of meiosis, and often the term meiosis used casually may refer to this unique period only. As in mitosis, meiotic prophase begins when chromosomes begin to be visible as distinct strands in the light microscope. But the meiotic prophase, unlike mitotic prophase, is complex and can be subdivided into descriptively named states. These are, in chronological sequence: leptotene, zygotene, pachytene, diplotene, and diakinesis. Some cytologists and cytogeneticists (Rhoades, 1961) insert a preleptotene stage before leptotene, as the beginning of prophase I. Preleptotene is being referred to more and more often and so is included here.

Preleptotene. Nebel and Ruttle (1936) stated, "Any account of meiosis commencing with leptotene is thus incomplete," and Hiraoka (1941) agreed. Preleptotene has been observed in a number of plant species such as *Sagittaria* (Shinke, 1934), *Allium* (Bonnevie, 1911), *Tulipa* (Newton, 1926), *Fritillaria, Trillium, Vicia,* and *Psilotum* (Hiraoka, 1941), *Lilium* (Taylor and McMaster, 1954), but especially *Lilium* (Walters, 1970). It has been described as the "premeiotic spiral prophase" or as small chromomere-like accumulations of DNA. Moens (1964) described it in tomato as pachytene-like; Brown (1954) described it as a "chromatic phase" in *Luzula camperstris*. In animals it has possibly been described as the "prochromosome stage." This is a highly controversial "possible" stage, and its meaning and structural significance are unclear. If it occurs, the chromosomes subsequently "uncoil" into the more elongate and thinner threads of leptotene. The chromosomes at preleptotene are probably not double, but certainly a number of cytologists have reported doubleness of presumably unpaired leptotene chromosomes (Oehlkers and Eberle, 1957). Since DNA replication seems to extend throughout preleptotene (Taylor, 1959) and into leptotene (Callan, 1968), it may well be the period of structural doubling. Preleptotene may be late interphase rather than prophase, if such a distinction is possible.

Leptotene. The first generally accepted stage of meiosis is leptotene, or leptonema. The concept of the nucleus at leptotene is that the chromosomes are very thin and the homologues are probably randomly distributed and unpaired. In many, especially animal, meiocytes the leptotene threads are not randomly arranged; they are polarized in the nucleus, having all ends attached to the nuclear envelope (Moens, 1969) on a small area with the chromosomes themselves looped across the nucleus, the whole arrangement forming what is described as the *bouquet*. It is very likely that the leptotene chromosomes are never randomly scattered, although that is certainly the appearance in meiocytes not having the polarized bouquet arrangement.

There are many tiny swellings, called chromomeres, scattered along the leptotene threads, although some cytologists have described them as localized aspects of late leptotene coiling that later spread uninterruptedly the length of the pachytene chromosomes. Leptotene terminates when the presumably single threads pair (synapse) with their homologues.

Zygotene. The term zygotene specifies the

coming together side by side of the single or double but unpaired homologous strands. The pairing or synapsing homologues, according to the classical scheme, are "attracted" to each other from their random distribution and are moved by some unknown "force" (see Rhoades, 1961, for discussion). In the nucleus with a diploid number of, say, 40 chromosomes, one can imagine the turbulence created by 40 long chromosomes writhing among one another, each being attracted to its homologue, like 20 supersexed pairs of 20 species of earthworms contained in a quart bottle, *and moving sideways!* It is generally stated that the pairing usually starts at chromosome ends if there is a bouquet configuration. Otherwise, pairing may start anywhere, wherever homologues make first contact.

The study of zygotene pairing by light or electron microscopy is very difficult. Chromosome threads are assumed to show pairing if a segment is seen in which two run closely parallel, or in which some strands are thick and presumably paired but other strands are much thinner and seemingly single. As in leptotene, many tiny swellings or irregularities in thickness, the chromomeres, can be seen at zygotene. Whether they are small masses of heterochromatin, or gyres of a coiled thread, or some presently unknown trivial or important aspect of the chromosomes is unknown. Irregularities show in electron micrographs but mean nothing so far.

It is during the pairing of homologues that the synaptinemal complex (SC) is put together. It is obvious, therefore, that some well-paired regions probably have completed the SC whereas other regions are still unpaired. Because zygotene, like the earlier leptotene, contains some unpaired strands at the same time that other regions have reached the next stage, pachytene, some cytologists have called it amphitene, that is, both. There is some shortening of the chromosomes, the details of which are unknown. Nor is there any evidence that this shortening of meiotic chromosome length is different from or the same as mitotic prophase shortening.

Zygotene is also the time of *synizesis*. That term applies to an apparent collapse of the chromosomes into a dense mass of chromatin, usually near the nuclear envelope and the nucleolus; the latter is very often flattened against the inner surface of the nuclear envelope. Synizesis is very common, especially in plants, and the term *synapsis* was originally applied to it. Now synapsis refers to the act of intimate association of homologous chromosomes achieved during zygotene. Synizesis may well be an artifact of fixation as the contents of plant nuclei at zygotene are very difficult to fix even with the new fixatives for electron microscopy. By pachytene, however, such collapse no longer occurs.

Although it is usually homologous chromosomes that pair side by side, nonhomologous pairing of whole chromosomes, as in haploids of maize (McClintock, 1933) and *Antirrhinum* (Rieger, 1957), or of inverted segments of otherwise homologous chromosomes (McClintock, 1933; Maguire, 1966) is known. In the light microscope the pachytene pairing of nonhomologous regions seems to be as intimate as that of synapsed homologues; but no chiasmata are formed, and the nonhomologues fall apart during diplotene-metaphase I. They have paired but not synapsed.

There are various conditions present at early prophase I that affect the number of chiasmata per bivalent at metaphase I. Some of these are discussed later in this chapter with meiotic genes and asynapsis. But in addition to gene and chromosomal aberrations affecting pairing, it is well known that hybridization usually affects pairing and/or reduces chiasma frequency. For example, interspecific hybrids in the cotton relative *Cienfuegosia* had chromosome pairing as in the parents but a much lower chiasma frequency (Wilson and Fryxell, 1970). There are also many reports that environmental conditions affect pairing, that pairing is different among genetically very similar plants, and that pairing conditions of the same plant in the same environment can be different on different collecting dates (Douglas and Brown, 1971).

Pachytene. During pachytene (meaning thick) the chromosomes are shorter and thicker than previously, about one fifth as long as at leptotene (Rhoades, 1961). It now seems likely that both ends of each chromosome are attached to the nuclear envelope, not only in the evident bouquet (Smith, 1956; Woollam et al., 1966; and many others) but also as demonstrated by nuclear reconstruction from serial electron microscope sections in an insect spermatocyte (Wettstein and Sotelo, 1967) and in plants (Moens, 1969; Stack, unpublished). Since zygotene and perhaps leptotene also show this condition in the bouquet arrangement, it is possible that chromosome ends are always attached to the nuclear envelope (meiosis and mitosis) from previous telophase.

Synaptinemal complex. Pachytene (Fig. 6-3) is also the stage of the completed and functional synaptinemal complex (SC) (Moses, 1956, 1968). This is a unique cellular structure present nowhere except during zygotene and pachytene of meiosis. It is now considered to be excellent evidence of meiosis. Although the early development of the SC has not yet been adequately worked out, it seems likely that each unsynapsed chromosome forms a longitudinal protein filament called a *core* during or before zygotene (Moens, 1969; Westergaard and von Wettstein, 1970; Stack, 1969). As the homologues synapse during zygotene, the cores of homologues move toward each other and become the parallel *lateral elements* of the SC. They become "connected" by strands that extend from the externally located chromosomes, "through" the lateral elements, to some sort of unknown but possibly very important association, called the *central element*, between the two lateral elements. The paired lateral elements and probably the central element are proteinaceous and extrachromosomal and are later discarded. They seem to be produced for, or they themselves bring about, synapsis. In fact synapsis can now be defined by the configuration of paired homologues with a synaptinemal complex between them.

Westergaard and von Wettstein (1970) from detailed electron microscopy of the ascomycete *Neottiella* have concluded that the lateral elements of the SC are formed between the closely associated sister chromatids of each homologue (Fig. 6-4) but the central region, including the central component, is formed within or upon the nucleolus (Fig. 6-5). Later they get together to form the complete pachytene complex.

Recent study of the synaptinemal complex by whole mount spreads of meiotic chromosomes that were untreated or treated with deoxyribonuclease, ribonuclease, trypsin, or other chemicals and examined by electron microscopy has revealed very significant details; Comings and Okada (1970, 1971) and others have concluded that the lateral and central elements are probably completely protein (no DNA) as are also the fibers connecting the central element to the lateral elements. In fact these connecting fibers loop from the two lateral elements, and where they come together they actually form the central element. Externally from each double lateral element there are extensive chromatin loops (up to 7μ) that compose the two homologous chromosomes, each of two chromatids (Fig. 6-6).

The synaptinemal complex is not always correlated with effective crossovers and chiasmata. There are a few cases known of normally formed pachytene synaptinemal complexes followed during late prophase to metaphase by achiasmatic bivalents, as in male mecopterous insects of the genus *Panorpa* (Gassner, 1967) and in the male achiasmatic mantid *Bolbe nigra* (Gassner, 1969). On the other hand achiasmatic flies do not have the SC (Meyer, 1964). Furthermore, recombination seems likely without the SC, as the reported cases of somatic recombination indicate. It must be emphasized, however, that it has not been proved that there is no SC at somatic crossing-over; it is merely unlikely. But these examples hardly prove or even strongly suggest that the SC is not an evolved structure to produce large amounts of meiotic crossing-over, 500 or more times as much as somatic crossing-over. As Peacock (1971) stated, "The complex appears to be a necessary prerequisite for meiotic recombination, but it is not essential for homologous pairing and orderly progression through meiosis (e.g. Drosophila males), nor does its presence ensure that crossing-over will occur."

It seems likely that the SC should bring paired homologous chromosomes that are in the proper structural condition into crossover alignment that is adequate for hundreds or thousands of effective exchanges to be formed

Text continued on p. 84.

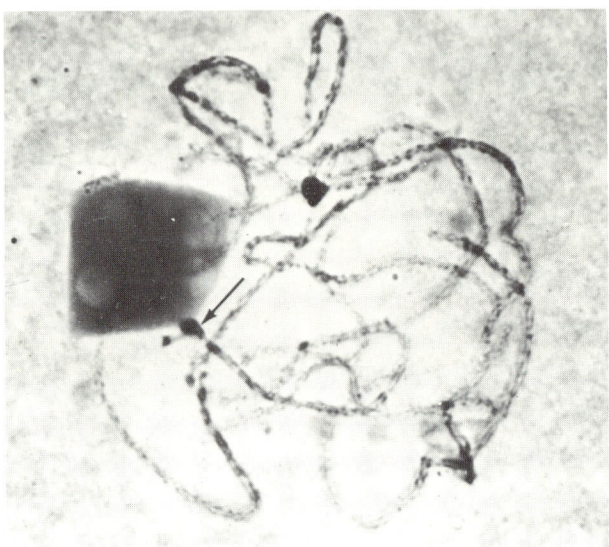

Fig. 6-2. A typical pachytene squashed cell of *Zea mays*. The nucleolus is the dark mass, with the nucleolus organizers of the two synapsed homologues appearing as a dark swelling (arrow) in contact with the nucleolus. The small remainder of the chromosome is evident and at mitotic metaphase would produce a very small satellite. (Courtesy Dr. M. Maguire, Department of Zoology, University of Texas at Austin.)

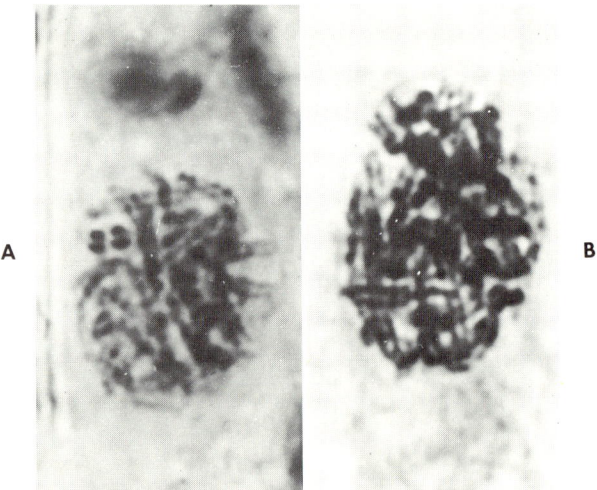

Fig. 6-3. Photomicrographs by optical microscopy of pachytene nuclei of the ascomycetous fungus *Neottiella*. The paired and parallel homologous chromosomes are evident. Such nuclei provided material for the electron micrographs of pachytene of this fungus that follow (Figs. 6-4 and 6-5). The measurement across homologous pairs and the space between is about 0.72μ, as it is in electron micrographs at pachytene (Figs. 6-7 and 6-8.) (From Westergaard, M., and D. von Wettstein. 1970. C. R. Trav. Lab. Carlsberg **37**:239-268.)

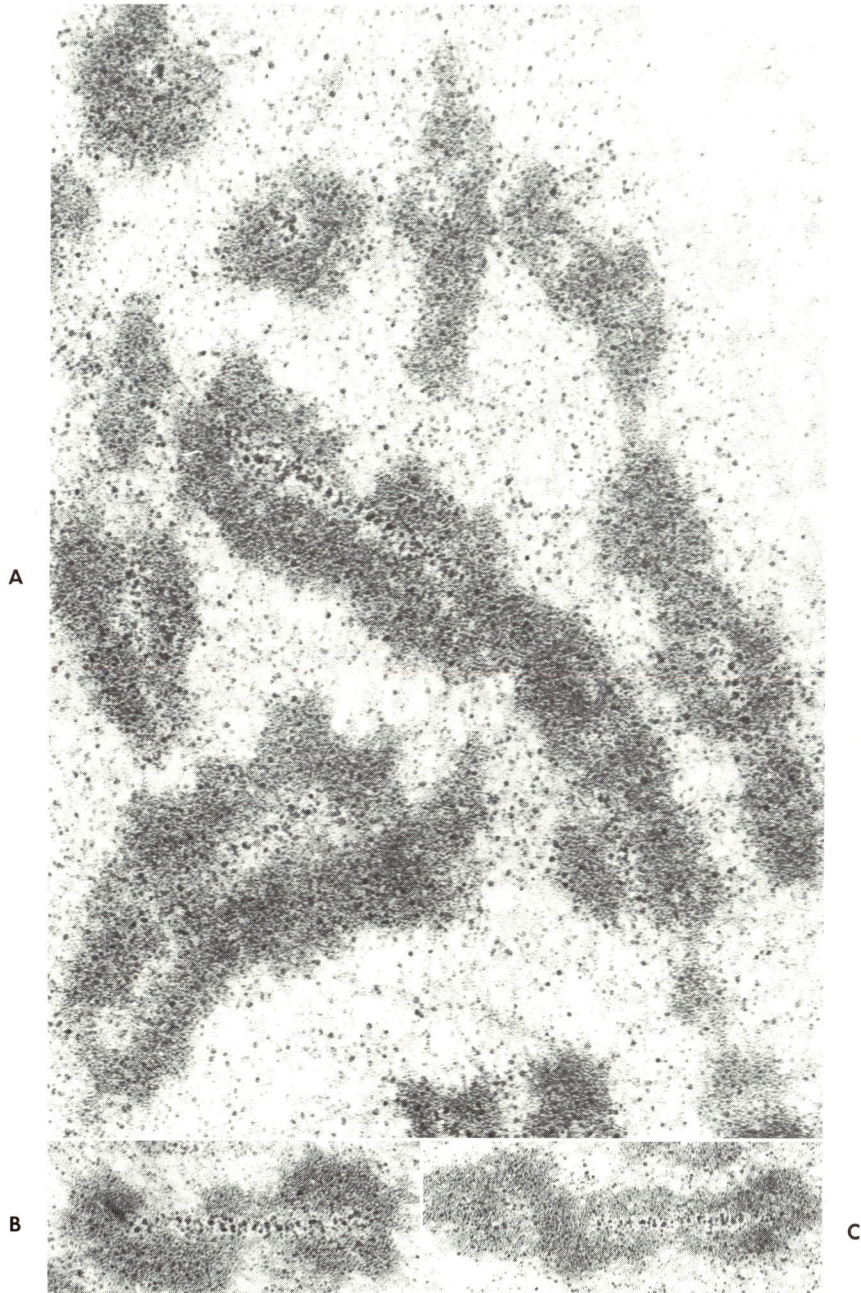

Fig. 6-4. Development of the synaptinemal complex in *Neottiella*, origin of the *lateral elements* between sister chromatids of unsynapsed homologues. This is indicated by the silver grains within clear regions between sister chromatids. It had been determined that the lateral elements precipitate metallic silver in situ from ammoniacal silver solution. (From Westergaard, M., and D. von Wettstein. 1970. C. R. Trav. Lab. Carlsberg **37**:239-268.)

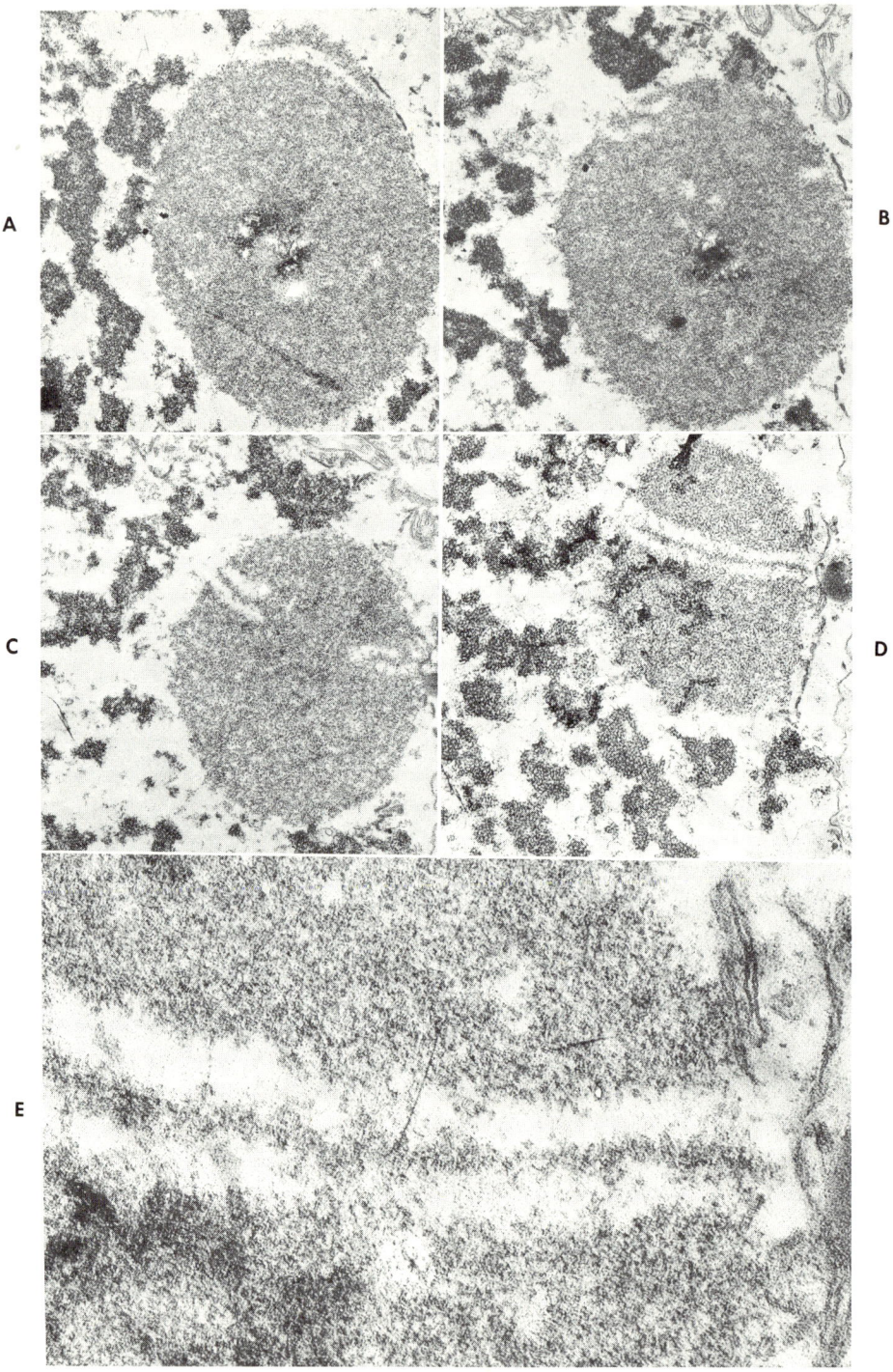

Fig. 6-5. Development of the synaptinemal complex in *Neottiella*, origin of the *central region* on the peripheral component of the nucleolus. **A** to **D,** Sections of a series cut through the same nucleolus of a fusion (presynaptic) nucleus after ribonuclease treatment. **E** is a higher magnification of part of **D.** The central regions appear as two parallel light bands. The true central component is missing in these figures due to treatment. If these are central regions, they are less wide (400 to 600 Å) than in the complete synaptinemal complex, where the central region is 1,200 to 1,500 Å wide. (From Westergaard, M., and D. von Wettstein. 1970. C. R. Trav. Lab. Carlsberg **37:** 239-268.)

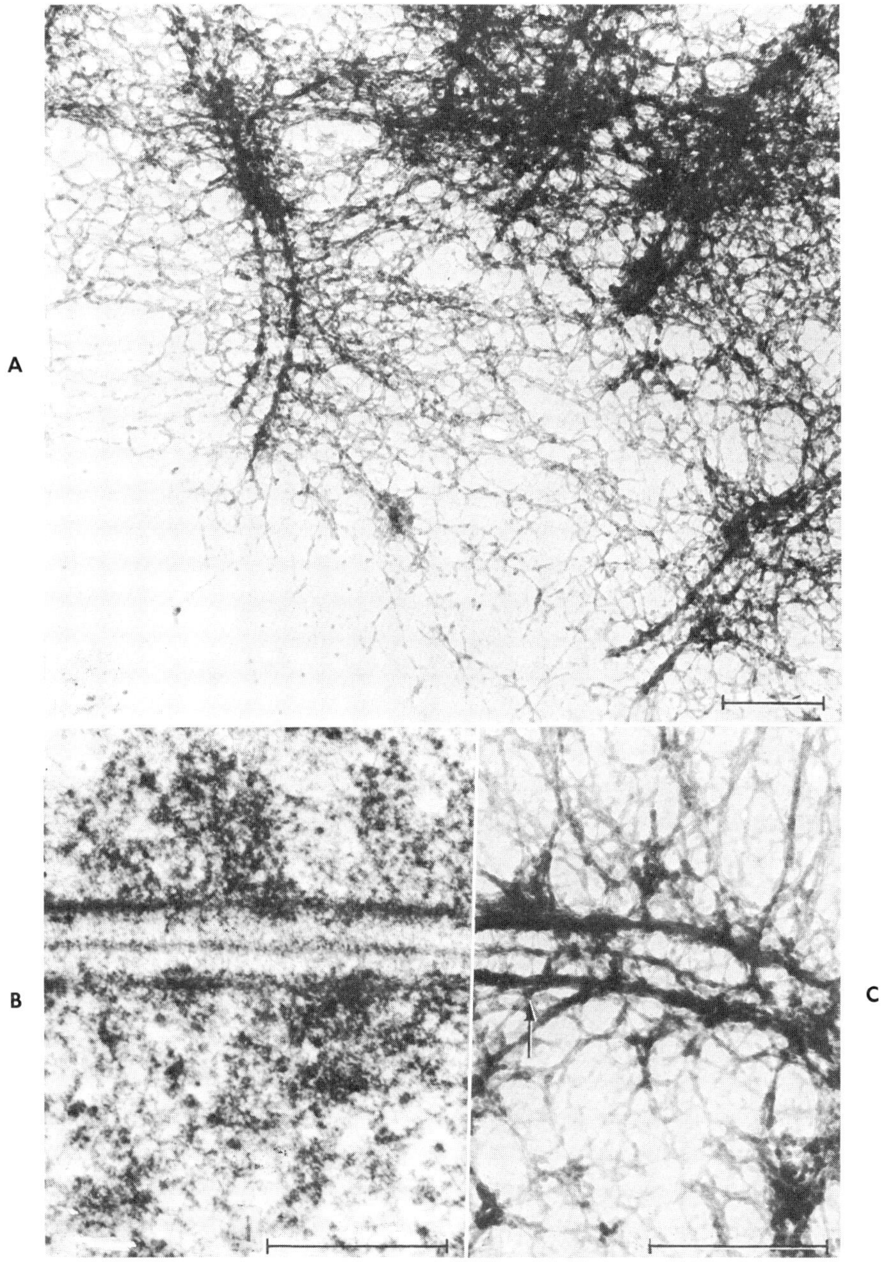

Fig. 6-6. The synaptinemal complex in section, **B**, and water spread, **A** and **C**, by electron microscopy. **A,** A water spread of pachytene chromosomes of a bird (quail) testis. The lateral and central elements of the SC are evident and the homologous chromosomes are outside the lateral elements. A section, **B,** and a water spread, **C,** of pachytene chromosomes of mouse testes at the same magnification for comparison. Markers represent 5,000 Å. (From Comings, D. E., and T. A. Okada. 1970. Chromosoma 30:269-286, Fig. 4, Springer-Verlag New York, Inc., New York.)

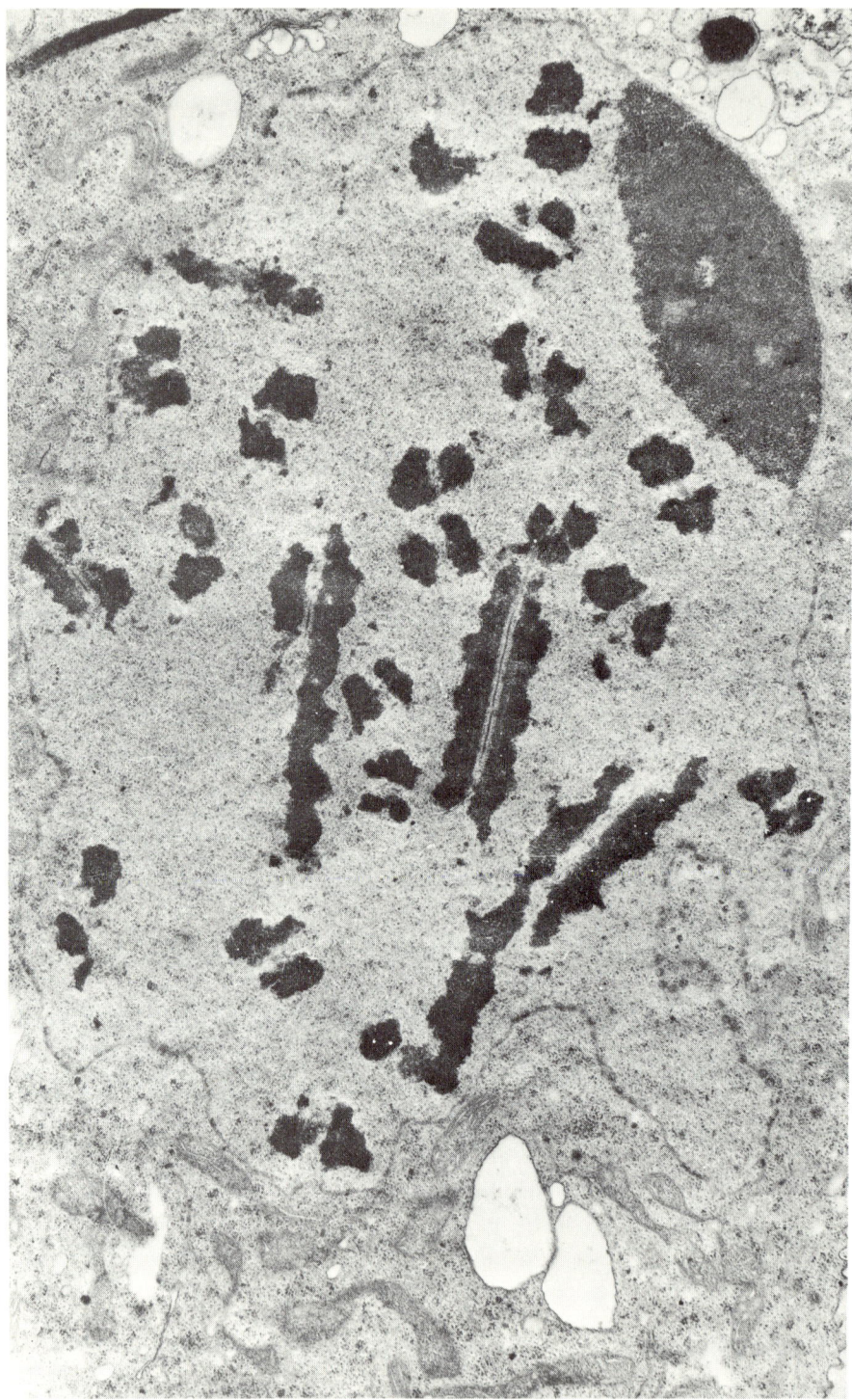

Fig. 6-7. Synaptinemal complexes (SC) at pachytene in a *Neottiella* ascus. Cross sections show the two synapsed homologues but hardly the SC. Some longitudinal or diagonal sections of the bivalents do show the central element as a line in the clear central region between the homologues. See Fig. 6-8 for higher magnification of a longitudinal section. Since the SC is a plane structure, some longitudinal sections do not show it. (From Westergaard, M., and D. von Wettstein. 1970. C. R. Trav. Lab. Carlsberg **37**:239-268.)

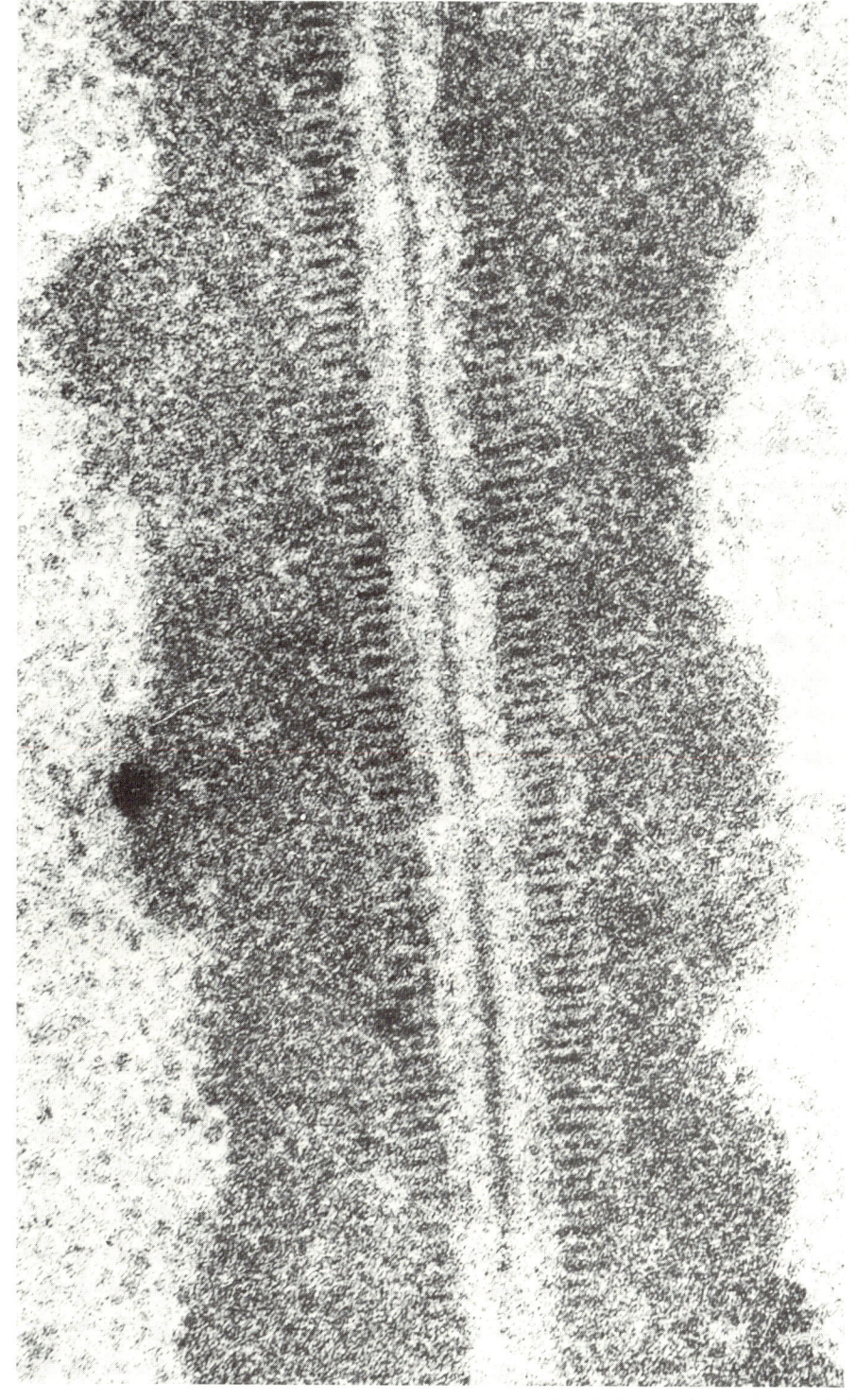

Fig. 6-8. The fine structure of the synaptinemal complex in *Neottiella*. The lateral elements are embedded one in each homologue and consist of parallel short rods. The central region is about 1,200 Å wide and "keeps" the lateral elements at that distance. The central element is evident as a dark line in the center of the central region. 125,000×. (See Fig. 6-9 for details and dimensions.) (From Westergaard, M., and D. von Wettstein. 1970. C. R. Trav. Lab. Carlsberg **37**:239-268.)

at pachytene. Yet the number of observable chiasmata actually present at later stages of meiosis I are usually in the tens, one or two per homologous pair or, rarely, from three to six per pair. Another indication of large numbers of pachytene exchange is the amount of "repair" DNA synthesized at zygotene-pachytene. It indicates in *Lilium* a number of exchanges of the order of magnitude of 10^4, yet only about 3.6×10^1 actual chiasmata are counted (Stern, 1969). The reason for this difference is unknown.

Since somatic interphase and meiotic pachytene crossing-over seem to be the same *genetically*, and if there is nothing like the synaptinemal complex in somatic cells to achieve somatic pairing, then it may be that somatic and meiotic crossing-over are the same *cytologically*, and the function of the SC is merely to draw the already paired homologous chromosomes close enough together (quite uniformly 1,200 Å) so that many more homologous regions are brought together or into contact *surrounding* the SC than can occur somatically during interphase. The considerable variation in details of lateral and central elements of the SC but the uniformity of the distance between lateral elements indicates that the distance between lateral elements (and, therefore, between homologous chromosomes) may be the *only* important aspect of the SC. In fact, the 30 to 60 Å proteinaceous central filaments may be contractile and pull the lateral elements (and therefore the homologous chromosomes) closer together (from 2,000–3,000 Å to 1,200 Å) by their contraction.

This concept would not make necessary any specially evolved, unique, and complex structural aids within the SC for the actual formation of meiotic crossovers. Nor does it predict the probable formation of thousands of pachytene crossovers per chromosome pair in contrast to the very few (from 1 to 10) actual diplotene chiasmata observed. There may be, indeed, only one mechanism of crossing-over, whether somatic or meiotic, and the difference between them is merely quantitative due to the absence or presence of the SC. This scheme fits the evolutionary assumption that meiotic crossing-over is not something new but is the same as somatic crossing-over. It has evolved directly from somatic crossing-over between somatically paired homologous chromosomes. The SC evolved to achieve a quantitative increase in amount of crossing-over.

This scheme does not attempt to explain why somatic and meiotic crossing-over almost always seems to occur between or within homologous genes of homologous chromosomal segments. But no proposed scheme of the mechanism of crossing-over has been able to do that.

Meiosis in an autopolyploid under the optical microscope demonstrates that any number of the homologous chromosomes can form chiasmata among themselves and multivalents result. But at any particular locus only two homologues can be united by a chiasma. Electron microscopy of autopolyploid *Lilium* microsporocytes has shown that at any locus the SC is associated with only two homologues, as is expected, but the SC can switch along its length to be associated with different homologous chromosomes (Moens, 1970). Moens concluded that "The cores may be associated with the recombinationally active hereditary material during meiotic prophase."

Pachytene is the stage at which crossing-over produces genetic recombination, according to the best evidence (Mitra, 1958; Henderson, 1966; Rhoades, 1968; Lindsley et al., 1968; Peacock, 1968, 1971). The *logical* time for crossing-over is during DNA replication; but there is much less acceptable evidence that it does occur then. But science must accept tentatively the *best* evidence at any particular time. In fact there is considerable acceptance that the sole function of the synaptinemal complex is to produce the meiotic crossing-over and the resulting recombination during pachytene. But see the section on asynapsis later in this chapter for failure of chiasma formation.

Crossing-over is proved by genetics, but the physical act is still hypothetical. The majority of geneticists and cytologists consider the evidence very strong that it occurs at pachytene and weak that it occurs during DNA replication during interphase, the only alternative considered. It is also generally agreed that when crossing-over occurs there is breakage

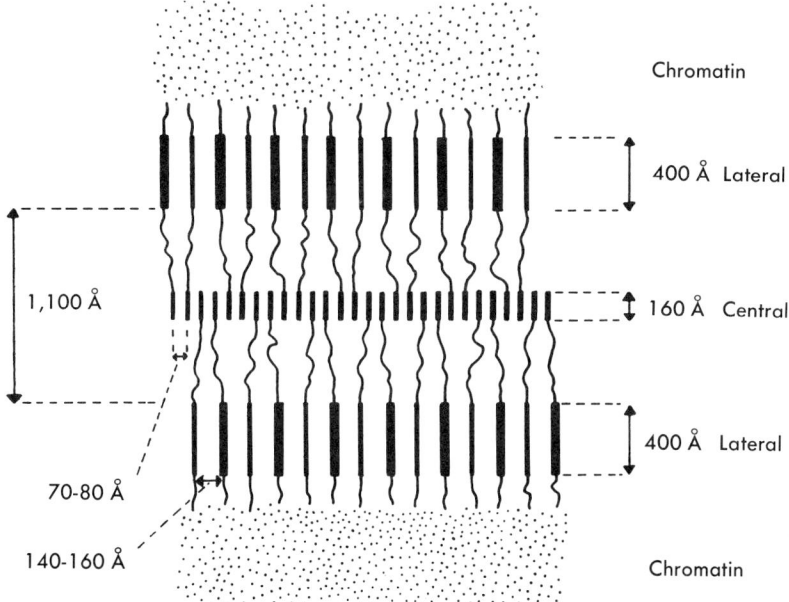

Fig. 6-9. Diagram of the components of the synaptinemal complex between two homologous chromosomal segments. The chromosomes (dotted regions) are adjacent to the two lateral elements and connected to them by chromatin filaments. Chromatin filaments extend from the alternating thick and thin rods of the lateral elements, through the central region, to the central element, which also consists of short parallel rods. The distance between lateral elements is typically about 1,200 Å. Compare this model of the fungus *Neottiella* (*Neurospora* and other ascomycetes are the same) with other models, such as in Fig. 6-11, that are more typical of other types of organisms. (From Westergaard, M., and D. von Wettstein. 1966. C. R. Trav. Lab. Carlsberg **35**:261-286.)

and reunion of chromatin strands, as originally proposed by Janssens in 1909 and 1924 in part from cytological study of grasshopper spermatocytes, the *partial chiasmatype theory*. Although neither he nor anyone else has ever seen the event at zygotene or pachytene, the chiasmata at diplotene are best interpreted as the result of breakage and fusion. The proposal in 1916 by Bridges, that crossing-over occurs at the four-strand stage and that only two non-sister strands are involved, fits well with the partial chiasmatype concept of Janssens, of breakage of two nonsister strands at the same locus and reunion with nonsister strands, and also fits the diplotene cytology as known.

The best cytological proof of the partial chiasmatype theory of crossing-over was provided by Brown and Zohary (1955). They studied heteromorphic homologues, one of which had a visible terminal deficiency. They found that the percentages of those arms having one crossover between the centromere and the deletions at pachytene equaled the number of anaphases showing dyads having one deleted and one normal monad. Such equivalence is predicted by the partial chiasmatype theory.

The recent evidence that repair DNA is synthesized at breaks also fits well with the evidence of Hotta et al. (1966) that there is a small amount of special DNA synthesized at pachytene to repair the recombination breakage.

Crossing-over (Fig. 6-12) must occur when each homologue is double structurally according to all genetic evidence (Bridges, 1916). Therefore, the homologues must double early in pachytene unless, as seems more likely, they double at some earlier stage, back to the S period or even earlier. But certainly the work on maize cytogenetics (Rhoades' excellent essay of 1955) strongly indicates double homologues at pachytene, but actual double strandedness cannot be seen before the next phase, diplotene. At the present time there is no acceptable scheme by which eukaryotic pairing,

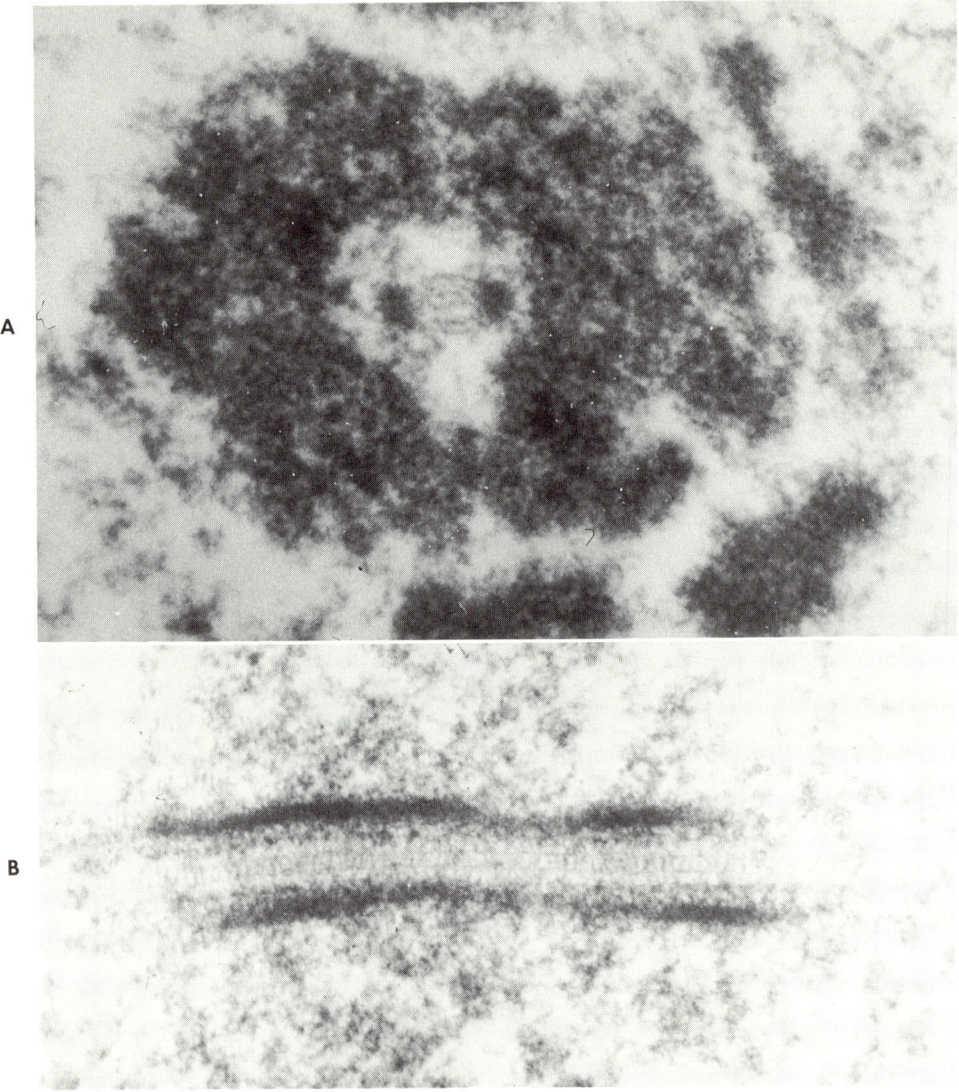

Fig. 6-10. Synaptinemal complex and synapsed homologues of the insect *Panorpa neptialis* by electron microscopy. **A,** Cross section showing the lateral elements in the center of a clear zone within or between the synapsed homologues that form a ring around the SC. A number of fibrils connect the two lateral elements, which in turn are connected to the masses of the chromosomes by numerous fibrils. **B,** Longitudinal section of the flat, ribbonlike synaptinemal complex. There are a number of faint longitudinal fibrils between the lateral elements and, therefore, within the central region. For interpretation see Fig. 6-11. Crossing-over could occur above and below the SC where the filaments of the homologues are in contact. (Courtesy Dr. G. Gassner, U.S.D.A., Agric. Res. Service, Metabol. Rad. Res. Lab. Fargo. From Brown, W. V., and E. M. Bertke. 1969. Textbook of Cytology. The C. V. Mosby Co., St. Louis.)

synapsis, and crossing-over are achieved with respect to chromosome fine structure, although many persons have tried (see Rhoades, 1961, for discussion). That is, the paired and synapsed chromosomes are not just two straight strands of DNA as in prokaryotes. A zygotene (Wolfe and John, 1965) or pachytene (Ris and Chandler, 1963) chromosome consists of a great many parallel chromatin strands, extending longitudinally, of each homologue, or it consists of zigzag or coiled strands. Other studies indicate that each pachytene chromosome consists of many appressed loops, like a 1,000μ lampbrush chromosome shortened to about 50μ. Studies of the (pachytene) synaptinemal complex (Moses, 1968) indicate massive looped chromosomes external to the lateral elements (Figs. 6-6 to 6-11); and it is well known that the amount of DNA in a chromosome if straightened out would be from 1,000 to 2,000 times the length of the pachytene chromosome (DuPraw, 1968). How can gene-for-gene pairing, which geneticists have demonstrated as occurring, be achieved in such an apparent mess?

In normal meiosis, however, out of pachytene come "bivalent" chromosomes showing in particularly good material the four-stranded chromosomes with chiasmata. Chiasmata are the physical basis of genetic crossing-over (Rhoades, 1961) and are first clearly seen at diplotene.

Diplotene. At the end of pachytene the synaptinemal complex is discarded, the chromosomes have acquired their chiasmata (crossovers), and genetic recombination of chromosomal segments of homologues has been achieved. Each such diplotene bivalent consists of two pairs of strands or chromatids, and the pairs appear to repel each other but are restrained from separating by the chiasmata. The repulsion is probably due to the chiasmata present because, in the same sporocyte, homologues closely paired but without chiasmata showed no such repulsion (Price, 1955). The number of chiasmata is usually small, and why this should be so is unknown. It would seem that two synapsed pachytene homologues could form dozens, even hundreds of cross-

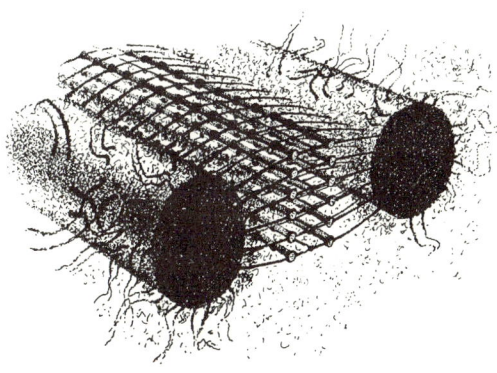

Fig. 6-11. Interpretive diagram of the synaptinemal complex of *Panorpa neptialis* (Fig. 6-10). The two large rodlike lateral elements are interconnected by six fibrils (internal strands) at each level along their lengths. In the center of the central region is the central element, which consists of six pairs of longitudinal rods that are united in pairs and to the lateral elements by the internal strands. External strands connect each lateral element to its adjacent chromosome. (Courtesy Dr. G. Gassner, U.S.D.A., Agric. Res. Service, Metabol. Rad. Res. Lab., Fargo. From Brown, W. V., and E. M. Bertke. 1969. Textbook of Cytology. The C. V. Mosby Co., St. Louis.)

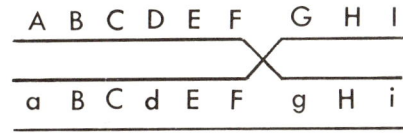

Fig. 6-12. Diagrammatic representation of crossing-over. The two pairs of homologous chromatids, two above and two below, vary for dominant and recessive conditions of various genes. After the crossing-over between the F and G genes, there are four genetically different chromatids. Two, the topmost and the bottommost, are unchanged, but the two in the middle have the sequences of alleles altered by the crossover so that one is now *ABCDEFgHi* and the other is *aBCdEFGHI*. These are new combinations not present before crossing-over occurred. (From Brown, W. V., and E. M. Bertke. 1969. Textbook of Cytology. The C. V. Mosby Co., St. Louis.)

overs. Furthermore, genetics claims that the presence of one crossover interferes (positive interference) with the existence of another close by, in the same chromosome arm, or even across a centromere. Rhoades (1961) stated, "Especially instructive would be a detailed comparison of genetically determined interference with cytologically measured chiasma interference in the same species."

Hypotheses have been published to try to explain the nature of the so-called diplotene repulsion. Darlington (1937) proposed two kinds of "electronegative" repulsion: one forces the pairs of chromatids apart and a second, centromere repulsion, causes chiasmata to move toward the ends of the chromosome arms (terminalization). Swanson (1957) proposed that the change from many small coils of the pachytene chromosomes to the fewer larger gyres of diakinesis produced the apparent repulsion. But it is true that chiasmata usually do move toward the chromosome ends and decrease in number.

Diplotene chromosomes are often stopped in their progression toward diakinesis. Since conversion to an interphaselike, possibly metabolic, condition is common, some cytologists have named the condition the *diffuse stage*. The long-arrested (up to years) condition in human oocytes is related to chromosomal aberrations, such as the 21-trisomic condition, in children of older mothers. The lampbrush chromosomes of amphibian oocytes are famous for their gene amplification and RNA transcription. In most species, however, diplotene may progress without interruption, by thickening and shortening, to diakinesis.

The result of physical crossing-over can be "seen" at diplotene and "explains" the genetic results (Fig. 6-12). That is, at any crossover one of the two strands of a homologue breaks and unites with one of the nonsister strands of the

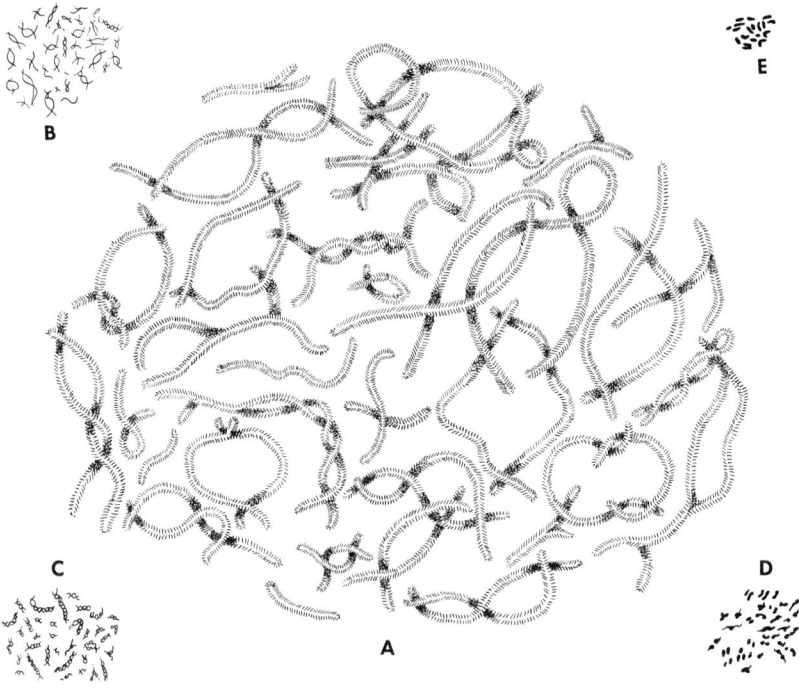

Fig. 6-13. Lampbrush chromosomes, **A,** in the egg of the shark *Pristiurus* at diplotene and drawn to the same scale as late diplotene, **B** and **C,** after they have contracted to the "normal" pachytene condition. During diakinesis, **D,** and metaphase I, **E,** the chromosomes remain typical of chromosomes during those phases. Thus, the lampbrush condition is very abnormal in the great extension of the chromosomes and the thousands of lateral loops all along them (Fig. 6-14). (Redrawn from Rückert, J. 1892. Anat. Anz. **7:**107-158.)

other homologue. As a result, two allelic genes, one on either side of the crossover, are recombined. If one homologue has alleles *A B* and the other homologue *a b*, a crossover between them would produce two recombined monads, one *Ab* and the other *aB*, as well as two nonrecombined monads, *AB* and *ab*. Actually, long segments of chromosomes, between consecutive crossovers, recombine. Such recombination mixes up segments of chromosomes and also permits the construction of genetic maps of chromosomes. The farther apart two genes are, the more frequently a crossover will occur between them. But because crossing-over may be and usually is more frequent in the distal regions of chromosome arms than in the heterochromatin near the centromeres, genetic linkage maps do not correspond with chromosomal or cytogenetic maps based upon cytological markers, as discussed in Chapter 3.

Lampbrush chromosomes. During meiosis in eggs of amphibia, such as the newt, *Triturus*, and some other vertebrates (Fig. 6-13), during the diplotene stage of long duration the bivalent chromosomes become extremely extended and diffuse. Each chromosome consists of the expected four strands showing chiasmata (Fig. 6-14); and along each double strand are many (about a thousand) tiny, darkly stained points, called chromomeres. Extending laterally from the thousand or so chromomeres of each bivalent chromosome are very long (up to 50μ, Gall, 1956) and thin loops of chromatin (Fig. 6-14). Evidently, each loop is a strand of DNA covered with ribonucleoprotein that includes polymerase molecules. The two ends of each loop are different for the amount of ribonucleoprotein (RNP) on the chromatin fiber. One end, called the *thin insertion end,* has only a thin covering of RNP; the other end of the loop, where it "reattaches" to a chromomere, is called the *thick insertion end* and is covered with a

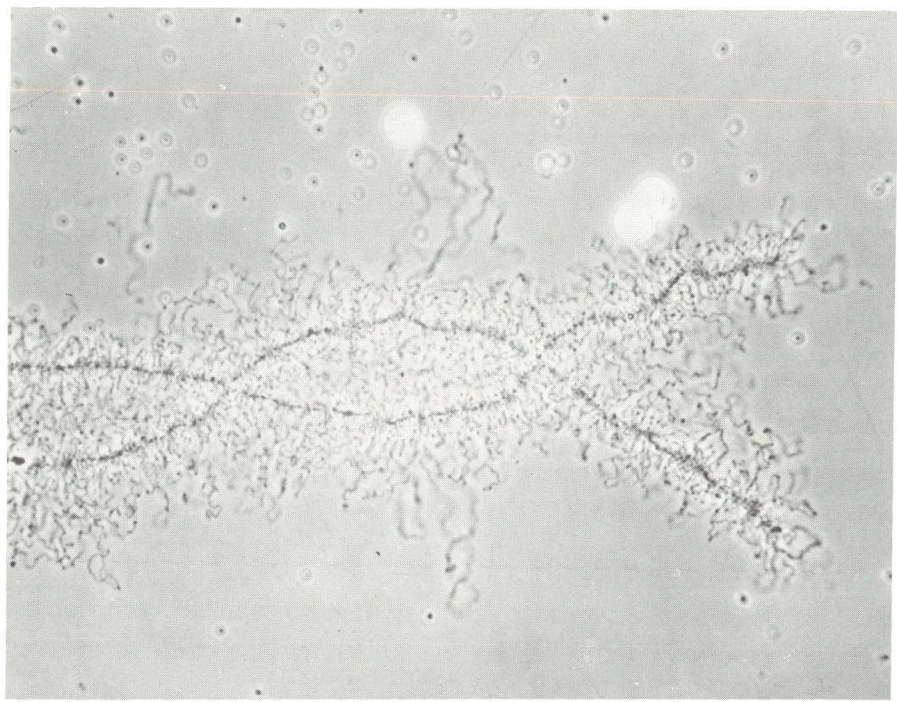

Fig. 6-14. A portion of a lampbrush chromosome of the amphibian *Triturus viridescens* made by phase-contrast microscopy showing the bivalent nature of the diplotene chromosome with two loci of crossovers evident. All along both homologues there are lateral chromatin loops of various lengths extending laterally from the series of chromomeres that form the longitudinal axes of the bivalent. (Courtesy Dr. J. G. Gall, Department of Biology, Yale University. From Brown, W. V., and E. M. Bertke. 1969. Textbook of Cytology. The C. V. Mosby Co., St. Louis.)

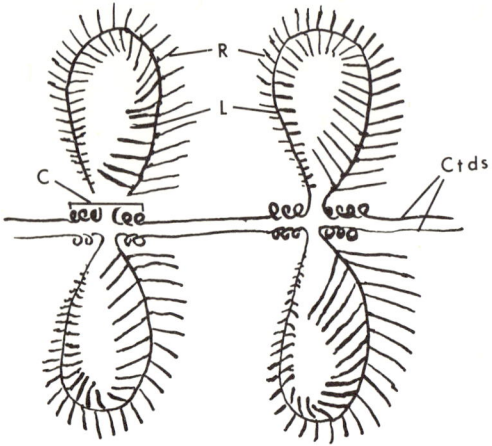

Fig. 6-15. Diagrammatic representation of the main axis (horizontal) and two pairs of lateral loops of two chromatids (Ctds) of one short section of one homologue of a bivalent lampbrush chromosome. Each continuous line represents a chromatin strand of a double DNA helix plus proteins. At the chromomere (C) the two chromatids separate and extend outward as a pair of loops (L) that return to the same chromomere. Along the loop RNA transcription (R) occurs (see Fig. 6-16) so that the RNA polymers vary regularly in length along each loop. Presumably most of the RNA filaments produced from the thousands of various loops present in the nucleus are messenger RNA and are probably being translated into protein either while still attached to the loop, or after separation from the loops, or both.

thick layer of RNP (Fig. 6-15). The lengths of loops and amount of RNP on them vary considerably from loop to loop. Presumably the RNP, or some of it, eventually leaves the loops and, therefore, the chromosomes. At least some of the RNP is stored for use in protein synthesis in early embryogenesis (Davidson et al., 1966; Crippa et al., 1967). Presumably, at least some of the RNA formed is messenger RNA. It is concluded that a very large number of loops (and genes) are very active during this lampbrush period. Before and after this highly modified diplotene period, the chromosomes are very much smaller and of typical form and size (Fig. 6-13). Callan (1963) has reviewed these matters.

Recently Miller et al. (1970b) have visualized the transcription of m-RNA along portions of lampbrush chromosome loops, near the thin insertion end, of the egg of the amphibian *Triturus viridescens*, the common spotted newt of eastern North America. The micrographs they take (Fig. 6-16) show presumptive RNA polymerase molecules along the chromatin strand of the loop, and extending outward from such molecules are many strands of m-RNA. Apparently these m-RNA-protein fibrils increase greatly in length. They report fibrils "tens of microns in length" at intermediate points along loop axes but have not yet been able to estimate the lengths of fibrils near the thick insertion end. RNA molecules tens of microns long can encode large amounts of genetic information.

The lampbrush condition seems to be a form of chromosome that permits a great deal of replication of RNA's of many genes for immediate and later use. It is not unlikely that a less spectacular "lampbrushing" occurs to diplotene chromosomes in meiocytes of many animals in addition to some vertebrates (some fish, amphibians, reptiles, and birds) such as grasshoppers (Hsu, 1948) and cockroach (Lewis and John, 1957), and plants (Grun, 1958). Those of some vertebrate eggs are merely the extremes of the general condition.

It is also from the lampbrush chromosomes that the amplification of r-DNA occurs in *Triturus* and has been seen by Miller et al. (1970b) as discussed in Chapter 2. It can be anticipated that similar processes occur in RNA puffs of salivary gland cell chromosomes (Chapter 3).

Diakinesis. During diakinesis chiasmata of many species continue to terminalize, and the bivalent chromosomes shorten and thicken so that the double nature of each half of the bivalent is undetectable. In many species the chromosomes become arranged against the nuclear envelope and widely separated. The nucleolus is fading by loss of its material, some of which apparently accumulates on the chromosomes (McClintock, 1934).

Aside from its value for cytological studies of terminalization, chromosome numbers, asynapsis, multiple pairing, etc., this phase of meiosis is not cytogenetically significant.

Prometaphase

The nucleolus eventually disappears, the nuclear envelope breaks down, the spindle

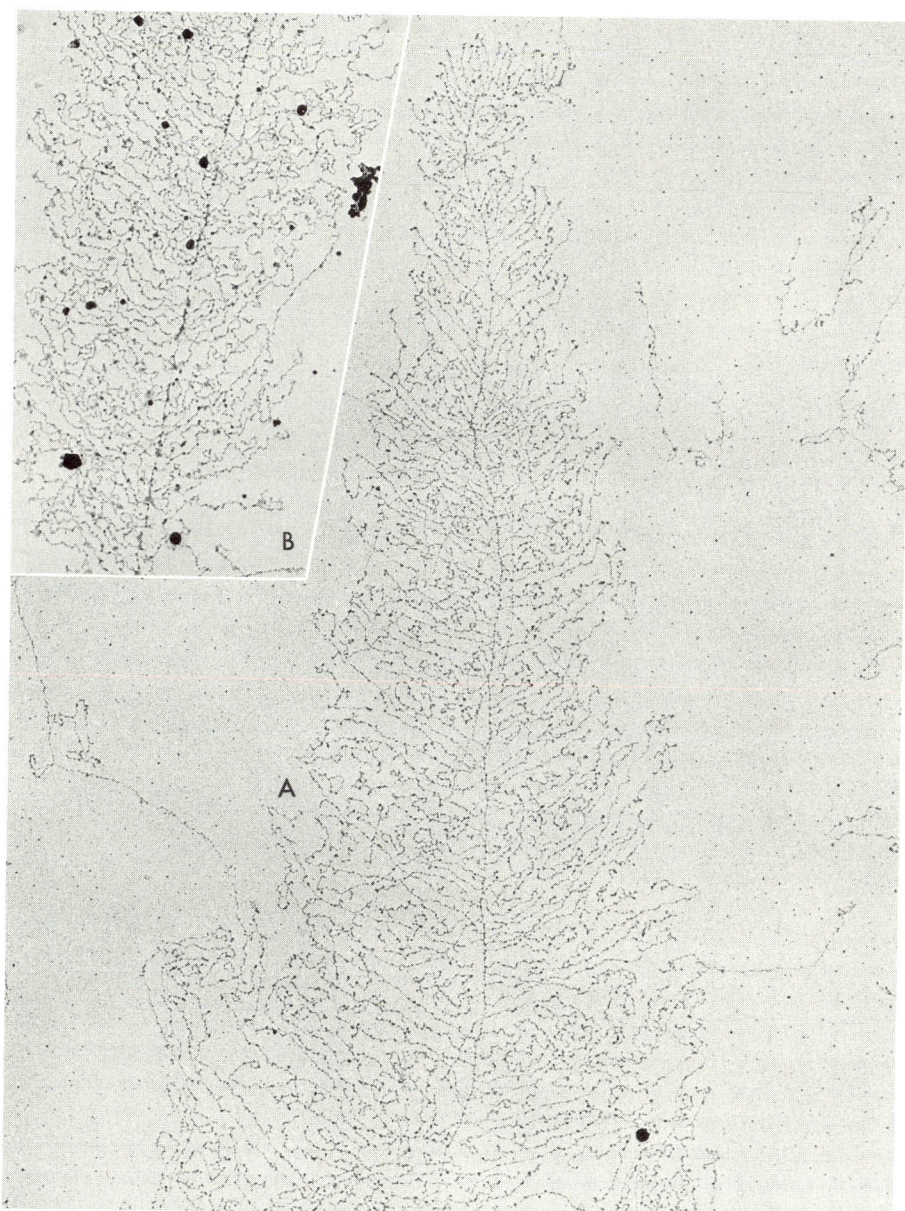

Fig. 6-16. Segments of loops of lampbrush chromosomes of eggs from the amphibian *Triturus viridescens* following treatment and observation with electron microscopy. **A,** A segment of a loop showing the progressive increase in length of the RNA polymers that extend laterally from the single chromatin strand of the loop. The sum total of all of the lateral ribonucleoprotein filaments on a loop is called *matrix*. 29,000×. **B,** Portion of a loop and filaments of matrix. 41,000×. Presumably, there is a molecule of RNA polymerase where a lateral filament "attaches" to the chromatin strand of the loop. The 100 Å chromatin strand of a loop apparently consists of a single uncoiled double helix (about 30 Å in diameter) covered with protein. This is another figure showing genes in action. (From Miller, O. L., B. R. Beatty, B. A. Hamkalo, and C. A. Thomas. 1970. Cold Spring Harbor Symp. Quant. Biol. **35:**505.)

forms, and chromosomes move during the short prometaphase. This stage terminates with all of the chromosomes on the equatorial plate at metaphase I.

Metaphase I

Prometaphase ends when the bivalent chromosomes are arranged on the metaphase plate of the spindle. At that time the chromosomes are short and thick. By removal of the matrix it is evident that each chromatid is formed of a spiral strand that itself is spiraled. Whereas in mitosis each centromere lies on the metaphase plate, at metaphase I each bivalent has two centromeres, one above and one below the plate. Two sister chromatids (except for crossovers) are attached to each centromere of the bivalent. Of course, at this time each of the two centromeres is connected by chromosomal fibers to the nearer pole. The two halves of the bivalent are still joined by one chiasma or more, and the shape of the bivalent is determined by the number and positions of chiasmata in each arm. As metaphase continues, the centromeres appear pulled toward opposite poles; but because of the resistance to separation caused by the one chiasma or more, the chromosomes adjacent to the centromere appear thin and stretched.

Metaphase, like diakinesis, is a good stage for observing chromosome numbers and such conditions of chromosomal pairing as univalents, bivalents, trivalents, quadrivalents, etc. of natural and hybrid organisms.

Almost without exception the bivalents are arranged at random with respect to which spindle pole is related to the paternal or maternal centromeres. That is, a random number of maternal (or paternal) centromeres go to each pole. This is the basis of Mendel's laws of random or independent assortment of nonlinked genes. There are a few mutant cases known in which a particular chromosome ends up in the egg (rather than a polar body or disintegrating spore) more often than random. This, if determined during meiosis, is called *meiotic drive* (Sandler and Novitski, 1957). In *Drosophila* Novitski (1951) found that the shorter member of a heteromorphic pair of homologues was regularly included in the egg.

Rhoades (1952) reported that "abnormal chromosome 10" of maize is included in the egg more often than random, perhaps because of the "neocentromeres" it produces in heterozygous dyads. Roman (1948) has also reported in maize that a chromosome containing a translocated supernumerary chromosome is included in about 60% of eggs. It is likely that orientation of the spindle-metaphase bivalent is the basis of meiotic drive (see Chapter 7).

Anaphase I

Anaphase I is a very important cytogenetic phase of meiosis because the final effects of some structural changes and crossing-over become visibly apparent at that time. As the homologues separate from one another as dyads, the doubleness of each dyad, which was unobservable at metaphase I, becomes startlingly evident (Fig. 14-7). Very often the arms of the two chromatids of each such anaphase I dyad chromosome spread apart as far as they can except that they are attached at the undivided centromere. They assume the form of X's and V's. If separation is free and complete, there is nothing else unusual about anaphase I except that each anaphase group is reduced.

There are, however, certain preceding conditions that result in chromosomal or chromatid connections between the two separating chromosomes, which are called *bridges*. Such bridges are often accompanied by fragments of chromosomes that, because they lack a centromere, are called *acentric fragments*. Also because they lack a centromere, they do not move toward either pole: they lag on or near the metaphase plate. These are discussed in Chapter 14.

Another occasional aberration of anaphase I of genetic significance is *nondisjunction*. The term disjunction refers to the anaphase separation of the homologues at anaphase. If for any reason the two homologues do not separate and/or both go to the same pole, it is a case of nondisjunction and is detectable genetically.

If synapsis of any number of homologues failed for any reason such as an asynaptic gene, which would produce no synapsis at all or only of a few pairs, or lack of homology between some chromosomes as in some hybrids, the unpaired chromosomes usually reach the metaphase I plate. However, if it is an abnormal condition, the univalents usually divide as in mitosis. Often, however, such univalents behave unusually such as not dividing, not getting onto the metaphase plate, lagging, or dividing oddly (Upcott, 1937; Endrizzi, 1958). That is, the two chromatids may or may not separate at anaphase. This indicates that the centromeres of normal two-chromatid anaphase I chromosomes may actually be double but do not separate perhaps because of unilateral attachment to the spindle fibers. Of course, if the univalent divides at anaphase I, it goes into metaphase II single and so cannot divide at anaphase II (Upcott, 1937; Endrizzi, 1958). There are, of course, envolved systems, such as male *Drosophila,* in which the homologues do not synapse or form chiasmata but remain paired loosely and "disjoin" regularly without division at anaphase I. There are many unusual but, for the particular species, normal cytological conditions known at anaphase I including *alternate* or *adjacent* disjunction, disjunction of sex chromosomes, etc., some of which are discussed later.

Telophase I and interkinesis

Among organisms telophase I varies from a mere beginning or none at all in many animal eggs (Sharp, 1926) and the plant *Trillium* (Rhoades, 1961) to complete. In animal eggs the late anaphase I chromosomes become directly the late prophase II chromosomes. Such a condition can occur without any adverse results because normally or always there is no DNA replication or chromosomal duplication between meiosis I and II. Usually in meiosis, however, there is a telophase leading to a uniquely modified interphase called interkinesis. This "quickie" interphase is almost unique and, in a sense, *is* reduction. It is of short duration and does uniquely reduce the DNA content to the haploid (1C) amount by telophase II and must have evolved for that purpose.

In general, telophase I does lead to interkinesis; but with respect to cytokinesis it varies. Usually cytokinesis does occur, but in many flowering plants both in microsporogenesis and megasporogenesis and in Ascomycetes and Basidiomycetes and animal eggs there may be no telophase I cytokinesis. In certain species of flowering plants the four nuclei (the four megaspores) all take part in the formation of the embryo sac (female gametophyte).

An unusual cytogenetic condition in *Drosophila* is called "sex ratio," which produces nearly all female progeny. In the sex ratio male the Y chromosome usually degenerates, but the X chromosome divides at *both* anaphase I and II (Sturtevant and Dobzhansky, 1936). If the description is correct, it is not known when the extra DNA replication for these two divisions of only one chromosome, a univalent, might occur: during mid-prophase as in *Chlamydomonas,* during interphase, or at some other time.

Prophase and metaphase II

Prophase II, whether it arises directly from telophase I or from interkinesis, is usually described as a rather typical mitotic prophase. This is more or less true except that the dyad chromosomes are already double from at least anaphase I and can be seen as such in maize (Rhoades, 1961) where they appear in prophase as X's, very much as they appeared at anaphase I.

Metaphase II is hardly different from a mitotic metaphase except that the chromatids of each dyad seem to be more separate than the two chromatids of a mitotic metaphase.

Anaphase II

Anaphase II, like anaphase I, sometimes shows the results of abnormal chromosomal aberrations and crossing-over, such as certain crossovers within inversion loops, from the much earlier pachytene. Thus, bridges or bridges and fragments may appear at anaphase II and indicate to the cytogeneticist what did happen at pachytene. Otherwise, anaphase II is not unusual.

Telophase II

Telophase II is the end of meiosis with each nucleus haploid and recombination having been achieved. As far as chromosomes or nuclei are concerned, there is nothing unique about it. In some organisms such as *Neurospora* and some other ascomycetes, *Chlamydomonas reinhardi*, etc. a mitotic division may immediately follow telophase II. In *Neurospora* there is no cytokinesis until the eight-spore nuclei have been formed. In some flowering plants there is no cytokinesis throughout meiosis until telophase II in anthers or ovule. In some embryo sacs there is no cytokinesis until after a mitotic division produces the eight nuclei of the mature embryo sac.

The cells produced by meiosis vary. In male animals they are usually spermatids that subsequently differentiate into spermatozoa. In female animals there is one large ootid and two tiny polar bodies, the ootid being the ovum. In plants, fungi, etc. they are called spores.

Among various species there are many modifications of the process of meiosis, especially in parthenogenetic and apomictic (asexual) species, some of which are discussed later.

THE NEOCLASSICAL SCHEME

One of the most significant challenges to part of the classical theory of meiosis is not new, although it is only recently that it has received strong support (see Brown and Stack, 1968, for detailed discussion). This concept is about as old as the classical. The classical scheme claims that homologous chromosomes are not paired before the beginning of zygotene. This contrary idea proposes that there is adequate indirect cytogenetic (Maguire, 1965; Grell and Chandley, 1965; Feldman, 1966) and direct cytological evidence from plants (Moens, 1964; Brown and Stack, 1968; Stack and Brown, 1969a), animals (S. G. Smith, 1942), and fungi (Singleton, 1953; Rossen and Westergaard, 1966) that homologous chromosomes pair in the condensed condition at some time before leptotene. In fungi pairing occurs just after fusion of diploid "gamete" nuclei. In animals and plants pairing occurs at anaphase of the last premeiotic mitosis or earlier. The earliest it can occur is in the zygote of the individual (*Yucca*, Watkins, 1935). In Diptera it occurs soon after, in early cleavage (Metz, 1916a). In some flowering plants it occurs gradually during the formation of the flower (Brown and Stack, 1968; Stack and Brown, 1969a; Stack, 1969). In lice it occurs in the primary spermatogonium, six divisions before meiosis (Hindle and Pontecorvo, 1942). In other animals it seems to occur in gonial cells up to the last gonial division (S. G. Smith, 1942).

In Diptera and plants such pairing in somatic tissue divisions has been called *premeiotic pairing*. Regardless of what it is called or when or where it occurs, it is assumed that it is an essential preparation for meiosis. As S. G. Smith stated in 1942, "The meiotic pairing consumated at pachytene is initiated at the latest by the telophase of the last premeiotic division."

In the classical scheme the terms pairing and synapsis are synonymous, and the event is supposed to occur only during zygotene. The developing neoclassical scheme uses the term pairing for somatic and premeiotic pairing, but the term synapsis is reserved for the zygotene process involving the synaptinemal complex, which is an *entirely different process*. But, according to this scheme, synapsis is impossible without previous premeiotic pairing. It is assumed that by the anaphase before meiosis the contracted homologues are paired, they lie closely side by side. During the telophase before meiosis they decondense but remain side by side with their ends attached to the nuclear envelope (Wettstein and Sotelo, 1967; Moens, 1969). During interphase and DNA replication they are closely associated, whatever that means in regard to fine structure. They are paired during preleptotene and leptotene (Moens, 1969; Stack, 1969). During leptotene, perhaps, but certainly during zygotene each homologue produces part of the synaptinemal complex (Moens, 1968; Stack, 1969; Westergaard and von Wettstein, 1970), and the complex formation "pulls" the homologues into final synaptic position. Either the position achieved or the SC itself produces the crossing-over.

This scheme avoids the "impossibility" of zygotene pairing, of the need for "pairing forces acting at a distance," and "explains" the rarity of interlocking of bivalents (Feldman, 1966; Brown and Stack, 1968) (Fig. 6-17). The obser-

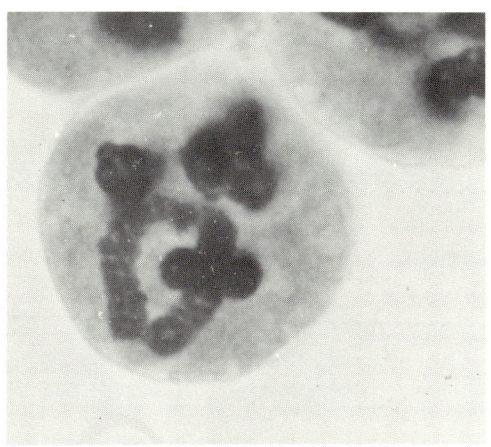

Fig. 6-17. Interlocking of two bivalent chromosomes at meiotic metaphase I. One chromosome of each bivalent lies between the homologues of each pair, thereby producing a configuration like two links of a chain. Interlocking is not common and may represent a weak sort of exchange quadrivalent. (From Brown, W. V., and E. M. Bertke. 1969. Textbook of Cytology. The C. V. Mosby Co., St. Louis.)

vational cytological evidence for premeiotic pairing is much more convincing than that for zygotene pairing. It is important that this concept does permit the cytological observations of zygotene pairing because it does *not* specify that all homologues are paired closely (about 1,200 Å) *throughout their lengths*. It does specify that *at least the chromosome ends* are intimately paired during leptotene. Other parts may be much more widely separated at leptotene, and so pairing more intimately during zygotene can occur and be observed. Regardless of the looseness of their pairing, there can be no other nonhomologous chromosomes between the pairs unless there was previous premeiotic pairing of homologous parts of two pairs of homologues and/or homeologues. Thus, interlocking represents evidence of translocation and is a loose "quadrivalent" association (Brown and Stack, 1968).

Another heretical concept is that most of the so-called meiotic prophase I is *not* comparable to mitotic prophase but might better be considered as a highly modified G_2 stage of interphase (Swanson, et al., 1967).

First, if homologous chromosomes are really paired by premeiotic interphase, the connotations of the descriptive terms leptotene, as single strands, and zygotene, as a period of pairing, are unnecessary and misleading. Preleptotene, leptotene, zygotene, and pachytene seem to represent in part stages in a continuous development and function of the synaptinemal complex.

Second, although the fine details of neither mitotic nor meiotic chromosomal condensation are known, it seems likely that they may be structurally different, in part because they seem physiologically different. Since somatic and meiotic recombination are genetically identical (Lindsley et al., 1968), it can be assumed that the physical conditions of the chromosomes when somatic and meiotic recombination occur are similar. The mitotic G_2 period is more likely than mitotic prophase as comparable to meiotic prophase, although this has not been established.

Third, throughout mitotic prophase there is neither DNA replication nor RNA transcription (D. D. Brown, 1966; Comings, 1966). But DNA replication extends through preleptotene into leptotene (Taylor, 1959; Callan, 1968), and RNA transcription extends into pachytene and beyond. There is even some DNA replication in pachytene (Hotta et al., 1966). In some oocytes of insects and amphibians there is considerable RNA transcription and substantial DNA replication in a modified diplotene. Physiologically, then, early meiosis is more like S and G_2 of interphase than mitotic prophase.

Fourth, the development and functioning of the synaptinemal complex seems more appropriate to interphase activity than to prophase inactivity. Of course, it is unique to meiosis, and it is not known whether mitotic prophase cells or chromosomes could do it.

Since it is likely that the formation and functioning of the synaptinemal complex is the raison d'etre for meiotic prophase, as a mechanism for a great deal of recombination, it is proposed that the period from the beginning of synthesis of the synaptinemal complex, the cores seen by Moens (1968), Stack (1969), and Westergaard and von Wettstein (1970), to the end of pachytene when the synaptinemal complex is discarded as stripped cores (Moses, 1968) be called synaptitene. Synaptitene would start probably in leptotene, would include zygotene, terminate at the end of pachytene, and be considered as a very special G_2 period. Ordinary diplotene might be considered as a modified prophase capable of great modification in certain meiocytes such as the diffuse stage of amphibian eggs with their lampbrush chromosomes. Perhaps such a diffuse stage and early diplotene generally should be considered as late G_2. Tentatively, meiotic prophase I would start somewhere within diplotene and include diakinesis.

Since this scheme is a classification of the premeiotic interphase, the period from the end of the premeiotic telophase to synaptitene, including G_1 and S_1, would be prosynaptitene. If the early part of diplotene is not prophase, it should be called telosynaptitene. Prosynaptitene, synaptitene, and telosynaptitene can be considered as substages of synaptiphase. This scheme also assumes that homologues are paired at least by premeiotic telophase and are paired throughout synaptiphase. These two terminologies and schemes can be summarized and compared as follows:

Neoclassical	Classical
Telophase (last premeiotic)	
Homologues paired	Homologues not paired
Synaptiphase	
	Premeiotic interphase
Prosynaptitene	G_1
	S
	G_2 (preleptotene?)
	Prophase I
	Preleptotene?
Synaptitene	Leptotene
	Zygotene
	Pachytene
Telosynaptitene	Early diplotene
	Late diplotene
Prophase I	Diakinesis
Prometaphase	Prometaphase
Metaphase I	Metaphase I

This scheme and terminology are not presented in an attempt to change the existing useful terminology but as an hypothesis to open up the thinking of cytogeneticists to a possibly useful concept. One of the dangers in science in general and cytology in particular is that hypotheses become accepted as "facts" without adequate testing, and as facts they strongly affect thinking. As someone once said, "Thinking it makes it so." For example, Sturtevant stated the following in 1951 about the classical scheme of meiosis that owes much of its establishment to Darlington during the 1930's:

Darlington has developed a detailed theory of the process of meiosis and its relation to ordinary mitosis. This is a beautiful and satisfying scheme, which has been very generally adopted by geneticists; it has almost come to be considered the very core of the chromosomal interpretation of genetics. It is with great reluctance, therefore, that I have found myself forced to discard large portions of the scheme—a reluctance made even greater by the fact that I see no wholly satisfactory substitute.

Great progress has been made in cytology and genetics since 1950. Perhaps the present understanding and lines of ongoing study of meiosis discussed in this chapter may be producing progress toward what he called "a more coherent and satisfactory picture" that he, Sturtevant, could not see in 1951.

EVOLUTION OF MEIOSIS

There is little literature directly concerned with the evolution of meiosis. It has been generally accepted that meiosis did not always occur among organisms but evolved from some preexisting conditions and characteristics of cells (Klingstadt, 1939). The old names "heterotypic mitosis" and "homeotypic mitosis" for the two meiotic divisions indicate the general acceptance that the two meiotic divisions evolved from "premitosis" (Sagen, 1967).

Wilson in his classic of 1925 wrote one paragraph on the subject (p. 629), from which three sentences are quoted: "The origin of the internal phenomena of meiosis is still wholly unknown.... It is easy to offer hypothetical teleological answers to such questions.... If the cytological problems here involved remain unsolved, they are non the less real and offer an interesting field for further inquiry."

Aside from discussions of differences between mitosis and meiosis, such as by Darlington (precocity theory of 1937), Gustafsson (1939), etc., the only two publications known specifically stating evolutionary concepts are by Cleveland (1947) and Stack and Brown (1969b).

The precocity theory of meiosis by Darlington (1937) was based on the highly questionable assumption that chromosomes must be double except from anaphase to the next prophase. In mitotic cells the chromosomes are already double by replication at the beginning of prophase, and so the requirement for doubleness is satisfied. As a premeiotic cell enters meiotic prophase, it was assumed that the chromosomes are single and satisfy the requirement for doubleness by homologous chromosomes pairing. Later, in pachytene, each homologue was assumed to double. This also satisfied the doubleness requirement; but since they were already double, there was the tendency for the homologues to repel each other, as they seem to do at diplotene.

This hypothesis was generally accepted but seems now to be relegated to history, especially since the probable doubleness of leptotene chromosomes seems to be likely. Furthermore, it does not consider all of the presently known differences between mitotic and meiotic divisions, time of DNA replication, the existence of the synaptinemal complex, etc.

Sax and Sax (1935) had earlier proposed that early stages of meiotic prophase progress more slowly than do comparable stages of mitotic prophase, which allows meiotic chromosomes to uncoil completely. Since they claimed that homologous chromosomes have at all times an attraction for each other, the uncoiled condition provided an opportunity for synapsis. The coiled condition of mitotic prophase chromosomes prevents pairing.

Beasley (1938) also assumed that homologues always have an attraction for each other but the great enlargement of the meiotic nucleus, the long time-span of prophase, and the decreased viscosity of the nucleus permitted pairing. He reported in the plant *Gasteria* that the mitotic nucleus increased about 1.5 times from interphase ($772\mu^3$) to prophase (about $1,200\mu^3$) whereas the meiotic nucleus increased 4 times, from 894 to $3,597\mu^3$. These hypotheses do describe probably real differences between mitosis and meiosis but are somewhat out of date now that we know about the synaptinemal complex, nor do they really deal with the evolution of meiosis from mitosis, only differences.

Cleveland (1947) in a short paper attempted to base an evolutionary scheme of the origin of meiosis from mitosis on nothing but centrioles and life cycles of symbiotic flagellate protozoa. On the basis of his own observations entirely, he proposed that "Evolution of meiosis has been direct: haploidy to diploidy; diploidy to haploidy and vice versa; fusion of nuclei; cytoplasmic division producing gametes." In all of these steps except the first, to diploidy, and the last, formation of gametes and sexuality, there seem to be no advantages of recombination or heterozygosity since the autogamously fusing nuclei are genetically identical or diploidy arises by endomitosis. The sequence does include an interphase without chromosomal duplication, pairing of homologous chromosomes, and numerical reduction. Cleveland was more concerned with life cycles and centriole-chromosomal asynchrony than pairing, synapsis, and recombination.

Stack and Brown (1969b) also assumed that meiosis evolved from mitosis. The biological importance of sexuality and recombination is to produce variation within the species as a basis for rapid evolution (Muller, 1932). If meiosis has evolved independently in numerous evolutionary lines of organisms but has always converged to give essentially the same meiotic system, then there must be some property of cells or chromosomes or both that made such convergent evolution inevitable. The only appropriate property of chromosomes seems to be homologous pairing, assumption of an inherent tendency for homologous segments of DNA or chromatin to pair. There is adequate evidence that chromosomes do pair or associate homologously, and paired homologues produce inevitably some crossing-over. Recombination in viruses and bacteria implies pairing of naked DNA genophores (Chapter 20). As Westergaard (1964) proposed, it is likely that homologous pairing is the primitive condition of chromosomes when diploidy arose. Therefore, it is assumed that homologues will pair unless inhibited, such as by the diploid pairing gene or genes of wheat (Riley and Chapman, 1958; Sears and Okamoto, 1958; Feldman, 1966).

Before there was meiosis and sexuality, there

was recombination-producing parasexuality, as occurs at least in asexual and ameiotic fungi (Sinha and Ashworth, 1969), a system that obviously requires somatically paired homologues in order to produce the somatic recombination observed. As sexuality arose as a genetically controlled system of plasmogamy and karyogamy out of the more random cell and nuclear fusion of parasexuality (like modern cell hybridization), the randomness of parasexual chromosome number reduction (Käfer, 1961; Hastie, 1967) also became a gene-controlled system also based upon somatic pairing. If homologues are paired at telophase and there is a gene-controlled inhibition of DNA replication during interphase, as during interkinesis, as in pupal gut-lining cells of mosquitoes (Berger, 1938; Grell, 1946) and in cycad roots (Storey, 1968), then at anaphase homologues separate from one another and an orderly reduction of genomes is achieved. Thus, sexuality and chromosome reduction without recombination had evolved as an adequate, orderly system, and there was also *somatic* recombination.

Genetics has demonstrated, however, that there is a great deal of recombination during the first meiotic division, many times as much as is achieved by somatic recombination (Pontecorvo, 1958). It is assumed that more or less random somatic recombination evolved into an orderly meiotic recombination division or divisions by gene-controlled formation of the synaptinemal complex, which, however, requires already "somatically" paired homologues. This "recombination division" could have occurred during any division of the life cycle; but if it occurs just before the previously evolved reduction division, the greatest amount of variability of gametes is achieved. So natural selection would put it there. But since this new first meiotic recombination division also reduces the chromosome number, the earlier evolved reduction division ceased to be that; but its unique interphase, without DNA replication, remained as a necessary part of the reduction process, as interkinesis.

The final aspect of the evolution of meiosis was the accumulation of meiotic genes that generally inhibit the "somatic" pairing of homologues in somatic cells of the organism, such as the 5B genes in wheat (Sears, 1969). Such nonpairing genes must themselves be inhibited or turned off at some time in cell lineages leading to the meiocytes. Thus, the necessary premeiotic pairing of homologues before meiosis is theoretically "accounted for"; but actual study of premeiotic pairing led in part to the synthesis of this scheme of the evolution of meiosis (Brown and Stack, 1968; Stack and Brown, 1969a; and finally Stack and Brown, 1969b).

As was just stated, the inherent characteristic of chromosomes to pair, often nonhomologously if homologues are not present, is the basic foundation of this scheme. At the present time there is no acceptable model for such pairing. It would seem likely that pairing of all sorts is achieved by chromosomal proteins rather than directly by DNA. But the mechanism of pairing is one of the many problems for the future.

MEIOTIC GENES

The whole meiotic process is, of course, genetically controlled, having probably evolved from mitosis (Stack and Brown, 1969b). Some aspects of meiosis may be modified by environmental conditions, especially the important effect of heat shock on amount of crossing-over (for example, Grell and Chandley, 1965; Henderson, 1966; Peacock, 1968). Cool temperature (9° C) during the S period caused asynapsis by diplotene in *Ornithogalum virens* (Church and Wimber, 1971). Synapsis and recombination can also be affected by structural differences between otherwise homologous chromosomes, such as inversions (Maguire, 1966), translocations (Roberts, 1966b), duplications (Roberts, 1966a), and, significantly, even by allelic heterozygosity (Holliday, 1964).

Genetic control of meiosis is made evident by abnormalities of the meiotic process. These are most evident in those characteristics that differentiate meiosis from mitosis. Although the mitotic process is also under genetic control, it is generally far less sensitive to genetic imbalance than is the meiotic process. But even the mitotic process has become genetically modified through evolution in some organisms such as predictable chromosomal elimination, chromosomal reduction, initiation of polyploidy, etc. Indeed, there are many forms of mitosis such as in fungi and dinoflagellates that hardly fit the definition or general concept of "mitosis." But the more complex process of meiosis seems to be easily distorted or modified by genetic change, such as results from inbreeding, hybridization, evolution of apomeiosis in many asexual forms, and single gene mutations.

Essentially all of the unique characteristics of meiosis seem to be easily modified such as premeiotic pairing, synapsis, the synaptinemal complex, recombination, bivalent chromosomes, chromosomal condensation that is different from mitotic, condensed sex bivalents, nondisjunction of centromeres at anaphase I, reconstitution nuclei, the "quickie" interkinesis with no DNA replication, normal bipolar spindles, parthenogenetic development, etc.

There are genes that affect the amount of genetic recombination in various ways (Lindsley et al., 1968, for general discussion and specific examples). One of the most evident is the normal gene-controlled lack of synapsis at meiosis in one sex such as in most male *Drosophila* and many other achiasmatic insects. Mutant genes are also known, called asynaptic genes, in many species of plants and a few insects (Soost, 1951; Rees, 1961) that result in all or some chromosomes being unpaired at metaphase I.

All meiotic genes do not produce complete or even partial asynapsis; some affect other aspects of meiosis. They may affect amount of "recombination" as in certain mutants of *Neurospora* (Catcheside, 1968), one of which increases 10 times the amount of recombination within the "histidine 1" locus only (Jessop and Catcheside, 1965). Bridges (1929) described a segregating progeny in which some individuals had about one-third normal recombination in the third chromosome of *Drosophila*. Lindsley et al. (1968) described some genes in *Drosophila* that promote nondisjunction (which may be partial asynapsis and univalents) and/or reduce exchange. They proposed that these mutants stop pairing along a chromosome before it is complete, thereby reducing the amount of exchange. If pairing begins near the centromere, as it is claimed to do in *Drosophila*, crossing-over is reduced in the distal regions of the arms and vice versa. "Sex ratio" in *Drosophila* (Sturtevant and Dobzhansky, 1936) that affects X-chromosome distribution during meiosis is a gene (or genes) acting only during meiosis.

It is not unlikely that regions of concentrated recombination may be in part genetically determined, that is, close to the centromere, in distal regions of arms, or randomly throughout. Emsweller and Jones (1934) studied segregation of regions of chiasmata in an *Allium* hybrid. *Allium fistulosum* has chiasmata localized in the proximal regions of the chromosome arms, near the centromere, whereas *A. cepa* has chiasmata randomly distributed along the arms. In the F_1 hybrid all chiasmata were randomly distributed, as though that condition were dominant. In the F_2 from a backcross to the "recessive," *A. fistulosum*, approximately one half had proximal chiasmata whereas the other half had the chiasmata randomly scattered, as would be expected for simple mendelian inheritance.

Lesley and Frost (1927) reported a recessive gene in the plant *Matthiola* that increases the lengths of chromosomes during meiosis and also decreases the degree of pairing.

In wheat there is a "gene" or segment on chromosome 5B which, in normal number, prevents the pairing of homeologous chromosomes but not of homologous chromosomes (Riley and Chapman, 1958; Sears and Okamoto, 1958). This is an inhibitory "gene" so that when it is present in excess it prevents pairing of both homologues and homeologues (Feldman, 1966). But *Aegilops speltoides*, an ancestral diploid species of hexaploid wheat, has a dominant gene that suppresses the effect of the 5B gene on pairing of homeologues (Riley and Chapman, 1964). These come under the category of "bivalent pairing" genes that prevent the expected multivalent associations of many polyploids so that the polyploid possessing such genes has as good bivalent pairing as a diploid. How common such genes are is unknown; but that they do exist is important knowledge, and they represent a class of important meiotic genes.

It is not unlikely that some species, such as in *Oenothera*, may have genes for alternate disjunction in translocation rings at anaphase I (see

Chapter 13). Although alternate disjunction in rings of four or more translocated chromosomes may be structurally determined, as in part by central location of the centromere, the high predictability in some species indicates the likely involvement of genes. For example, in maize alternate disjunction is random from ⊙ 4 occurring almost exactly 50% of the time in a translocation between chromosomes No. 8 and No. 9, which have centromeres so located as to produce long and short arms (Rhoades, 1955). Nevertheless, genes from alternate disjunction have been proposed repeatedly (Cleland, 1936).

In many apomictic plants and parthenogenetic animal species, not only are genes for asynapsis common but genes for restitution nuclei at anaphase and telophase I or anaphase II are often present to prevent chromosome number reduction. In other such species different gene-controlled modified meiotic divisions have been described (White, 1954; Gustafsson 1946-47) such as a premeiotic chromosome number doubling to the 4n condition so that normal meiosis produces diploid eggs in certain species of parthenogenetic animals. The difference between egg development only after fertilization or as parthenogenesis is clearly under genetic control.

Among insects, such as coccids and cecidomyiids, gene-controlled abnormalities of meiosis have been reported including nondisjunction, unipolar spindles, chromosomal elimination, and unusual chromosomal movements (White, 1954). In the *Rosa canina* complex a similar unusual unipolar spindle and chromosomal movements occur (Gustafsson and Hakansson, 1942; Gustafsson, 1944). The unusual "prometaphase stretch," common in meiosis of many insect spermatocytes but not in mitosis, must be gene-controlled. The unusual but typical "quickie" interkinesis that occurs without DNA replication is typical of all standard meiosis. The heterochromatic "sex bivalent" during meiosis in many animal spermatocytes, gene amplification, lampbrush chromosomes, facultative heterochromatin, etc. are all gene-controlled.

Genetic control of meiosis is made evident also by such abnormalities as "neocentromeres"
in the presence of abnormal chromosome No. 10 of maize (Rhoades and Vilkomerson, 1942) and, although it occurs just after meiosis, the "polymitotic" gene reported in maize by Beadle (1931).

A number of cytologists have expressed the opinion that genetic conditions can produce breakage of meiotic chromosomes (Marquardt, 1952; Walters, 1956; Rees, 1961; Magni et al., 1964). In fact Lewis and John (1966), Vig (1968), and Tai (1970) have proposed that anaphase I bridges without or with fragments, or fragments only, may be the result of such chromatid breaks and reunions rather than some sort of exchanges associated with inversion loops. Tai's study was made on two "mutant" plants of the diploid species *Agropyron cristatum* that also produced multiple spindles and other related abnormalities. Vig reported on one plant of *Aloe vera* that showed at metaphase I asynapsis in 6.5% and stickiness in pollen mother cells (PMC's), and at anaphase I: 4.7% bridges without fragments, 6.7% fragments without bridges, 1.4% irregular distribution of chromosomes, 7.7% with lagging chromosomes, and 3.9% showing difficult separation. Walters found in some plants of *Paeonia californica* that some PMC's contained more than 50 fragments of various sizes. The plants were heterozygous for a number of translocations, but there was no correlation between number of translocations and fragmentation.

Vasek (1962) and Tai (1970) reported "multiple spindles" in *Clarkia exilis* and *Agropyron cristatum*, respectively. Genetic analysis revealed that a single recessive mutation when homozygous in *Clarkia* determines this condition in from 50 to 65% of microsporocytes. Most such PMC's contained two spindles but as many as five rarely. Segregation of numbers of chromosomes on these spindles was very variable; essentially all possibilities were found for the 2n = 18 chromosomes on from one to five spindles. Pollen sterility corresponded to the percentage of multiple spindle-containing PMC's.

Clark (1940) reported a single recessive gene mutation, *dv* in maize that produces "divergent spindle" at the first meiotic division in anthers. The result of the homozygous condition of this gene is that nearly all PMC's have abnormal spindles that usually flair out in the polar region rather than converging. Other conditions found were long, curved or multiple spindles. This gene points up the importance of converging spindles, to pack the chromosomes tightly together at telophase so that only one nucleus is formed at each pole. With the divergent meiotic spindle condition, numerous small nuclei form at each pole at telophase. This mutation is similar to the *claret* of *Drosophila simulans* that has an abnormally wide first meiotic spindle in eggs that produced from 4 to 12 nuclei at telophase II. Spindle formation seemed to require the presence of chromosomes No. 2 and No. 3 in the same nucleus (Wald, 1936). Divergent spindle does not occur in ovules of maize. The *dv* gene of Clark is a different mutation from the "polymitotic" mutant of maize (Beadle, 1931) that causes a series of mitotic divisions after completion of meiosis in anthers but apparently without interphase DNA or structural replication of chromosomes. Therefore, the numbers of chromosomes in the numerous nuclei become lower and lower.

Some other known meiotic genes are for failure of cytokinesis (Beadle, 1932a) and for sticky chromosomes (Beadle, 1932b) in maize, for precocious centromere division by the end of meiosis I in tomato (Clayberg, 1959), for coenocyte formation prior to meiosis from failure of cell wall formation in barley (L. Smith, 1942) and a somewhat similar condition in maize (Lebedeff, cited by Clark, 1940), and a gene *em* in maize for numerous effects on spindle and desynapsis during meiosis (Suto, 1955, cited by Vasek, 1962). The effect of the mutant chromosome No. 10 (K 10) of maize in increasing crossing-over in both normal and structurally rearranged chromosome No. 3 (Rhoades and Dempsey, 1966) is well known.

It is evident that there are numerous meiotic genes that can mutate to the recessive condition and, when homozygous, upset one or more components of the overall process from premeiosis to postmeiosis. None of these genes affects the mitotic divisions of the organism; but because such genes begin to act during the last one or more premeiotic mitoses, Rees (1961) assumed the change from mitosis to meiosis is achieved by a cytoplasmic change that builds up gradually over a number of cell generations. Brown and Stack (1968) and Stack and Brown (1969b) considered that genes for meiosis are turned on, or inhibitors are turned off, gradually over a number of premeiotic cell generations, as already discussed in this chapter.

Asynapsis

The lack of synapsis at meiosis is called asynapsis or desynapsis. It is either a normal and evolved condition or a result of a mutation. It is not evident until diplotene, but from then to metaphase it is evident by homologues not being united by chiasmata. There are two sorts of evolved and, therefore, normal asynaptic conditions, (1) no crossing-over in males of particular species and (2) some forms of apomeiosis in some asexual species. In *Drosophila* and many or all (White, 1954) of the short-antennae (Brachycera) flies there is pairing but asynapsis and no chiasmata (called achiasmatic) in the males, although crossing-over does or can occur in *Drosophila ananassae* males (Mukherjee, 1961). In the mosquito *Culex tritaeniorhynchus* there is no genetic recombination in the female, at least of the X chromosome when the mutant *golden* is present on the X chromosome, although under similar conditions there is recombination in the male (Baker and Rabbani, 1970). Both males and females have two X chromosomes.

Homologues stay together in such achiasmatic spermatocytes or oocytes and move reductionally at anaphase I. Needless to say, there is no synaptinemal complex, at least in *Drosophila* males (Meyer, 1960), although Gassner (1969) reported synaptinemal complex in the achiasmatic spermatocytes of the mantid *Bolbe nigra*, and Stack (unpublished) has found quite typical SC in the completely asynaptic triploid form of the wild onion *Allium amplectans*. Asynapsis functions in some recently evolved apomictic plants and parthenogenetic animals are one aspect of the overall mechanism of asexual reproduction (Chapter 18).

Asynaptic mutants arise occasionally in many species of plants (Beadle, 1930; Prakken, 1943; Rees, 1961; Soost, 1951; Miller, 1963) and animals (White, 1954). In some species there may be pairing but no synapsis (or at least chiasmata), in others not even pairing, and, as an extreme, a typical mitotic division may replace meiosis I. The place and time of action of asynaptic genes can be at pairing, at synaptinemal complex formation, or at crossing-over.

Soost (1951) classified reported cases of asynapsis into five groups: (1) asynapsis in species hybrids, (2) asynapsis caused by the loss of a chromosome pair, (3) asynapsis in apomictic organisms, (4) asynapsis induced by external conditions, and (5) asynapsis due to the action of a gene or genes.

Asynapsis need not involve all chromosome pairs of a meiocyte (Price, 1955); the amount of asynapsis may vary among meiocytes (Beadle, 1930), and among plants (Miller, 1963), or may regularly occur in only one pair or only in the long chromosomes of a grasshopper (Rees, 1957); often the sex chromosomes do not form chiasmata. An example of asynapsis induced by external conditions was reported by Roth and Parchman (1967). They found that when microsporocytes of *Trillium* were extruded into culture solution just prior to leptotene, in pre-leptotene, a synaptinemal complex formed and pachytene chromosomes appeared normal until diplotene, when it was evident that no chiasmata had formed. They also reported "that if protein synthesis is inhibited at late zygotene with cycloheximide there is no visible effect on synapsis, but achiasmatic chromosomes are produced at diplonema. Thus, even though the SC is formed, protein inhibition effectively mimics the transplantation phenomenon." Inhibition of the special DNA synthesis at leptotene and zygotene in extruded microsporocytes also prevents chiasma formation by inhibiting complete synthesis of the synaptinemal complex (Roth and Ito, 1967). It is not surprising, then, that there are numerous reports of the desynaptic condition being caused by unusual environmental conditions. Roth and Parchman concluded, "therefore, that the events of chiasma formation, although dependent on the presence of the SC, are separately mediated by the presence of further protein synthesis that is required in addition to the formation of the synaptinemal complex." Thus, it cannot be assumed that because no chiasmata are present after pachytene that synapsis and a synaptinemal complex did not occur. It is for this reason that some cytogeneticists prefer the term desynapsis to asynapsis.

A form of asynapsis has been reported in salivary gland chromosomes in hybrids of culicine mosquito species and is discussed in Chapter 3.

Pseudoassociations

Price (1955) reported in *Secale montanum* that had been X-rayed the common occurrence of "pseudoassociations" along with normal chiasmatic bivalents. The pseudoassociations consisted of homologues lying side by side through diakinesis but with no chiasmata. The ends were stuck together by matrix, there was no repulsion of homologues at diplotene, and separation into univalents occurred by metaphase I. Similar "quasibivalents" have been reported in plants by Ostergren and Vigfusson (1953), Walters (1954), and Person (1955).

CONCLUSION

Meiosis evolved in part as a chromosomal number reduction process and in part to produce recombination. It evolved out of mitotic divisions one or more times but has remained very sensitive to environmental conditions as well as to mutations that act at various times and to other upsets such as hybridity may create. Many variations of it have evolved as "normal" for certain species, such as various forms of apomeiosis (see Chapter 18), nondisjunction, directed chromosomal eliminations, multiple or unipolar spindles, etc. And the study of meiosis constitutes a large part of the subject matter of cytogenetics.

CHAPTER 7

Sex determination

The observations cited above, as well as a multitude of others that cannot here be reviewed, render it certain, however, that sex as such is not inherited.
E. B. Wilson, 1906a
(Probably written in 1905 or 1904)

It seems possible that the differential chromosomes may perform a definite and special function in sex production without being in themselves specifically male-determining and female-determining.
E. B. Wilson, 1906b

... the view advanced by McClung ('02) and first shown to be correct in principle by the work of Stevens and myself, that half the spermatozoa are male-producing and half female-producing.
E. B. Wilson, 1909

In Chapter 6, sex chromosomes were introduced as those that have evolved as the sex-determining mechanism in most species of animals and some plants. This knowledge was determined early in the twentieth century, shortly after the 1900 rediscovery of mendelian genetics by Correns, De Vries, and Tschermak, and indeed made an essential contribution to the formation of the science of cytogenetics.

Before 1900 sex was considered to be determined by environmental factors. It is now known that environmental conditions do often trigger sexual reproduction, but in rare cases only does it determine the sexes of individuals. During the next 10 years geneticists tried to explain sex determination in general by a pair of mendelian alleles such that there is one sex homozygous and one heterozygous and that one or other sex is dominant and the other recessive (Castle, 1909; Bateson, 1909). At the same time cytologists were demonstrating that one sex has two similar chromosomes (XX or ZZ) but the other has only one of that chromosome (X or Z) although it may have also a unique chromosome (Y or W). Both concepts fitted the results of genetic experiment; the difference is that the chromosomal explanation is polygenic and neither sex is dominant to the other. Both permitted the concept that each sex has attributes of the other because both have a common embryonic condition from which they develop.

McClung (1901) was the first to propose that the so-called accessory chromosome (now recognized as the single X) in male grasshoppers might in some way be related to sex determination since it does not divide at first meiotic anaphase but does so at the second. Therefore, two sorts of sperms are produced in equal numbers, half with and half without the accessory chromosome. The female (XX) at metaphase I of meiosis, however, has a normal X bivalent rather than the univalent accessory chromosome of the male, and all eggs are chromosomally equivalent. Thus, the equal numbers of two sorts of sperm cells could produce the equal number of two sexes. The bug (Hemiptera) *Protenor belfragei* was found by Montgomery (1901) and Wilson (1906b) to have the same single accessory chromosome in the male. By 1910 this XX female/XO male scheme of sex determination was established, the XX sex being homogametic, producing only one type of gamete.

By 1920 the XX-XY scheme of sex determination in *Drosophila* and some other organisms had also been demonstrated, for *Drosophila* by Stevens (1908), Wilson (1913), and Metz (1914). In this scheme, too, as in the XO scheme, it is usually the male that is digametic or heterogametic, producing equal numbers of two kinds of gametes. The two schemes differ only by the presence or absence of the Y, although the Y is only slightly different from no chromosome at all. In fact, by 1920 it was generally accepted that the XO condition is derived from the XY by the evolutionary loss of the Y. During that same decade the heterogametic female condition was described in the Lepidoptera (moths, butterflies, etc.) birds (Figs. 11-2 and 11-3), etc., the female being XY (or ZW) the male XX (or WW) or some more complicated scheme.

During the subsequent 50 or 60 years many species of animals and dioecious plants have been examined cytologically for sex chromosomes or groups of chromosomes (see White, 1954, for animals; Allen, 1954, for bryophytes; Allen, 1940, and Westergaard, 1958, for angiosperms; and Mittwoch, 1967, for all groups, but especially mammals). Many species have been reported to have chromosomal sex-determining mechanisms more complicated than the XO or XY schemes. Some are discussed later.

DETERMINATION OF SEX

It is obvious that if two different sex chromosomes are to be recognized cytologically, they must be structurally different. It seems evident that many X and Y chromosomes are essentially identical cytologically and may differ for only one or a few alleles. This seems to be true of many species and genera of the suborder of Diptera called the Nematocera (White, 1954). For example, the mosquitos all have $2n = 6$, and in some genera there are no recognizable sex chromosomes. In fact the sexes of *Aedes* and *Culex* are determined by alleles of one gene; the males are heterozygous, Mm, and the females are homozygous recessive, mm (Gilchrist and Haldane, 1947; McClelland, 1962; R. H. Baker, 1968). Philip (1942) could find no detectable chromosomal sex differences even in salivary gland chromosomes in two species of *Chironomus*, and White (1954) considered the whole family Chironomidae as lacking sex chromosomes. Mosquitos of the genus *Anopheles*, however, do have heteromorphic X and Y chromosomes. Whether the chromosome carrying a single sex gene should be called a sex chromosome or not is debatable. Then there are many species, especially dioecious angiosperms (pistillate plants and staminate plants) with no recognizable sex chromosome differences. In fact it is remarkable, rather, that X and Y chromosomes are so often heteromorphic. Why should they differ in size so often?

The answer to this question is related to gene content and the genetic system that *produces* males, or, alternately, females, in contrast to the mechanism that *determines* maleness or femaleness. For example, a single allelic difference may determine a male rather than a female in the mosquitos *Aedes aegypti* or *Culex tritaeniorhynchus*, but many developmental genes must be involved in actually producing an adult male or adult female organism. The single gene or chromosome may act merely as a trigger to initiate a specific male-producing or a different female-producing chain of gene actions, the genes located on any or all chromosomes of the set. Yet, with respect to sex production, the genes involved seem to be peculiar. Dobzhansky and Schultz (1934) could not pin down any individual sex-producing genes on the X chromosome of *Drosophila*. Rather, they concluded that it is the X chromosome *as a whole* that appears to be female-producing. And it is not the Y chromosome but the No. 3 and No. 4 that contain the male-producing genes (Pipkin, 1960). Thus, the trigger, the male- or female-determining chromosome, such as the Y, need not carry many genes. In fact the trigger can be merely the lack of an X chromosome as in the XO scheme.

The above is true of *Drosophila* and perhaps those species having the *Drosophila type* of sex determination in which it is the ratio of the number of X chromosomes to the number of autosome sets (A) that determines sex. Thus 2X/2A is female, 1X/2A is male whether XY or XO, 2X/3A or 3X/4A are intersexes, and XXY/2A is female (with a Y) whereas XO/2A is male (without a Y), so the Y chromosome does not determine maleness. In contrast is the *Melandrium type* of sex determination, as in the plant *Melandrium*, in man, etc., in which maleness almost always results if the organism has even one Y regardless of the number of autosome sets or X chromosomes. The Y is male-determining whether XY, XXY, XXXY, or XXXXY.

In both types, however, the Y chromosome is almost always largely heterochromatic with few, if any, detectable genes, or it has been discarded completely as in XO types. Yet the Y does usually contain a portion of its length homologous to part of the X so that the two can pair (if not synapse) at meiosis so that gametes produced by the heterogametic sex will not have both sex chromosomes or neither of them. Thus the Y chromosome usually consists of such a *pairing region* and also a short or long *differential region*. The latter is unique to the Y and makes it the Y; segments of the pairing region, if there is crossing-over between the X and the Y, change back and forth between X and Y chromosomes each generation. Of course in many XY species there is no synapsis between X and Y, and in some species there is not even any pairing, as in the milkweed bug *Oncopeltus*. But even in abnormal female *Drosophila* in which the pair of X chromosomes do not form chiasmata, the X's disjoin normally at anaphase I, one to each pole (Cooper, 1945). This is merely another demonstration that chiasmata are not always necessary for normal disjunction at meiosis.

Whereas the Y chromosome usually contains few, if any, detectable genes (in man the only likely one is for "hairy ears," which is common and esteemed among men of India), the X has usually, but not always (rodents seem to have none), many sex-linked genes that have nothing to do with sexual dimorphism. That is, the X is usually a more typical chromosome, the Y a rather atypical one. Because of its low gene content and heterochromatic condition it can and usually has become quite small and, therefore, morphologically recognizable.

The Y chromosome in most species of *Drosophila* has fertility factor (Brosseau, 1960); it contains a nucleolus organizer and is important in the production of nucleolar RNA (Ritossa and Spiegelman, 1965); some Y chromosomes of *D. paramelanica* can suppress X-chromosome-borne "meiotic drive" phenomena (Stalker, 1961); the Y of *melanogaster* can suppress various X-chromosome lethals (Hess, 1963); it plays a role in variegated position effect (W. K. Baker, 1968); the number present (XO, XY, or XYY) affects the salivary gland chromosomes, especially the form of the ends (Schultz, 1947); and in *D. melanogaster* and *hydei* the number of Y's affects the lengths of sperm remarkably (Hess and Meyer, 1968). In *D. melanogaster* the XO condition produces sperm 1.2 mm long; XY^S, 1.3 mm; XY^L, 1.5 mm; XY, 1.8 mm; and XYY, up to 3.5 mm. In *D. hydei* the XY condition produces sperm 6.6 mm long; XY plus partial duplication produces intermediate lengths; and XYY, from 13 to 14 mm.

Furthermore, Y chromosomes vary in size among species of *Drosophila* such as *D. pseudoobscura* and *persimilis* (Dobzhansky, 1935, 1937), *D. affinis* (Miller and Roy, 1964), and *D. athabasca* (Miller and Voelker, 1968). In *affinis* the Y ranges from a large J to complete absence, and in *athabasca* the size variation is continuous.

Species of XY type in which XO individuals or populations occur, in addition to *Drosophila affinis* and some others, have been reported in Hemiptera such as *Metapodius femoratus* (Wilson, 1910) and *Auchennorrhyncha* (Halkka, 1959). Smith (1953, 1960) has listed a number of XY species of Coleoptera with some XO individuals, and Baker and Hsu (1970) found an XO species of bat.

Although X chromosomes are more typical chromosomes than Y's, even the X often behaves atypically, especially with respect to heterochromatization. When a chromosome or segment of a chromosome is heterochromatic as during interphase or in a postmitotic cell, the genes within the heterochromatin are supposedly inactive. It is also well known that cells can cause sex chromosomes to be heterochromatic when autosomes and even other sex chromosomes are euchromatic. The commonest case is in female mammals in which, in many cells, only one X is heterochromatic as the Barr body (also called drumstick or the sex chromatin) (see Chapter 19). Condensed sex chromosomes also occur during meiosis in male spermatocytes. It is very often seen that the single X of grasshoppers is heterochromatic from before meiosis begins until metaphase I (White, 1954). Even in many XY species of animals the synapsed "sex bivalent" is heteropyknotic (in the male only) from premeiotic interphase to metaphase I, as in man (Evans and Swezy, 1929) and mammals in general (Ohno et al., 1961). The assumed result of one X only being euchromatic in interphase somatic cells of female mammals is for *dosage compensation*, to make them equal to the single X of males with respect to gene activity. The function served by heterochromatic sex chromosomes during meiosis in males only when synapsis and crossing-over are taking place may be to suppress crossing-over, although chiasmata are formed in many species. If other chromosomes are transcribing RNA up to diplotene, which seems likely though not generally demonstrated, the heterochromatic sex chromosomes might not be able to do so.

Such sex chromatin (Barr) bodies in interphase nuclei have been reported as general in mammals, questionable in reptiles, in one bird, not in amphibians, none in Mollusca or Chilopoda or Diplopoda, in few Arachnida and Crustacea, in all Lepidoptera (♀!), in few Hemiptera and Homoptera but rare in Insecta generally (Habert and Beckert, 1971).

EVOLUTION OF SEX CHROMOSOME MECHANISMS

It is likely that sex chromosomes began evolving at the same time that males and females of multicellular organisms were evolving their distinctive primary and secondary sex differences. Probably the first differentiation was the distinction of the primary sex organs in different individuals. (Of course hermaphrodites do not have sex chromosomes, White, 1954.) In such primitively sexual species a few genes would be different in the two sexes so that sperm- or egg-producing gonads (or the plant equivalents) would form. Such a primitive system would be very delicate and easily upset genetically. White cites the conversion of male heterogamety involving one gene or a few in the fish *Lebistes* to female heterogamety (for details see Mittwoch, 1967). Bellamy (1936) and Gordon (1947) have reported that in the fish genus *Platypoecilus* one species has male heterogamety but in another it is the female that is heterogametous. White proposed in fish in general, and probably amphibians also, that X and Y chromosomes are little differentiated from autosomes.

In flowering plants the hermaphroditic condition, with anthers and pistils in the same flower, is most common and so is considered to be primitive. Therefore, the dioecious condition, with some individuals staminate and other sporophyte individuals pistillate, has evolved as the derived condition in numerous families. In all such dioecious species the abortion of stamens produces the pistillate condition, and abortion of ovaries produces staminate individuals. This is a delicate balance, and perfect flowers are often formed. It is also likely that one or few genes are involved in the inhibition. It is not surprising, then, that Allen (1940) reported no cytologically detectable heteromorphic sex chromosomes in 46 dioecious species of 32 genera in 25 families.

On the other hand, 69 dioecious species in 30 genera of 26 families do have X and Y chromosomes of visibly different sizes, and in some species evolution has advanced to the level of multiple sex aggregates of three or more sex chromosomes.

Arrhenotokous parthenogenesis (haploid males) is another genetic and environmental sex-determining mechanism that has arisen in some animals, especially the Hymenoptera (bees, wasps, sawflies, etc.), without typical sex chromosomes. In this system it should be remembered that little more than the germ line of so-called haploid males is really haploid, that the somatic tissues become diploid, often polyploid, so really the males are not haploid.

Although sex determination in Hymenoptera is determined by fertilization (which produces a diploid and a female) versus no fertilization (which gives a haploid and a male) (Newell, 1915), it is now known that diploid males *can* occur and "sex chromosomes" *are* involved

(Whiting and Whiting, 1925; Whiting, 1935, 1939, 1943, 1945). Apparently an XX female X male system is in operation. One particular chromosome, the X, of a set has a series of alleles of a specific gene so that if only one of these alleles is present, even if in duplicate, a male is produced, but if any two different alleles (on the two X's) are present in an organism, it is female. Among social bees (Apidae) sex seems to be determined by a set of "maleness" genes, which have no additive effect, in balance with a group of "femaleness" genes, which do have additive effect. Both sorts of genes may be major or minor in strength (Kerr, 1969).

This is another example of the confusion of genes and chromosomes in sex determination that has been going on since 1902 merely because the genetic and cytological are merely two ways of describing the same cause and effect. Ultimately, of course, genes are fundamental.

Sex determination, as in the annelid *Bonellia*, in which distance from a female and concentration of some male-producing or female-inhibiting molecule determines sex, is another example of a trigger mechanism that initiates the development of a large complex female or a tiny parasitic male.

The XY condition, usually with the male heterogametic, was probably the next evolved condition because simple gene differences lead directly to it as in related genera of mosquitos. Almost always it is the Y chromosome that has decreased in size and function; it is very often the smallest of the set and has probably little genetic function. In *Drosophila* nearly 60% of species have the Y equal in size to the X, in about 33% of species the Y is somewhat smaller than the X, in 9% of species the Y is tiny, and in less than 2% there is no Y. The species *annulimania*, *longala*, and *orbospiracula*, and a few collections of *affinis* are XO. *D. miranda* is $X_1X_2Y_1Y_2$. But often the Y is smaller than the X and among the smallest of the set.

Nevertheless, when the Y is present it is usually necessary, except as in *D. affinis*, which has populations that can get along without it although in cage competition the XY forms completely displaced the XO forms (Voelker, 1970). In *D. melanogaster* (Brosseau, 1960) and in *D. hydei* (Hess and Meyer, 1968) the Y is necessary for fertility of sperm. Hess and Meyer reported in spermatocytes that the Y produces five pairs of loops, like those of lampbrush chromosomes, which are probably the "Y fertility factor" of Brosseau. These loops are all necessary for full fertility (determined by various deletions) by their synthesis of unique RNA found in testes. Therefore, Y-less species probably have these few essential elements, or genes, translocated onto other chromosomes of the set.

The X chromosome is a more typical chromosome; it is rarely, if ever, one of the smallest. In man, dog, cat, etc. it is longer than average; in many *Drosophila* species, the mammal *Chinchilla lanigera* (Galton, et al., 1965), etc. it is the largest of the set; and in the rodent *Microtus agrestis* (Hansen-Melander, 1965) it is not only the longest of the set but it is a giant 14μ long and is equal to about 30% of the haploid autosome set, and it is one of the longer chromosomes known among mammals.

At least one case of apparent reversion from XY male is known, in the common housefly, *Musca domestica* (McDonald, 1971). The species is typically XY in the male, but the Gainesville strain has *apparent* XX males, and no morphological Y can be detected cytologically. However, one member of the third pair of autosomes is always transferred from male parent to its male progeny (holandric inheritance) "as though it were linked to a male determiner or a Y chromosome." Furthermore, since there is no crossing-over in male houseflies, the third chromosomes are functionally and morphologically similar X and Y chromosomes. In this strain the holandric third chromosome (= Y) never receives genes or alleles from the other third (X) chromosome and so is free to mutate independently of the homologous third chromosome. In a sense, then, the X chromosomes have become autosomes, and the third chromosomes have become sex chromosomes.

Wahrman (1966) found in some beetle genera the change from XY to XO (or the reverse), which seems to have occurred commonly and independently. He stated, "These two systems

[XO and XY] often occur side by side in the same genus. In fact they coexist in 8 genera, including almost all of those where more than one or two species have been studied. This appears to indicate that the changes from XY to XO or from XO to neo-XY have occurred repeatedly and independently."

From the XY condition, then, reduction to elimination, as in *Drosophila*, leads directly to the evolution of the XO condition. The XO condition is common or universal in grasshoppers and Odonata, is very common in the Heteroptera, and occurs in other groups.

The XO condition is not the end of the evolutionary line of sex chromosome evolution, but further evolution from XO, or XY, leads to more complex sex mechanisms. It involves translocations with autosomes, thereby decreasing the number of autosomes as they become whole or parts of sex chromosome complexes. For example, *Drosophila prosaltans* I has XY and four autosomes, but the derived *prosaltans* II has four of the six chromosomes as sex chromosomes, $X_1 X_2 Y_1 Y_2$ (Patterson and Stone, 1952).

For example, a translocation between one X and an autosome of an XO species means that one of the involved autosome pair becomes a Y chromosome; and two translocated chromosomes, an X and autosome, become X_1 and X_2 (Fig. 7-1). The old autosomes then must be replaced by the new X_1 and X_2 in the population. The new form has all the genetic material of the old X and autosome pairs, and bivalents are formed in the female. But at meiosis in the male the old autosomal arms of the new X_1 and X_2 pair with one intact old autosome which now is the Y and is restricted to males. The Y can begin to evolve genetically and morphologically independently of the female sex. Such an $X_1 X_2 Y$ condition is typical of mantids and occurs in some species of numerous groups of animals and in a few plants.

Translocation between a Y and an autosome can produce the $XY_1 Y_2$ condition (Fig. 7-2). At meiosis the two Y's go to the same pole and determine male progeny. The $X_1 X_2 Y$ condition of a species means that the female has one more chromosome in diploid cells than the male, whereas the male of an $XY_1 Y_2$ species has one chromosome more than the female. *Drosophila miranda* and the Rhodesian pygmy mouse

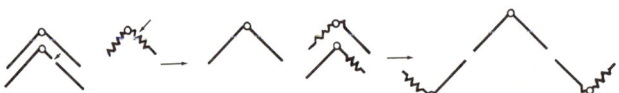

Fig. 7-1. Evolution of a tripartite $X_1 X_2 Y$ sex chromosome mechanism from an original XO scheme. A reciprocal translocation between one of a pair of autosomes (A_2) and the X chromosome (arrows) yields $Y(=A_1)$, X_1, and X_2 chromosomes. At meiosis in the male each arm of the $Y(A_1)$ pairs with its homologous arm, now on the two X's. By alternate disjunction the X_1 and X_2 go to one pole at male meiosis and the Y goes to the other. In the female there is a pair of X_1's and a pair of X_2's; the A_1, now the Y, is lost.

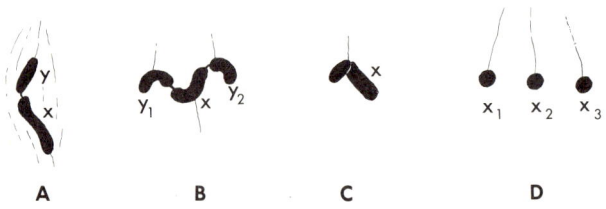

Fig. 7-2. Some simple sex chromosomal arrangements at meiotic metaphase in the heterogametic sex. **A,** Example of the XY bivalent (see Fig. 19-8 for such an association in man). **B,** The tripartite $XY_1 Y_2$ association with alternate disjunction indicated. The $X_1 X_2 Y$ would appear similar. **C,** The univalent X chromosome of the XO type. **D,** The three X's of the $X_1 X_2 X_3 O$ type all somehow go to the same pole. (From Brown, W. V., and E. M. Bertke. 1969. Textbook of Cytology. The C. V. Mosby Co., St. Louis.)

(Matthey, 1965) are examples of the X_1X_2Y condition; and *D. americana*, the gerbil, *Gerbillus gerbillus* (Wahrman and Zahavi, 1955), and a number of bats (Baker and Hsu, 1970) are examples of XY_1Y_2. White (1954) lists other examples of these and more complex schemes. Complexities of the XO scheme are designated as X_nO, and types up to $X_1X_2X_3X_4X_5O$ are known. Complexities of the XY scheme are designated as X_nY, and examples up to $X_1 \ldots X_8Y$ have been reported, especially in vertebrates. In the beetle *Blaps lusitanica* at meiosis of the male the $X_1X_2X_3X_4$ form a complex like a compound chromosome that is "paired" to the Y. At anaphase the Y goes to one pole while the whole group of four X's goes to the other (Wilson, 1925).

Complex sex chromosome schemes of other sorts are also known, especially in mammals. Ohno et al. (1963) reported that the "creeping vole," *Microtus oregoni*, is a gonosomic mosaic. It is mosaic because there are two chromosomal conditions in each individual, and it is gonosomic because the germ line is different from the soma. The male somatic tissue has 18 chromosomes, $16A + XY$, but the male germ line has 17 chromosomes, $16A + OY$; presumably, the X is eliminated from the male germ line by selective disjunction. The female soma has 17 chromosomes, $16A + XO$, but the female germ line has 18 chromosomes, $16A + XX$; presumably, the extra X is acquired by selective nondisjunction (see below). XY zygotes produce males, but XO zygotes produce females.

Another mammal with unusual sex chromosome behavior is the Indian mongoose, *Herpestes auropunctatus* (Fredga, 1965). In this species the female is $34A + 2X$; the male is $34A + XO$. At meiosis in spermatocytes one end of the single X is paired with one end of one autosome of a particular pair. Fredga proposed that originally there was a Y in the male but it was largely translocated to an autosome and the remainder lost. Now the X pairs with the pairing segment of the Y on a tip of one arm of one autosome to form an odd trivalent at metaphase I. How this translocated chromosome remains confined to the males in spite of crossing-over between the centromere and the X-pairing tip is unknown.

In addition to a variety of normal, evolved sex chromosome schemes for the determination of sex in animals, there are also aberrations that are found in a few individuals only of a species. Such aberrations in man are discussed in Chapter 19, and similar aberrations of sex chromosomes should occur in other species. But because so few and usually "normal" individuals of such species are examined it is hardly likely that aberrants will be often found. Nevertheless, in mice, which have been examined considerably, Cattanach (1962) and others have found an XO in apparently normal and fertile females, which would compare to Turner's syndrome in a human being. A few XXY apparently normal but sterile males have been reported by Cattanach (1961) and others. These compare to human beings with Klinefelter's syndrome.

Matthey (1966) reported in a study of a species of an African mouse that four out of seven female individuals had only one normal metacentric X. The other X, apparently the second X, had a deletion.

There are two famous aberrations of sex chromosomes in *Drosophila* that are worthy of mention, namely *primary nondisjunction* and *attached X*.

Primary nondisjunction is an irregular segregation of X chromosomes during meiosis in oogenesis so that both X chromosomes are present in the egg, or no X is present there. Thus, after fertilization the zygotes are XO, YO, XXY, or XXX. Of these the XXX usually dies and the YO always does. This phenomenon was discovered when white-eyed and red-

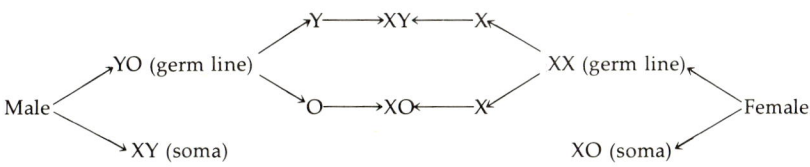

eyed drosophilas were crossed in various ways. White eye is recessive to red eye, and the gene is in the X chromosome. Usually when a white-eyed female *(ww)* is crossed to a red-eyed male *(W)*, all males are white-eyed *(w)* and all females are red-eyed *(Ww)*. But Bridges found exceptional white-eyed females (about 1 in 2,500) and exceptional red-eyed males (about 1 in 1,200). Bridges proved that these exceptional individuals were the result of primary nondisjunction so that white-eyed females were XXY and red-eyed males were XO, the X derived from the father.

Attached X was discovered by Morgan (1922) as a similar case of apparent X-chromosome nondisjunction, but it was 100%. Cytological examination revealed that the male is normally XY but the female has a Y and also an abnormal X. This attached X consists of two partial X chromosomes attached together, apparently by translocation close to the centromere, so that it is a metacentric chromosome with two almost identical arms. Thus, it is nearly the equivalent of two X's, at least for certain genes such as white eye. A cross, then, between a white-eyed attached-X female (attached X plus Y) and a red-eyed normal male (XY) gives: (1) attached X plus X = sterile abnormal female, red-eyed; (2) attached X plus Y = white-eyed female; (3) XY = red-eyed male (Y from mother, X from father); and (4) YY = lethal.

More complex X chromosomes have been produced and studied, especially the production of ring-X's and the effects of crossing-over within such rings (Sandler, 1965). He mentions a "tandem metacentric compound X chromosome," which is composed of two complete X chromosome elements completely reversed with respect to each other and attached to a single centromere, and a "tandem acrocentric compound X chromosome," which is two X chromosomes attached to each other in tandem and to a single subterminal centromere. Sandler found that all sorts of rings are unstable during meiosis even when no crossing-over occurs in the presence of the asynaptic gene, C3G.

PLANT SEX CHROMOSOMES

Sex chromosomes in Bryophyta (mosses and liverworts) have been discussed by Allen (1935, 1945). In contrast to that of animals, the life cycle of plants is complicated by alternation of haploid and diploid generations. The diploid is the sporophyte, and meiosis occurs to produce either one sort or two sorts of haploid spores. If only one sort of spore is produced, the haploid plant that grows from it is probably monoecious. That is, it produces both male (sperm) and female (egg) gametes. If, however, two sorts of haploid spores are formed by meiosis, two sorts of gametophytes develop; one type of spore grows into a sperm-producing (male) gametophyte, and the other type develops into the egg-producing (female) gametophyte. Union of egg and sperm forms the diploid zygote that grows to be the next, asexual, generation sporophyte. The function of sex chromosomes is to determine male and female gametophytes, and, if dimorphic, the X and Y can be recognized as a heteromorphic pair at meiosis in the diploid sporophyte. In the Bryophyta the gametophytes are the plants generally called mosses and liverworts.

In all dioecious Bryophyta having sex chromosomes the scheme is XY, with the Y smaller than the X except in two species. The X is found in female gametophytes and the Y in male. These sex chromosomes are usually wholly or partially heterochromatic, especially the X, during prophase.

Occasionally meiosis (in the sporophyte) fails, and diploid (2A + X + Y) gametophytes are formed. Thus, the special characteristics of the gametophyte are not determined by the haploid condition. Similarly, diploid gametophytes (2A + 2X and 2A + 2Y) can be formed by direct chromosome doubling from haploid gametophytes, and tetraploid sporophytes can arise from sexual reproduction.

In *Sphaerocarpos donnellii,* one of the most studied of bryophytes, partial destruction of the X chromosome by irradiation resulted in males even though no Y was present. It was assumed that male-determining genes are borne on the autosomes but female-determining genes are on the X. The X, even though "totally heterochromatic," also carries genes for structure.

Sex chromosomes are also found in the flowering plants, but in this group they determine the "sex" of the "asexual" diploid sporophyte. Thus, while most species of flowering plants bear both ovaries and stamens on each plant, numerous species in various families consist of two sorts of sporophytes: staminate (producing pollen) and pistillate (producing ovules and embryo sacs). Such species are described as being *dioecious*. It is only in dioecious species that sex chromosomes are to be expected and are reported (Allen 1940).

Many dioecious angiosperms (flowering plants) have no heteromorphic chromosomes but probably do have "X" and "Y" chromosomes of similar form. They may differ for one or a few genes only. But sex chromosomes have been found in a number of species—69 species in 30 genera of 26 different families were listed by Allen. Thus, the condition must have arisen independently many times. As is generally true, the XY scheme occurs in nearly all species, such as poplars, willows, *Cannabis sativa*, *Lychnis (Melandrium)*, *Elodea*, *Smilax*, etc.

There are a few, more complex and derived systems. In the mistletoe, *Phoradendron*, the female is 2A and the male is 2AY. The hop species *Humulus japonicus* is XY_1Y_2 but *H. lupulus* is $X_1X_2Y_1Y_2$. A number of species of *Rumex* are XY_1Y_2. *Atriplex hymenelytra* is reported to be X_1X_2Y.

A few species of plants having sex chromosomes have been able to achieve polyploidy but to still maintain chromosomal sex determination. In *Rumex*, *R. hastatulus* is $2n = 8(6A,2X)$ in the female but $2n = 9(6A,XY_1Y_2)$ in the male. *R. acetosa* is $2n = 14(12A,2X)$ in the female and $2n = 15(12A,XY_1Y_2)$ in the male. Thus, the autosomes are tetraploid in *acetosa*, but there has been no evolutionary change in the sex chromosomes. Sex in *R. acetosa* is determined by the A:X ratio as in *Drosophila*; 2A,2X is female, but 2A,X is male without any Y. The classical case in plants of the Y determining the male sex, as in man, is *Melandrium dioicum* (Warmke, 1946). Warmke found that XYY, XY, XXY, and XXXY are all male, but 4XY is hermaphrodite. In this species the Y is considerably larger than the X. Tetraploid plants XXXX female and XXXY male were found to be fertile and gave equal numbers of male and female plants, whereas XXXX crossed with XXYY gave mostly males because nearly all gametes produced by the male plant were XY more commonly than random, about 89% (Warmke and Blakeslee, 1940).

SEX RATIO

In 1928 Gershenson in *Drosophila obscura* and in 1936 Sturtevant and Dobzhansky in *D. pseudoobscura* and *persimilis* found an X-linked genetic condition that, when present in the *male*, leads to a very high percentage of female progeny. This condition, where only X-bearing sperm seem to function, is called *sex ratio*. Sturtevant and Dobzhansky reported that in male spermatocytes carrying this condition the X divides at both meiotic divisions, but Novitski et al. (1965) found meiosis regular. Both groups reported the full 128 spermatids and sperms in each bundle in the testes. Therefore, they proposed the *functional pole* hypothesis to "explain" sex ratio—that there is "preferential movement of the X chromosome to the functional pole at anaphase." Policansky and Ellison (1970), however, by electron microscopy, also found in normal genotypes more than 100 (112 ± 7.4) spermatids per bundle but in the mutant only about 55 ± 8.2 sperms per bundle. They assume, therefore, that meiosis is normal but Y-containing spermatids do not produce sperm. They called it *spermiogenic failure*.

MEIOTIC DRIVE

The sex-ratio condition is an example of a broader condition known as meiotic drive, which Sandler and Novitski (1957) defined as a condition of the meiotic divisions that modifies the breeding structure from the expected ratios by altering the frequencies of alleles in a population. That is, the alteration in meiosis "drives" the expected ratios to some unexpected ratios. In the sex-ratio condition the 1 male:1 female ratio is "driven" (perhaps not by meiosis as *now* understood) or altered to almost all females. Other cases of nonrandom segregation of genes or chromosomes or both to the gametes are known. One in maize is determined by abnormal chromosome No. 10. Abnormal chromosome No. 10 (Longley, 1938; Rhoades, 1955) possesses an addition on the distal end of the long arm at pachytene about equal in length to the short arm. In addition to a rather typical knob near its attachment to the end of the long arm, it has a long segment of heterochromatin visible at pachytene. The distal end of this additional segment is euchromatic. These abnormal No. 10 chromosomes were found in maize grown by Navajo, Pueblo, Mescalero Apache, and Ute Indians. Longley (1937) also found this abnormal No. 10 in the close maize relative, annual teosinte. In the heterozygous condition the abnormal chromosome No. 10 produces at anaphase I "neocentromeres" from knobs on the chromosomes (Rhoades and Vilkomerson, 1942). The inclusion of abnormal No. 10 (and other knob-bearing chromosomes also, Longley, 1945) into the particular megaspore that produces the embryo sac and egg occurs about 70% of the time (Rhoades, 1942).

This is comparable to a similar condition in maize where the translocation chromosome containing most of a B chromosome also gets into the egg from 66 to 75% of the time (Roman, 1948). In *Drosophila* heterozygous for a heteromorphic pair of chromosomes, the shorter chromosome gets into the egg nucleus more often than the longer. In the mouse the chromosome containing "tail-less" alleles gets into functional sperm as much as 95% of the time.

Experimental genetic-environmental drive is known in the housefly, *Musca domestica* (McDonald, 1971). Individuals homozygous for the *tsl* gene can be reared at 25.6° C but die in the late larval or pupal stage if reared at 33.3° C. This gene is on the third chromosome. By using the Gainesville strain, in which the third chromosomes have become the sex chromosomes, "X" and "Y" (see earlier in this chapter), and by having the new holandric "Y" with the normal gene but the "XX" female homozygous for *tsl*, when reared at 33.3° C only males emerge. McDonald proposed this system as a cheap method of producing only males for control of the pest by the sterile-male principle.

Such genetic and chromosomal abnormalities, many of which are determined, "drive" the population toward an unexpectedly high number of individuals possessing the abnormality. The factors producing meiotic drive are not always (or perhaps usually) aberrations of meiosis. Somewhat similar factors occur with respect to *preferential distribution* of supernumerary chromosomes as discussed in Chapter 9.

Another unusual sort of meiosis occurs in *Dysdercus koenigii* (Ray-Chaudhuri and Manna, 1952). The X and Y go to opposite poles at anaphase I of spermatocytes quite typically; but at second division they do not divide as usual so that instead of two spermatids containing X's and two containing Y's, one contains an X, one a Y, and two neither. These sex-chromosomeless spermatids do not function. The X and Y finally divide in the zygote division. This is an unusual case of a chromosome going through two divisions without its chromatids (or centromere) dividing and separating.

Makino (1951) also reported a case of unusual segregation of sex chromosomes during meiosis in a field mouse. Almost universally in XY species having unicentric chromosomes, the two X chromatids at anaphase I go to one pole and the two Y's to the other, called *prereduction*. But in the field mouse *Apodemus* an X and a Y go to each pole at anaphase I and the X's and Y's separate at anaphase II. This is called *postreduction* and is generally typical of species with diffuse centromeres.

CONCLUSION

It is evident that sex chromosomes often behave in unusual ways but can also determine sex in many different ways. The XY scheme is basic; but from it by loss of the Y and by translocations, simpler and more complex schemes have evolved in plants and animals. Although chromosomes may determine sex, the differentiation of the sexual organism itself requires many genes that are often located in autosomes.

CHAPTER **8**

Mitotic metaphase chromosomes

No phenomena in the history of the cell more clearly indicate the existence of a morphological organization which, though resting upon, is not to be confounded with, the chemical and molecular structure that underlies it; and this remains true even though we are wholly ignorant of what that organization is.

E. B. Wilson, 1909

The study of mitotic metaphase chromosomes, as the physical and observable aspects of the genome, is a significant part of cytogenetics. Such chromosomes often represent by their numbers, sizes, sets, and forms morphological representations of species, hybrids, and evolutionary relationships and directions. Aberrations such as one chromosome too many or too few, exchanges of pieces between two nonhomologous chromosomes, observable deficiencies, excessive length, etc. are of significant value, as in human cytogenetics. Studies of karyotypes, chromosome numbers, basic numbers, fundamental numbers, ploidy, etc. contribute significantly to systematics. Thus, knowledge of metaphase chromosomes is essential to the general cytogeneticist.

The term chromosome was first applied during the last quarter of the nineteenth century to those bodies that appeared in the prophase nucleus of eukaryotic cells when stained with certain dyes and that later underwent the movements characteristic of cell divisions. They were believed then to *disappear* at telophase, only to *reappear* in the same forms and numbers at the next prophase. The concept has since been extended to a changeable structure that exists at all times including interphase, to the chromatin material of eukaryotic nuclei, and even to the DNA alone, including (Chapter 20) the genophores of prokaryotes, mitochondria, plastids, and even viruses. Here, however, chromosomes in the original sense, those evident at metaphase of higher eukaryotes (plants and animals), are discussed.

It is true that such chromosomes do disappear at telophase and do reappear at prophase. It is assumed that at telophase they swell or expand and lose their visible identity but maintain their relative positions and genetic integrity (Fig. 4-3). Expressed another way, they change at telophase from heterochromatin to euchromatin. At prophase the reverse occurs; the loose, diffuse, and mostly uncondensed structure changes to the highly condensed and dense heterochromatin. It is likely that changes in RNA and protein occur as part of these conversions. Furthermore, DNA and histone double during the S period of interphase which is necessary for the structural doubling of one telophase chromosome into a subsequent metaphase chromosome of two equal and identical chromatids.

GROSS STRUCTURE

Each chromosome at mitotic metaphase consists of two side-by-side sister chromatids that

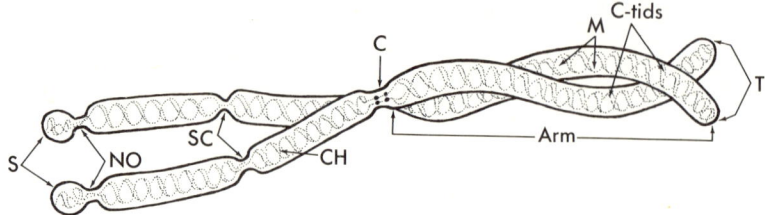

Fig. 8-1. Mitotic metaphase chromosome. The centromere, C, or primary constriction, is median and contains (according to some cytologists) four centromeric granules. Each arm consists of two chromatids, C-tids. Each chromatid may consist of two coiled or folded chromonemata, CH. Some chromosomes have a nucleolus organizer, NO, that is often called the secondary constriction. The short, spherical region of the arm distal to the NO is called a satellite, S. The ends of arms may contain a structure called a telomere, T. Some cytologists have reported constrictions other than the primary and nucleolus organizer, and these should be called secondary constrictions, SC. It is likely that the chromonemata are embedded in special proteinaceous material called matrix, M. Compare with Fig. 8-7. (From Brown, W. V., and E. M. Bertke. 1969. Textbook of Cytology. The C. V. Mosby Co., St. Louis.)

are somehow attached to one another at the centromere, as though the centromere were still single (Fig. 8-1). These sister chromatids are morphologically and genetically identical, except for very rare errors, from end to end. The centromere is also called the *primary constriction* and is the point of reference in discussing chromosome morphology. Chromosomes are characterized by the type, number, and location of centromeres. If the centromere is diffuse and spread from end to end of the chromosome, the chromosome is described as *polycentric* or *holokinetic* as in plants such as sedges and rushes and in bugs and Lepidoptera (Bauer, 1967). A chromosome with no centromere is acentric and is often called a fragment. Most chromosomes of higher plants and animals have a single centromere and are described as unicentric or monokinetic. Occasionally, abnormal chromosomes are observed having two centromeres and are called *dicentric;* for example, in human beings with Bloom's syndrome (Figs. 19-14 and 19-16). Acentric and dicentric chromosomes are abnormal and behave abnormally at anaphase, although Sears and Camara (1952) studied a transmissible dicentric in wheat. Subsequent discussion is limited to the common unicentric type.

Centromeres

The location of the single centromere along the chromosome determines types of unicentric chromosomes (Levan et al., 1964) (Fig. 8-2).

If the centromere is truly at one end, a *telocentric* chromosome results. A centromere not quite at one end produces a chromosome with one long arm and one very short arm, and the chromosome is *acrocentric*. A *metacentric* chromosome results from a centrally located centromere, and the two arms are equal in length. A *submetacentric* chromosome has unequal arms and is intermediate between an acrocentric and a metacentric. These adjectives are often used as nouns, and one speaks of a metacentric or acrocentric. Telocentrics are rare or unknown in plants but are evidently not rare in animals.

Chromosomes can be designated more exactly than merely as submetacentric or subacrocentric by using actual measurements and formulas. The system used in human cytogenetics is used here. The long arm length = q, the short arm length = p, and total chromosome length = p + q.

$$RL = \frac{p + q \times 1{,}000}{\text{length of haploid set}}$$

$$AR = \frac{q}{p}$$

$$CI = \frac{p}{p + q}$$

The *arm ratio* (AR) equals the length of the long arm divided by the length of the short arm. The *centromeric index* (CI) equals the short arm divided by the total length of the chromosome. The *relative length* (RL) equals 1,000 times the length of the particular chromosome divided

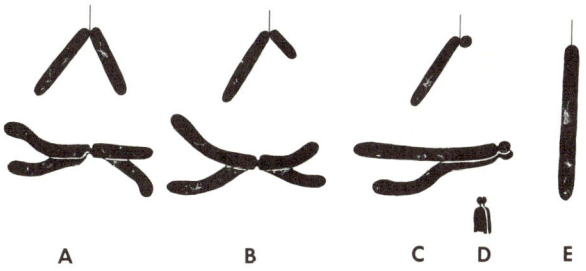

Fig. 8-2. General forms of chromosomes based upon position of the centromere. **A,** Median position produces a metacentric chromosome. **B,** If both arms are nearly but not quite equal, the chromosome is described as submetacentric. **C,** If the short arm is approximately spherical and has the diameter of the long arm, it is subacrocentric. **D,** If the short arm can be seen and has a diameter clearly less than the diameter of the long arm, it is an acrocentric chromosome. **E,** The telocentric has no observable short arm whatsoever. The main problem is distinguishing between telocentrics and acrocentrics.

by the sum of the lengths of the haploid set of chromosomes including the X chromosome.

These characterizations often reveal small differences between specific chromosomes either in constructing normal idiograms or karyotypes or in detecting small aberrations such as pericentric inversions, deletions, or translocations.

Telocentrics. Telocentric chromosomes are produced artificially from unicentric chromosomes by transverse breakage through the centromere. Thus, each unicentric chromosome can produce two entirely different telocentrics, and there can be twice as many telocentrics as unicentrics. In wheat ($n = 21$) nearly all possible of the telocentrics have been found and are maintained in culture (Morris and Sears, 1967; Sears, 1969). These telocentrics can be described as semipermanent. They are more permanent than the maize telocentrics (McClintock, 1932) and do persist in culture with help. This may result from the wheat telocentric centromeres being more than just half centromeres. This concept would imply that complete terminal centromeres might be as permanent as intercalary centromeres.

The wheat telocentrics have been of considerable value in mapping genes in that polyploid species. If a plant contains a telocentric and a particular genotype, that genotype will not be transmitted through the pollen if it is on the telocentric because telocentrics are rarely passed on via the pollen in competition with a homologous whole chromosome. Thus it is determined that the gene determining that genotype was located on that telocentric and, therefore, on that particular arm of a particular chromosome from which the telocentric was derived. Of course, most of the known or available telocentrics have to be involved in a number of hybridizations before the correct one is found by trial and error.

Telocentrics of wheat were also used in a study of somatic pairing in root tips because one or a few can easily be identified among the rest of the normal metacentric chromosomes of the set (Feldman et al., 1966).

Telocentrics, also called telosomes, have been used in mapping genes in cotton by Endrizzi and Kohel (1966), who found that map distances from the centromere are reduced when a telocentric is used because of reduced crossing-over near the centromere.

Telocentric chromosomes are generally considered to be rare in nature in plants, although Marks (1957) and others have thought otherwise. In animals, on the other hand, truly telocentric chromosomes seem to be more common, such as in mouse tumor cells in culture (Gimenez-Martin et al., 1965; Comings and Okada, 1970e), in a parasitic crustacean, *Ulophysema oresundense* (Melander, 1950), in all chromosomes of the fish *Esox masquinonge* (McGregor, 1970), in grasshoppers (John and Hewitt, 1966; Southern 1969), and in many others. Melander considered that the telocentrics he studied have complete terminal centromeres at metaphase. Each such complete centromere of each chromatid contains two centromeric centrioles, so that there are four for each metaphase chromosome. Gimenez-Martin et al., on the other hand, found only one centromeric chromomere per chromatid in the terminal centromeres of mouse chromosomes, and Comings and Okada found the electron microscopic equivalent for the same material. They considered that there may be two or more kinds of terminal centromeres, those in mouse chromosomes being one half of ordinary internal centromeres which they called monochromomeric, and those reported by Melander and others that are dichromomeric. It is also possible that the monochromomerics are produced by a break between the two centromeric chromomeres whereas a break between the arm and the nearer chromomere would produce a terminal dichromomeric centromere. The nonpersisting maize terminal centromeres produced by McClintock (1932) were probably monochromomeric. On the other hand, the wheat telocentrics are described as semipermanent, normally functioning, and complete, so they are doubtless dichromomeric. Nevertheless, the monochromomeric terminal centromeres of mouse, and probably other animals, function normally.

The fine structure of centromeres is discussed later in this chapter.

Nucleolus organizer

Another standard structural modification of some but not all metaphase chromosomes is the nucleolus organizer (Fig. 8-1). Every nucleus probably has at least one, as in a haploid cell. In diploid cells there is one pair or more,

Fig. 8-3. Various positions of satellites and nucleolus organizers in plants. They are usually on short arms, but a few species have satellites on the ends of long arms. The centromeres are indicated by open circles.

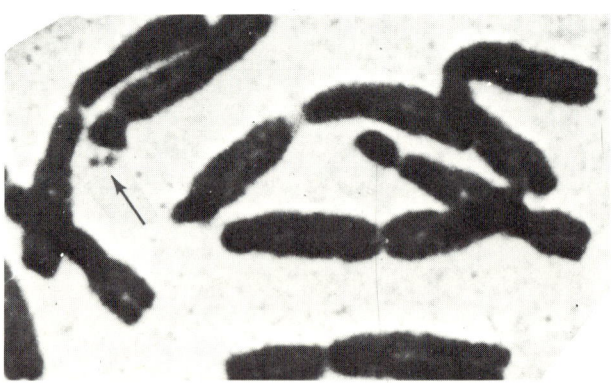

Fig. 8-4. Photomicrograph of some metaphase chromosomes of *Paeonia suffruticosa* clearly showing primary (centromere) constrictions that seem to vary in length and two small satellites (arrow) on the short arm of one chromosome. (Courtesy Dr. A. H. Sparrow, A. Rogers, and S. Schwemmer, Biology Department, Brookhaven National Laboratory.)

and polyploid cells also have one pair or more. The number is not related to the number of chromosomes in each nucleus of a species but is usually species-specific. *Haplopappus gracilis* (2n = 4) has one pair as does *Zea mays* (2n = 20). On the other hand, nucleolus organizers are somewhat like salivary gland chromosome puffs in that they may or may not form. Wheat, for example, has six chromosome sets and eight theoretical nucleolus organizers. Typically, however, only two pairs form and function. If these normally functioning organizers are lacking from a nucleus, as in a micronucleus, then the usually suppressed organizers will form and function.

Nucleolus organizers, as evident in metaphase chromosomes, are regions of chromosomes as are centromeres. At metaphase, therefore, the organizer appears as a short or long constriction in one arm or more of particular chromosomes (Figs. 8-3, 8-4, and 8-7). Almost always they are not terminal in plants but are often terminal in animals and are usually in short arms of submetacentric chromosomes (Longwell and Svihla, 1960); for example, in human beings they are in the short arms of the D and G groups of chromosomes but never in all at once. If the region is short, a mere constriction appears, called the *secondary constriction* in animals. If the nucleolus organizer is long or very long, then the constriction appears as an unstained region up to a number of microns in length, and the short, stained part of the arm distal to it appears as a small ball called a *satellite* (Fig. 8-4). Thus, nucleolus organizers are markers of certain specific chromosomes of the set.

Variations

Chromosomes also vary greatly in size, both length and thickness, although thickness is usually directly correlated with length. The longest known mitotic metaphase chromosomes are those of the plant genus *Trillium*, some of which are more than 30μ long (Fig. 8-5). Average length is about 6μ, and the shortest, often holokinetic chromosomes, are considerably less than 1μ as in fungi, sedges and rushes, and some animals. Not only do chromosome lengths vary among organisms (Fig. 8-6) but they often vary within the set of chromosomes of a species. In man, for example,

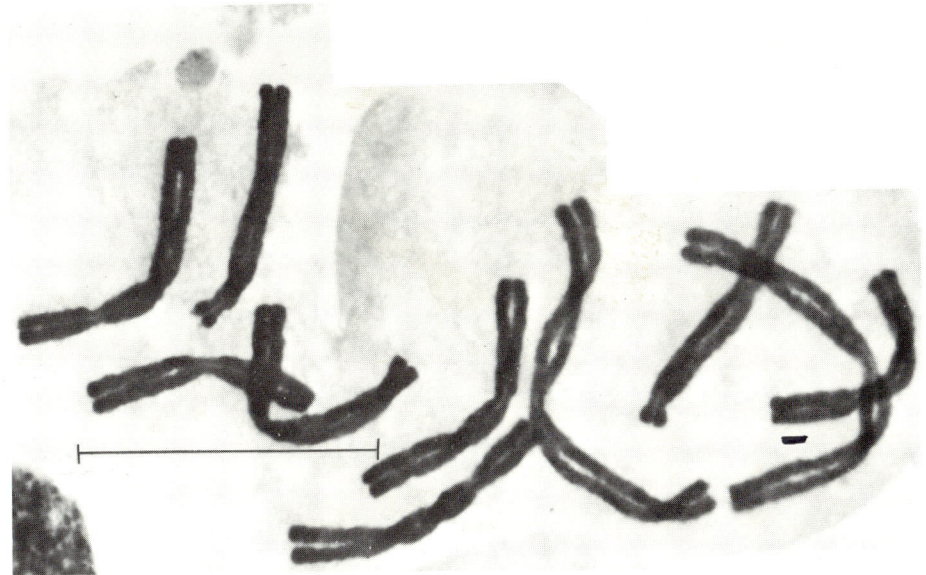

Fig. 8-5. Examples of the largest known metaphase chromosomes, of *Trillium erectum* (2n = 10). These chromosomes range in length from about 22μ to 36μ. The bar indicates 30μ. (Courtesy Dr. A. H. Sparrow, R. C. Sparrow, K. H. Thompson, and L. A. Schairer, Biology Department, Brookhaven National Laboratory.)

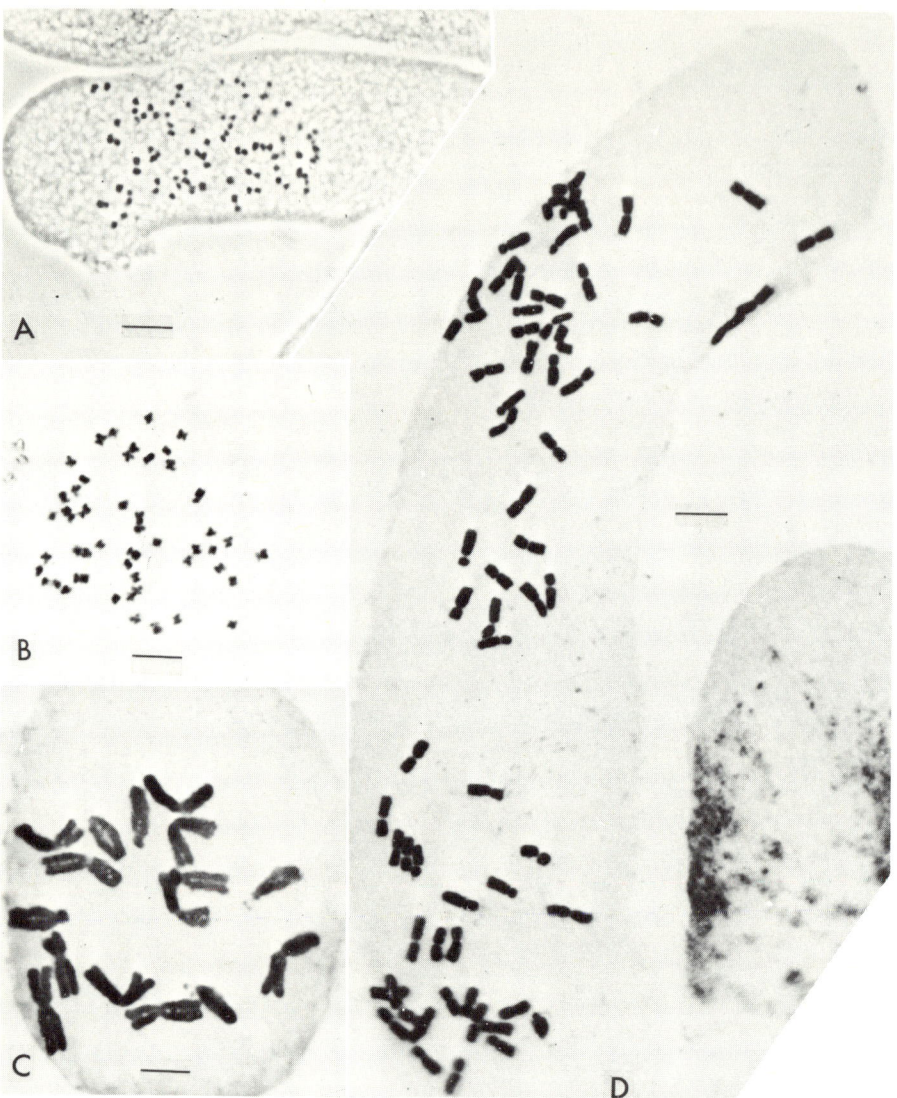

Fig. 8-6. Examples of small (**A** and **B**) and medium-sized (**C** and **D**) plant chromosomes. **A,** *Acer rubrum.* **B,** *Coleus blumeri.* **C,** *Clematis jackmanii.* **D,** *Chrysanthemum arcticum.* Scale, 5µ. Bars indicate 5µ. (Courtesy Dr. A. H. Sparrow, Biology Department, Brookhaven National Laboratory.)

they vary from about 2 to 10µ forming a graded series (Chapter 19). In many forms, such as birds, lizards, the plant *Yucca*, etc., there may be two distinct size classes, large and small. There are also species in which all chromosomes of the set, regardless of size, are approximately the same length. Difference in chromosome sizes between sets of polyploid plants can indicate evolutionary origin. Tetraploid cottons of the New World have one set of small chromosomes from Old World diploids and one set of smaller chromosomes from New World diploids (Skovsted, 1934).

Another variation of metaphase chromosomes is the number in the nucleus or species (Chapter 10). It is generally true that the chromosome number is constant within a species other than the diploid-haploid alteration caused by meiosis and fertilization. Of course, there are aberrations in certain cells or individuals such as an extra chromosome No. 21 in mongoloid human beings or in experimental plants such as the 21 monosomics and the 21 trisomics of wheat. The chromosome number of a gamete, the haploid number of a diploid species, can be called the *basic number* (x). Accordingly, a tetraploid would have 4x chromosome sets, a hexaploid, 6x, etc. (Chapter 12).

Chromosome numbers of species vary greatly, from 2n = 4 to 2n = about 1,200 (Brown and Bertke, 1969). Groups seem to have a restricted range of chromosome numbers; fungi, Diptera, families of flatworms, etc. all have low numbers, whereas ferns, mammals, birds, etc. have high numbers. Within such groups, however, there is variation, usually even within genera such as 2n = 6 to 14 in *Drosophila*. In contrast, some genera are remarkably uniform such as the plant genus *Sagittaria*, all of whose species that have been examined, have not only 2n = 22 chromosomes but all have morphologically the same set or idiogram (Brown, 1946; Baldwin and Speese, 1955).

Although most species have a constant chromosome number, more and more exceptions are being found. Of course, polyploid facultatively apomictic "species" of plants, such as Kentucky bluegrass *(Poa pratensis)*, have many chromosome numbers. But there are sexual species also that vary. The plant species *Claytonia virginica* has chromosome numbers from 2n = 12 to 190 (Rothwell and Kump, 1965; Lewis et al., 1967). Some species of animals also have variable numbers, such as the mole rat *Spalax ehenbergi* with 2n = 52, 54, 58, and 60 (Wahrman et al., 1969), the common shrew *Sorex araneus* with 2n = 22 to 27 (Ford et al., 1957), etc. Some of these, such as *Spalax*, are incipient species, called sibling species, that is, a number of species evolving from one ancestral species.

It is evident that chromosomes and chromosome sets vary remarkably in form, size, numbers, and variability as studied by optical microscopy. Study of such characters provides data for taxonomic and evolutionary conclusions.

FINE STRUCTURE

The fine structure of metaphase chromosomes, from very fine to what can be seen in the optical microscope, is studied by electron microscopy. There has been far less contribution to knowledge from electron microscopic study of chromosomes than was hoped for 15 years ago, but there has been some. At the finest level some observers have seen in sectioned material filaments from 20 to 30 Å in thickness, which represent Watson-Crick double helices. At the next level of size are reports of filaments from 50 to 500 Å in thickness. Probably the best characterized of such filaments are those about 250 Å, as discussed by DuPraw (1968) and others, for burst cells (Fig. 2-2). Such 250 Å filaments must represent coiled 20 to 30 Å filaments plus more protein. These filaments can be called chromatin.

Each metaphase chromosome consists of a mass of such chromatin filament in the form of a metaphase chromosome of two chromatids. Presumably—but certainly not proved—each chromatid consists of one very long chromatin strand. There is disagreement, however, whether this strand extends back and forth from one end of the chromosome to the other many times (Wolfe and John, 1965) or whether it makes only one such trip but zigzags and coils many, many times from side to side in its progress from one end to the other. DuPraw (1965a, 1968) seems to present both concepts in his so-called folded fiber model. He stated that "the chromosomal unit is regarded as a single DNA-protein fiber which is repeatedly folded back on itself both longitudinally and transversely."

Although whole mount spreads of chromosomes examined by electron microscopy (DuPraw, 1965a; Comings and Okada, 1970b, 1970e) do show details impossible to see by other techniques, they do produce rather messy tangles of chromatin strands. How much of this appearance is artifact and how much indicates reliably the original in vivo fine structure of the chromosomes is very uncertain. Strands seem to run in all directions but mostly longitudinally.

On the basis of such technique Comings and Okada (1970e) have concluded that vertebrate chromatids at metaphase are single, that they are not divided into two subchromatids. Neither they nor anyone else has settled this problem of possibly subdivided chromatids yet.

Comings (1970) has discussed subchromatid aberrations (breaks) and subsequent anaphase side-arm bridges and chromatid versus chromosome breaks based upon the conflicting concepts of single-stranded in contrast to multistranded chromatid structure. He and Okada (1970e) have concluded that the single-stranded structure of DuPraw's (1965a) folded fiber organization could account for the findings (mostly chromosome breaks) quite adequately.

They did demonstrate an important concept of diagrammatic representation, namely, that no model of a eukaryotic chromosome should be represented by a straight line. Rather, it should be represented in all its known structural complexity of form to be meaningful and to provide valuable concepts.

Light microscopy and protein digestion has contributed a third picture of metaphase chromosomes (Trosko and Wolff, 1965). This scheme was derived from the chromosomes of the plant *Vicia faba* (Fig. 8-7). It does not show the mass

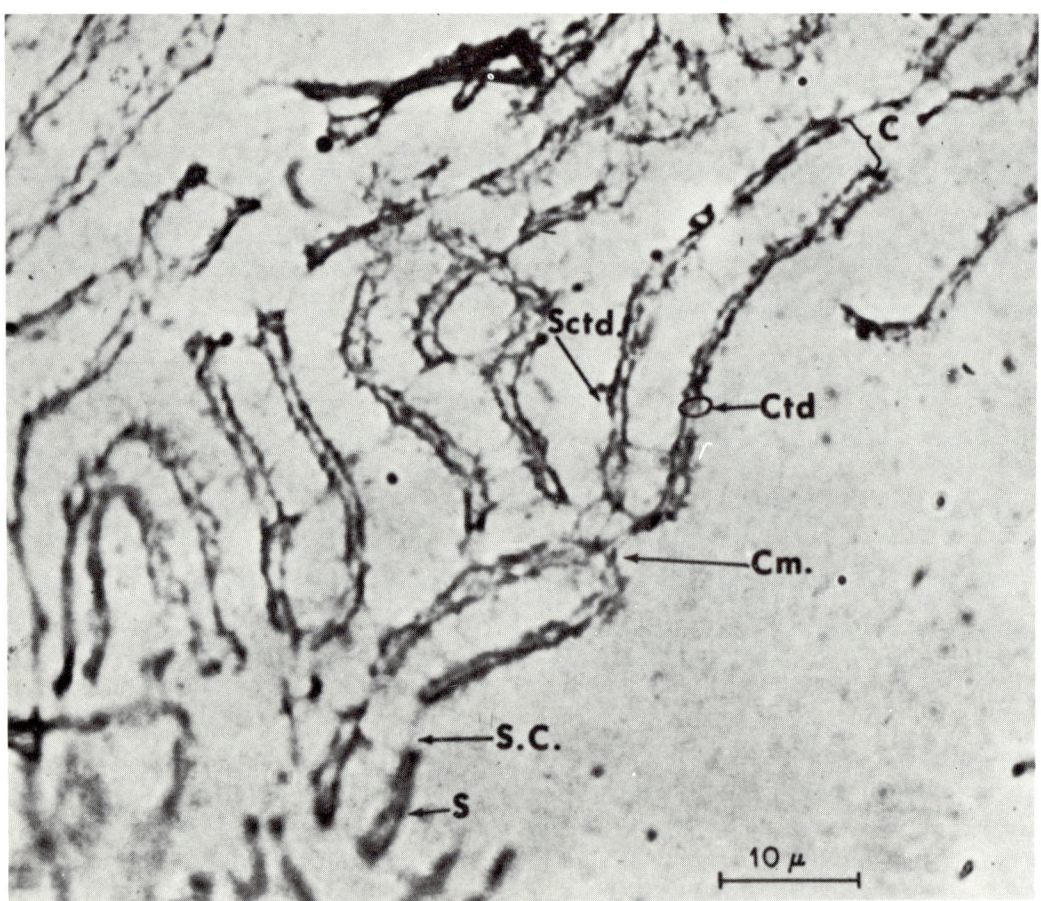

Fig. 8-7. Specially prepared chromosomes of *Vicia faba* that seem to show strandedness not seen in chromosomes as usually prepared. These were removed from the cell, spread on a slide, treated with trypsin to remove proteins, and Feulgen-stained to reveal DNA. Notice especially the long M chromosome that extends from lower center to upper right. In this preparation the M chromosome is almost four times as long as normally observed. The M chromosome is submetacentric, *Cm.*, and has a nucleolus organizer, *S.C.*, and satellite, *S*. Each arm, *C*, consists of two chromatids, *Ctd*, and each chromatid of two subchromatids, *Sctd*. (From Trosko, J. E., and S. Wolff. 1965. J. Cell Biol. **26**:125-135.)

of chromatin filament but rather four or eight strands of the chromosome composing the chromatids, subchromatids, and possibly sub-subchromatids. This latter appearance corresponds well with ideas of chromosome structure of plant chromosomes from more typical light microscopy and other methods of electron microscopy.

In all of these images derived from burst cells and nuclei or protein digestion studied by electron and light microscopy, it must be understood that much of the protein of the chromosome is lost, at least the matrix protein that fills in among the chromatin filament loops and gyres. It is also possible that all three of these models of chromosome fine structure are correct since they were based upon three different sorts of chromosomes. DuPraw studied unicentric animal chromosomes (human and honeybee), Wolfe and John studied the polycentric and meiotic chromosomes of the bug *Oncopeltus*, but Trosko and Wolff studied protein digested plant chromosomes. Cytologists assume that all chromosomes should have essentially the same general fine structure, but perhaps that assumption is false. Certainly there *are* other very distinctive forms of chromosome fine structure such as those of the dinoflagellates, which are quite different (Dodge, 1963a; Zingmark, 1970). But it is too soon yet, there are too few good observations, to close our minds by accepting any present model of chromosome fine structure as dogma. In any case, whatever the arrangement of the chromatin fibril, at metaphase it is in a highly condensed heterochromatic form. Additional aspects of chromosome structure or implications of possible structure are presented in Chapter 5.

Recently two new methods have been developed that stain heterochromatin only; the euchromatin is not stained, especially of metaphase chromosomes. The quinacrine procedure gives fluorescent heterochromatic regions with ultraviolet light, whereas the Giemsa method produces ordinary microscopic stain of heterochromatin (see Chapter 19 for further discussion).

Both methods produce staining on both sides of the centromere (the centromeric heterochromatin) as well as any blocks of heterochromatin in the arms of the chromosomes. The extent and locations of the heterochromatin can vary between homologous and between otherwise indistinguishable nonhomologous chromosomes among individuals of a species and among species of a genus, much like the allocyclic (heterochromatic) clear (H) regions of *Trillium* chromosomes following cold treatment (Kurabayashi, 1958; Darlington and Shaw, 1959) and the salamander *Triton* (Callan, 1942). These new methods have been used so far only with cultured animal chromosomes, for the identification of particular chromosomes of the set (Miller et al., 1971), for evidence of chromosomal polymorphism within the (human) species (Craig-Holmes and Shaw, 1971), and as studies of chromosomal evolution in *Drosophila* (Barr and Ellison, 1971).

Centromeres

The fine structure of the centromere has been determined by optical (Lima de Faria, 1950, for review) and electron microscopy (a recent discussion by Hoskins, 1969). Electron microscopy of centromeres uses two basically different techniques. One is by the standard embedding and then sectioning, such as by Nebel and Coulon (1962), Brinkley and Stubblefield (1966) (Fig. 8-8), Jokelainen (1967), Dietrich (1968), Wilson (1968), and others. The other method (Hoskins, 1965, 1969) (Figs. 8-9 and 8-10) involves tissue-cultured cells that by means of micromanipulation have the prometaphase or metaphase spindle pulled out of the cells with centromeres and condensed chromosome arms attached, a process called *micrurgy*. These masses of spindle and mitotic chromosome are dehydrated, dried, and examined unstained with the electron microscope. This method often separates centromeres from arms since the chromosomes break more easily than do the centromere-spindle unions. As a result, whole centromeres (although dehydrated and dried) are examined, often with attachments to spindle fibers and to chromosome (or chromatid) arms evident. This method is more similar to optical microscopy than is thin sectioning and perhaps provides the best details of structure (Fig. 8-9).

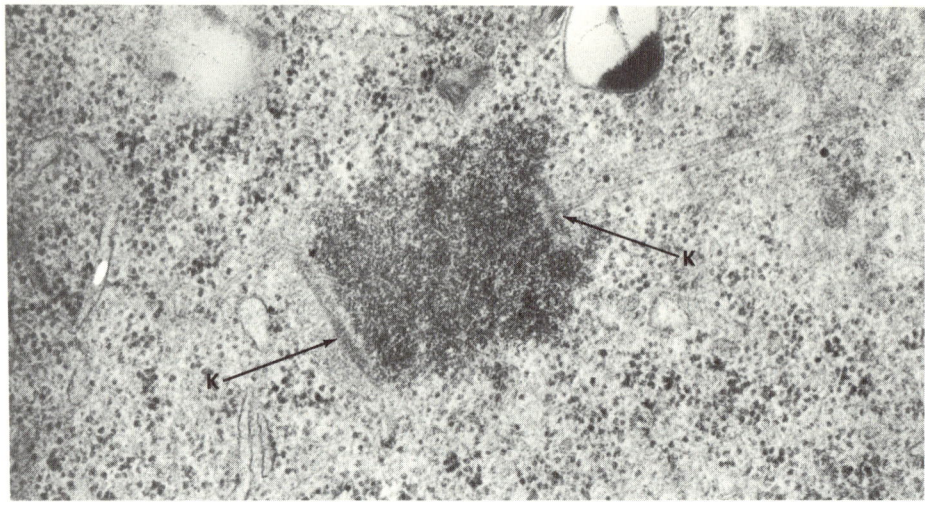

Fig. 8-8. Electron micrograph of a cross section of a centromere from sectioned material. Microtubules can be seen extending from left and right to the two plates (kinetochores, K), one on each side of the chromosomal material of the centromere. It seems that during mitosis the centromere must form the kinetochores temporarily for chromosomal movement. Microtubules of the spindle seem to attach to the two kinetochores of each centromere. (From Brinkley, B. R., and E. Stubblefield. 1966. Chromosoma 19:28-43.)

The tentative model of metaphase centromeres derived from all methods of study and as presented by Hoskins (1969) follows as the best documented at this time. The nature of the centromere during interphase is unknown; it may form as a structure during prophase and more or less cease to exist during interphase.

In optical microscopy the centromere appears as a constriction, and, therefore, it is somehow different from the structure of the arms. Hoskins was able to take pictures of nearly spherical, somewhat heart-shaped centromeres about 0.8μ in diameter that are probably the proteinaceous amorphous "centromeres" that produce the constriction. These are evident because they are not stained as is the nucleoprotein of the arms. This structure he called "matrix," preferring not to give it a formal name. At late metaphase this matrical body divides itself in half in a plane perpendicular to the axis of the spindle. At early anaphase these two halves separate from each other. Attached to each side of the matrix are two spindle fibers from each spindle pole, each spindle fiber consisting of numerous microtubules. Thus, there are four fiber attachments to each matrix that correspond to the four stained granules (Ostergren, 1947b; Lima-de-Faria, 1950; Bajer, 1965) called

Fig. 8-9. Centromeres as whole, pulled-out objects, unsectioned and unstained, prepared by micromanipulation under constant optical microscopic observation, photographed by electron microscopy, printed at two densities to reveal different details, and interpretations drawn (**1b, 2b, 3c,** and **4c**). 2a to 2c and 4a to 4c are of human chromosomes with 8,000 Å diameter centromeres C. **1a** to **1c** is of a mouse 4,000 Å centromere, and **3a** to **3c** is of a Chinese hamster 8,000 Å centromere. The centromeres (C) are more or less hemispheric or valentine-shaped (**3c, 4c**). Various fibers are attached to the centromeres: *CRF*, chromosomal fibers; *SF-1* to *SF-4*, bundles of spindle fibers; *cs-1* to *cs-4*, centromere-spindle fibers; *DBC*, dense black cuff surrounding some *SF* (spindle fiber bundles). The base of the chromosome arm is connected to the centromere by *CRF*, and the centromere is connected to the spindle fiber bundles (*SF*'s) by *cs*'s. The complete structural model is illustrated in Fig. 8-10. (From Hoskins, G. C. 1969. Caryologia 22:229-247.)

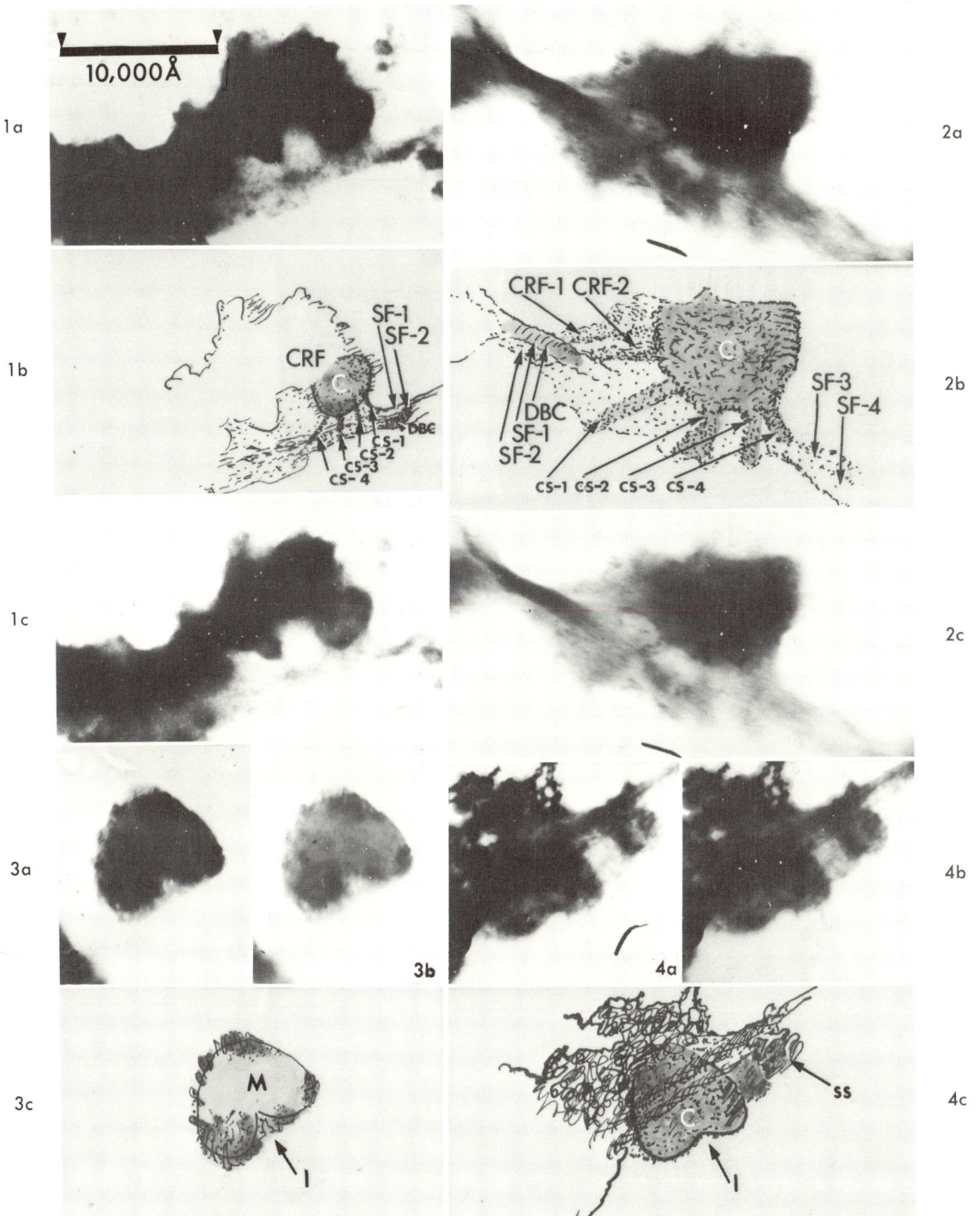

Fig. 8-9. For legend see opposite page.

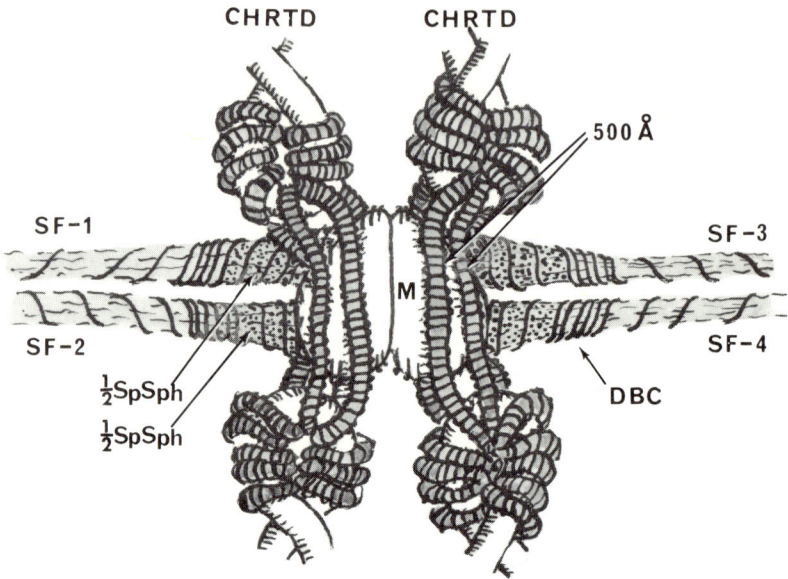

Fig. 8-10. The Hoskins model of the centromere. The chromosome arms (CHRTD) are indicated as extending upward and downward, whereas the spindle fibers (SF-1 to SF-4) extend from the centromere to the left and right. The chromonemata (indicated as 500 Å) extend across the centromere from one arm of one chromatid to the other arm of the same chromatid. The two valentine-shaped centromeres in Fig. 8-9, 3c, called matrix (M), here are present as a pair, one for each chromatid, associated base to base. Where the spindle fiber bundles (four to each of the pair of centromeres) make contact with the matrix, there is a swelling ($\frac{1}{2}$SpSph) called the spindle spherule. The four $\frac{1}{2}$SpSph's associated with each of the two M's form one spherule that is just visible in the optical microscope. At beginning anaphase the two M's would separate horizontally in the figure. (From Hoskins, G. C., 1969. Caryologia 22:229-247.)

spindle spherules or *centromeric chromomeres*. At anaphase two spherules go to each pole as part of each anaphase chromosome and as part of each half of the divided metaphase matrix (Fig. 8-10).

Extending across the surface of or running through the matrix are probably eight strands of 500 Å thick chromosome that are connected to and continuous with the strands of chromonemata of the two chromatids of each metaphase chromosome. That is, each chromatid has four strands of chromosome connecting across each half of the partially divided matrix. This implies that each anaphase chromosome may consist of four separate and distinct cytogenetic strands (also called chromonemata or just strands), as has often been proposed (Sparvoli et al., 1965; Trosko and Wolff, 1965; and many others).

Hoskins interpreted his material as showing other fine fibers, especially 100 Å thick fibers in the region of and attached to the matrix. At the attachment of the spindle fibers to the matrix he believed that these fibers form a collar or cuff of unknown function around the spindle fibers.

This model does lend itself well to the concept that transverse, rather than longitudinal, division of the centromere, or so-called matrix, would produce two functional telocentric chromosomes or two isochromosomes. The isochromosome would still have two spindle fibers attached to a rather normal half matrix and would, therefore, be a rather normal "anaphase" chromosome. The telocentric chromosome would have, however, only one spindle fiber to each pole and would otherwise be abnormal. This might explain the rarity of normally functioning, experimentally produced telocentric chromosomes produced by a transverse break across the middle of the matrix. A break between the centromere and an arm, however, should produce a rather normal centromere and so function normally, as in wheat.

Comings and Okada (1970e) compared the truly telocentric and median centromeres of mouse metaphase chromosomes and concluded that the telocentric represented one half of the median centromere. Sister chromatids are united by many fibers of chromatin, but there is little evidence of centric chromomeres.

Other models of centromeres, such as the one having two *kinetochores*, one on each surface of each metaphase centromere (Fig. 8-8), might represent the special attachment regions of spindle fibers to the matrix and the cuffs reported by Hoskins. This model of a definite and highly complex structure of the localized centromere does make difficult any conception of the nature of the "diffuse centromere" (Buck, 1967) that, presumably, evolved from the localized centromere and even more difficult a conception of the gene-controlled "neocentromeres" of maize.

FUNCTION

The function of the metaphase chromosome is at most only slightly genetic. There is certainly no DNA replication and essentially no or absolutely no RNA transcription detectable by incorporation of radioactive nucleotides. The only replication likely would be of genes in the partially condensed centromere, or, by anaphase, in the nucleolus organizer regions. Certainly nucleoli start to form in *Luzula purpurea* during anaphase (Castro et al., 1954). The formation of the protein kinetochore plates or matrix occurs during late prophase when the chromosomes are inert, but the RNA for their synthesis (and synthesis of the monomers for the spindle fibers) may well have been completed during interphase. But it does seem likely that there must be some sort of prometaphase and metaphase control of spindle formation and function. Nuclear envelope breakdown, nucleolar "disintegration," centriolar movement, and mitochondrial arrangement in some cells also occur during the period of division when chromosomes are condensed and essentially inert.

As far as it is known, however, the only function of condensed chromosomes during division is orientation, disjunction, and movement.

KINDS OF CHROMOSOMES

In addition to the variations in size, shape, and number as just discussed are other sorts of chromosomes. One sort consists of *sex chromosomes or heterosomes* (Chapter 7). These can vary from being no different from the typical *autosomes* as in culicine mosquitos (Breland and Gassner, 1961) in which sex is determined by a single pair of alleles (Mm = male; mm = female) (Baker and Rabbani, 1970) to chromosomes that behave very differently from autosomes, such as the sex chromatin body in female mammalian nuclei and the heterochromatic condition of sex bivalents just before and during meiosis. In general, X chromosomes contain many genes, but Y chromosomes usually carry few. As a result, Y chromosomes can be evolved out of existence to give the XO condition.

Supernumerary chromosomes (Chapter 9) constitute a more distinctive class, which is common among flowering plants but is rare in animals (Brown and Bertke, 1969). Ostergren (1947a) proposed that supernumerary chromosomes may be parasites having genes for their own survival but may be damaging to the host organism only if present in excessive numbers. Whether they are parasites or not, their characteristics are more like those of parasites than of normal cell organelles. They are not necessary to the species, none are known to be beneficial to their host, they occur in only a few individuals or populations of a species, they are not harmful in the usual small numbers per cell (from one to three) but in greater numbers (from five to fifteen or more) they greatly reduce vigor and reproduction, and in various ways they have adapted to their hosts for their own survival and increase (see Chapter 9).

On the other hand, they look like small chromosomes, have centromeres, and behave like chromosomes. They go through the mitotic cycle like heterochromatin, they replicate and disjoin at anaphase, during meiosis they pair if not synapse among themselves but not with normal chromosomes, but they can have translocations with normal chromosomes and can evolve into two or more different morphological types.

Their adaptation to their host (flowering plants and at least one flatworm) is to its life cycle. They often eliminate themselves from somatic tissues, such as roots, but remain in the germ line of animals and stems of plants. In the pollen grain divisions they double their numbers by nondisjunction so that all, in double number, get into the male gametes.

Some cytologists studying mechanisms by which chromosome numbers (and centromeres) can be increased have postulated that supernumerary chromosomes might be a source of centromeres, by translocation, for such chromosome increase. Although such translocation has been observed in maize, there is no proof that centromeres from supernumerary chromosomes have ever contributed the required centromeres for chromosome number increase. Rather, it is generally assumed that supernumerary chromosomes originated as remnants of chromosomes, consisting of a centromere and unnecessary heterochromatin that remain following translocation of the necessary gene-containing segments to other chromosomes.

SILENT REGIONS

It is well known from cytogenetics that there are long and short regions of chromosomes where *detectable* genes are very few or completely lacking, in spite of intensive efforts to find them. Such "geneless" regions are often described as "silent" although essential genes are located in them, as deficiencies within silent regions have demonstrated (Rhoades, 1968; Khush and Rick, 1968). In maize and tomato such silent regions seem to be most common in distal regions of chromosome arms. Phillips (1969) reported silent regions at the ends of maize chromosome arms, especially long arms, such as 3L, 5L, 6L, and 7L, and 5S. The silent region in the long arm of 7L amounts to 38 cross-over units and about 24% of the cytological length of the arm. The silent region in 6L occupies approximately the terminal third of the arm. Khush and Rick (1968) and Rick (1971) found silent regions in tomato especially in terminal regions, and chromosome No. 12 is almost completely silent. Chromosome No. 8 of maize has very few genes for its cytological length. Cytogenetic maps of the chromosomes of *Drosophila melanogaster* also show genes very irregularly and nonrandomly distributed with some short and some rather long silent regions, such as near the end of one arm of chromosome No. 3. Nonrandom distribution of genes on chromosomes and among chromosomes of the set seems to be a real and general condition in nature.

As Rick (1971) emphasized, some reports of silent areas result from failure to detect genes located there or from the lack of markers suitable for locating genes on chromosomes, but certainly there are regions carrying important and vital genes that cannot be detected and characterized by specific phenotypic effects.

It is evident that so-called silent regions are not necessarily regions of heterochromatin. Most extensive silent regions are located in distal regions of chromosome arms whereas extensive regions of heterochromatin are usually located in proximal regions, and heterochromatin does not seem to contain vital genes, as silent regions do.

Differential mutability

Rick (1971) has proposed the hypothesis of differential mutability to "explain" a correlation observation in tomato and other organisms. The hypothesis states that "regions of higher marker density are subject to increased mutation rate." For example, of the known genes in the proximal regions of genetic maps of arms, a higher percentage exists there of multiple alleles. He relates mutability of genes to heterochromatin and cites an unpublished finding by M. M. Green in *Drosophila* that when gene W is moved away from its normal position close to some interstitial heterochromatin, the gene loses its high mutability rate. He concludes that the nonrandom gene distribution reflects some factor in addition to differential recombination.

So-called silent genes, genetic loci that produce no demonstrable products, seem likely in organisms. Boyer and associates (1971) have proposed a silent gene in apes and man, termed *hemoglobin* ^3L, which has mutated in at least one chimpanzee and one gorilla to a condition in which it *can* produce an unusual L chain of hemoglobin. This locus is assumed to be separate from the locus for A-α and to have persisted for millions of generations.

CONCLUSION

Metaphase chromosomes are tightly packaged linkage groups of the various sorts of genes in an inactive form for ease of distribution, with very little error, to the two daughter nuclei during mitosis. Like all other structures of organisms, they are susceptible to gene action and evolution. Thus, they can be made heterochromatic or not and can change in gross structure and number in evolutionary time. Although a picture of the fine structure of interphase, mitotic, and meiotic chromosomes is slowly developing, cytology has far to go before a model in accord with the semiconservative scheme of replication will be synthesized from the apparently disordered strand organization presently available. Nevertheless, mitotic chromosomes are useful cytogenetic structures, and such uses are discussed in detail in subsequent chapters.

CHAPTER 9

Supernumerary chromosomes

> The decisive factor for the selction of them would be whether they were *useful to themselves*. They are probably in many cases leading an exclusively *parasitic existence* in the plant.
>
> G. Ostergren, 1947a

As Longley (1927) pointed out, from about 1910 to 1925 the chromosome number of *Zea mays* (maize) was uncertain because 2n numbers from 20 to 25 or more and n numbers from 10 to 15 were reported. Longley, as had some of the immediately preceding cytologists, believed that $n = 10$ was the true chromosome number but in some strains a few plants had more. He found in his 1927 study that $n = 10$ to 15 occurred, but he considered those chromosomes in excess of 10 as supernumerary. There was, however, no implication that they were any different from regular chromosomes at meiosis; they were about comparable in size to the smallest regular ones. He did show that the number of these supernumerary chromosomes varied in progenies from the number expected from the number in the parents of crosses.

Crosses of $n = 11 \times n = 11$ gave 6 F_1 plants with 20, 19 with 21, 13 with 22, 30 with 23, and 5 with 24 chromosomes. That is, some progeny had fewer supernumeraries (20 and 21 chromosomes) than the sum of those in the parents, some had more (23 and 24). There were also more progeny with greater numbers of supernumerary chromosomes from reciprocal crosses if the pollen parent had the supernumerary chromosomes.

Blackwood (1956) confirmed these results in maize. Smith (1968) reported in a species of *Haplopappus* that the cross 0 supernumeraries (female) × 1 in the male gave equal numbers of 0 and 2, whereas 1 female × 0 male gave equal numbers of 0 and 1, but 6 × 6 gave plants with 1, 2, 3, 4, and 6.

In 1928 Randolph examined the chromosomes of maize and indicated that the supernumeraries were different from the true, or A, chromosomes. Depending on size he called the supernumeraries B-F chromosomes. Subsequently, these peculiar chromosomes have often been called "B-chromosomes" (Rhoades, 1955; Battaglia, 1964; Blackwood, 1956; and others), "accessory chromosomes" (Bosemark, 1956; Frost, 1959b; others), and numerous other names (Battaglia, 1964). Thus, supernumeraries achieved unique status as a special kind of chromosome, not only in maize but subsequently in many species of plants and a few animals. For example, Randolph (1941) could find no genetic evidence that the maize supernumeraries contained any of 46 genes that were known to occupy loci in 17 of the 20 arms of the A-type chromosomes. He further discovered what has developed as a rather general characteristic of supernumeraries, that they often, at specific times or in specific divisions, regularly undergo nondisjunction.

Randolph (1941) reported that in maize this

nondisjunctional division occurs in the pollen grain when the generative nucleus divides to form the two male gamete nuclei, the second pollen grain division. Blackwood (1956) has confirmed this. By this irregular division of supernumeraries two sorts of male gametes occur in one pollen grain. Thus, a plant with one supernumerary produces eggs (haploid) with one supernumerary but male gametes with two or none. Therefore, progeny should have one or three. Breeding results of maize by Longley, Randolph, and Blackwood indicated that nondisjunction occurs to some extent also at anaphase I. Blackwood cited for a plant with 2 B's at anaphase I in 103 microsporocytes: 84% had 1 and 1 segregation, 13% had 0 and 2, 2% had 1 and 2, and 2% had 2 and 2. Therefore, with some nondisjunction at anaphase I and a great deal at the second pollen grain division, male gametes from a plant with two will have from zero to four supernumeraries and F_1's will have from zero to six or more. Longley and Blackwood reported data that show that F_1's derived from a cross with the pollen parent having the supernumeraries contain plants with greater numbers of supernumeraries than if the female parent contains the B's. The implication is that nondisjunction occurs only or more frequently during microsporogenesis than during megasporogenesis. Thus, it is possible naturally or experimentally for numbers of supernumeraries to increase and otherwise change from generation to generation. In maize, for example, plants with as many as 34 supernumeraries per cell were produced (Randolph, 1928; Rhoades, 1955), in *Tradescantia edwardsiana* (Fig. 9-1) as many as 9 were accumulated (Brown, 1960), and in *Haplopappus validus* (Smith, 1968) as many as 9 also accumulated.

Battaglia (1964) considered the term nondisjunction inadequate to describe these irregular divisions that seem to tend to increase supernumerary chromosomes or to keep them in the germ line (or plant equivalent) but eliminate them from the soma. He suggested the term "preferential distribution," which can be mei-

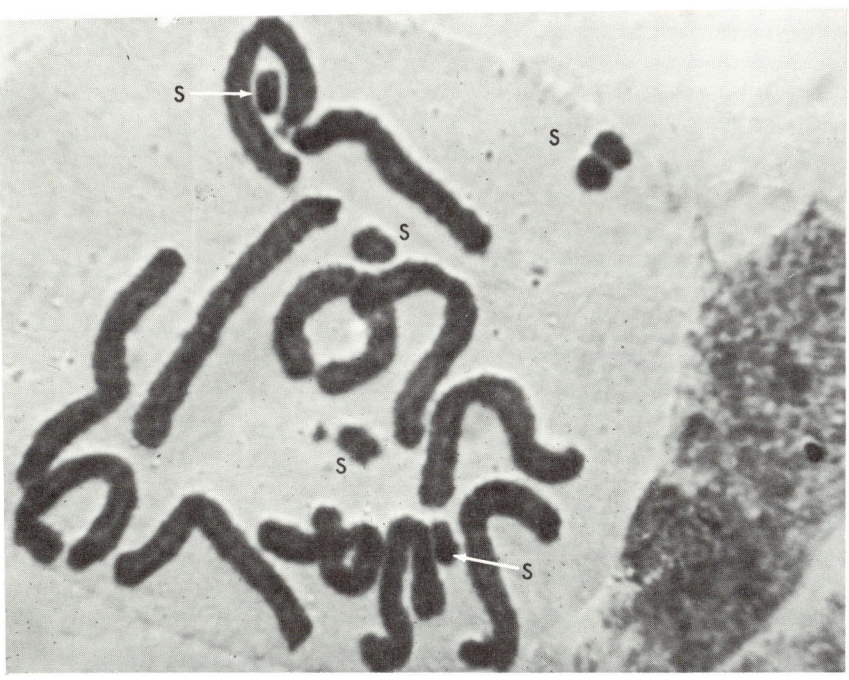

Fig. 9-1. Supernumerary chromosomes in a root tip cell of *Tradescantia edwardsiana*. Among the twelve normal, long, metaphase chromosomes are a number of small supernumerary chromosomes (S); the one in the middle of the figure has a definite acrocentric constriction. (From Brown, W. V. 1960. Southwest. Nat. **5**:49-60.)

otic or postmeiotic. As examples of meiotic preferential distribution he cites *Lilium callosum* (Kayano, 1962), *Trillium grandiflorum* (Rutishauser, 1956), and *Plantago serraria* (Frost, 1959a). In these cases the nondisjoined supernumerary passes to the anaphase I pole of the megasporocyte division so that two are present in from 75 to 85% of eggs rather than 50%, which would occur if they moved randomly at anaphase I.

In the flatworm *Polycelis tennis* there is an extra division of the supernumerary in the ovary during meiosis that doubles the number in the egg, thereby tending to increase the number each generation.

Examples of postmeiotic preferential distribution are more common and occur in the pollen grain or embryo. The most common subgroup, present in at least 10 species, Battaglia called the Secale type. The first pollen grain (or microspore) division produces the vegetative and generative nuclei (Fig. 15-5), and the generative divides later as the second pollen grain or pollen tube division resulting in the two male gametes present in each pollen grain or tube. In Secale-type nondisjunction at the first pollen grain divisions, the double supernumerary preferentially goes to the generative nucleus. Thus, each male gamete after the second pollen grain division receives double the number present in the microspore, and both gametes are the same.

Battaglia's Zea type has already been discussed, consisting of nondisjunction at the second pollen grain division and resulting in different gametes, one with few or no supernumeraries and one with an increased number. This alone is not enough to increase the number of B's in the population; but in maize, at least, a further effect of the supernumeraries is *preferential fertilization*. That is, the gamete having the supernumeraries unites more often with the egg than does the sister gamete. Roman (1948) reported about 60% of fertilizations were accomplished by the supernumerary-containing gamete, and Blackwood (1956) also reported this phenomenon.

Preferential distribution also occurs in the development of the embryo in a number of plant species and in a flatworm. Battaglia calls such cases "mitotic elimination," but they are also preferential distribution since they are regular in occurrence. In some plants supernumeraries may not be found in any roots or even leaves although they are present in stems and flowers. Examples are: *Sorghum purpureum-sericeum* (Janaki-Ammal, 1940; Darlington and Thomas, 1941) and *Xanthisma texanum* (Berger et al., 1955); but in *Poa alpina* the supernumeraries are present in primary roots and stems but as the plant grows they are eliminated from the leaves and adventitious roots (Muntzing, 1948).

A so-called boosting mechanism has been reported in *Crepis capillaris* (Rutishauser and Rothlisberger, 1966) and probably also in *C. pannonica* (Frost, 1960). A seedling that starts out with a few (from one to three) supernumeraries per cell as it nears flowering is found to have more and more cells with twice as many B's as in the seedling. In the receptacle of the flower head 80% of cells showed the doubled number, and in the florets the percentage of cells with the boosted number was found to be 90%. This is another mechanism for increase of these peculiar chromosomes.

In *Tradescantia edwardsiana* Brown (1960) found various numbers of supernumeraries in root tip cells and microsporocytes (Fig. 9-2). Crossing plants with low numbers (from one to three) gave F_1 and F_2 plants with varying numbers to nine. In this species the behavior was irregular even for supernumeraries. Vernon and Witkus (1970) found in the Liliaceous species *Puschkinia libanotica* that 72% of plants had no supernumeraries, 19% had 1, 6% had 2, 1.5% had 3, 0.7% had 4, and 0.7% had 5.

The gist of all of these examples is that supernumerary chromosomes as a group are able to exploit nondisjunction in various critical cells or structures of the plant or animal to boost their numbers during reproduction. But since they do not seem to be present in many plants of a species nor in great numbers in any, it must be assumed that there must be some means of eliminating them if only that they are somewhat detrimental to their host, or, as Smith (1968) proposed, in great numbers they adversely affect the ability of pollen tube growth in competition with pollen tubes with

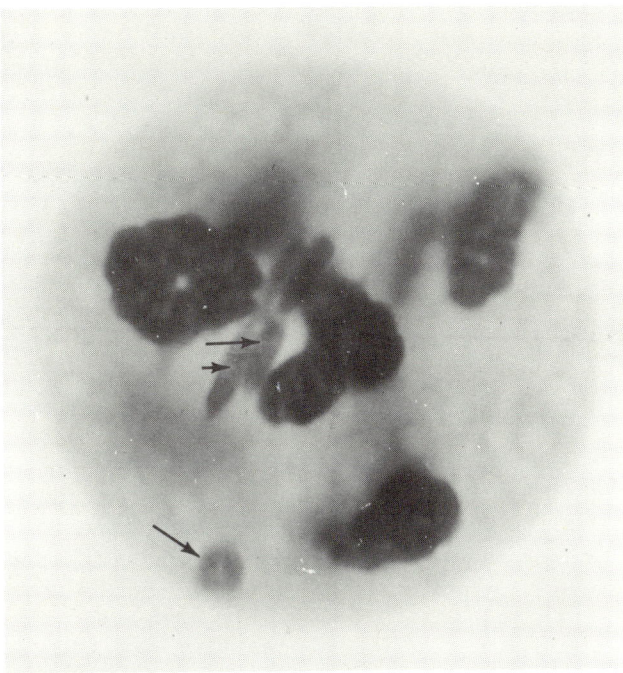

Fig. 9-2. Supernumerary chromosomes during meiosis of *Tradescantia edwardsiana*, at metaphase I of a microsporocyte. These chromosomes are much thinner and paler than the normal chromosomes. (From Brown, W. V., and E. M. Bertke. 1969. Textbook of Cytology. The C. V. Mosby Co., St. Louis.)

small numbers or none at all. This is certainly true in large numbers such as in *Tradescantia edwardsiana* (Brown, 1960) and in maize (Rhoades, 1955), but even in very small numbers (Ostergren, 1947a) they seem to have adverse effects.

MECHANISM OF NONDISJUNCTION

The method by which nondisjunction is achieved has been studied in a few plants, especially in rye, *Secale cereale*. In that species a few different sorts of B chromosomes occur, and the smaller forms are derivatives of the common "standard B." This standard B possesses a knob near the end of the long arm. If this knob is not present in a cell, the derived forms of B's in the cell cannot perform nondisjunction. But if the cell contains such a knob, all B's in that cell, whether they are knobbed or knobless, are able to nondisjoin (Lima-de-Faria, 1962). Similar presence of a mass of heterochromatin in the supernumerary for nondisjunction has been reported in *Festuca* and *Phleum* (Bosemark, 1956a and b), and Battaglia concluded that the large amount of heterochromatin in the B chromosome of maize is required for nondisjunction. Bianchi et al. (1961) concluded that the heterochromatin of the B is probably capable of inducing nondisjunction of even A chromosomes in the same cells, and Hanson (1962) reported that B's influenced crossing-over in chromosome No. 3 of maize.

Thus, it seems that a large amount of heterochromatin (and supernumeraries are usually mostly heterochromatin) is related somehow to nondisjunction of themselves and others within the same cell. It is unknown how heterochromatin per se is able to achieve this unless, as Lima-de-Faria (1952) found in rye, at metaphase the centromeres have divided but the sister chromatids are "stuck together" at a proximal region (of heterochromatin) on each arm. With respect to such boosting mechanisms, as has been reported in *Crepis* (Rutishauser and Rothlisberger, 1966), in which doubling of accessories increases during ontogeny, there may be a relationship between that process and the progressive "meiosization" during ontogeny of A chromosomes, as reported by Stack (1969, 1971).

EFFECTS OF SUPERNUMERARIES

Generalization about supernumerary chromosomes must be cautious because of their diversity. Some may be little more than trisomics whereas most are, and all should be, unique. Most are partially or largely heterochromatic whereas others are described as euchromatic. Some, as in maize, rye, etc., are probably centric fragments of standard types, and certainly they must vary greatly in genetic content and phenotypic effect. But their phenotypic effects must be related to the genomes with which the B's are associated. Thus, results vary among tests of the same plant species. Almost all results, however, indicate that in small (from one to three) numbers B's are without detectable effect or they produce slightly unfavorable effects. Almost universally, however, bad effects result from abnormally great numbers in experimental material; such numbers are not found in nature.

A few reports of questionably favorable effects have appeared, such as Fröst's (1959b) with *Centaurea*, Rutishauser's (1956) with *Trillium*, and Bosemark's (1957). These are all borderline or questionably favorable (Battaglia, 1964).

Unfavorable effects of few supernumerary chromosomes are equally few (Ostergren, 1947a) but there are many reports of deleterious effects of abnormally great numbers. Battaglia (1964) listed eight of such cases. The classical case of deleterious effects of increased numbers of B's is in maize (Rhoades, 1955). The so-called normal number in strains is from 1 to 3, but as many as 10 or 15 seemed not to affect the vigor or reproduction. From 15 to 25 there is a direct correlation between increasing number and decreasing vigor, seed set, and fertility. Although a few plants with from 30 to 34 were found, they were of very low vigor and were completely sterile. There is no other species known that can endure such great numbers as maize. Rarely in other species can more than from 5 to 10 be found or perhaps tolerated, and

they affect plant vigor, pollen fertility, and seed setting. Smith (1968) reported that some plants of *Haplopappus validus* subspecies *graniticus* had one or two supernumeraries. He was able to increase them because of nondisjunction at the first pollen grain division. As many as six supernumeraries in a plant apparently had no morphological effects, but any increase in number greater than two per cell did affect fertility as judged by pollen stainability. He proposed that pollen with greater numbers (from three to nine) were at a disadvantage in competition with pollen tube growth of pollen with few or no supernumeraries. Ostergren (1947a) considered that supernumerary chromosomes, like well-adapted parasites, occur only in such small numbers that they and their host are adapted to each other and the host is little, if any, affected by them, that they "injure the plant as little as possible."

OCCURRENCE

In animals supernumeraries have been reported in a very few groups except in grasshoppers, a couple of species of flatworms, a snail, a few bugs, a few beetles, a few other insects in 6 orders and 17 families. More than half of all insect species having supernumeraries are short-horned grasshoppers of the family Acrididae (White, 1954; Battaglia, 1964).

In the flatworm *Polycelis tennis* the supernumeraries are eliminated from the soma, and there is an extra division during meiosis that increases the number; but even a few are detrimental to the individuals.

In mosses it is possible that there may be supernumeraries, but observations are difficult and are confused by the m (for minute) chromosomes, which are very common. These m chromosomes are too constant within species to be supernumeraries but they and sex chromosomes are probably confused.

Supernumeraries have been reported in no other group of plants except the angiosperms (flowering plants), in which they are *very* common.

Supernumerary chromosomes have been reported in more than 475 species of 163 genera of 42 families of angiosperms (Brown and Bertke, 1969). It is possible that if enough individuals of all species of angiosperms were examined supernumerary chromosomes would be found in most of them. Of course, they occur in only a few individuals of any species, and doubtless there are species in which they do not occur.

Longley (1938) reported in maize a relationship between presence of supernumerary chromosomes and number of knobs visible at pachytene. The number of knobs in maize varies from none to 16 in single plants and among plants in 22 different positions (Rhoades, 1955). Longley found that, in strains from 33 North American Indian tribes, in those strains having more than seven knobs, only 6.2% of plants had B chromosomes whereas in strains having fewer than seven knobs, 20% of plants examined had B's. Knobs, like B's, consist of heterochromatin; therefore, there was a reciprocal relationship between these two forms of heterochromatin. The meaning and significance, if any, of this relationship is unknown.

ORIGIN OF SUPERNUMERARIES

It is *assumed* that supernumeraries represent remnants of typical chromosomes that have persisted and perhaps evolved. There is very little evidence for this sort of origin but none for any other origin. Perhaps the best evidence is in *Drosophila*, a genus devoid of supernumeraries except in the only $n = 7$ species, *D. trispina*, and in $n = 5$ *D. putrida II* (Patterson and Stone, 1952, p. 138 and 179). These "free" centromeres are largely heterochromatic. In *D. putrida II* from Florida ($n = 5$) there is a pair of small chromosomes that are only slightly longer than the common "dot" chromosome in the genus; but this pair is lacking in *putrida I* from Texas ($n = 4$). It is assumed that *putrida* arose from an ancestor with $n = 5$ (4 large plus dot). By centric fusion resulting from breaks near the centromeres of two acrocentric chromosomes, two new chromosomes were formed. One is represented by a long metacentric pair, and the other is the pair of very small "supernumerary" chromosomes of *putrida II*, each of which consists of a centromere and two very short heterochromatic arms. This so-called supernumerary was presumably lost in the evolution of *putrida I*. Patterson and Stone also consider that the only known case of chromosome increase from the primitive number for the genus, $n = 6$, to $n = 7$ is in *D. trispina*. It was achieved by a similar centric fusion followed, however, by the loss of the *large metacentric* and so, essentially, achieved the introduction of that supernumerary into the original $n = 6$ set. By many definitions these small chromosomes in these two species of *Drosophila* are true supernumeraries. They are small, they are heterochromatic, they may carry few detectable genes, and since they do not occur in *putrida I* they may be unnecessary. However, since they occur in all individuals of *trispina* and all of *putrida II*, they must be considered as not typical supernumeraries.

Wedberg et al. (1968) also proposed a relationship between translocations and supernumerary chromosomes in *Clarkia williamsonii*. Populations in yellow pine forests have low frequencies of translocations (from 0.0 to 0.1%) and less than 15% of individuals have a supernumerary chromosome, whereas plants at lower elevations and in foothill woods have a higher frequency of translocations (from 0.3 to 0.4%) and more than 15% of individuals have a supernumerary chromosome.

The accumulated evidence from *Drosophila*, *Clarkia*, *Haplopappus*, etc. does certainly indicate that the most likely source of supernumeraries is that they are the nonessential centric fragments remaining after unequal translocations.

Hovin and Hill (1966), however, reported in a number of species and hybrids of *Lolium* that at meiosis in PMC's few or no supernumeraries were seen at diakinesis whereas at metaphase I and anaphase I they were present. They proposed, therefore, that "they originated by misdivision of A-chromosomes during prometaphase I." They also reported progeny with supernumeraries from parents that lacked them, as though they could be produced frequently from one generation to the next in species and hybrids essentially devoid of translocations. They observed further that they could be lost from one generation to the next. These conclusions need to be reinvestigated since they conflict with most other conclusions.

Estimates of when supernumerary chromo-

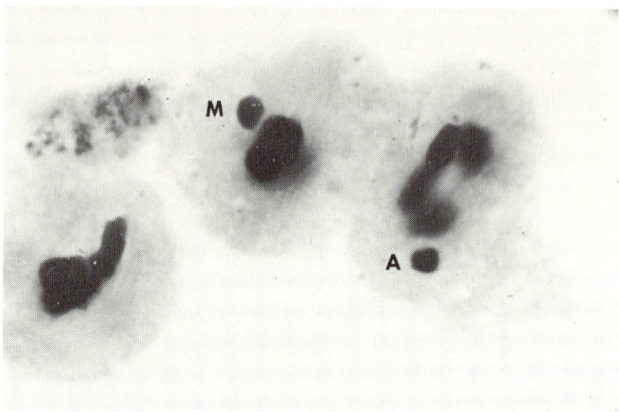

Fig. 9-3. Supernumerary chromosome at metaphase-anaphase I of *Haplopappus gracilis* (n = 2). The supernumerary at metaphase *(M)* is undivided, and at anaphase *(A)* it goes to one pole of the spindle.

somes arose in a particular species or genus vary from previous to the origin of the species, for example *Anthoxanthum aristatum* (Ostergren, 1947a) and rye (Muntzing, 1957), to very recently, as a common product of a common phenomenon. The phenomena usually credited with producing B's are (1) centric fusion (reciprocal translocation) as already discussed in *Drosophila* and by Jackson (1962) in the origin in *Haplopappus gracilis*; (2) misdivision and fragmentation producing tiny telocentrics of isochromosomes (Cleland, 1951, in *Oenothera*; and Whitaker, 1936, in *Tradescantia*); and (3) subsequent modification of a normal species into which B's are introduced by hybridization (Li and Jackson, 1961). Subsequently, evolution of such remnants that persist could produce the parasitic nature proposed by Ostergren (1947a).

Certainly the discussion of supernumeraries in this chapter does show that even if they are not parasites according to some definitions, many of their characters are parasitelike and true parasites by other definitions. They do seem to have "genes" for their own survival, and they are not beneficial even in small numbers to their host. But they may play a part in chromosome evolution by providing centromeres for increase in chromosome number.

CHAPTER **10**

Chromosome numbers

Every species of plant or animal has a fixed and characteristic number of chromosomes, which regularly recur in the division of all of its cells; and in all forms arising by sexual reproduction the number is even.

E. B. Wilson, 1896

As the excerpt from Wilson indicates, a generalization about the numbers of chromosomes and species was possible before the end of the nineteenth century. Wilson listed the chromosome numbers of 21 animals (including man with 2n = 16!) and three species of plants. During subsequent years the chromosome numbers of tens of thousands of species of organisms of all groups from even viruses to mammals and flowering plants have been determined. From time to time lists of the chromosome numbers of species of particular groups have been prepared (Makino, 1951, all animals; Hsu and Benirschke, annually, 1967-1970, mammals; Darlington and Wylie, 1956, flowering plants; Moore, annually, 1956-1968, all plants and fungi).

The topic of chromosome number was briefly introduced in previous chapters, where it was indicated that 2n numbers range from 4 to more than 1,000 but usually lie between 10 and 100. Very large chromosomes, such as those of salamanders, grasshoppers, *Tradescantia*, *Trillium*, have 2n ranges from about 10 to 24, but small chromosomes range from few (fungi and mosses) to many (decapod Crustacea). Viruses and bacteria have only one and might be considered to be haploid, but the n = 1 form of *Parascaris equorum* really seems to have compound chromosomes. There is no species of Eukaryota with n = 1.

RECOMBINATION INDEX

Chromosome number is, of course, correlated to the number of genetic linkage groups. Presumably, the more linkage groups, the greater variability in the species due to sexuality and crossing-over, and, therefore, the greater the evolutionary potential. This relationship led Darlington (1937) to propose the *recombination index*, which is "the sum of the haploid number of chromosomes and the average chiasma frequency of all the chromosomes in a meiotic cell." The higher the recombination index, the larger the number of "blocks" of genes segregating at meiosis. For example, in a species with a haploid chromosome number n = 3 and three chiasmata per bivalent the recombination index would be 3 + 9 = 12. On the other hand, in a species with n = 50 and an average of one chiasma per bivalent the index would be 50 + 50 = 100. Presumably, the species with a recombination index of 100 would mix up blocks of genes eight times as much as the species with an index of 12, from meiosis alone.

It should not be assumed, however, that there is in general a correlation between rapid evolution and high recombination index. For example, birds with high chromosome numbers and flies with low are both evolving rapidly, whereas horsetails (*Equisetum*, n = 108) with high chromosome numbers and mosses with low are both evolving slowly.

AMOUNT OF DNA

Chromosome numbers are not necessarily related to the amount of DNA contained in a cell. *Trillium*, for example has 10 chromosomes totaling about 280μ in length by about 1.5μ in diameter, whereas *Eleocharis ovata* has 10 chromosomes totaling about 20μ in length by less than 1μ in diameter. Many similar extreme comparisons can be made, such as salamanders and frogs, or lungfish and teleosts; but even among closely related genera or among species of a genus (excluding polyploidy) various amounts of DNA have been recorded for the same chromosome number. Rather exactly the amount of DNA is related to the volume of chromosomes as revealed by general chromosome staining procedures (Miksche, 1967; Rees et al., 1966).

In the plant genus *Aster*, Huziwara (1959) illustrated *A. tripolium* (2n = 18) with chromosomes averaging 9μ long by 1.6μ thick, whereas *A. modestus*, which is also 2n = 18, has chromosomes 3μ long by 0.8μ thick. McLaren et al. (1966) found two "sibling forms" of the copepod *Pseudocalanus*, both of which have 2n = 16. The chromosomes of the large form are much longer and thicker than those of the small form and nuclei have seven times as much DNA. The authors considered this a form of polyteny. In the gymnosperm genus *Pinus*, Miksche (1967) reported 138 picograms of DNA per cell in *P. resinosa* but only 75 picograms of DNA in *P. contorta*, both species having the same chromosome number. In the toad, *Bufo*, Ullerich (1966) reported about 50% more DNA in *B. viridis* than in *B. calamita* in spite of their having the same chromosome number.

Martin (1968) has demonstrated great differences in DNA amount within the genera *Vicia*, *Microseris*, and *Plantago*. *Plantago lanceolata* has 3.6 times as much as *P. bigelovii*, and *Microseris lacininata* has 2.6 times as much DNA as *M. douglasii* although both have the same chromosome number. In 12 species of the genus *Vicia*, using the amount of DNA in *V. faba* as 100, the relative amounts of DNA went down 83, 60, 52, 28, 22, 18.5, and 17 (5 omitted) although all have 10 or 12 chromosomes. The species *V. amphicarpa* has two chromosomes fewer than *V. faba* but one sixth as much DNA. Martin cites many other cases of such DNA variation, such as 20:1 in the genus *Crepis*. Such intrageneric variation among animals and plants is common rather than rare (Rees et al., 1966).

The implication drawn from such data is that there is lateral polyteny (Martin and Shanks, 1966) or sequential redundancy of whole DNA chromosome strands. But until better knowledge of chromosome fine structure has been acquired, it is best to keep such proposals as the hypotheses that they are and to not accept them as proved facts.

INTERNAL ALTERATIONS

In addition to DNA variations among related species of the same chromosome number there are almost always structural alterations. These can be various sorts of inversions, minor deletions, and duplications, which alter the chromosomes rather little. Large pericentric inversions and translocations, on the other hand, can greatly alter chromosome forms of the set. Such changes are usually homozygous within a species but between "good" species and even "sibling" species may be markedly different. Accumulation of such differences accompanied by gene mutations gradually produces internal sterility barriers as two or more species derived from one common ancestor evolve without necessarily changing the chromosome number. The reverse, two species forming one by hybridization, *recombinational speciation*, can also occur although probably rarely (Grant, 1966a, 1966b). That is, two related species were hybridized (*Gilia malior* × *G. modocensis*) and the hybrid, without polyploidy, was able to persist in culture through a period of high sterility and low vigor, eventually forming in the F_9 generation an intermediate, vigorous new species that is isolated by sterility from its parental species. In the wild the weak diploid hybrid would rarely be able to persist until a vigorous form could arise.

ROBERTSONIAN CHANGES

It is becoming increasingly evident in vertebrates that changes in chromosome number occur within species, between sibling species, and among fully developed species by (1) more or less metacentric chromosomes breaking crosswise through the centromere to produce a greater number of telocentric (or acrocentric) chromosomes and/or (2) the reverse, by which one more or less metacentric chromosome is formed by the union of centromeres of two telocentrics (or acrocentrics) (Fig. 10-1). By either of these types of change the number of chromosome arms (the fundamental number, FN) remains the same although the number of chromosomes (and, therefore, linkage groups) is changed. That is, one metacentric chromosome has two arms and is equal to two acrocentric chromosomes. There may be a loss of some short arms during a series of such centric fusions; but in general the FN changes little, while the chromosome number changes a great deal.

In the rodent species *Gerbillus pyramidum* three cytological forms that may be sibling species have been found (Wahrman and Zahavi, 1958). Although the 2n chromosome numbers are 40, 52, and 66, the numbers of arms are respectively 78, 74-76, and 76 and the number of metacentrics goes down from 38 to 22-24 to 10 as the number of acrocentrics increases from 2 to 28-30 to 56 as represented below:

Population	2n	Metacentrics	Acrocentrics	Arm number
Algeria	40	38	2	78
Coastal Israel	52	22-24	28-30	74-76
Negev	66	10	56	76

Another example of Robertsonian change occurs in the mole rat, *Spalax* (Fig. 10-2). Four sibling species with 2n = 52, 54, 58, and 60 have been found in and near Israel (Wahrman et al., 1969). Wahrman and associates assume that evolution has progressed from 52 to 60 by centromere fission. Again, as the number of metacentrics decreases the number of telocentrics increases. Other examples are the common shrew, *Sorex araneus* (Ford et al., 1957), in which the FN is constantly 36 but the chromosome number ranges from 22 to 25 in the XX female and from 22 to 27 in the XY_1Y_2 male; and the marine snail *Thais lapillus* (Staiger, 1956), in which the following three chromosomal forms are found: (1) with 2n = 36, all acrocentric chromosomes, occurring at low-tide level (FN = 36); (2) with 2n = 26, five pairs of which are metacentric (FN = 36), at high-tide level; and (3) with 2n = 27 to 35, having various numbers of metacentrics and acrocentrics (FN always = 36) between the low- and high-tide

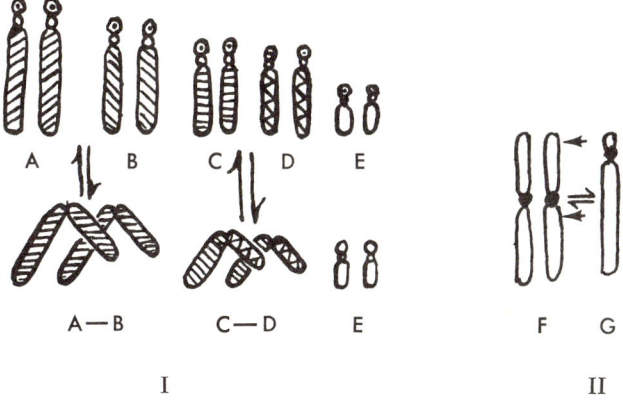

Fig. 10-1. **I**, Robertsonian-type chromosomal changes, from acrocentrics to metacentrics by centric fusion, very common, or from metacentrics to acrocentrics by centric fission, apparently rare. The fundamental number (FN) in this example is 10, because there are 10 long arms in the set. **II**, F-G change of a metacentric to an acrocentric by a large pericentric inversion. Arrows against F indicate points of breakage, followed by inversion and then reunion to produce G. The reverse can also occur.

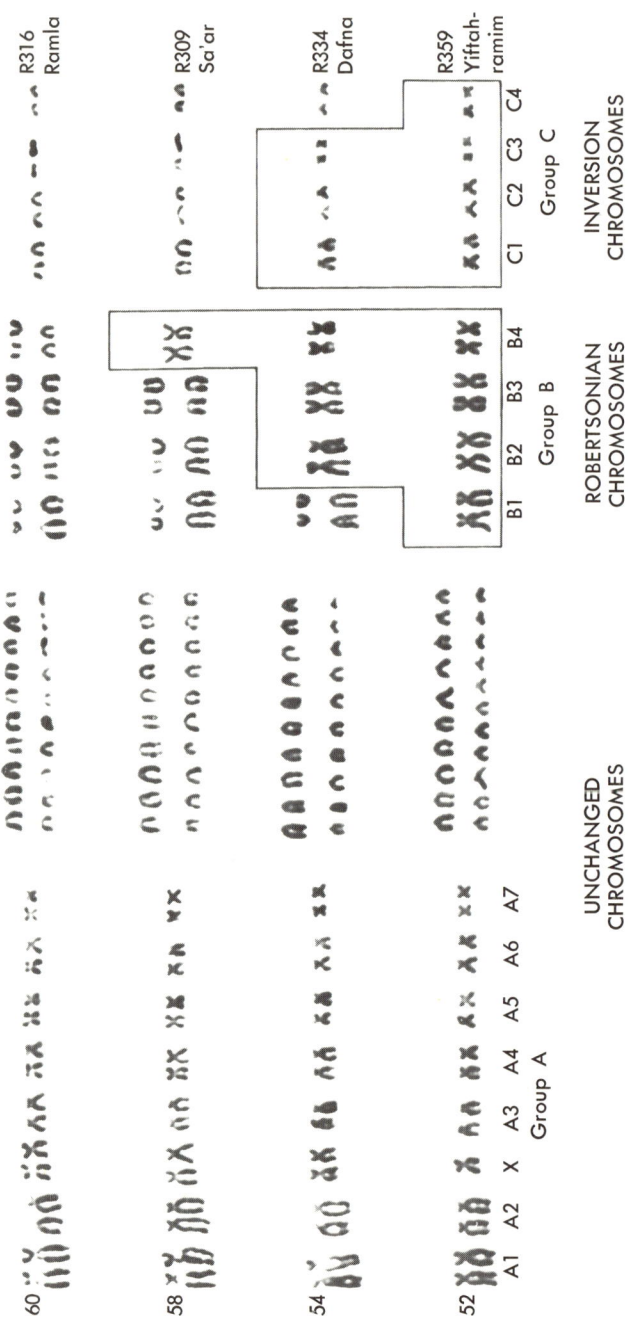

Fig. 10-2. Chromosomal polymorphism within the species of mole rat, *Spalax ehrenbergi*, of Israel. The chromosomes that show Robertsonian change are the B-group, boxed at the lower right. There are four chromosomal types of populations that are essentially indistinguishable morphologically. They have diploid chromosome numbers (left) of 52, 54, 58, and 60. Hybrids seem to occur rarely, and the chromosomal types in general occupy different regions. The chromosomes of interest here are the B-group. The 52-chromosome individuals and population have eight more or less metacentric B-group chromosomes. In the 54-chromosome population chromosome B1 is replaced by two telocentrics or acrocentrics (outside the box). The 58-chromosome population has only B4 left as a metacentric, and in the 60-chromosome individuals all metacentrics have become converted to acrocentrics or telocentrics by whole-arm Robertsonian change. The number of chromosomes has increased, but the number of long arms has remained constant. It is conceivable that evolution may have gone in the opposite direction by chromosome fusions. (From Wahrman, J., R. Goitein, and E. Nevo. 1969. Science **164**:82-84. Copyright 1969 by the American Association for the Advancement of Science.)

levels. Numerous other species are known illustrating Robertson's law such as *Mus minutoides* from the Congo in which three forms have been found (1) with 34, all acrocentric autosomes, (2) with 32 chromosomes but one pair of metacentrics, and (3) with 30 chromosomes but two pairs of metacentrics (Matthey, 1964). Bianchi et al. (1969) reported in the rodent *Akodon molinae* some individuals with $2n = 42$ with a large metacentric and others with $2n = 43$; the latter have two new elements derived from the large metacentric by centric fission. Badr and Badr (1970) found in a wild population of rats, 14% with $2n = 38$ and two long metacentrics, 54% with $2n = 42$, and 32% with $2n =$ about 54 (50 to 60) with no metacentric chromosomes.

An example of Robertsonian change in plants is recorded by Hair and Benzenberg (1958) in the gymnosperm family Podocarpaceae. The authors "read" the evolution in the direction from $2n = 40$ acrocentric chromosomes, by repetitive centric fusions to the "saturated" condition of 20 metacentrics. Robertson (1916), who proposed the concept of constancy of the number of the chromosome arms although the number of chromosomes may have changed during evolution, based this concept on studies of Orthoptera, especially short-horned grasshoppers. Considerable subsequent work on grasshoppers has been done, and White (1954), who himself had worked on them, discussed the changed chromosomes during evolution in, especially, the Acrididae. There are two subgroups that can be designated the 10-chromosome group and the 12-chromosome group. In the 12-chromosome group all conditions exist (but not in an evolutionary sequence), from 12 pairs of acrocentrics to 6 pairs of metacentrics. In some species the condition has not stabilized, for example in *Chortophaga viridifasciata* and in *Hesperotettix* where most individuals have 12 pairs of acrocentrics but as many as 4 pairs of metacentrics are observed, with, of course, a corresponding decrease in the number of acrocentrics. These changes have resulted from translocations that have produced centric fusions.

Matthey (1949, 1951) proposed the concept of the *fundamental number* (FN), which is the total number of arms of the autosome chromosomes of the 2n set.

There are, of course, taxa with variation of chromosome number but not of the Robertsonian type. An obvious example is the genus of deer mice, *Peromyscus* (Hsu and Arrighi, 1966). In this genus the reverse of the Robertsonian constancy of the fundamental number is found but variation occurs in chromosome number. In *Peromyscus* the chromosome number remains constant at 48 but the FN ranges from 56 to 96. Species like *P. collatus* and *P. eremicus* have all biarmed (no acrocentric) chromosomes whereas the subspecies *rowleyi* of *P. boylei* has almost all (40) chromosomes of the acrocentric or telocentric type. Exchanges and pericentric inversions have presumably achieved these results.

It must not be assumed that centric fusion and possibly centric fission are the only mechanisms of chromosome number change or alteration of chromosome form. Acrocentrics can become metacentrics by internal changes, called "homosomal change," such as pericentric inversions, as in *Podisma sapporoensis* (Helwig, 1942). Small heterochromatic chromosomes that may remain after one or more translocations of its euchromatin to other chromosomes can be lost without adverse effect on the species as in aneuploid (or dysploid) decrease. And polyploidy is common in plants as a significant chromosome number change.

CENTRIC FUSION

There is wide agreement among cyto-evolutionists that one of the commonest methods of chromosome number change is *centric fusion*. This is a particular subclass of reciprocal translocations (see Chapter 13) and is rather limited to exchange between short arms of two acrocentric chromosomes (Fig. 10-1). It is generally assumed, and often correctly, but perhaps not always, that two nonhomologous chromosomes are broken, one close to the centromere in the very short arm and the other very close to the centromere in the long arm. Following reunion one chromosome consists mostly of a centromere and two short arms, a "free centromere," and the other consists of a centromere and two nonhomologous long arms. The very small chromosome would be composed of a centromere and all, or mostly, of proximal heterochromatin. It might be considered to be a supernumerary and could be lost without serious effect on the species. Presumably, after many generations, the long metacentric chromosome can replace the two equivalent acrocentrics so that in an evolutionary sense a pair of metacentrics have replaced two pairs of acrocentrics. A couple of examples of that sort of centric fusion, in addition to grasshoppers, are set forth by Ford et al. (1957) and Nogusa (1960). Ford et al. found such chromosomal polymorphism in the common shrew, reporting chromosome numbers from 22 to 27 in the species. Nogusa found polymorphism in the fish *Acherlognathus rhombea*. In the testes of one individual he saw spermatogonial divisions with 44 chromosomes including 4 metacentrics, 46 chromosomes with 2 metacentrics, and $2n = 48$ with no metacentrics. Such cases as these suggest that some mechanism other than translocation may indeed be operating.

White (1948a, p. 162) made the observation that in grasshoppers "It is interesting that nearly all of these centric fusions have taken place between chromosomes of similar length, so that the V's produced have approximately equal arms." He cited cases of only long acrocentrics combining to form large metacentrics and only small acrocentrics uniting to form small V's. Recently Cohn and Clark (1967) were struck by the supposedly derived metacentrics of five species of crocodilians having exactly equal arms. They, however, felt forced to consider such metacentrics to be isochromosomes, a questionably valid conclusion. It is, of course, true that exchanges and such aneuploid chromosome number reduction does also occur between other sorts of chromosomes. Smith (1965) found in the ladybird beetles *(Chilocorus)* that *C. orbus* has $2n = 22$; *C. tricyclus*, $2n = 20$; and *C. hexacyclus*, $2n = 14$. He reported that all three species have the same number of euchromatic (E) chromosome arms but have 16, 12, and 0 heterochromatic (H) arms. H arms can be lost by centric fusion without affecting viability as the chromosome number decreases. Numerous other examples of centric fusion are cited throughout the remainder of this chapter.

There is a possibility that many animal chromosomes are really telocentric rather than acrocentric. The mechanism of centric fusion between two telocentrics, having centromeres on the tip of the chromosome, if such exist, might be different from centric fusion of acrocentrics, which have short arms beyond the centromere. Either the two centromeres could fuse directly to make a metacentric with a "double" centromere and no "free centromere" to be lost, or else two breaks, through two such terminal centromeres, followed by fusion would again produce a metacentric and a truly free centromere having no arm at all. This mechanism has been proposed by Marks (1957), John and Hewitt (1966), Southern (1969), and Comings and Okada (1970e). Comings and Okada's conclusion was based upon electron microscopic comparison of terminal and median centromeres of mouse chromosomes; the latter seemed to be equivalent to two of the former (see also discussion of telocentric chromosomes in Chapter 8).

CENTRIC FISSION

Centric fission is the opposite of centric fusion. A chromosome with two long arms (a metacentric or submetacentric) has the centromere split transversely so that two truly telocentric chromosomes result. Rhoades (1940) and McClintock (1941b) have studied such broken chromosomes in maize and concluded that they are unstable and are lost during a number of cell generations. Therefore, it is generally assumed that there may not be any truly telocentric natural chromosomes. Marks (1957), on the other hand, has claimed that truly telocentric chromosomes do exist in nature; and in recent years telocentric chromosomes have been reported in a number of plants, such as Podocarpaceae (Hair and Benzenberg, 1958), Cycads (Khoshoo, 1960; Marchant, 1968), *Tradescantia* (Jones and Colden, 1968), and *Nigella* (Strid, 1968). Jones (1970) stated recently that "the frequency of such cases will increase when the telocentric receives the recognition it deserves." It is also likely that many animal mitotic metaphase chromosomes prepared by the new techniques (see Chapter 19) may well be telocentric and stable. Considerable study on this matter is urgently needed because, if centric fission is able to produce stable telocentrics, evolution of the Robertsonian sort can be "read" in either direction or both. Now it is assumed to be possible only by centric fusion. It seems that the presently available evidence that telocentrics do not or cannot occur in nature is completely inadequate for a sweeping generalization. The examples reported by Staiger (1956) and by Ohno et al. (1965) as well as others such as Nogusa (1960) strongly indicate that both true centric fusion and fission not involving exchange may occur rapidly among individuals of a species and within a single individual. In fact, Ohno et al. have proposed a mechanism for such true centric fusion and fission found in tissues of the rainbow trout, *Salmo iridens*, that the number of arms (the FN) remained constant at 104 but the chromosome number varied from 58 to 65. They proposed that in early embryogenesis fissions and fusions occur to produce some metacentrics from telocentrics and telocentrics from acrocentrics. They proposed that there is heterochromatin at the end of the telocentric that has strong attraction for heterochromatin at the ends of *nonhomologous* telocentrics, but that attraction between heterochromatin at ends of *homologous* telocentrics is nonexistent. This seems to be attributing rather too much specialized character to heterochromatin. Perhaps terminal centromeres only are involved.

Staiger (1956) found similar Robertsonian change in the marine snail, *Purpura lapillus*. The 18-chromosome form with no metacentrics occurred at low tide, the 13-chromosome form with five pairs of metacentrics occurred at high-tide level, but between tide levels various combinations of metacentrics and telocentrics occurred. His anaphase I illustrations indicated that probably telocentric rather than acrocentric chromosomes occurred. Furthermore, it is implied that only 5 pairs of metacentrics or the 10 pairs of telocentrics are involved in these alterations. The remaining eight pairs of acrocentric, not telocentric, chromosomes are unchanged in all forms. Rhoades (1955, p. 137) reported the increase of the chromosome number in an experimental line of maize from $n = 10$ to $n = 11$ by the formation of two differently sized metacentric isochromosomes from one submetacentric by transverse misdivision of a centromere.

An interesting and spectacular chromosome number change downward has recently been recorded in a primitive subfamily of Asiatic deer, the Mutiacinae (Fig. 10-3). Among deer in general, 13 species have $2n = 56$ to 70 and the fundamental number (FN) of 70 to 74 (Wurster and Benirschke, 1970). The fundamental number is the total number of arms of the autosomes of the 2n set (Matthey, 1951). The species *Muntiacus reevesi*, however, has the lower chromosome number of $2n = 46$ and the FN of 46. The spectacular chromosome number is in the related and similar species, *M. muntjak* of India, that has only $2n = 6$ in the female and 7 in the male (Wurster and Benirschke)! The three haploid mitotic chromosomes of the female are: one 17μ metacentric, one 10μ acrocentric, and the third 13μ subacrocentric. Thus the chromosomes are very large for vetebrates. The X chromosome is probably derived from centric fusion between a telocentric X and a small autosome. In the male such an original telocentric X, a subacrocentric X similar to the two X's of the female, and a very short metacentric Y are present. The authors cite reports that *M. reevesi* and *M. muntjak* hybridize to produce fertile hybrids and that a third species, *M. rooseveltorum*, is intermediate. How the chromosomal difference between these two species arose, probably by spectacular reduction, is unknown. Meiotic pairing in a hybrid should be very interesting. Dr. Wurster (personal communication) has found subsequently that the Javan muntjac has $2n = 8$, in contrast to $2n = 6$ in the Indian subspecies. Thus, the three cytologically known forms in the genus are: *Muntiacus reevesi* ($2n = 46$), *M. muntjak* ($2n = 8$), and *M.m.* subspecies *vaginalis* ($2n = 6$). Relative amount of DNA per cell is greatest in *reevesi* and least in subspecies *vaginalis*, as would be expected.

Somewhat less spectacular drops in chromosome numbers are known, such as from $2n = 18$ to 37 to $2n = 8$ (*Graphipterus serrator*) in carabid beetles recorded by Wahrman (1966); and Saez (1957) found that the grasshopper *Dichroplus silveiraguidoi* has only 12 chromosome arms although all other species have from 19 to 23 arms.

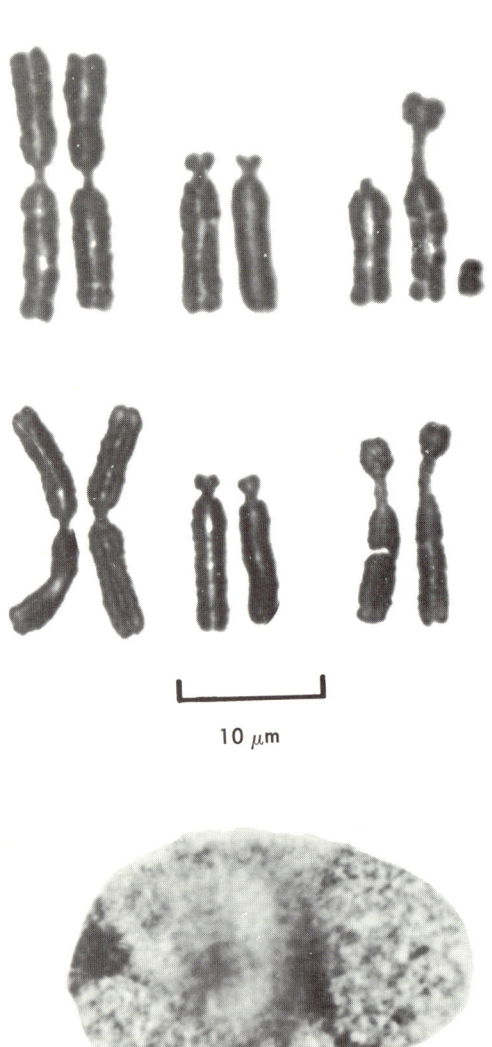

Fig. 10-3. The chromosomes of the Indian muntjac deer, *Muntiacus muntjak* (male 2n = 7; female 2n = 6). The male seems to have one long X (13μ long), one long Y, and one short Y. The longest pair are metacentric and about 17μ long, very large chromosomes for mammals or any organism. The chromosome number of the most closely related species, *M. reevesi*, is 2n = 46. It is most likely that spectacular evolutionary chromosomal reduction has occurred from 2n = about 50 to 2n = 6. (From Wurster, D. H., and K. Benirschke. 1970. Science **168**:1364-1366. Copyright 1970 by the American Association for the Advancement of Science.)

INTRASPECIFIC VARIABILITY (POLYMORPHISM)

Although it is generally true that all individuals of a species have the same chromosome number and that the number is even (Wilson, 1896), there are exceptions. Invariability of chromosome number of an extreme sort occurs when all species of a genus or most genera of a family have the same chromosome number, as in some families of gymnosperms and horsetails. Among flowering plants a few genera have all species at one diploid level: *Sagittaria*, all n = 11; *Paeonia*, all n = 5; *Ribes*, all n = 8. An example of trivial intergeneric chromosomal variability was reported by Wohnus and Benirschke (1966) in the primates called marmosets. Four species in three genera had chromosomes so similar that they could all be matched into one composite idiogram.

There are many genera of plant that have both diploid and polyploid species (Chapters 12 and 16). There are also numerous genera composed of species with two or more diploid chromosome numbers, such as *Drosophila*, n = 6,5,4,3, and one 7; *Crepis*, n = 5,4,3,6, and 7; and many others. Such cases demonstrate intrageneric chromosome number decrease and increase from a known or hypothetical ancestral basic number. Examples of this sort of change are discussed in Chapters 13 and 15.

There are numerous animal and plant species within which aneuploidy is common, with or without polyploidy. Among plants, *Claytonia virginica* is exceptional for its extreme aneuploidy combined with polyploidy, thereby having plants with chromosome numbers from 2n = 12 to 2n = 191 (Rothwell, 1959 Rothwell and Kump, 1965; Lewis et al., 1967). Actually, the whole genus *Claytonia* is exceptional, showing aneuploid variation at the diploid level (3 subspecies, n = 5; 7 subspecies, n = 6; 14 subspecies, n = 8). Euploidy based on these diploid numbers and aneuploidy at the polyploid level in a number of species other than *virginica* is known; for example, *C. lanceolata* has 2n = 16, 24, 32, 36, 37, 44, 48, 52, 54, 64, 74, 76, 84, and 90.

Another chromosomally variable plant "species" is the grass *Poa pratensis* (Kentucky bluegrass) that is both sexual and apomictic. Thus, it is an "agamic complex" rather than a species. It is based on a diploid chromosome number of $2n = 14$, but its chromosome numbers range from about 28 to 140. Most individuals have multiples of 7 (are euploid), but nearly all aneuploid numbers have also been reported.

Among animals, especially vertebrates and insects, intraspecific variations in chromosome number are being reported more and more commonly. An unusual example is the orthopteran *Euscirtus hemelytrus* of Japan (Ohmachi and Ueshima, 1957). Not only do individuals differ in chromosome number, from $2n = 17$ to 24 (although 60% have $2n = 19$); and have variations of chromosome number among spermatogonia within an individual such as 19 and 20, 17 and 19, 19, 19, 20, or 21 and 22; but in meiosis at metaphase I the number of bivalents varies from 5 to 9 and of univalents from 1 to 8, and in spermatogonia the number of V-shaped chromosomes varies from 4 to 7. The authors made no attempt to explain these variations.

In some species of deer mice *(Peromyscus)* intraspecific chromosomal polymorphism has been observed (Hsu and Arrighi, 1966; Sparks and Arakaki, 1966). The latter authors reported the following variations in *rubidus* and *gracilis*:

Form	Acrocentrics	Metacentrics	Chromosome number
P. rubidus	18, 19, 20	30, 29, 28	48
P. gracilis	8, 9, 10	40, 39, 38	48

Wahrman and Goitein (preprint) have found that the Near Eastern spiny mouse, *Acomys cahirinus*, consists of two chromosomal forms; one is $2n = 38$, including four telocentrics, whereas the other is $2n = 36$ in which the four telocentrics are represented by two metacentrics. Apparently one whole-arm Robertsonian change occurred. In natural hybrids ($2n = 37$) the two telocentrics pair with the two arms of the metacentric at meiosis.

INTRAINDIVIDUAL VARIATION

Variation in chromosome number, other than meiosis, within an individual is usually euploid, or the multiplication of whole sets of chromosomes such as polyploid cells in stems of tomato (McMahon, 1956), polyploid cells in the liver of mammals, etc. There are very many such cases known. But there are known cases of aneuploid increase and decrease, especially in certain tissues of the organism. For example, Nygren et al. (1968) reported that the fish, pike ($2n = 50$), has cells in the kidney with chromosome numbers from $2n = 18$ to 75, and in the salmon ($2n = 58$) the kidney has from $2n = 47$ to 63. Even more extreme is the report of Ambühl (1953) that chromosome numbers in young embryos of the salmon relative, *Corigonus lavaretus*, ranged from 4 to 590 ± 10!

In the previous section the great variability of chromosome numbers in *Claytonia virginica* was discussed. But variation within plants also occurs, called *multiple genotypes* by Lewis et al. (1971). They found in plants having $2n = 28$ chromosomes in roots that $2n = 29$, 30, and 31 were very common in shoots and in microsporocytes. If such plants do have different genotypes in roots and shoots, because these organs have different chromosome numbers, the difference must be trivial in the light of such great chromosome differences between plants, all of which are essentially the same morphologically and reproductively.

In Chapter 5 chromosome reduction in plants is discussed, especially the condition in a certain cell layer of cycad roots in which reduction from about $2n = 20$ to one or two chromosomes *per cell* occurs.

Hegwood and Hough (1958) found a pistillate apple variety that had a constant chromosome number of $2n = 34$ in very young leaves but a pollen parent variety had only 50% of cells with $2n = 34$, 50% had chromosomes from 28 to 40, the diploid hybrids had $2n = 27$ to 45, and a triploid hybrid had $2n = 45$ to 61 with 36% of cells showing the triploid number of 51. The researchers cite other cases in plants of such a random chromosome mosaic condition; mitotic irregularities were commonly seen and were the evident cause. Chromosome reduction from the diploid to the haploid in certain asexual fungi, such as *Aspergillus* and *Verticillium*, apparently by gradual chromosome reduction has also been discussed earlier.

Euploid increase is, of course, very common within individuals of plants and animals, such as mosquito larval gut, various tissues of the water strider, tetraploid, octoploid, etc. spermatogonia of the salmon, the somatic tissues of male Hymenoptera, numerous cells of plant roots, stems, tapetum, etc. Such cases are discussed in Chapters 11 and 15.

Another type of intraindividual chromosome variation is the mosaic (also called chimera). These are very common exceptional individuals; normal chromosomal variation, which is very common, is not always considered as the mosaic condition. Among human beings mosaics for sex chromosomes are far from rare, such as XO/XX, XX/XY, XX/XXXX, XO/XX/XXX, etc. (see Chapter 19). In *Drosophila* and other animals, especially insects, gynandromorphs have been reported which are half male and half female. In the silk worm, *Bombyx mori*, genetic control of the fertilization of the egg and polar nuclei occurs to form gynandromorphs predictably (Goldschmidt and Katsuki, 1931). Occasionally bee colonies have been noted with many genetically determined gynandromorphs and mosaics (Morgan, 1916), the male parts resulting from unfertilized nuclei, the female parts from diploid nuclei. There are also many cases of mosaic plants in which various concentric or sectorial regions are tetraploid in a basically diploid stem (Derman, 1947, 1953; Stewart and Derman, 1970). Lima-de-Faria and Jaworska (1964) reported two mosaic plants of *Haplopappus gracilis*; one was 1.5% haploid and the other was 77% haploid, yet the plants grew as well as the fully diploid plants. Mosaics (Hyde and Powell, 1916; Crew and Lamy, 1938), including gynandromorphs (Oliver, 1934), have been reported many times in *Drosophila*. A chicken embryo of 63% triploid and 37% diploid has been reported (Bloom and Buss, 1966).

Other interesting examples of such intraspecific chromosomal alteration without change of chromosome number have been reported in the white-throated sparrow (Thorneycroft, 1966) and a phasmatid (Craddock, 1970). When idiograms of 35 white-throated sparrows were compared, it was evident that there was a variety of combinations of two pairs and a single "M" chromosome (Chapter 17). Each bird has four of these five elements but in at least five different combinations. The second longest pair is designated pair 2 consisting of subacrocentrics. Pair 3 consists of acrocentrics, and the M chromosome is nearly metacentric. Sixteen birds had 2,2, 3,3 and no M; 15 individuals had 2, 3,3, M; 3 had 2,2, 3, M; one had 2,2,2, 3; and one had 2,2,2, M. These same birds are apparently nullisomic for M, or 3, or monosomic for 2 and M or 3 and M, or trisomic for 2.

The Australian stick insect, *Didymuria violescens* (Craddock, 1970), was found to have chromosome numbers ranging from $2n = 26$ to 40 and three sex chromosome systems. The primitive sex chromosome mechanism in phasmatids and in some individuals of this species is XX-XO. From that, by translocations with autosomes, two different XX-XY systems were detected. Variation in autosomes seems to be reciprocal among long metacentrics and acrocentrics or telocentrics. Other examples of this sort have been discussed and related to evolutionary mechanisms by White (1968).

Thus, it is obvious that all cells of individuals of many, if not most or all, species do not necessarily have the same number of chromosomes.

BASIC NUMBERS

For years it has been a popular sport among some plant cytological evolutionists to try to determine a so-called basic chromosome number for various taxa, such as genera, subfamilies, families, and the whole angiosperm class. The same game had been played for some animal groups. The reliability of such conclusions varies greatly so that some are very convincing with little disagreement whereas others are highly controversial.

The genus *Drosophila* is one of the most completely studied of animal genera, and chromosome numbers from 3 to 7 have been recorded (Patterson and Stone, 1952) among about 215 "forms" ("at least 198 species"). Patterson and Stone reached the following conclusions with respect to chromosomal evolution (p. 160); (1) The basic chromosome number is $n = 6$, consisting of 5 rods and a dot. (2) Only one increase, to $n = 7$, has been found, *D. trispina* from near Needles, California. (3) A new species *group* usually evolves from single founder *species*. And (4) differences in chromosome number within a species group must usually have evolved subsequent to the formation of the group, and from the common ancestor with the highest represented number (or a higher one) or from a mixed group that contained a member with the highest number. Gains in chromosome number are relatively rare. Most rearrangements have been centric fusions, by which two acrocentric chromosomes form one metacentric, and by both paracentric and pericentric inversions. Translocations, other than centric fusions, have been uncommon in chromosome evolution in *Drosophila*.

The basic chromosome number of a genus in which all species have the same number, such as the plant *Sagittaria* (all species studied have n = 11), is obvious. The same is true of the grass subfamily or super tribe Festucoideae, in which essentially all genera have chromosomes that are exact multiples of seven.

Less certain are many other genera and groups. The plant genus *Crepis* has been very thoroughly studied, and chromosome numbers from 3 to 7 have been reported (Babcock, 1947). In his 1947 monograph Babcock stated (p. 8) that "The most primitive chromosome numbers in *Crepis* are 6 and 5." However, later (1949, 1950) he revised this opinion and considered the chromosome numbers 5 and 4 (of section VI) as most primitive, with the possibility of extinct, 6-chromosome, tap-rooted forms as even more primitive than section VI. Fig. 10-4 diagrams the likely chromosomal evolution of *Crepis* with 4 or 5 or 6 the possible original (x) basic number. The symbol "x" is used to indicate *the* basic number of the original taxon of the group being considered.

Fig. 10-5 presents schemes for the chromosomal evolution in *Plantago* (x = 6), *Callitriche* (x = 5), and *Aphanostephus* (x = 4). These are simple genera and have simple and likely schemes. The genus *Aster*, on the other hand, has two "logical" schemes for the evolution of chromosome number based upon two possible basic numbers, 5 or 9 (Fig. 10-6). Most workers on the chromosome numbers of *Aster* and related genera accept x = 9 as the most likely basic number, chiefly because most species have n = 9 and many polyploid multiples of that number exist (Huziwara, 1959; Solbrig et al., 1964). On the other hand, the assumption that x = 5 or 4 or 6 is the most likely basic num-

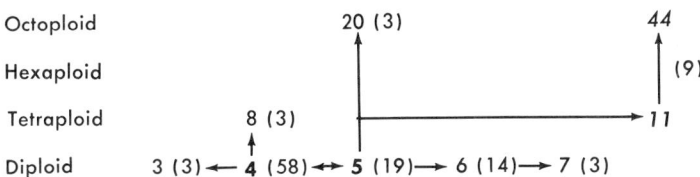

Fig. 10-4. Hypothetical scheme of chromosome changes during evolution of *Crepis*. Numbers in parentheses are numbers of species. Italicized numbers are of the sexual and related apomictic species (x = 11). Because the highest polyploidy is based upon n = 5, it is considered that n = 5 is the most likely basic number, in agreement with Babcock's conclusion, although n = 4 is equally likely.

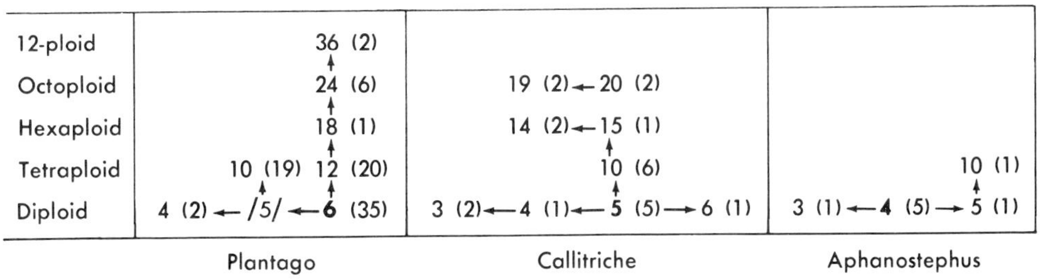

	Plantago	Callitriche	Aphanostephus
12-ploid	36 (2)		
Octoploid	24 (6)	19 (2) ← 20 (2)	
Hexaploid	18 (1)	14 (2) ← 15 (1)	
Tetraploid	10 (19) 12 (20)	10 (6)	10 (1)
Diploid	4 (2) ← /5/ ← 6 (35)	3 (2) ← 4 (1) ← 5 (5) → 6 (1)	3 (1) ← 4 (5) → 5 (1)

Fig. 10-5. Schemes of chromosomal evolution in three genera expressed in haploid (gametic) chromosome numbers. Numbers in parentheses are numbers of species having that chromosome number. That n = 5 in *Plantago* is hypothetical; no such species is known.

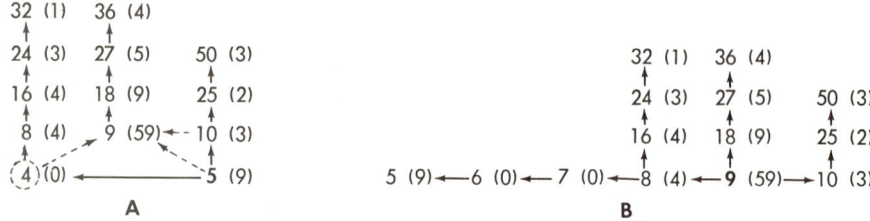

Fig. 10-6. Two possible schemes for chromosomal evolution in the genus *Aster*. In **A** the basic number is 5; in **B** it is 9 (which is the commonest haploid number in the genus). In **A**, n = 4 is hypothetical.

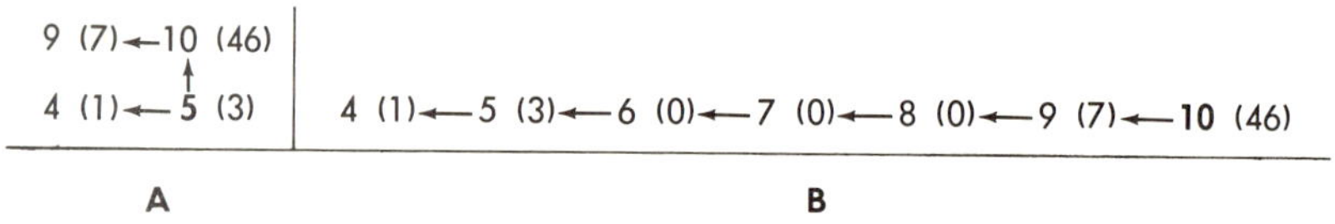

```
      9 (7)←10 (46)
         ↑
      4 (1)← 5 (3)      4 (1)← 5 (3)← 6 (0)← 7 (0)← 8 (0)← 9 (7)← 10 (46)

           A                               B
```

Fig. 10-7. Two possible schemes for chromosomal evolution of genera of the grass tribe *Andropogoneae*, based upon n = 5 in **A** or n = 10 in **B**. The commonest basic number in the tribe (46 genera) is n = 10.

ber of genera and tribes of the Compositae, rather than 9, was proposed by Turner et al. (1961), Turner et al. (1962), Turner and Horne (1964), Turner and Lewis (1965), Turner and Flyr (1966), Turner (1967), etc. Turner and associates have assumed the low basic number from study of many genera and tribes of the Compositae, including the Astereae, and have repeatedly noticed, as in *Aster* (Fig. 10-6), that within a genus or tribe chromosome numbers of 4, 5, and 6 as well as 9 and 10 occur but there are no species with n = 7 or 8 and often 6. Such a gap of one, two, three, or even four chromosome numbers is far too common to be chance.

The evolutionists who assume x = 9 or 10 for genera consisting of species with also, for example, n = 4 and 5, have proposed a continuous evolutionary decrease, 9-8-7-6-5, but they assume the species with n = 8, 7, and 6 have died out. This is conceivably possible occasionally; but when such a gap occurs as often as it does, one is compelled to discard such an assumption. The case of the grass tribe Andropogoneae is such an extreme example (Fig. 10-7). Most genera are based upon n = 10 and there are a few with n = 9. But then there are three genera (including *Sorghum*) with n = 5 and one with n = 4. In this tribe there is a gap of n = 8, 7, and 6. Why? Certainly there is no evidence that n = 8, 7, or 6 would be at any evolutionary disadvantage because of chromosome number per se and so have become extinct. The frequency of these gaps in the range of 7 and 8 or 6, 7, and 8 is so common that an assumption of a high basic number of 9 or 10 and continuous aneuploid decrease followed by extinctions to create the gap is untenable.

A logical alternative is a low basic number close to x = 5 followed by tetraploidy to establish n = 10. A simple aneuploid reduction could establish n = 9 as is diagrammed for the Andropogoneae (Fig. 10-7), and another could establish n = 8 (and aneuploid decrease seems to be a very common evolutionary change). Thus, the gap is "explained." Another possible increase directly to n = 9 can be postulated by hybridization between an n = 5 with an n = 4 species after an aneuploid reduction from 5 to 4 had occurred, as is indicated as possible for *Aster* in Fig. 10-6. Establishment of apparent n = 8 could really be based upon n = 4. Certainly this latter scheme is simpler and both cytologically and logically more sound. But, as someone once stated, "Logic is a way of being wrong with confidence."

The best ways of determining which of these schemes is the more sound are (1) determination of the chromosome numbers of the assumed most primitive species and genera and (2) examination for secondary association of bivalents at metaphase I of meiosis. Solbrig et al. (1964), after assuming that x = 9 in the

Table 1. Haploid chromosome numbers of the most primitive genera in the tribe Astereae

Genera	Number of species	Species counted	Haploid numbers; number of species in parentheses
Olivaea	2	2	6(2)
Xanthocephalum	8	5	4(4), 6(1)
Grindelia	50	23	6(15), 12(8)
Machaeranthera	30	22	4(14), 5(5), 6(1), 8(1), 9(1)
Haplopappus	150	63	2(1), 3(1), 4(22), 5(8), 6(11), 8(3), 9(13), 12(1), 18(3)

Astereae, then stated that the most primitive genera in the tribe are *Haplopappus* (and *Machaeranthera*), *Olivaea*, *Xanthocephalum*, and *Grindelia*. Table 1 lists the haploid chromosome numbers in these genera. They all seem to have basic numbers of 4 and 6, not 9. Studies have not been made of secondary association in the Compositae, but in a number of other genera having apparently high basic number such as $x = 10$ or 12 the association (if it is real and if it really does imply residual homology) has indicated a more ancient low basic number in the range of 6 ± 1 (Nandi, 1936, for rice; Catchside, 1937, in *Brassica* for example). The presence of bivalents and secondary association at meiosis in haploids has also been used as evidence of residual homology and a lower basic number (see Kimber and Riley, 1963, for examples).

BASIC NUMBER OF THE ANGIOSPERMAE

A few studies of chromosome numbers have been made of all angiosperms as a group, and the likely original basic number has been proposed. In all cases, regardless of method, the concluded basic number is 7 or 8. Most of such conclusions have been based upon the commonest basic number in all species or genera that have been studied or in selected (such as primitive or woody) genera. Avdulow (1931) graphed total numbers of species against basic chromosome numbers and recorded the highest peak at $n = 8$ with a secondary peak at $n = 12$. Stebbins (1938a) graphed numbers of *genera* according to basic haploid numbers and found a distinct high peak at $n = 8$ for mostly herbaceous plants and a secondary peak at 12, the latter present mostly in woody plants. Grant (1958) reported peaks at $n = 7$ to 8 for herbs and $n = 11$ to 14 for woody forms from about 12,000 woody and herbaceous dicotyledons. If number of *families* is plotted against the single lowest chromosome number known in each family, the peak is at $n = 6$ and 7. Raven and Kyhos (1965) and Ehrendorfer et al. (1968) studied chromosome numbers in "primitive" woody families and concluded that the likely basic chromosome number of the original angiosperms was $x = 7$. There is at least unanimity of opinion on this topic.

Jones (1970) in an essay on chromosomal changes in plant evolution has warned against applying too generally cytological conclusions from one genus or a few with respect to the direction of evolution, or of basing the direction of evolution on chromosome changes alone. This is good advice. Chromosome evidence is very valuable; but, if unsupported by other evidence, it is often unreliable.

CHAPTER **11**

Haploidy and aneuploidy

> The appearance of haploid individuals among organisms normally diploid is a very rare occurrence.
> It is of interest that such [aneuploid] forms usually differ in appearance from the normal diploid much more than do the tetraploids, triploids, or other euploid forms.
>
> L. W. Sharp, 1926

Although most of Chapter 12 is devoted to a discussion of polyploidy, and that mostly in plants, in order to achieve completeness and an adequate basis of comparison nonpolyploid conditions such as haploidy, diploidy, and aneuploidy will be treated in this chapter. Diploidy, of course, is the basis of comparison for the other conditions.

In animals polyploidy is perhaps autopolyploidy and is very often associated with asexual reproduction called parthenogenesis (White, 1954; Takenouchi, 1970). In plants, on the other hand, polyploidy is usually associated with hybridization and sexual reproduction. Polyploidy in animals is uncommon but in plants is very common. In animals polyploidy plays almost no part in evolution; in plants it is a very significant aspect of evolution and speciation. Not only do plants seem able to tolerate polyploidy better than do animals but they seem to be able to tolerate better other variations from the normal diploid condition such as haploidy and aneuploidy. Therefore, most of this chapter is, by necessity, botanically

oriented. Because of the intimate relationship of polyploidy and plant evolution, there will be considerable overlap between this and the next two chapters.

HAPLOIDY

An organism is haploid if it has the chromosome number of a gamete of the species. Of course there can be haploids of a polyploid, but these should be called *polyhaploids* (Chapter 12). Haploids of diploid species can be classified in two categories: (1) natural and normal, as an evolved condition; and (2) abnormal, fortuitous, occasional individuals that should but do not have the diploid chromosome number.

Normal haploidy

Haploidy as a natural and normal condition occurs in all groups of organisms. Viruses, bacteria, blue-green algae, etc. are essentially always haploid; each individual contains one "chromosome." They are asexual and have no meiosis. Among animals, including protozoa, the haploid condition is generally restricted to gametes; the organisms are diploid. Exceptions do occur among animals as in some insects, such as all males of Hymenoptera, some Hemiptera, etc. (Schrader and Hughes-Schrader, 1931; White, 1954) where males arise by parthenogenesis and are haploid, at least in the germ line. Among algae and fungi there is usually a life cycle one portion of which is at the haploid level. The relative extent of the haploid condition can be permanent, as in many imperfect fungi, it can include the whole life cycle except the diploid zygote, it may be about equal to the diploid phase, or it may occur only in the gametes. Such haploid phases arise in plants from development of haploid spores produced by meiosis. Among the bryophytes there is normally no parthenogenesis; the haploid (gametophyte) generation, the moss plant, is haploid, having developed from the haploid spore. The diploid (sporophyte) generation, which develops from the fertilized egg, is a short-lived parasite on the gametophyte. Among vascular land plants the haploid generation ranges from obscure but independent organisms in the cryptogams, such as *Psilotum, Equisetum,* and *Lycopodium;* to reduced but independent in *Selaginella, Marsillia,* etc.; to bulky "parasites" within the ovules and pollen grains of the gymnosperms; and finally to tiny eight-nucleate, seven-celled embryo sacs and the two- or three-celled pollen grain of the angiosperms.

Expressed another way, animals generally have meiosis tied to gamete production whereas land plants have meiosis followed by haploid spores, which produce haploid gametophytes that finally form the gametes.

Abnormal haploidy

Abnormal haploidy occurs occasionally in nature and in culture or may be produced experimentally in the laboratory. It is the asexual development of a form that should be diploid, as a haploid. In animals and many plants the asexual development occurs by development of an unfertilized egg, called parthenogenesis. Among animals haploids rarely develop very far toward maturity, although haploid *Drosophila* specimens have been reported rather often (Bridges, 1925a; Castle, 1934). Haploid sporophytes of flowering plants, if given special care, very often do reach maturity, usually as dwarfed plants, and they are fairly common (Kimber and Riley, 1963).

One of the first-reported haploid angiosperms was in *Datura stramonium* (Jimson weed) among cultures of the species being studied by Blakeslee et al. (1922) and Belling and Blakeslee (1927). Of course, meiosis in a haploid of a diploid species is very irregular because there are no homologous chromosomes to pair and synapse. As a result, at metaphase I in *Datura* ($n = 12$) there were 12 univalents, each chromosome, of course, consisting of two chromatids. In about 88% of microsporocytes (pollen mother cells) the 12 two-chromatid univalents, without dividing, moved randomly at anaphase I to the two poles: 6 and 6, 27%; 7 and 5, 38%; 8 and 4, 19%; 9 and 3, 12%; 10 and 2, 3%; 11 and 1, 1%. They reported no 12 and 0 segregation. At the second meiotic division these "reduced" cells and chromosomes did divide, but mostly dead and empty spores were produced. In 125 cases, however, the 12 univalents were seen to divide as in mitosis at anaphase I to produce two nuclei, each having 12 chromosomes. There was no second meiotic division, and normal haploid pollen grains were formed. Presumably, the same sort of division occurred in a few ovules because these haploid plants did produce some seeds each of which, following self-fertilization, contained a completely homozygous diploid embryo. Thus, haploid plants can be partially fertile.

By 1929 haploid plants had been found also in *Nicotiana tabacum, Triticum compactum, Crepis capillaris, Solanum nigrum, Matthiola, Oenothera,* and *Lycopersicon esculentum* (tomato). The main difference between haploid *Datura* and tomato is that the latter produces only about 1% of apparently good pollen. But when pollinated with good pollen from a diploid plant, some fruits were set and a few good diploid seeds were formed (Lindstrom, 1929).

During the 1930's haploids of diploid ($n = 7$), tetraploid ($n = 14$), and hexaploid ($n = 21$) wheats were studied (Krishnaswamy, 1938). In the diploid *Triticum monococcum* ($n = 7$) Kihara and Katayama (1933) recorded in pollen mother cells (PMC's) of haploids almost always seven univalents (98%) but there was rarely a bivalent. Kihara (1936) found in a few PMC's of a haploid of tetraploid wheat as many as three bivalents. Gaines and Aase (1936), Kihara (1936), and Krishnaswamy (1938) studied meiosis in PMC's of hexaploid wheat haploids (polyhaploids, n = 21) and found from 48 to 86% of cells had 21 univalents, from 10 to 37% had 19 univalents and 1 bivalent, from 2 to 14% had 17 univalents and 2 bivalents, and from 0.1 to 5% had 15 univalents and 3 bivalents. Only rarely were more than three bivalents or any trivalents observed. The variation among these three reports is typical of different studies of haploids since the researchers used different strains of *T. vulgare* and external conditions do affect the amount of pairing in such haploids (Kimber and Riley, 1963). Actually, as is now known, the presence in hexaploid wheat of a locus on the 5B chromosome prevents the pairing of homeologous chromosomes, and this probably explains the low number of bivalents in the polyhaploid of hexaploid wheat.

In 1952 McGinnis and Unrau also studied meiosis in a haploid of hexaploid *Triticum vulgare* (n = 21). They found about 58% of PMC's with from 1 to 3 bivalents, a few cells with 42 univalents, and one with 21 bivalents. When pollinated with normal pollen from a normal hexaploid, this haploid plant produced a few seeds. In subsequent F_2 and F_3 generations most plants had 21 bivalents at meiosis, but irregularities, especially trivalents and quadrivalents, were common. Such abnormalities were considered to have been produced by crossing-over in the haploid between only partially homologous chromosomes during meiosis, which would be equivalent to reciprocal translocations. Plants with fewer or more than 42 chromosomes in the F_2 and F_3 resulted probably from eggs in the haploid plant with fewer or more than 21 chromosomes. As is usual with plants with unusual meiosis of this sort, each generation became more normal (see diploidization in Chapter 12).

Haploids of maize (*Zea mays* n = 10) are of interest (Chase, 1949, 1952). It was found that pollen from certain strains stimulates haploid parthenogenesis in ovules irrespective of pistillate strain, and that certain pistillate strains produce monoploid (haploid) embryos more than others, regardless of pollen source. This was confirmed by Coe (1959) and Sarkar and Coe (1966). By breeding, monoploid-producing strains were developed by Chase that yield about one haploid per 1,000 plants, but Coe (1959) found a strain that, *as pollen parent*, would induce up to 3% monoploids of pistillate parthenogenetic origin!

About 10% of untreated maize monoploids yield self-progeny because some somatic chromosome doubling occurs in the tassels and ears. This is another example of haploids *as haploids* being sterile but by various means (*Datura*, wheat, etc.) producing diploid progeny. Maize haploidy is an example of genetic control of parthenogenesis and, interestingly, of the effect of pollen on rate of parthenogenesis even though the pollen genes are not present in the developing egg. It is reminiscent of parthenogenetic fish (*Poeciliopsis*) requiring fertilization by males even of other species (Schultz, 1969). Later Chase (1963) reported that about one monoploid maize plant in 80 originates from a nucleus from the pollen, androgenetic or patroclinous, rather than from an egg nucleus, gynogenetic or matroclinous. Kermicle (1969) found in maize a gene called "indeterminate gametophyte" which, among other effects, if present in the pistillate parent, results in nearly all of the monoploid progeny starting from a male nucleus of the pollen tube; they are androgenetic. And Coe (1959) and Sarkar and Coe (1966) reported a strain of maize that produces up to 3% monoploids.

A method of producing very many androgenetic haploids of various species of *Nicotiana*, including tobacco, by removing microspores (or uninucleate pollen grains) from anthers and growing them in culture has been developed (Nitsch and Nitsch, 1969). In culture some of the immature pollen grains develop into haploid embryos.

By 1963 haploid angiosperms had been reported in at least 71 species of 39 genera in 16 families (Kimber and Riley, 1963), but mostly of species and genera of or related to crop plants, such as *Hordeum distichum* (Tometorp, 1939), wheats, *Nicotiana*, tomato, peppers, maize, etc.

The value of haploids in breeding programs is great. Since a haploid must be hemizygous, when its chromosomes are doubled either naturally or by heat shock, colchicine treatment, or decapitation, a *completely homozygous diploid* results. Repeated inbreeding may ap-

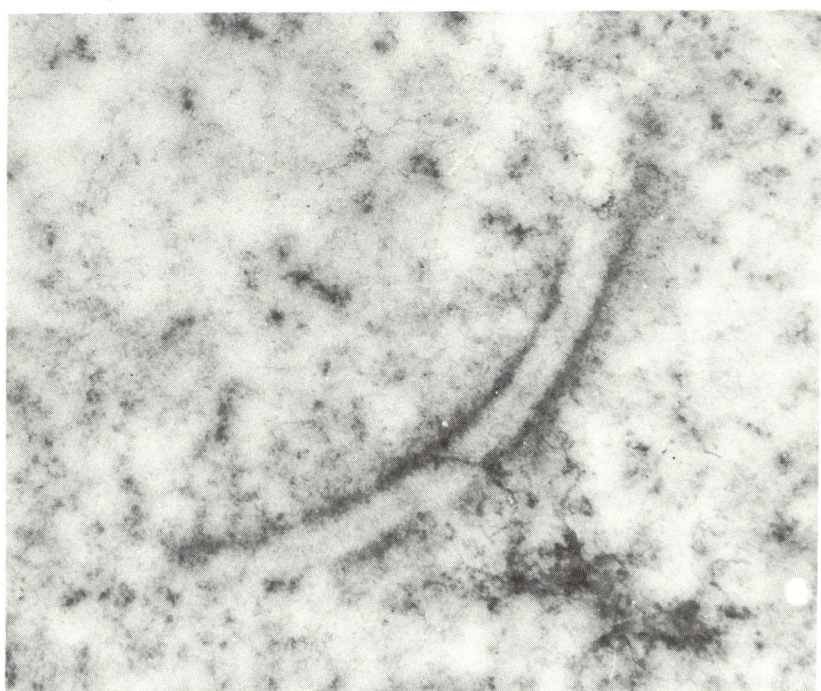

Fig. 11-1. Short segment of synaptinemal complex in a microsporocyte of haploid tomato (n = 12). These profiles were rarely observed. Presumably, they represent synapsis of homologous segments in a true diploid, or they may represent residual homology in a nearly completely diploidized descendant of an ancient amphidiploid from a cross n = 6 × n = 6 or nonhomologous synapsis. A small amount of pairing or synapsis is often reported in haploids of diploids with chromosome numbers above n = 7. (From Menzel, M. Y., and J. M. Price. 1966. Amer. J. Bot. **53**:1079-1086.)

proach homozygosity after many generations, but doubled haploids can be produced in one generation. The advantage of the monoploid method of producing completely homozygous strains is obvious (Chase, 1949).

The study of meiosis in haploids has made tentative contributions to ideas of basic chromosome numbers in a number of genera. The presence of secondary association and of bivalent pairing and even some synapsis at meiosis in haploids (Fig. 11-1) implies some lower basic numbers than that of the presumably diploid species being studied (Kimber and Riley, 1963).

However, pachytene pairing of nonhomologous chromosomes in haploids of *Hordeum, Secale*, diploid wheat, tomato, rice, even with synaptinemal complexes in *Petunia, Antirrhinum, Zea*, and tomato (Fig. 11-11) has been reported (Sadasivaiah and Kasha, 1971). It seems likely that pachytene synapsis between nonhomologues is possible, but crossovers do not form and bivalents are rare at metaphase I.

DIPLOIDY

The diploidy condition is the "normal" for animals and for plant sporophytes. It is the condition to which haploidy and polyploidy are compared. It is the level at which most evolution occurs, but of slow evolution compared to the sudden origin of species by polyploidy. Evolution at the diploid level consists of the slow drifting away from each other with respect to various characters of varieties, or sibling species, and true species by accumulation of mutations and chromosomal alterations, with some form of isolation being necessary. The sudden origin of species by allopolyploidy occurs by the bringing together in a hybrid the genomes of two related species that had already separated by diploid evolution or by previous allopolyploid evolution.

A diploid individual or species has two sets of chromosomes that consist of pairs of homologues. At meiosis these homologues pair and synapse to form good bivalent pairing. However, if there is a closely related form with one half or one third or one quarter as many chromosomes, the presumed diploid may not be really diploid but polyploid. There are also obvious polyploids that act in every way like diploids, called *amphidiploids,* such as tetraploid and hexaploid wheats. The chromosome number of a gamete of a true diploid is described as haploid (n) and may also represent the original or basic chromosome number (x) of the genus or higher category in plants, such as $x = 7$ in wheats, and in some animals, such as $x = 6$ in *Drosophila*.

ANEUPLOIDY

The aneuploid condition occurs when a cell, organism, or species has a number of chromosomes other than an exact multiple of the basic (x) number or of the haploid number of a true diploid. That is, the term aneuploid is used in contrast to the term euploid. The latter is a cell, organism, species, or genus that does possess an exact multiple of the basic number. A haploid (n) has the basic number multiplied by one. A diploid (2n) has the basic number multiplied by two. Polyploids such as triploids, tetraploids, hexaploids etc. are also euploid. Aneuploids, on the other hand, are, for example, $2n + 1, 2n + 2, 4n - 1, 6n - 3$, etc. A $2n + 1$ is also called a *trisomic,* and many trisomics or trisomes are known. A $4n - 1$ or $6n - 1$ can be considered as *monosomic* because it is lacking one homologue of a pair; but monosomics exist in culture or nature usually only at the polyploid level. The monosomic condition of a diploid is usually lethal, and only one such is known in the human species except for sex chromosomes (XO). XO mice are female, fertile, and apparently normal (Morris, 1968). XO's have also been found in *Drosophila* as have also various aneuploids derived from triploids and tetraploids (Bridges, 1922, 1925a, 1939). Aneuploidy upsets the physiological processes of the cell, such as the rate and manner of DNA synthesis during the S period and the rise and fall of certain enzymes during the G_1, S, and G_2 periods of aneuploid (heteroploid) cultured cells of Chinese hamster (Klevecz, 1969).

Aneuploid plants are usually present in F_1 progenies derived from crosses of diploids with triploids. The triploids provide gametes of various chromosome numbers and combinations, and the survival of F_1 individuals depends upon the viability of each combination in the haploid plant spore and/or gametophyte as well as in the 2n zygote or sporophyte. Species differ considerably in the range of F_1 aneuploids recovered (Table 2). In general, aneuploid individuals with from one to three chromosomes more than the diploid number are most abundant, as in petunia (Rick, 1971) and maize (McClintock, 1929; Punyasingh, 1947); but those two species rather unusually also tolerate some aneuploids ranging up to or nearly to the triploid number. Tomato (Rick, 1971) and most other diploid plant species (Khush, 1970) seem to tolerate very few additional chromosomes. Petunia is an extreme and

Table 2. Percentages of plants with aneuploid chromosome number in progeny from $3n \times 2n$ crosses in petunia, maize, and tomato*

Chromosome number	2n	1	2	3	4	5	6	7	8	9	10	11	12
Petunia (n = 7)	9	32	35	14	3	5	2	0 = 3n					
Maize (n = 10)	2	12	30	26	20	11	3	1	0	0	0 = 3n		
Tomato (n = 12)	38	43	16	1	0	0	0	0	0	0	0	0	1 = 3n

*Aneuploids ranged from diploid to triploid in petunia, and nearly the whole range occurred in maize, but in tomato almost no plants above $2n + 2$ were recovered. Few individuals in petunia and maize were diploid whereas many diploid tomato plants were present in the progeny. (Data from Rick, C. M. 1971. Stadler Genet. Symposia, **1** and **2**:153-174.)

unusual species in its tolerance of aneuploidy, and Rick further pointed out that "the viability of the most unbalanced plants was not noticeably reduced. So little morphological effect was wrought, moreover, that it was impossible to distinguish phenotypically between the aneuploids or between them and the diploids." Two other diploid species that also show indifference to the aneuploid condition in the range between the diploid and triploid chromosome numbers are *Clarkia unguiculata* (Vasek, 1956) and *Collinsia heterophylla* (Dhillon and Garber, 1960). This extreme unresponsiveness to the aneuploid condition as in petunia is unusual and can be contrasted to the condition in *Datura* and tomato trisomics discussed in the next section. Khush (1970) has recently reviewed the subject of aneuploidy at length and in depth.

TRISOMY

The commonest form of aneuploidy is the primary trisomic condition (2n + 1) in which the extra chromosome is a normal chromosome (Fig. 11-2). In a particular species there can be as many different trisomics as there are chromosomes in the gametic set. If a species has chromosomes A, B, C, D, and E there can be five primary trisomics: AAA, BBB, CCC, DDD, and EEE. In hexaploid wheat (n = 21) all 21 trisomics have been produced. Trisomy for sex chromosomes is common in man and other animals, but trisomy for autosomes in man is limited to a very few of the smallest autosome chromosomes, such as the 21 trisomic of Down's syndrome.

In plants trisomics are rather common. The classical example is the series of trisomics of *Datura stramonium*. This species has n = 12. At first certain "mutant" forms were thought to be gene mutants of a peculiar sort (Blakeslee and Avery, 1919), but later they were found to be strains that were trisomic for each of the 12 chromosomes of the set. The phenotype of each trisomic was distinctive and named (Blakeslee and Belling, 1924). By that time Blakeslee had also acquired a number of secondary trisomics that have the extra chromosome in the form of an isochromosome (Belling and Blakeslee, 1924). At metaphase I there are pairing arrangement differences between primary and secondary trisomics that indicate the sort of trisomic (Fig. 11-3) (Belling and Blakeslee, 1924; Rhoades, 1933). The difference between the primary and secondary trisomics is that in the primary the extra chromosome is normal, A-B, but the isochromosome of the secondary trisomic has two identical arms, A-A or B-B. Since in meiotic pairing A must pair with A, and B with B, the metaphase I configurations are different.

Blakeslee and co-workers derived most of their aneuploid types from irregular segregation out of a triploid crossed to a diploid plant. The triploid had been produced by a cross of a diploid with an autotetraploid. They also recorded many other aneuploids such as: 2n + 2 (tetrasomic), 2n + 1 + 1 (double trisomic), 2n − 1 (monosomic), 3n + 1, 3n − 1, 4n + 1 + 1, 4n + 1 − 1, etc. (Blakeslee and Belling, 1924). The tetrasomic has four of one kind of chromosome whereas the doubly trisomic individual is trisomic for two different chromosomes in the same organism. Levin (1968) found two plants of *Liatris* in a hybrid swarm of *L. aspera* (2n = 20) and *L. spicata* (2n = 20), one of which

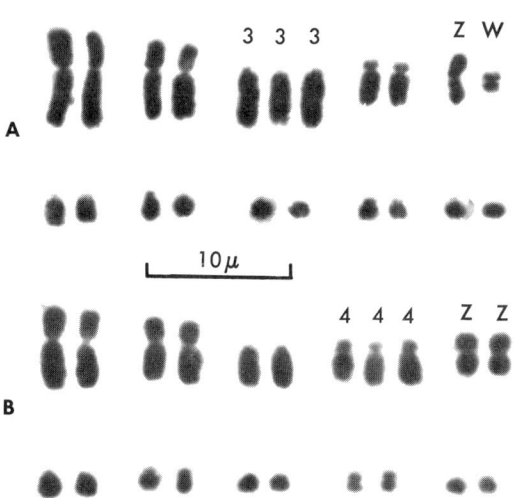

Fig. 11-2. Trisomy in chick embryos that were near death at 4 days of incubation. The 10 largest pairs of chromosomes are represented. **A**, Trisomy for chromosome No. 3 in a ZW sex chromosome female. **B**, Trisomy for chromosome No. 4 in a ZZ male. (From S. E. Bloom. 1970. Science **170**:457-458. Copyright 1970 by the American Association for the Advancement of Science.)

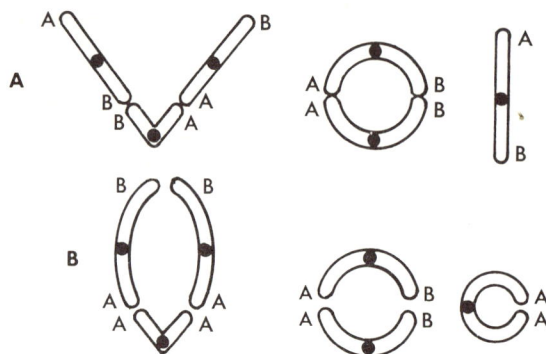

Fig. 11-3. Comparison of the commonest chromosomal configurations in primary and secondary trisomics. Trivalent association (**A**) in the primary is a chain of three; in the secondary it is a ring of three. In the bivalent-univalent configuration of **B** the two ends of the univalent isochromosome pairs together to form a small ring that cannot be produced by the normal univalent in the primary trisomic.

was trisomic for four different chromosomes since it often formed four trivalents at meiosis; the other was trisomic for six different chromosomes (up to six trivalents at meiosis). Such plants could be segregates from a cross of a diploid with a triploid. Roy and Gould (1971) reported similar aneuploidy in natural hybrids of diploid *Bouteloua hirsuta* × *B. pectinata*, both species being 2n = 20. As in *Liatris*, hybridity in nature has produced aneuploid-containing hybrid swarms. In *Bouteloua* most hybrids were aneuploid.

At meiosis in a trisomic there are three of one kind of chromosome, and during synapsis only two can synapse at any particular site. But at different places along the chromosomes different pairs may synapse. Therefore, at metaphase I various configurations are found. The two commonest are trivalents of various sorts, most frequently a chain of three, or a bivalent with a univalent. The chain of three usually orients itself at metaphase I so that at anaphase I the middle chromosome goes to one pole while the two end chromosomes of the chain move to the other pole. Almost always, therefore, two sorts of gametes are formed for that particular chromosome, some with two and some with one of the three homologous chromosomes.

Such segregation, if a known gene is located on that chromosome, gives distinctive genetic ratios in subsequent generations. For example, if the trisomic is AAa it produces the following gametes: 1AA: 1a: 2A: 2Aa. If test crossed to a normal diploid homozygous recessive, aa, the 5:1 phenotypic ratio results. Such an AAa would be the trisomic resulting from the cross AAA × aa.

When a trisomic ratio can be assigned to a particular chromosome that can be seen to be trisomic and can be identified as chromosome No. 1 or No. 2, etc., the trisomic method of identifying individual chromosomes with their respective linkage groups is employed. It has been used successfully in maize (McClintock and Hill, 1931), tomato (Lesley, 1928; Rick and Barton, 1954; Rick et al., 1964), *Antirrhinum* (Rudorf-Lauritzen, 1958), *Datura* (Avery et al., 1959), *Spinacea* (Janick et al., 1959), and *Hordeum* (Tsuchiya, 1959).

In plants trisomics are common. A recent addition is *Arabidopsis thaliana*, all five primary trisomics of which have been produced and identified with known linkage groups (Sears and Lee-Chen, 1970). This species of flowering plants can have about 12 generations a year. In man, *Drosophila*, and most animal species, however, trisomics are rare. The only viable *Drosophila* trisomic is for the minute chromosomes No. 4. Chicken embryos trisomic for chromosomes No. 3 or No. 4 die at about 4 days (Bloom, 1970) (Fig. 11-4).

Rick and Barton (1954) and Rick et al. (1964) have identified all 12 trisomics of tomato with pachytene chromosomes and linkage groups by trisomic ratios, except that not one of 60 genes tested was located on chromosome No. 12. They concluded (1964) that "spontaneous mutant genes are not randomly distributed among the tomato chromosomes," a condition also reported in maize (Rhoades, 1955).

Trisomics for all 10 chromosomes of maize have been identified by trisomic ratios (Rhoades, 1955). Rhoades found that in some ovules of the 2n + 1 plants the extra chromosome is eliminated so that normal haploid eggs are formed. Thus, the actual trisomic ratio observed is about 4:1 rather than 5:1. But it is still very distinct from the 1:1 disomic test

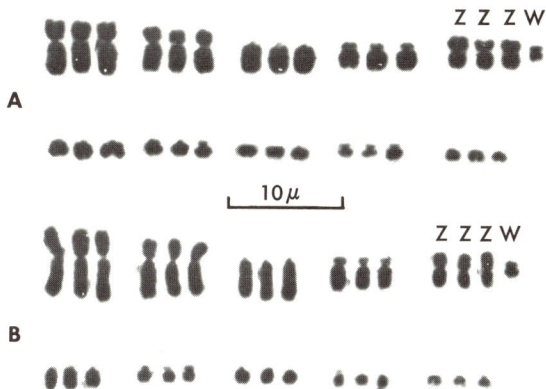

Fig. 11-4. Chromosomes (the 10 largest only) from a triploid female chick embryo that appeared normal at 4 days of incubation. Each chromosome except the W is present in triplicate in each of the two cells (**A** and **B**) represented. (From S. E. Bloom. 1970. Science **170**:457-458. Copyright 1970 by the American Association for the Advancement of Science.)

cross ratio. In maize the only secondary trisomic found is an isochromosome for the short arm of chromosome No. 5. Pairing was as reported earlier for secondary trisomic *Datura*.

Rhoades (1955) also stated that numerous tertiary trisomics of maize have been reported. A tertiary trisomic has the extra chromosome composed of parts of two nonhomologous chromosomes. These have been derived from normal diploid plants heterozygous for large exchanges as follows: At synapsis the two normal chromosomes and the two translocated ones form a typical group of four. Rather typically the two normal chromosomes that lie opposite each other in the ring of four at metaphase I go to one pole at anaphase I and the two translocated chromosomes go to the other pole. But if the two normal chromosomes and one translocated chromosome go to one pole, the gametes resulting will be n = 11 and after union with a normal n = 10 gamete will produce a n = 21 tertiary trisomic plant.

Rick and Notani (1961) proposed an interesting generalization about trisomics of various sorts, that primitive or wild forms possess greater tolerance to one or more extra chromosomes than do highly cultivated species or forms. They demonstrated much greater tolerance to and transmission of extra chromosomes by the more nearly wild-type, Red Cherry, variety of tomato than a typical large-fruited commercial variety, Marzano. They also gathered data on other wild and cultivated trisomics for tolerances, vigor, and transmission of trisomes. In all cases the wild forms were superior to the cultivated. The authors proposed that "Primitive or wild forms undoubtedly possess such tolerance as part of their plasticity and ability to withstand various unfavorable situations" and this plasticity permits them to tolerate extra chromosomes.

MONOSOMY

Monosomy is the 2n − 1 chromosomal condition. A nullisomic is 2n − 2, and the two missing chromosomes are homologues. The terminology is similar to that for trisomy. Because a monosomic is deficient for a whole chromosome, monosomics are rare in diploid species and even in tetraploid wheat but are occasionally reported, as in *Datura* (Blakeslee and Belling, 1924) and the G monosomic reported in man (Al Aish et al., 1967). Monosomy of sex chromosomes in diploid animal species, however, is not uncommon, as in *Drosophila*, man, and mouse already mentioned. Monosomics are not uncommon, however, in cultures of polyploid plants such as cotton and wheat.

There can be as many monosomic types as there are chromosomes in the gamete. In wheat (2n = 42) 21 are possible, and in the American tetraploid cottons (2n = 52) there are 26 possible. In wheat all 21 possible 2n = 41 monosomics have been produced at least once (Sears, 1944) and all 21 monosomics of the wheat variety, Chinese Spring, are maintained as useful cultures (see later). All 21 nullisomics (2n = 40) were also produced by Sears and maintained. Both monosomics and nullisomics (lacking both homologues of a pair) are variably more sterile than diploids. Nullisomic wheats can persist by doubling a pair of homeologous chromosomes so that the nullisomic is also tetrasomic and has 42 chromosomes. That is, nullisomic 5B is tetrasomic 5A or 5D. 5A, 5B, and 5D are homeologous chromosomes that can functionally more or less compensate for one another. 5B is the chromosome No. 5 of the

B genome, and 5A and 5D are the chromosomes No. 5 of the A and D genomes.

Monosomics, like trisomics, can be used for determining linkage groups (Knott, 1959; and many others) by the chromosome substitution method. The production of all 21 aneuploid monosomics of hexaploid wheat (Sears, 1954) has opened up a new approach to the genetic study of a polyploid species as well as providing material useful in other cytogenetic studies. One use of these monosomics is in the production of varietal *substitution lines*. One donor variety of wheat is crossed to a recipient monosomic line, and the progeny are repeatedly back-crossed to the monosomic. Eventually a line is produced that has the 20 pairs of chromosomes of the monosomic line plus one substituted pair of the donor line in the place of the original pair that was half missing in the monosomic. This process can be and has been repeated for all 21 monosomics of the wheat variety Chinese Spring, and in Europe the plan of the European Wheat Aneuploids Cooperative is to produce all 252 substitution lines for four selected varieties, each to have 21 substituted chromosomes of the other three (Sears, 1969). As Sears wrote, "The interaction of any chromosome with its genetic background can be determined by first crossing the proper substitution line with both the recipient variety and the donor variety and then comparing the two F_1's with each other, with the substitution line, and with an F_1 between the two varieties." Askel (1967) used this method successfully.

If the donor of the substituted chromosome is a different species, even of a different genus, such as rye, *Agropyron, Aegilops*, or other species of *Triticum*, a so-called alien substitution line is produced. This is possible because all these genera have homeologous chromosomes (Sears, 1969). These alien substitution lines have been deliberately produced (Anderson and Driscoll, 1967) or have been detected in progeny from interspecific hybrids (Knott, 1964). The usual method of producing such alien substitution lines is to derive from the hybrid an "additional line" with 42 wheat chromosomes plus a pair from the alien parental species. This, when crossed to the proper wheat monosomic, has the alien pair of chromosomes inserted in place of the monosome (the single chromosome of the monosomic).

The alien chromosome replaces or compensates for one of the three homeologous chromosomes of wheat better than the others, which indicates that although each chromosome has two homeologues in hexaploid wheat, these homeologues have evolved considerable differences. Furthermore, the alien chromosome often behaves differently than the wheat chromosomes, being more subject to asynapsis, misdivision, or loss (Sears, 1969). Crossing-over between the wheat homeologues and the substituted alien chromosome can be made to occur by having the wheat also nullisomic for the 5B chromosome. The 5B chromosome carries a gene locus that inhibits homeologous pairing. When it is lacking (in the nullisomic-5B condition) homeologues, even between different genera, can pair, synapse, and recombine.

The term aneuploidy is also used in connection with changes of basic chromosome number either upward or downward. A much better term for such change is *dysploidy*. For example, it has been demonstrated in *Crepis* that the $n = 3$ species *C. fuliginosa* has been derived by aneuploid (or dysploid) decrease from an $n = 4$ form similar to *C. neglecta* by translocation (Tobgy, 1943). This form of evolution is discussed in Chapter 13.

Another form of evolution can arise by hybridization between two species having different chromosome numbers. The F_1 hybrids produced by Tobgy between *C. neglecta* ($n = 4$) and *C. fuliginosa* ($n = 3$) had $2n = 7$, a chromosome number that is aneuploid relative to either of the parental species. Such hybrids are usually highly sterile because of genetic imbalance and chromosomal structural differences.

It can be concluded that abnormal haploids and aneuploids, although genetically useful and cytologically interesting, are unsuccessful in nature and of little evolutionary significance.

CHAPTER 12
Polyploidy

Any cell, organism, or species that has a chromosome number exactly three or more times the gametic number of a proved diploid or an hypothetical diploid is a polyploid. To some extent aneuploids of such polyploids are also considered to be polyploid, as can be also a haploid of a polyploid. Such examples occur commonly in plants. Here, however, will be discussed polyploids as just defined, that is, *euploids* rather than aneuploids. Haploidy and aneuploidy have been treated in Chapter 11. Euploids are named for the number of sets of chromosomes present just as are monoploids (haploids) and diploids. Examples of euploidy are triploids (3n), tetraploids (4n), pentaploids (5n), hexaploids (6n), octoploids (8n), 10-ploids, 12-ploids, 38-ploids, etc.

Most polyploid species are tetraploid, hexaploid species are less common, etc.

A few examples of high polyploidy are: a species of *Kalanchoe*, 2n = about 500 (Baldwin, 1938); the New Zealand grass, *Poa litorosa*, approximately 38-ploid with 2n = 265 and x = 7 (Hair and Benzenberg, 1961) (Fig. 12-1); an *Hibiscus* hybrid with 2n = 216 (x = 18?) is about 12-ploid (Menzel and Wilson, 1963) (Fig. 12-2); *Galium grande* with 2n ± about 220 (x = 11) is about 29-ploid, (Dempster and Stebbins, 1968); the radiolaran (Protozoa) *Aulacantha* with from 1,000 to 2,700 chromosomes (Grell and Ruthmann, 1964); and, of course, some ferns and fern allies, such as *Schizaea dichotoma* (2n = 1,080) and some plants of *Ophioglossum reticulatum* with about 1,260 chromosomes must be highly polyploid although the basic number is unknown. Such high polyploids are often aneuploid because a few chromosomes more or less make no difference to a plant so highly redundant for genomes and genes.

Triploids and pentaploids, because they have an odd number of sets of chromosomes, have irregular meiosis and are unstable; they do not persist in nature or culture. Polyploids with even numbers of sets of chromosomes have meiosis from very irregular to as regular as any diploid. Polyploids can also be categorized by similarity of their chromosome sets, from autopolyploidy to extreme allopolyploidy. Naturally, there is a graded series beginning with four or six *identical* sets but fading to levels of nearly complete incompatibility, where hybridization becomes impossible. Those polyploids with four or six or eight identical to very similar sets are *autopolyploids*, whereas those with sets from rather similar to very dissimilar are *allopolyploids*. Characteristics and dissimilarities of these contrasting types and intermediates are discussed in this chapter.

TRIPLOIDY

Triploids, like haploids, are unstable and do not persist in sexual forms. They do persist as asexually reproducing species such as triploid potatoes. They are certainly more common in

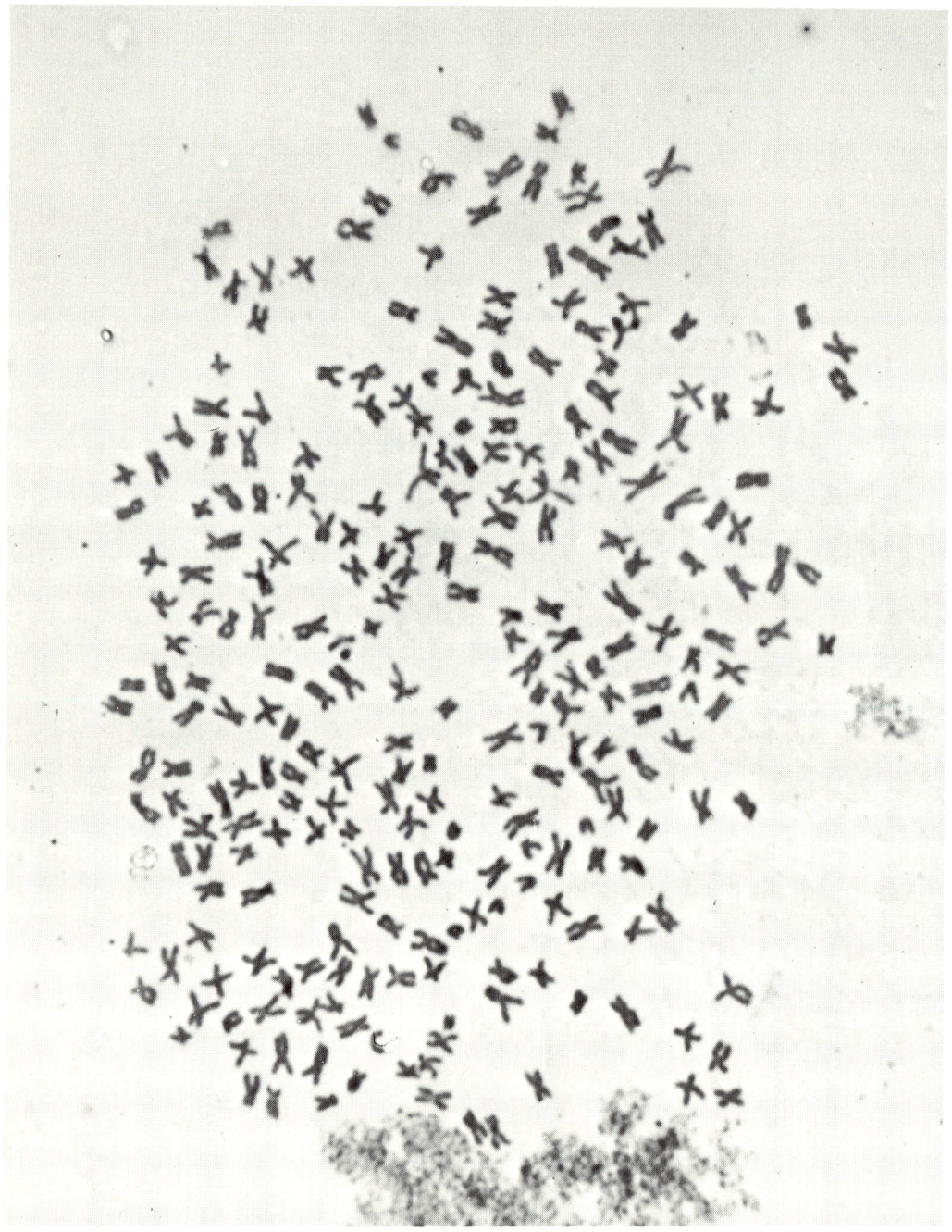

Fig. 12-1. An example of high polyploidy, in the New Zealand grass *Poa litorosa*. The basic number in *Poa* is x = 7, so this species with about 265 chromosomes is 38-ploid. (From Hair, J. B., and E. J. Benzenberg. 1961. Nature **189:**160.)

Fig. 12-2. Meiosis in a high polyploid hybrid between tetraploid *Hibiscus radiatus* (2n = 72, x = 18) × octoploid and decaploid *H. diversifolius* (2n = 144 and 180). At metaphase I in this hybrid most chromosomes were paired and the chromosome number was found to be 2n = 216, an allododecaploid, or 12-ploid. (From Menzel, M. Y., and F. D. Wilson. 1963. J. Hered. **54**:55-60.)

plants than in animals, mainly because polyploidy is more common in plants; but numerous parthenogenetic animals are triploid hybrids (see later). In *Drosophila* until 1939 (Muller, 1939) triploids had been reported repeatedly, but studied genetically only, such as by Bridges (1922) and Metz (1925). Morgan (1925) and Dobzhansky and Bridges (1928) studied triploid intersexes. Morgan's example had two "attached X" chromosomes. Redfield (1930), Mather (1933), Kikkawa (1934), and Beadle (1935), for example, studied crossing-over in triploid drosophilas. Redfield (1932) compared crossing-over in triploid and diploid *Drosophila*. Schultz and Dobzhansky (1933) produced sterile female allotriploid hybrids from a cross of triploid *melanogaster* × diploid *simulans*. The hybrid had two sets of *melanogaster* chromosomes and one of *simulans*. Bridges (1925b, 1928) and Bedichek (1938) studied the relation of sex to chromosome sets, and Schultz (1934) studied the manifestation of dominance in the triploid.

Triploidy in man is lethal. No triploids reach birth, but they have been reported in studies of naturally aborted foetuses. Triploid newts are viable and may be of either sex (Fankhauser, 1938), but haploids seem to be female (Fankhauser, 1937). Triploid and tetraploid Uriodeles are found in natural populations of diploids (White 1948a), and triploids have been reported in various insects. Polyploid parthenotes (parthenogenetic animals) are not rare in animals such as triploid and hexaploid weevils (Takenouchi, 1970), triploidy being more common than other polyploid stages in parthenogenetic weevils (Suomalainen, 1969).

Triploid plants, on the other hand, quite regularly reach maturity and are somewhat fertile. They have been found or produced in most of the species discussed with haploidy in Chapter 11. They are occasionally formed by a diploid when an unreduced egg (2n) is fertilized by the haploid sperm. Rhoades (1936), however, reported a triploid maize plant derived from a haploid egg and a diploid sperm, and Beadle (1930) recovered many triploids from a cross of his "asynaptic" mutant with a normal diploid. Most triploids are produced experimentally by crossing a diploid and a tetraploid. Hundreds of triploids have been produced in this way in the relatives of wheat, cotton, potato, tobacco, etc. in genome analysis. Another experimental importance of triploids is that they produce useful aneuploids, such as various trisomics, when crossed with a diploid. They are quite unimportant in nature, however.

Triploid species or varieties can persist without sexual reproduction if they have some sort of natural form of vegetative, parthenogenetic, or apomictic reproduction or are perpetuated artificially by man. There are four triploid species of *Solanum* (x = 12) in the tuber-bearing group that includes the potato. The species *S. chaucha* has been studied by Lamm (1945). At meiosis it has mean numbers of eight trivalents, four bivalents, and four univalents per PMC. About 20% "good" pollen is formed, but the species is sterile. Good pollen means that the grains are full of protoplasm, but that does not mean they can function adequately in forming pollen tubes, achieving fertilization of the egg and polar nuclei, producing good endosperm, or producing a viable embryo. *Lilium tigrinum*, the commonly cultivated tiger lily, is

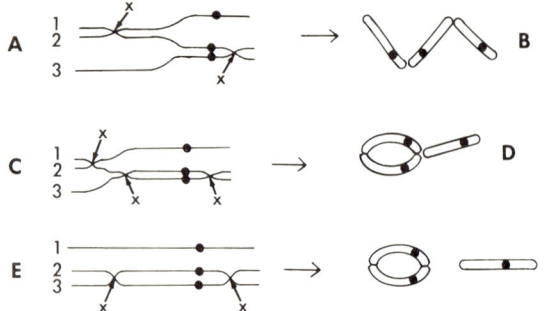

Fig. 12-3. Associations of three homologous chromosomes in meiosis, as in a trisomic or triploid. **A, C,** and **E** represent pachytene associations; **B, D,** and **F,** the resulting metaphase I appearances. **B** and **F** are the commonest associations. Crossovers at pachytene become terminalized by metaphase; solid circles represent centromeres.

triploid (2n = 36, x = 12) and reproduces by bulbils in the axils of the leaves (Noda, 1966). At metaphase I there are numerous trivalents and some univalents in addition to some bivalents.

Meiosis in an autotriploid is like that of a trisomic but for all the chromosomes of the three sets. That is, at metaphase each group of three homologous chromosomes is usually in an association of a trivalent or a bivalent and univalent (Fig. 12-3). From such configurations the segregation is two chromosomes to one pole and one to the other, but segregation is random for all the various chromosomes. As a result, one extreme is a few haploid and diploid cells, but usually each telophase I nucleus contains a number of chromosomes somewhere between the haploid and the diploid condition; that is, 1, 2, 3, etc. more than haploid, and 1, 2, 3, etc. less than diploid.

Not only is segregation apt to be irregular, but very often univalents and even bivalents lag on the metaphase plate and so are not included in the daughter nuclei (East, 1933; King, 1933). King reported in a triploid (2n = 18, x = 6) of *Tradescantia* the following distribution of chromosomes at anaphase I.

	Number of chromosomes at poles							Number of cells
	5	6	7	8	9	10	11	12	
None lagging			3	13	50	13	3		82
One or more lagging	2	2	26	48	31	7	2	1	119

Thus, most telophase I nuclei received 8, 9, or 10 of the 18 chromosomes. He estimated 47% good pollen, although Riley (1959) observed only 1.6 and 3.5% good pollen in two similar triploid plants.

Yarnell (1930) in a triploid (2n = 21, x = 7) of *Fragaria* recorded almost all metaphase I configurations as 10 bivalents and 1 univalent, or 9 bivalents and 1 trivalent. Since there are only seven sorts of homologues, there must be non-homologous pairing to produce 9 or 10 bivalents. East (1933) also considered some sort of non–gene-by-gene pairing in addition to homologous pairing. He emphasized the wide variation among individuals of progeny from a triploid selfed and crossed to a diploid of tobacco regardless of chromosome number.

Thus, there is great variation among cells produced by meiosis with respect not only to chromosome numbers but also to which chromosomes are present once or twice or absent. Since many of these combinations are lethal, triploids have low fertility. The progeny resulting from a cross between a triploid and a diploid often have chromosome numbers 2n, 2n + 1, 2n + 2, and perhaps 2n + 3 (Belling and Blakeslee, 1922, on *Datura*; Lesley, 1928, on tomato) although more of the intermediate types have been recorded (Yarnell, 1931, on *Fragaria*; East, 1933, on tobacco; van Overeem, 1921, on *Oenothera*). Furthermore, McClintock (1929) found that most fertilizations by pollen from a maize triploid contained only 10, 11, or 12 chromosomes; the others, evidently, could not compete successfully in the overall process of fertilization. The reciprocal cross produced many more kinds of aneuploids. As Rhoades (1955) wrote, "The different primary trisomics isolated by McClintock in the progenies of diploid by triploid crosses and their subsequent correlation with specific linkage groups marked the beginning of maize cytogenetics."

Interspecific hybrids between tetraploid and diploid species are discussed in Chapter 16.

Unlike triploidy, tetraploidy and other ploidy levels that are whole-number multiples of 2x are usually stable conditions, and a great many plant species have such chromosomal constitution. Tetraploidy is by far the most common. For example, of the more than 30 species of

Gossypium all are diploid except for 3 tetraploids.

PENTAPLOIDY

Pentaploid species, like triploids, can persist if they have asexual (Chapter 18) or other (Chapter 17) means of reproduction. There are at least three species of *Solanum* that are pentaploid and at least one of them, *S. curtilobum* (2n = 60), is sexually fertile. Lamm (1945) studied that species. It produces 40% good pollen and, when not selfed or crossed to other species such as the tetraploids *tuberosum* or *andigenum*, it sets abundant seeds. Meiosis is about as irregular as other polyploids of the tuberous section of the genus with varying numbers of univalents, bivalents, trivalents, quadrivalents, and pentavalents. When crossed to a tetraploid species, the progeny have chromosomes ranging from 51 to 62, mostly from 53 to 56, and the expected number is 54 (30 + 24); but the same numbers are found when it is selfed. This is expected since meiosis is irregular in all the polyploid species of the group. True-breeding pentaploid dog roses are discussed in Chapter 17.

POLYPLOIDY IN ANIMALS

Polyploid species of animals that are sexual used to be considered as very uncommon. Most polyploid animals are also parthenogenetic, and many are triploid, so polyploid animals will be discussed at this point. Of course, occasional autotetraploid individuals of normally diploid species are reported, as in *Drosophila* (Metz, 1922; Burdette, 1938; Bridges, 1925b, etc.), although there are no known polyploid species in the genus (Patterson and Stone, 1952). That is true of essentially all animal genera. The usual reason given (Muller, 1925) is that polyploidy in a genus of sexually reproducing male and female organisms, in which sex is determined by sex chromosomes, is intolerably mixed up by polyploidy. For example, if an XX-XY species became suddenly polyploid, it would be XXXX-XXYY. At meiosis in the male, if X and Y paired and were oriented randomly, the following sorts of sperm would be produced: XX, YY, and XY. Zygotes would be XXXX, XXYY, and XXXY if segregation in the female were regular, which it might not be. The XXXY individuals might be sterile or female. But Muller (1925) proposed that in the male X would pair with X, and Y with Y, so that almost all male gametes would be XY and the progeny XXXY and abnormal. Thus, a true-breeding tetraploid strain could not originate. Of course, some plants with sex chromosomes have become tetraploid *(Rumex)* at least for the autosomes, by simply keeping the sex chromosomes at the diploid level, XX-XY. Animal polyploids have rarely done this.

Another possible explanation of the lack of polyploidy in animals is that natural hybridization in animals has seemed to be very rare, and almost all known animal polyploids are *autopolyploids* (White, 1948a, p. 35) (but see later) whereas almost all plant polyploids are allopolyploids, the result of hybridization. Failure to hybridize among animal species in nature results from many isolating mechanisms not the least of which are sexual signals of many sorts that are species-specific, such as calls, movements, colors and patterns, odors, etc. that are often insurmountable barriers to hybridization, barriers not present in plants. Thus, lack of hybridization *and* chromosomal sex-determining mechanisms *and* almost universal bisexuality (most plants, on the other hand, are hermaphroditic) seem adequate to explain the lack of successful polyploidy in animal evolution.

Hybridization in animals, with or without polyploidy and/or parthenogenesis, is not as rare as used to be thought, at least in some groups. A few examples of natural hybridization are: in fish, especially freshwater forms (C. L. Hubbs, 1955; C. Hubbs, 1967), in toads (Blair, 1941), in salamanders (Uzzell and Goldblatt, 1967; Highton and Henry, 1970), reptiles (Lowe and Wright, 1966; Wright and Lowe, 1967), several species of the moth genus *Platysamia* (Sweadner, 1937), *Drosophila* (Patterson and Crow, 1940), a sawfly (Smith, 1941), grasshoppers (Klingstedt, 1939), weevils (Suomalainen, 1969), and snails (Boettger, 1922; Franz, 1928). The Hubbses considered hybridization among species and close genera of freshwater fish as almost inevitable. On the other hand, hybridization among marine fish species is rare. C. L. Hubbs (1955) mentioned only about six known cases. Natural hybridization among birds, mammals, and animals in general is probably less common, at least compared to plants; but Gray (1954) did list numerous mammalian hybrids, and the hybridity of the mule, cattle, dog-wolf, etc. are well known. Mayr (1969, p. 33) stated recently that there has been no positive identification of allopolyploidy or of hybridity in polyploid parthenotes in animals, but there is disagreement about this. Part of this difference between plants and animals is the ease of studying meiosis in plants but the difficulty of studying it in most animals.

Polyploids can be produced by various treatments, of which the colchicine method is best known and most productive. For example, Kawaguchi (1936) produced triploid, tetraploid, and hexaploid individuals of silkworms by centrifugation of developing eggs. Triploid and tetraploid males and females were more or less fertile.

The only known sexual allopolyploid animals include a synthesized tetraploid hybrid of the silkworms *Bombyx mori* and *B. mandarina* (Vereyskaya, 1965). This was achieved by crossing an autotetraploid parthenogenetic female *mori* (4n = 112) to a diploid male *mandarina* (2n = 56). The triploid (3n = 84) contained two genomes of *mori* and one of *mandarina*. By use of temperature shock, triploid females were made parthenogenetic, and they contained some 6n cells. Either a reduced hexaploid egg or an unreduced triploid egg was fertilized by a haploid sperm of *mandarina* to produce the allotetraploid, which was fertile and was maintained for at least six generations.

Wright and Lowe (1967) and Lowe and Wright (1966) proposed natural interspecific hybridization to produce allodiploid and allotriploid lizards, and Hubbs and Hubbs (1932) and Schultz (1969) have proposed a similar hybrid origin for allodiploid and allotriploid fish. But other polyploid animals, such as the fish *Carassius auratus* (a gold fish) and the carp *Cyprinus carpio*, that are tetraploid by chromosome count and DNA amount are most likely autopolyploid (Ohno et al., 1967). Schultz (1969) has proposed the following chromosomal scheme in the sexual diploid and parthenogenetic triploid (Cy, Cx, and "new") fish *Poeciliopsis*:

Lucidum	Cy	Cx	"New"	Monacha
2n(LL)	3n(LLM)	2n(LM)	3n(LMM)	2n(MM)
Sexual	Partheno.	Partheno.	Partheno.	Sexual

Similarly, Uzzell and Goldblatt (1967) proposed two sorts (species) of triploid parthenogenetic salamanders in the *Ambystoma jeffersonianum* complex:

Jeffersonianum	Platineum	Tremblayi	Laterale
2n(JJ)	3n(JJL)	3n(JLL)	2n(LL)
Sexual	Partheno.	Partheno.	Sexual

This conclusion is based upon morphology and analysis of serum protein electrophoresis.

The bug *Thyanta calceata* has the peculiar sort of ploidy called *agmatoploidy*. Smith (1941) did express the belief that the sawfly *Diprion similis* (Hymenoptera), which has n = 14, is an allotetraploid of n = 7, the chromosome number common in the rest of the genus.

There are a few species or populations of sexual animals that are naturally polyploid, such as anuran amphibians (Beçak et al., 1966; Beçak, 1969; Bogart, 1967; Wasserman, 1970; Beçak et al, 1970). Wasserman reported a tetraploid clasping-pair and complete tetraploid progeny, based on a sample of 19, as well as other reports of 2n = 48 as evidence for considering the frog species *Hyla versicolor* tetraploid relative to 2n = 24 species. There are four of each kind of chromosome at mitotic metaphase so these polyploids are probably autopolyploids with, however, regular bivalent pairing and regular chromosome segregation during meiosis (Fig. 12-4). Beçak et al. (1970) reported two tetraploid and one octoploid anuran amphibians with the basic chromosome number of x = 11 and 13. The tetraploid species *Odontophrynus americanus* (n = 13) had at meiosis 10 ⊙4 plus 6 bivalents, indicating that it is mostly an autotetraploid. Bogart (personal communications) states there are nine known (published and unpublished) polyploid species of anuran amphibians: five South American, three African, and one North American. Amphibians and the fish *Poeciliopsis* do not have heteromorphic sex chromosomes, which may be related to existence of polyploid sexual

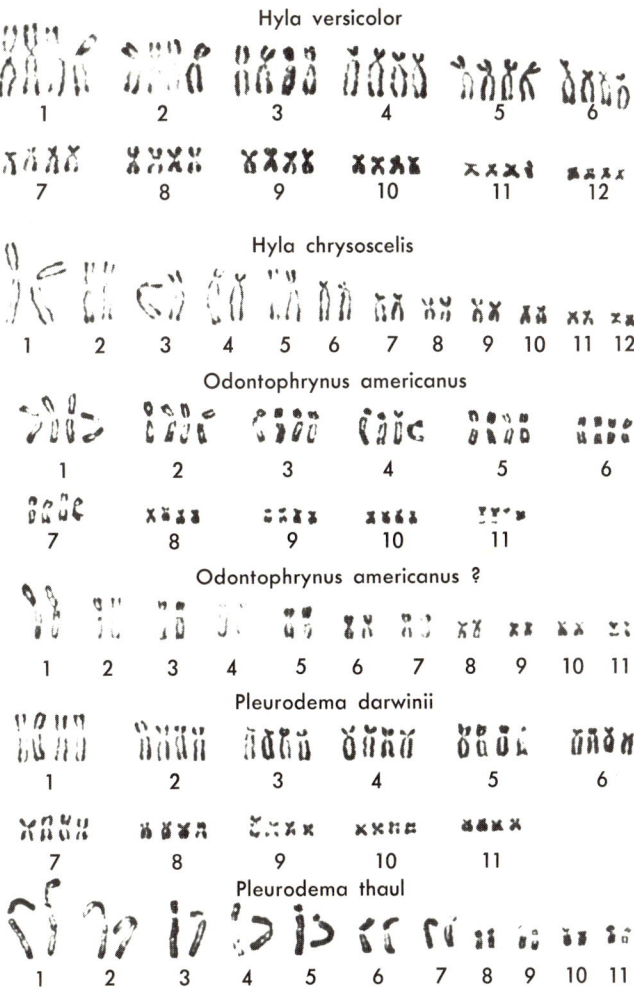

Fig. 12-4. Diploid and related tetraploid anuran amphibians. *Hyla versicolor* is a tetraploid relative of diploid *H. chrysoscelis. Odontophrynus americanus* is a tetraploid relative of diploid *O. americanus? Pleurodema darwinii* is a tetraploid relative of diploid *P. thaul.* Since the tetraploids have four of each sort of chromosome, they are doubtless autotetraploid. (Courtesy J. P. Bogart, Department of Zoology, Louisiana Tech. University, Ruston.)

species. White (1948a) cited Wilson as reporting the bug species *Thyanta custator* as n = 7 but *T. calceata* as n = 14 and tetraploid. Schrader and Hughes-Schrader (1956) have found that *T. calceata* has no more DNA than *T. custator*. This is then a case of agmatoploidy (see later).

Parthenogenesis

Polyploid parthenotes (parthenogenetic animals) are common (White, 1954, Suomalainen, 1940, 1969). The classical example of polyploidy associated with parthenogenesis is the brine shrimp, *Artemia salina* (White, 1948a). The bisexual condition exists at the diploid (2n = 42) level although a tetraploid (probably autotetraploid) sexual form has been reported. The parthenotes are diploid, tetraploid, and hexaploid. In all forms the first meiotic division is normal, but in the parthenotes there are two reconstitution (diploid) nuclei formed at the second meiotic division rather than four haploid spermatids. Again as in sexual polyploid amphibians, good bivalent pairing occurs at meiosis in spite of autopolyploidy (see below).

Parthenogenetic polyploid animals are not uncommon in many groups, such as in the Crustacea the isopod *Trichoniscus elizabethae* consisting of sexual diploids (2n = 16) and asexual triploids (2n = 24), the ostracod *Cypris fuscata* with sexual diploid and parthenogenetic triploid races, and *Daphnia pulax* (Schrader, 1925) with a hexaploid asexual race. One tetraploid parthenogenetic mollusc was mentioned by White (1948a). Many polyploid parthenogenetic insects are known in beetles, moths, wingless grasshoppers, phasmatids, etc. Parthenogenesis at the diploid level has also been found.

Parthenogenesis at the diploid level has already been mentioned in *Artemia salina*. In *Drosophila* there are a number of fractionally (facultatively) parthenogenetic diploid species (Stalker, 1954; Carson, 1967b), in which the percentage of parthenogenetic eggs can be increased by selection (Carson). Presumably, most cases of parthenogenesis in *Drosophila*, such as *D. funebris*, *D. tropicalis*, and *D. willistoni*, are incipient and polygenic, although *D. mangabeirai* is completely parthenogenetic. In no species of *Drosophila* has the parthenogenetic condition been accompanied by polyploidy. Suomalainen (1969) reported that more than 100 species of weevils in the two subfamilies Oliorrhynchinae and Brachyderineae have been studied in detail. Of these, 60 are diploid and bisexual and 42 contain parthenogenetic strains all of which are polyploid except one species, which is a parthenogenetic diploid. He reported 1 diploid, 25 triploid, 12 tetraploid, and 4 pentaploid parthenogenetic forms. Of these 42 species, 22 have both diploid bisexual and polyploid parthenogenetic races. Some of these races might be considered as sibling species. The parthenogenetic races almost always have wider geographical ranges than do the related diploids. He also reported examples of genome segregation in eggs of parthenotes at metaphase I. That is, whole sets of chromosomes were arranged on separate metaphase plates. Of such divisions 55% were 4n or irregular, 12% had one plate of 3n and a second plate of n chromosomes, 10% were 2n and 2n, and 22% were 2n, n, and n. He concluded that polyploid parthenogenetic forms have been derived from diploid sexuals via diploid parthenogenetic forms.

Parthenogenesis in vertebrates has been reported in a few fish (Hubbs and Hubbs, 1932; Schultz, 1969), two species of salamanders (Uzzell, 1964), and a few lizards (Darevsky, 1966; Maslin, 1962; Lowe and Wright, 1966; Thomas, 1965; Martens, 1960). Most of these are triploid but some are diploid and a few are tetraploid. Schultz (1969) considers the parthenogenetic fish *Poeciliopsis*, both diploids and triploids, to be species hybrids (allodiploid and allotriploid) as do Lowe and Wright (1966) their parthenogenetic lizards of the genus *Cnemidophorus*, and Uzzell and Goldblatt (1967) their triploid parthenogenetic salamanders.

Although most polyploid parthenotes probably arise from diploid parthenotes that have meiotic genes for parthenogenesis, it is also possible that polyploidy per se might produce parthenogenesis. That is, the polyploid condition in the egg might, by itself, initiate parthenogenesis, which anyhow is merely division of the zygote. In plants, too, apomixis is almost always associated with polyploidy. But meiosis in eggs of polyploid parthenogenetic animals is very regular whereas meiosis in autopolyploid plants is notoriously irregular, at least in recently synthesized forms.

Thus, aside from the interesting polyploidy in a few sexual amphibians and possibly a few other animals, polyploid species and subspecies of animals are asexual in the polyploid condition.

POLYPLOIDY IN PLANTS

Because polyploidy is so common in plants, there are many conditions of polyploidy most of which intergrade to produce a confusing array and terminology. The terminology of Stebbins (1950) is followed here, consisting of autopolyploidy, allopolyploidy, segmental allopolyploidy, genomic allopolyploidy, and autoallopolyploidy.

Autopolyploidy

At one extreme is homozygous tetraploidy, which is the clearest and least controversial form of polyploidy. That extreme condition occurs when the chromosome number of a single diploid plant doubles or is doubled. The result is a type with four sets of homologous chromosomes. Approximately the same condition may occur in a hybrid between two very closely related varieties or even species followed by chromosome doubling. That such types are autopolyploid can usually be determined by a study of meiosis. In fact, autoploidy is largely specified by a variable but significantly large number of quadrivalents, trivalents, and univalents at meiosis.

Meiosis in true autopolyploid plants is characterized by meiotic irregularity because, instead of only pairs of homologous chromosomes, there are more than two of each, as was discussed with meiosis in triploids. In general, in autotetraploids quadrivalents, trivalents plus univalents, and bivalents are the commonest chromosomal configurations at diakinesis and metaphase I. In autohexaploids a few hexavalents are usually present, but hexaploid autopolyploids are quite uncommon.

An autotetraploid produced in this way, directly from a diploid, has no qualitative differences from the diploid, although it may have some quantitative and/or physiological differences (Noggle, 1946) (Fig. 12-5). The polyploid has nothing new nor is it a new combination of characters as is a species hybrid.

Autotetraploid plants are quite easily produced from diploids by various treatments, especially heat shock at various critical phases such as meiosis or pollen grain division, but most easily and usually by colchicine or its synthetic equivalent, colcemid. In some plants, like tomato, if the stem is cut across (decapitation) and the cut surface coated with petroleum jelly, a callus forms. Because the stem contains some tetraploid cells, some of the plantlets that develop in the callus are autotetraploid. Some examples of true autopolyploids have been found in barley (Smith, 1951), maize (Rhoades, 1955), lettuce (Einset, 1947a), Asiatic cotton (Beasley, 1940), and very many other crop species and wild species (Stebbins, 1950).

Fertility of autopolyploids varies widely from about 90% seed set in maize (Randolph, 1941a) to almost complete sterility in cotton (Beasley, 1940). Such irregular meiotic pairing is true also in the few natural autotetraploid species known, except that they are usually quite fertile.

Galax aphylla is a wild monotypic species of eastern United States that exists as both diploid and autotetraploid indistinguishable individuals within the same range (Baldwin, 1941). *Sedum ternatum* (Baldwin, 1942) and *S. pulchellum* occur also as autopolyploids.

Giles (1942) found diploid, autotetraploid, and autohexaploid individuals of the *Tradescantia*-like *Cuthbertia graminea*. In the autotetraploid at meiosis most chromosomes are paired as bivalents and quadrivalents. In the autohexaploid bivalents, quadrivalents, and hexavalents are most common. The diploids are all restricted to a region just east of the Fall Line in North Carolina on Cretaceous sediments, and the tetraploids range widely from North Carolina to Georgia and central Florida. The two hexaploid plants, one from central Florida and one from South Carolina, seem to have arisen recently from the tetraploids among which they grew.

Randolph and Fischer (1939) reported the recovery of a number of diploid individuals by parthenogenesis from a few autotetraploid lines of maize. The parthenogenetic-arisen diploids were much more fertile than their tetraploid sibs. There are parthenogenetic genes in diploid maize, and perhaps the polyploid condition may have influenced them to a threshold of effectiveness. Certainly polyploidy per se

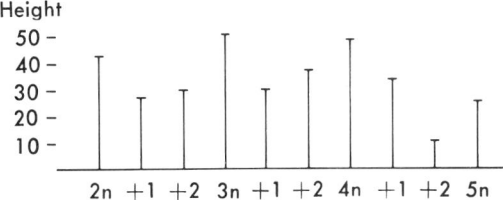

Fig. 12-5. Representation of heights of plants of *Luzula purpurea* (n = 3) in relative dimensions demonstrating that euploid plants (2n, 3n, 4n, and 5n) are more vigorous than aneuploid plants (2n + 1, 2n + 2, 3n + 1, 3n + 2, 4n + 1, 4n + 2) with unbalanced genomes, and that low euploids (3n and 4n) are often more vigorous than the diploid and usually than the higher polyploid (5n). (Data from de Castro, D., and T. Mello-Sampayo. 1954. Agron. Lusitana **16**:111-114.)

may influence parthenogenesis since there is a definite correlation (with some exceptions) between the two phenomena.

A somewhat different sort of autopolyploidy occurs when two ecotypes, varieties, sibling species, or very closely related species form a diploid hybrid that has almost regular meiosis and is almost as fertile as either of the parent forms. A tetraploid of such a hybrid or the result of a cross of two autopolyploids of the two diploid parental types can be considered an autopolyploid although it may have some new hybrid characters. It might possess hybrid vigor and eventually become completely fertile and possibly a new species. The process of becoming fully fertile with regular bivalents rather than multivalents at meiosis is called *diploidization*.

Stebbins (1950) describes how *Biscutella laevigata,* once thought to be an autotetraploid living with its diploid subspecies (Manton, 1937), may be thought to be a new species. Also the common pasture grass *Dactylis glomerata* is essentially an autotetraploid (Myers, 1943) although its diploid ancestors, *D. aschersoniana* and a *D. hispanica*-like form from Iran, may be considered species (Myers, 1948). Such examples lie between autopolyploidy and allopolyploidy.

Diploidization. The term diploidization refers to polyploids that start with irregular meiosis such as multivalents, bridges, and general heterozygosity and through selection, either natural or human, gradually settle down into as regular meiosis as is characteristic of good diploids. Often the term *amphidiploid* is applied to the final diploidlike polyploid. Of course, diploid interspecific hybrids (allodiploids) also may start out with meiotic and genetic instability and eventually achieve stability (Grant, 1966a, 1966b).

An example of at least the beginning of the diploidization process in an autopolyploid was the study performed by Gilles and Randolph (1951) with hybrid autotetraploid maize. Some grains of this heterozygous autotetraploid strain produced in 1938 were kept for 10 years while others were planted, and the plants were selected for "vigor, fertility, disease resistance, and other desirable traits," seed collected, planted, etc. each year for 10 years. In the eleventh year, 1948, the 10-year-old seed and seed of the tenth generation of selection were both planted and microsporocytes examined for number of bivalents and quadrivalents (trivalent-univalent combinations were rare). In general the 10 years of selection had reduced the average number of quadrivalents per cell by 1, thereby increasing the number of bivalents by 2, according to the percentages of cells with from 10 to 5 quadrivalents each:

Number quadrivalents	10	9	8	7	6	5	
1938 seed	11.5	44.2	30.7	7.8	5.8	0.0	100%
1948 seed	0.0	12.0	40.0	31.6	13.8	2.4	99.8%

The assumption is that selection for such characteristics continued long enough would result eventually in a tetraploid with all bivalents at meiosis.

Actually, achievement of all bivalent pairing and diploidization can be speeded up if a gene or genes for bivalent pairing appear. During the few thousands of years (and generations) of cultivated wheat, a gene for bivalent pairing appeared long ago (Frankel and Munday, 1962) that inhibits the pairing of homeologous chromosomes in the normal 42-chromosome hexaploid (Feldman, 1966). Thus, only homologous chromosomes pair as 21 bivalents and wheat is completely diploidized.

Another example of diploidization is reported by Gajewski (1946). *Anemone silvestris* (2n = 16) was crossed to *A. multifida* (2n = 32). The "triploid" hybrid was highly sterile, producing at meiosis univalents and a few bivalents. When the chromosome number was increased, meiosis in the allopolyploid was much more regular, with mostly bivalent pairing. In the second generation fertility was from 20 to 70% and increased in subsequent generations; it was becoming diploidized. Gajewski concluded that diploidization was achieved by alterations in chromosome structure with elimination of duplicate and deficient segments and with mutations promoting bivalent pairing and/or reducing asynapsis. Physiological adjustments can also be made.

Thus, diploidization applies to both autopolyploids and allopolyploids and the commonly applied cytological test for autopolyploidy, meiotic irregularities, would not be found in many old and diploidized forms.

Allopolyploidy

The prefix *allo-* indicates the involvement of two or more "good" species. That is, the polyploid is in part an interspecific hybrid. The prefix is used in animal studies to specify even a diploid hybrid, as an allodiploid (Schultz, 1969). Allopolyploidy is the commonest type of polyploidy in plants but the rarest in animals, although Schultz (1969) and Stebbins (1950) predicted it will be found more commonly in animals in the future, and Stebbins explains the paucity of identified animal hybrids by the difficulty of finding a small percentage of interspecific hybrids in a population of animals compared to a population of plants. Allopolyploidy is the antithesis of autopolyploidy, and the extreme contrast to "pure" autopolyploidy is a polyploid derived from diploid or lower polyploid species that are so different that they can just barely cross. Closer to and overlapping with autopolyploidy of intraspecific "hybridization" is the rather vague and ill-defined but common type called segmental allopolyploidy (Stebbins, 1950).

Segmental allopolyploidy. The prefix *allo-* specifies that a hybrid is derived from parents that are more distinctively "species" than the diploid forms of *Biscutella laevigata* and *Dactylis glomerata* already cited. Stebbins (1950), who coined the term, defined a segmental allopolyploid "as a polyploid containing two pairs of genomes which possess in common a considerable number of homologous chromosomal segments or even whole chromosomes, but differ from each other in respect to a sufficiently large number of genes or chromosomal segments, so that the different genomes produce sterility when present together at the diploid level."

Segmental allopolyploids occupy the middle of a continuous spectrum having completely homozygous autopolyploidy at one extreme and allopolyploidy derived from species so different that they can just barely form a hybrid at the other. Segmental allopolyploids are considerably more common than autopolyploids.

Cytologically, segmental allopolyploids have from quite regular bivalent pairing to some multivalents but vary greatly in fertility due to segregation for short-segment differences, gene differences, and genetic imbalance. But as defined, they are polyploid derivatives of "good" diploid species that produce diploid hybrids characterized by good bivalent pairing but which are highly or completely sterile. An example is *Delphinium gypsophilum* (Lewis and Epling, 1946, 1959; Epling, 1947).

The two diploid species *Delphinium recurvatum* and *D. hespericum* can be crossed to produce a diploid hybrid that has almost perfect bivalent pairing, but it is almost sterile. It is, however, so very similar to diploid forms of *D. gypsophilum* that the hybrid origin of the latter is indicated. The tetraploid form of *D. gypsophilum* probably arose from the diploid form by chromosome doubling and automatically achieved high fertility. Doubtless, both the diploid and tetraploid forms that arose naturally have diploidized since their origins.

Solanum tuberosum (potato) is another species that Stebbins (1950) considered a segmental allopolyploid of a different sort. It is definitely tetraploid ($2n = 48$, $x = 12$) and has rather irregular meiosis with a few multivalents, as high as hexavalents and octavalents, although most pairing produces bivalents. Since the species has been selected for quality of tubers and asexual reproduction rather than for sexual fertility, it has probably been diploidized not at all or even altered in the opposite direction.

According to Ugent (1970) the cultivated potatoes are genetically very diverse, having been derived from a variety of tuberous diploid species (or microspecies or subspecies) of a section of *Solanum*. The place of origin is the Peruvian-Bolivian region of west central South America, and its use arose as man modified the environment and at first gathered and then cultivated roots for food. At first about half a dozen diploid species were "camp followers" as weeds often growing together and hybridizing. Eventually man began selecting and planting, probably seeds, but hybridization continued because weedy and cultivated diploid forms grew together and belonged to the same species complex. Because different peoples lived in a variety of habitats and regions the cultivars were genetically diverse. Some diploid forms of *tuberosum* are still cultivated in the regions of origin.

Probably a number of times and in various places tetraploids arose, were selected, and became cultivated. It is possible that some tetraploids were autoploid; it is also possible that alloploidy occurred, with the most likely diploid parents *S. stenotomum* and *S. sparsipilum*. Perhaps about that time tuber eyes were planted rather than seeds. Even though tetraploidy cut down significantly on gene flow from diploid forms, triploids did arise and are cultivated, and the pentaploid *S. curtilobum* is close to tetraploid *tuberosum*.

Thus, the usually cultivated tetraploid *tuberosum* is a variable mixture of genetic and chromosomal contributions from many close and interfertile diploid "species." Cytological analysis of meiosis in the tetraploid potato does not provide clear evidence of model autopolyploidy or allopolyploidy. Certainly the fact that meiosis in haploid *S. tuberosum* shows from 11 to 12 bivalents (Kimber and Riley, 1963) indicates that it is closer to being an autotetraploid than an allotetraploid. It is probably an example of an intermediate condition with some meiotic characteristics of both.

A valuable reference is the "Inventory of Tuber-Bearing Solanum Species," Agricultural Experiment Station, University of Wisconsin, Bulletin 533, June, 1958, which lists all of the species of that section of the genus, the chromosome numbers of many, varieties, collections, plant introduction numbers, arrangement by ploidy level, hybrids, foreign varieties, relations to fungal and viral diseases, some insect pests, etc.

Two important differences between segmental and genomic allopolyploids derive from the characteristic of segmental allopolyploids to have some pairing between chromosomes from the two ancestral diploid species in the form of multivalents (heterogenetic associations). In each such multivalent nonhomologous chromosomes exchange segments and the result is subsequent segregation of some of the characteristics by which the two ancestral species differed from each other. Such segregation for disharmonious combinations of genes during generations after formation of the polyploid may lead to eventual failure of the line (Clausen et al., 1945) although not always (Kostoff, 1938b). The second difference between segmental and genomic allopolyploids is that the former often produce backcrosses that are partially fertile with any autopolyploid derivatives of either of the parental species. Barriers against this exist in genomic allopolyploids.

Genomic allopolyploidy. Genomic allopolyploids include most of the natural and artificially synthesized plant allopolyploids known. They are derived from diploids or lower polyploids that are so different genetically and chromosomally that when brought together in a hybrid there is essentially no pairing of chromosomes. Thus the hybrid is completely or almost completely sterile, and there is no exchange of segments between homeologous chromosomes. As a result when the chromosome number is doubled, good bivalent pairing occurs and there is little or no segregation in later generations. Thus, polyploids with good bivalent pairing are either segmental allopolyploids or other sorts that have been diploidized. Stebbins (1950) lists a number of examples of both cultivated and wild genomic allopolyploid species such as tobacco, New World cottons, and wheat as well as *Galeopsis tetrahit, Brassica napus, Iris versicolor, Madia citrigracilis, Poa annua,* and many others. Almost any polyploid species with regular meiosis is probably of this genomic sort unless proved otherwise by resynthesis from diploid or lower polyploid forms.

Further discussion of genomic allopolyploid plant species can be found in the next five chapters, especially Chapters 16 and 17.

Autoallopolyploidy

Autoallopolyploid plants must exist at the hexaploid level or above (Stebbins, 1950). As the name indicates, such a species is a combination of autopolyploidy and allopolyploidy and must be derived from at least two different species. If letters indicate chromosome sets or genomes, the two original diploid species can be represented by AA and BB. The autoploid of

AA is AAAA, and if that is combined with BB making an alloploid, the hexaploid autoallopolyploid can be represented as AAAABB. The A's might be somewhat different, even as different as in a segmental allopolyploid, and so could be represented by A_1A_1 and A_2A_2: such an autoallohexaploid would be $A_1A_1A_2A_2BB$. At high levels various combinations of alloploid and autoploid conditions can occur together.

Two examples of autoallopolyploids, *Phleum pratense* (Levan, 1941) and *Solanum nigrum* (Jorgenson, 1928), are discussed further under the topic of polyhaploidy where evidence of their polyploid constitution is clear.

Polyploid levels

Theoretically there is a very high level of ploidy possible, but tetraploidy is by far the most common level. Hexaploidy and octoploid species are less common, but known species above octoploidy are uncommon. Nevertheless, high levels are known, such as the 38-ploid New Zealand grass species, *Poa litorosa* ($x = 7$, $2n = 263$ to 265) (Hair and Benzenberg, 1961), the 20-ploid *Galium grande* ($x = 11$, $2n =$ about 220) (Dempster and Stebbins, 1968), and the dodecaploid hybrid between *Hibiscus radiata* × *H. diversifolius* ($2n = 216$) (Menzel and Wilson, 1963).

Saccharum, the genus of sugar canes, is used here as an example of an unusual and complex problem in polyploidy. *Saccharum* has been a cytological and reproductive problem for a long time because of high chromosome numbers (from 50 to 140), hybridity, chromosome number jumps, and aneuploidy. The basic chromosome number is $x = 10$, and most collections of *S. officinarum* have $2n = 80$ and of *S. robustum* mostly $2n = 60$ and 80. *S. spontaneum*, on the other hand, has a wide range of chromosome numbers (from 64 to 128) many of which are aneuploid; and many cultivated forms have high chromosome numbers and are aneuploid. Apparently most of the cultivated forms are of hybrid origin of unknown parentage but many involve all three species just listed. Both apomixis and parthenogenesis have been claimed but are unlikely (Price, 1959).

Although all details have not been worked out, aneuploidy can probably be explained by irregular meiosis and tolerance of aneuploidy by high polyploids. Large chromosome jumps seem to be explained by production of progeny having the somatic number of the pistillate parent plus the gametic number of the pollen parent ($2n + n$) following some crosses (Price, 1961). Price found that with few exceptions, *officinarum* × *officinarum* and *officinarum* × *robustum* produced progeny having chromosome numbers equal to the sum of the parental gametic numbers ($n + n$). But *officinarum* × *spontaneum* produced only $2n + n$ progeny! There was also one exceptional 134-chromosome plant from $2n = 80$ *officinarum* × $2n = 60$ *robustum* that probably was $2n + 2n - 6$.

It is evident that *S. officinarum* can contribute either the gametic or the zygotic chromosome number to progeny, depending upon the pollen parent. Apparently it does not accomplish this in the usual way by apomeiosis, such as restitution nuclei in the megaspore mother cell. Bremer, in numerous published studies including that of 1959, has concluded that meiosis is normal so that haploid megaspores are produced. In some ovules haploid embryo sacs and eggs are produced, but in others there is in the functional megaspore a postmeiotic endomitotic chromosome doubling so that the embryo sac and egg are diploid, but segregation is accomplished by the normal meiotic division. Price (1961), on the other hand, assumed that perhaps all species just listed fertilize both haploid and diploid eggs but only $n + n$ progeny develop from *officinarum* or *robustum* pollination and only $2n + n$ hybrids develop when *spontaneum* is the pollen parent.

Price (1959, 1961) has reviewed the literature on *Saccharum* reproduction that includes reports or proposals of parthenogenesis, apogamy, nuclear embryony, chromosome elimination en bloc by an odd meiotic division, postmeiotic fusion of two megaspore nuclei, partial endomitosis, etc. He cited some cases of aneuploidy that he could not explain. *Saccharum* is a peculiar, probably completely sexual, complex in the chromosome numbers of hybrid progeny produced. Since most cultivated clones are probably to some extent hybrid, many crosses do not produce progeny with expected chromosome numbers.

Polyhaploidy

Haploid plants of autopolyploid and allopolyploid species have been reported very often. Kimber and Riley (1963) listed 94 reports of polyhaploids, many, of course, within the same species, such as 15 of *Triticum aestivum (vulgare)* alone. Since they arise after meiosis by haploid parthenogenesis, each would contain one genome of each parent species if of an allopolyploid but would be a diploid if derived from an autotetraploid. Analysis of pairing in a polyhaploid gives indication of the nature of the polyploid from which it arose (Duara and Stebbins, 1952).

Solanum tuberosum, a tetraploid, has been claimed as both autotetraploid (Peloquin and Hougas, 1960) and segmental allotetraploid (Stebbins, 1950). In a potato polyhaploid meiosis showed a number of univalents as well as anaphase I bridge-fragment configurations (Ivanovskaya, 1939), indicating an allotetraploid composition. Peloquin and Hougas (1958), however, reported great cytological and fertility differences between two potato polyhaploid plants of different varieties. One regularly had 12 bivalents and was very pollen-fertile when crossed to other diploid species. The other polyhaploid produced almost no good pollen, and at meiosis less than 10% of microsporocytes contained 12 bivalents. If the potato is autotetraploid, it is likely to be nearly a segmental allotetraploid. On the other hand, if the species is a genomic alloploid, there should be no pairing or very little. But potatoes differ so greatly there is no cytogenetic "potato."

Solanum nigrum ($n = 36$, $x = 12$) (Jorgenson, 1928) and *Phleum pratense* ($n = 21$, $x = 7$) (Levan, 1941; Nordenskiöld, 1941) are auto-allohexaploids, and their polyhaploids show in *Solanum* 12 bivalents and 12 univalents and in *Phelum* 7 bivalents and 7 univalents. That is, in the normal hexaploid there are four sets alike (auto) and two different sets, AAAABB. In the polyhaploid, AAB, the A's will pair but the B must appear as univalents. The same type of pairing occurs in the *Phleum* polyhaploid

and the fertile polyhaploid of octoploid *Bromus inermis* Elliott and Wilsie (1948).

In the polyhaploids of hexaploid *Triticum aestivum* (wheat) that has been derived from three distinctive species, there are usually only one or two bivalents (Kimber and Riley, 1963), but a great deal of variation exists among the 15 polyhaploids studied. The very low number of bivalents between homologous chromosomes strongly indicates an allopolyploid, which it is known to be (see Chapter 16). It is now known, however, that the 5B gene inhibits the pairing of homeologous chromosomes. When polyhaploids lack this gene, many bivalents are formed, almost as would be expected in an autohexaploid. Hexaploid wheat is, therefore, a segmental allohexaploid. Interestingly, when Kaltsikes et al. (1969) produced AABB tetraploids from AABBDD hexaploid wheats, the tetraploids did not resemble the natural AABB tetraploid wheats morphologically but did have regular meiosis.

Raven and Thompson (1964) raised the possibility that polyhaploids might be fertile enough to start new evolutionary lines at a lower chromosome level; that is, that the evolutionary direction, diploid to polyploid, is reversible. Actually DeWet (1968) did produce a fertile polyhaploid in the grass genus *Dichanthium*, which is largely apomictic. Nevertheless, the conclusions of Jones (1970) that polyhaploids are nearly always so weak and sterile that they could hardly survive in nature and of Stebbins (1970) that they could hardly start a new evolutionary line since they possess nothing new strongly indicate that the direction, diploid to polyploid, is, indeed, as irreversible as any biological generalization may be.

AGMATOPLOIDY

As far as chromosome numbers are concerned agmatoploidy (Malheiros-Gardé and Gardé, 1951) is a form of polyploidy (including aneuploidy and euploidy), but with respect to amount of DNA it is not. In all previous forms of ploidy the addition of chromosomes increased the amount of DNA, and the subtraction of chromosomes depleted the amount of DNA. Agmatoploidy does not do this. The bug *Thyanta calceata* has twice as many chromosomes as its close relatives, but it has the same amount of DNA (Schrader and Hughes-Schrader, 1956). The best documented case of agmatoploidy is the plant genus *Luzula*, the wood rush (see later).

Another difference between other forms of ploidy and agmatoploidy is that regular ploidy occurs as an aspect of usual, unicentric chromosomes. Agmatoploidy, however, *requires* the unusual type of chromosome that seems to have the centromere spread along its whole length or, to put it another way, the whole chromosome is a centromere, a "diffuse" centromere. Regardless of the nature of the centromere, at metaphase-anaphase there are spindle fibers attached along its length so that at anaphase it moves sideways toward the pole. When a unicentric chromosome is broken anywhere except at the centromere, the portion lacking a centromere cannot move to a pole at anaphase, and it is lost. But when a chromosome with a diffuse centromere (a polycentric or holocentric chromosome) breaks, both parts function normally and these half-chromosomes in turn can break into quarter-chromosomes and still move normally.

Agmatoploidy is just that, the breaking in half of one or all chromosomes of the set. *Luzula purpurea* has $2n = 6$ holocentric chromosomes that are relatively large. A number of species have $2n = 12$ chromosomes each about one-half the size of that of *purpurea*. There are species with quarter-sized chromosomes with $2n = 24$ and one-eighth size with $2n = 48$ (Nordenskiöld, 1951, 1956). But all have the same amount of DNA (Halkka, 1964).

There are "aneuploid" agmatoploid species also, produced when one or a few chromosomes only break. Again, there is no change in amount of DNA. But there is also regular polyploidy in *Luzula*, in which case the DNA amount *is* increased. The genus *Luzula* has undergone all of these changes. It is likely that the genus *Juncus* in the same family with *Luzula*, the Cyperaceae (sedges) (Battaglia, 1954; Hakansson, 1954, on *Eleocharis*), the insect orders Homoptera and Hemiptera, the alga *Spirogyra* (Godward, 1954), probably Lepidoptera (Bauer, 1967), and certain scorpions such as *Tityus* (numerous papers of Piza, see Battaglia, 1955a) all have holocentric chromosomes and can have agmatoploidy, as seems to be true in *Thyanta*. Lorkovic (1941) found in a number of genera of the Lycaenidae that as the chromosome number increases the chromosomes become smaller and that the polyploidy is not strictly euploid. For example, haploid chromosome numbers in *Polyommatus* are 23, 45, and 90; in *Leptidea* 29, 54, and 104; and in *Erecia* 21 and 40. The aneuploid numbers probably result from a few chromosomes not having "broken," since Lorkovic reported variation in chromosome sizes. La Chance and Degrugillier (1969) reported the persistence of chromosome fragments through at least three generations in the bug *Oncopeltus*, which is known to have holocentric chromosomes.

In general, holocentric chromosomes of plants and animals are small, and plant genera having them tend to have a great deal of apparent aneuploidy. The large genus *Carex* has long been notorious for so much aneuploidy that euploidy and basic numbers cannot be determined. For different opinions of diffuse centromeres, holocentric chromosomes, and their divisions, see Battaglia (1955a).

Polyploidy, then, is a very important evolutionary process in plants including about two thirds of all grass species and one third of all angiospermous species. Nevertheless, the slow evolution at the diploid level continues. It is very likely, however, that the high diploid chromosome numbers of many groups (from 10 to 20) are really secondary diploids that once were tetraploids. Polyploidy is a mechanism for a very rapid origin of species and of evolution that has been and is being rigorously exploited by plants.

ENDOPOLYPLOIDY

Endopolyploidy is a common process, also called *endomitosis* (Geitler, 1937; Lorz, 1947), occurring naturally and normally in plants and animals. It is distinct from polyteny (see Chapter 3) in that polyteny increases DNA and genomes without affecting the number of chromosomes in the nucleus. Endopolyploidy increases the DNA and genomes and the number of chromosome sets in the nucleus. It is accomplished by DNA and chromatid replication and chromatid separation without effective anaphase disjunction. That is, the mitotic process is inhibited somewhere between prophase or even interphase and completion of anaphase separation. Just where mitosis is stopped varies among organisms. Furthermore, endomitosis may be repeated a few or many times to produce highly polyploid cells.

Endopolyploidy evidently accomplishes two results. One is the building up of DNA in cells for use of other cells. Presumably the endopolyploidy of tapetal cells in anthers of flowering plants (Brown, 1949) provides DNA (or at least nucleotides) for the use of microspores during pollen development since the tapetal cells break down completely at that time. The other likely function is to produce more than the normal two copies of each gene so that a higher rate of RNA transcription can occur. Thus it accomplishes the same result as the polytene giant interphase chromosomes.

The presence of polyploid cells in the livers of rodents and other vertebrates is well known. Recently 4n, 8n, and, rarely, 16n nuclei have been reported in livers of frogs, but mostly during the breeding season (Bachmann et al., 1966). Polyploid cells are regularly found in stem and other parenchyma of flowering plants when stimulated to divide (Bradley and Crane, 1955).

A classical example of endopolyploidy occurs in the gut lining cells of mosquitos (Berger, 1938; Grell, 1946). All mosquitos have $2n = 6$ and, as in all Diptera, homologues are always paired. During larval life there are apparently four endomitotic interphase divisions of each chromosome; the 16 chromosomes derived from each original chromosome are held loosely together in a bundle, but homologous bundles are still loosely paired. Thus there are three pairs of bundles of 16 chromosomes or a total of 96 chromosomes, although they cannot be seen as such until prophase of the first pupal divisions of these cells. This layer of large cells becomes the gut lining of more and smaller cells of the imago during the pupal stage by four quick "reduction" divisions within about 8 hours; that is the reason for the endopolyploidy.

Actually, the reduction divisions are far more interesting than the endomitotic divisions. During prophase of the first pupal division, the three pairs of bundles of 16 chromosomes, 96 in all, separate into 48 pairs of chromosomes by metaphase so that two telophase nuclei are formed, each of 48 chromosomes. There is a very quick interphase or the telophase goes directly into second prophase, very like meiotic interkinesis. At the second metaphase there are 24 pairs of presumably homologous chromosomes that form 24-chromosome telophase nuclei. Another quick interphase with, presumably, neither DNA nor structural replication leads to 12 pairs at metaphase and two 12-chromosome telophases. The fourth similar division produces two 6-chromosome telophases. Thus, one large 96-chromosome endopolyploid cell has produced 16 small diploid cells, each with 6 chromosomes, in about 8 hours with no interphase metabolism, even of DNA. These divisions are comparable to amphibian cleavage divisions where quick divisions and smaller cells result, but they do seem to have DNA replication during a short interphase.

Perhaps the best known case of endopolyploidy is that of the water strider, *Gerris* (Geitler, 1937), the male of which is XO with $2n = 21$. During interphase the single X remains visible as a small heterochromatic mass. Thus there is one sex chromatin mass (a prochromosome) for each 21 chromosomes, and the level of endoploidy is 2 for each prochromosome present. Geitler reported: muscle cells, tetraploid; sperm duct lining, octoploid; mid-gut epidermal cells, 16-ploid; seminal septum, 16-ploid; various regions of the Malpighian tubules, 16-, 32-, and 64-ploid; fat body, 64- and 128-ploid; and salivary gland cells, 1,024- and 2,048-ploid. A nucleus of *Gerris* that is 2,048-ploid would contain about 43,000 chromosomes!

Endomitotic divisions have also been reported in relation to meiosis, especially at the last premeiotic mitosis so that the meiocyte is tetraploid. That permits the production of diploid eggs (even though meiosis occurs) and permits the formation of diploid progeny by parthenogenesis (Omodeo, 1955, in earthworms).

POLYPLOIDY IN EVOLUTION

Polyploidy in evolution is beyond the scope of this book, but it has been suggested or indicated repeatedly in this chapter, in Chapters 16 and 17, and occasionally in others. (See Stebbins, 1950, and Grant, 1971, for extensive coverage of this topic.) It is clear that the majority of plant genera have evolved one successful polyploid species or more. In some instances such polyploid species are clear-cut and distinctive, but often they and their diploid progenitors intercross to some extent, thereby producing new polyploid taxa. As a result, polyploid complexes occur in many groups (see Grant, 1971, for extensive discussion of polyploid complexes in plant evolution). In many genera apomixis has arisen within such a polyploid complex to produce additionally an agamic complex (Chapter 18; Stebbins, 1950; Grant, 1971) as in the grass genus *Bouteloua*, which consists of sexual diploids, low polyploid sexuals, and high polyploid apomicts (Freter and Brown, 1955; Gould and Kapadia, 1962, 1964; Kapadia and Gould, 1964). Thus, generic evolution may occur solely at the diploid level, at the diploid and polyploid levels, or mostly at the polyploid level.

CHAPTER 13

Translocation

When structurally new types of chromosomes arise by interchange or inversion in natural populations, their maintenance largely depends on their effects upon two, sometimes conflicting, properties; first, the *survival* of individuals and, second, their *fertility* — the latter almost invariably being reduced in structural heterozygotes.

Rees, 1961

Two nonhomologous chromosomes' exchange of pieces is called translocation, interchange, or exchange (Fig. 13-2). All evidence indicates that the translocations observed are, with no known exception, reciprocal. Probably both simple and reciprocal translocations were formed; but in order to survive, a cell must have a balanced set of genes. A cell with a reciprocal translocation or exchange does have that (Fig. 13-2, *D*); but a simple translocation (Fig. 13-3, *C*), whereby a piece of one chromosome, Y of Y-Z, is attached to part of the other, W of W-X, but the piece X is not attached to the remainder of Z, would result in a deletion (X) that could be lethal since the acentric fragment would be lost. Even if the simple translocation could persist in a line of mitotic cells, the deletion would doubtless be lethal in the haploid cells following meiosis because of severe genetic imbalance. Thus the term *translocation* always implies the reciprocal type. A translocation between two homologous chromosomes would produce severe duplications and deletions and, following meiosis, would be lethal (Fig. 13-4).

Exchange of segments of chromosomes can occur only when breaks are formed. Only broken ends can unite; the normal ends of chromosome arms cannot. There seems to be something different at the end of a normal chromosome arm, the telomere, that prevents it from uniting with either another telomere or a broken end. Since the internal structure of a mitotic chromosome or the structure of an interphase "chromosome" is unknown, it is pointless to speculate on what "break" and "reunion" mean; but genetically the breaks and reunions occur within linkage groups, and the reunion seems to produce a perfectly normal chromosomal segment, probably by some DNA synthetic "healing." Just how a translocation break and reunion differs (if it does) from a somatic or meiotic crossover, which is also a break-and-reunion phenomenon, is not known. Of course, crossovers must occur at identical loci on homologous chromosomes, and meiotic crossovers seem to require the synaptinemal complex.

Simultaneous breaks in two nonhomologous interphase chromosomes that lie close together apparently occur in nature but can be produced in large numbers by radiation. Translocations often consist of exchanges of end segments of chromosome arms in which two breaks only are required. But sometimes a piece of one chromosome is inserted into another as part of the change so that four breaks are required. The first translocation

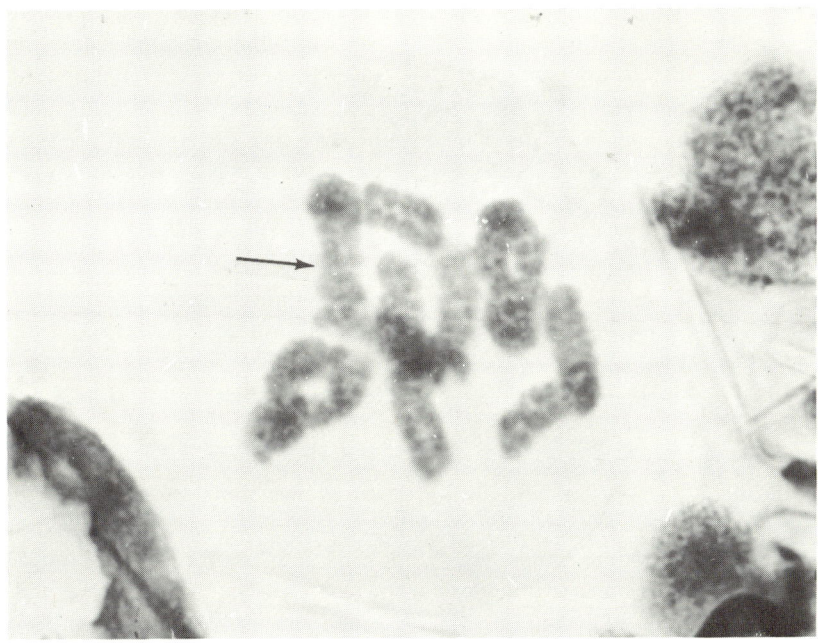

Fig. 13-1. A quadrivalent chromosome (arrow) at meiotic metaphase I of *Tradescantia* with four bivalents also. The four chromosomes of the quadrivalent are attached end to end by terminalized chiasmata to form a circle of four (⊙4).

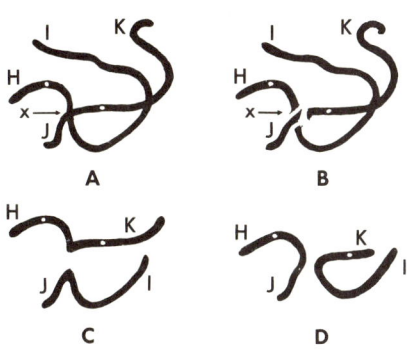

Fig. 13-2. Reciprocal translocations or exchanges are produced by two adjacent breaks in two nonhomologous chromosomes (**A** and **B**) followed by reunions (**C** and **D**) to produce new chromosomes. If, as in **C**, the broken ends unite to form a dicentric and an acentric, lethality follows and the abnormal chromosomes do not persist. **D**, Chromosomes *H-I* and *J-K* at **A** have become the translocated chromosomes *H-J* and *I-K*. (From Brown, W. V., and E. M. Bertke. 1969. Textbook of Cytology. The C. V. Mosby Co., St. Louis.)

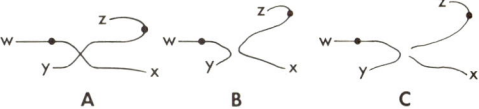

Fig. 13-3. A simple translocation, **C**, differs from a reciprocal translocation, **B**, by one pair of broken ends not uniting, **C**, to produce a deleted chromosome, *Z*, and an acentric fragment, *X*. Since this condition is lethal, it is not found persisting in populations.

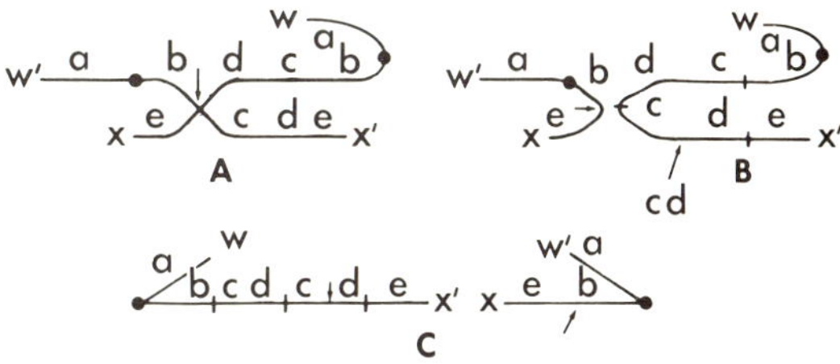

Fig. 13-4. **A,** Two *homologous* chromosomes broken at arrow; the break can occur anywhere on each chromosome. **B,** Reciprocal translocation between homologous chromosomes. One, *W'X*, has a deletion, *cd*; the other, *WX'*, a duplication, *cd cd*. **C,** At anaphase I one daughter nucleus will receive the duplication; the other, the deletion. Both are apt to be lethal and lost following meiosis.

found in *Drosophila* was an insertion (Bridges, 1923). Such breaks seem to be at random, and the result is that no two translocations are identical even if the same two chromosomes are involved. For example, in 1935 Anderson listed 90 different interchanges in maize of which only 14 were natural. By 1950 Longley was able to list 588 in that species, and others have been produced since (cited from Burnham, 1956).

The results of translocations are numerous. Burnham (1956) devoted six pages to a listing of "values and uses of interchanges" including 24 such values and uses, with 14 subtopics under one of the 24. These items are of cytogenetic and/or agronomic value. One is that new chromosomes with new linkages are formed. Such exchanges have been important in cytogenetics by providing visible markers for the cytological location of genes. Pachytene analysis of maize chromosomes and salivary gland chromosome studies in *Drosophila* have demonstrated that genetic maps determined by crossover analysis and by cytological observation do not exactly correspond to the positions of genes in the chromosome (Chapter 3). The linear sequence is the same, but cytological analysis places genes closer to the arm ends and the physical distance between genes does not always correspond to the crossover distances. It seems evident, therefore, that in most, but not all, species crossing-over is limited to or is most frequent toward the ends of chromosome arms, a condition not always revealed by crossover analysis.

Another effect of heterozygosity for translocation is that crossing-over is greatly reduced near the point of interchange (Rhoades, 1955). This reduced crossing-over is not a direct effect of the break and reunion per se but the result of failure of synapsis of homologous segments in the regions of all chromosomes near the exchange (Fig. 13-5).

Another effect of translocation in most species is some lethality of spores or progeny, as discussed in the following section. An interesting potential use of the lethal effect of translocation in insect pest control is presented by McDonald and Rai (1970) for the yellow-fever mosquito, *Aedes aegypti* ($2n = 6$). They produced two translocations, one between chromosomes No. 1 and No. 2 and another between No. 1 and No. 3, each with about 30% fertility when crossed to the normal (Fig. 13-6). But when the two exchanges were combined to form a double heterozygote, the cross of it to the normal wild type produced from 90 to 93% sterility. They proposed that release of such double heterozygous males into a population not only would greatly reduce progeny, as does the release of sterile males of screwworm flies (Knipling, 1959) or the mosquito *Culex* (Patterson et al., 1970), but would also release into the population the two single and the double translocations which would continue to cause lowered fertility in the population for a number of generations.

They also reported that crossing-over is increased considerably in one region close to a break point and that some of this crossing-over

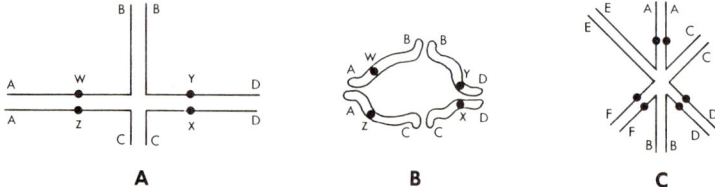

Fig. 13-5. Synapsis during meiosis in an organism heterozygous for one translocation or more. **A,** At pachytene homologous sections of the four different chromosomes synapse in a cross configuration. **B,** By metaphase any chiasmata have terminalized and a ring of four chromosomes results. With *alternate* disjunction A-B and C-D go to one pole and A-C and B-D go to the other, and each cell has a complete set of genes. With *adjacent* disjunction A-B and A-C can go to one pole while B-D and C-D go to the other. Each cell produced is duplicate and deficient, and the condition is lethal. **C,** Pachytene configuration, mostly theoretical, in an organism heterozygous for two translocations among three chromosomes. At metaphase I a ring of six chromosomes is found. (From Brown, W. V., and E. M. Bertke. 1969. Textbook of Cytology. The C. V. Mosby Co., St. Louis.)

has produced a new short chromosome consisting of parts of all three chromosomes as well as the two new chromosomes consisting of parts of two wild-type chromosomes. When the two single 1:2 and 1:3 exchanges are combined in the double heterozygote, all six chromosomes form a ⊙6 at meiosis.

At meiosis in an organism heterozygous for one or more translocations each meiocyte will contain, for the chromosomes involved, no homologous chromosomes, only homologous parts of chromosomes. If it is heterozygous for one translocation (Fig. 13-5), there are four nonhomologous chromosomes, two unchanged and two altered, each of which has parts homologous to parts of two other chromosomes. Thus each of the four chromosomes has no homology with one other. Usually the significant homologies for pairing and synapsis are located toward the ends of arms. Therefore, at pachy-

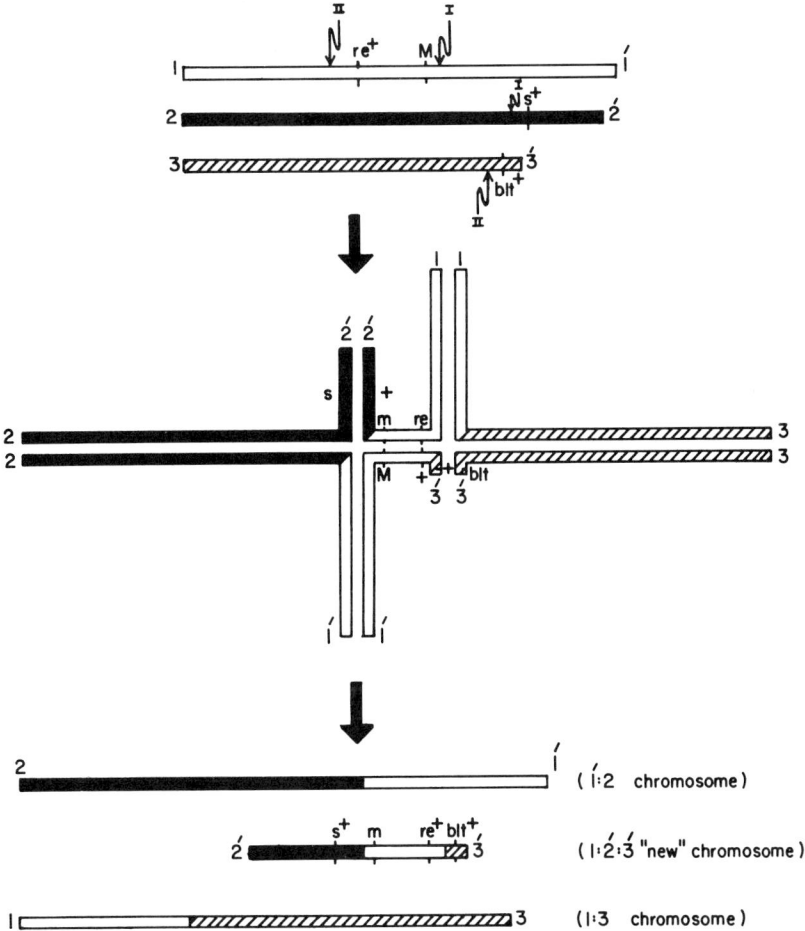

Fig. 13-6. Double translocation heterozygote pairing in the mosquito *Aedes aegypti* (n = 3). The three chromosomes, 1-1′, 2-2′, and 3-3′, have two translocations, I between 1-1′ and 2-2′ and II between 1-1′ and 3-3′. When the double heterozygote was formed, it had a normal 2-2′ and a normal 3-3′ but no normal 1-1′ as well as two different translocation chromosomes each of 2-1′, 1-3′. At pachytene the double cross-shaped hexavalent is presumed to have formed. Where the middle regions of 1-2′ and 1′-3′ are paired are located the alleles M and m for sex and rc and + for eye color. A crossover between those genes produced the three chromosomes at the bottom of the figure; the 1-2′-3′ "new" chromosome consists of portions of all three of the original three chromosomes figured at the top. (From McDonald, P. T., and K. S. Rai. 1970. Science 168:1229-1230. Copyright 1970 by the American Association for the Advancement of Science.)

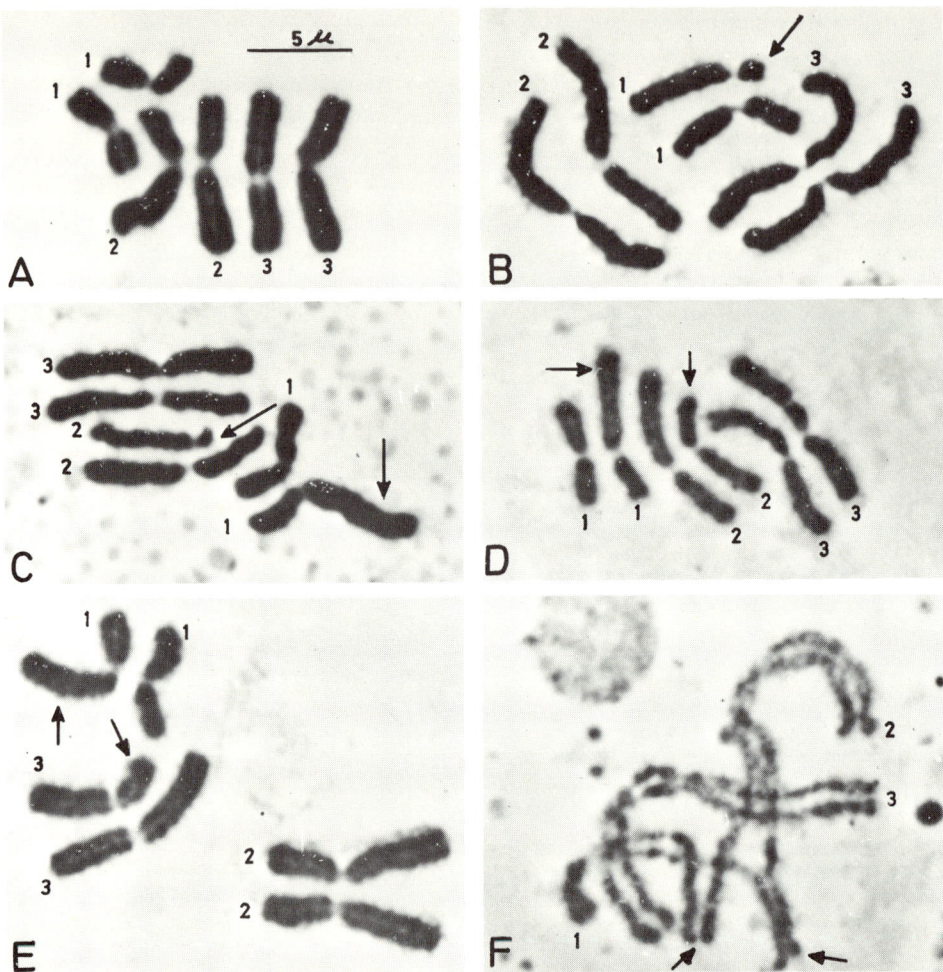

Fig. 13-7. Mitotic chromosomes of the mosquito *Culex tritaeniorhynchus* (2n = 6). Notice the somatic pairing of these dipterous chromosomes. **A,** The normal set; the three pairs are all submetacentric. **B,** An unequal pericentric inversion has moved the centromere of one chromosome No. 1 (arrow) toward the end. **C** to **F** show reciprocal translocation effects. **C,** A 1-2 translocation. **D,** A different 1-2 translocation. **E,** A 1-3 translocation. **F,** Mitotic prophase in which somatic pairing duplicates meiotic pachytene pairing of a 1-2 translocation. (From Baker, R. H., R. K. Sakai, and A. Mian. 1971. Science **171:**585-587. Copyright 1971 by the American Association for the Advancement of Science.)

tene all homologous regions will pair and a cross-formation results (Fig. 13-5). By metaphase I the chiasmata have terminalized and the group of four chromosomes forms a ring of four (⊙4) (Figs. 13-1 and 5). If there are two translocations involving completely different chromosomes, there will be at metaphase I, 2 ⊙4. But if the second translocation involves one of the four involved in the first exchange, then there will be an association of six chromosomes at pachytene (Fig. 13-5) and a ⊙6 at metaphase I.

Translocations can sometimes be observed at mitotic metaphase if the exchange of pieces alters observably the shapes of the two chromosomes involved. Reports of such have come from human cytogenetics (Chapter 19). More obvious cases are known such as in mosquitos (Fig. 13-7, *C-F*) and other Diptera in which homologues are always paired. Fig. 13-7, *F* even shows the pairing of the four homologous ends of the four chromosomes much as at pachytene, yet this stage is, according to reassurance by personal correspondence with

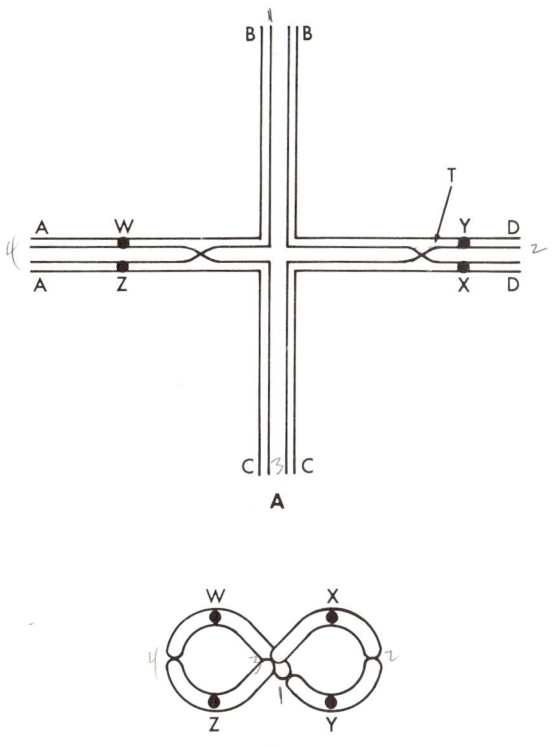

Fig. 13-8. Translocation heterozygosity at meiosis. **A,** At pachytene the cross configuration is seen if there is one exchange. Crossovers in the interstitial region as at T, between the centromere and the exchange point, can modify the effects of alternate and adjacent segregation with respect to viability. **B,** If the four chromosomes of the ring, for inherent structural reasons, assume such an alternate arrangement regularly, full fertility results because adjacent disjunction is prevented.

Baker, probably prophase in a spermatogonial cell. It could be, then, premeiotic pairing rather than somatic pairing, if there is any difference between those two categories in Diptera.

This figure also demonstrates the difference between chromosomal change due to a pericentric inversion (Fig. 13-7, B) involving only one chromosome, and exchanges (C, D, E, and F) that involve two chromosomes, No. 1 and No. 2 (C, D, F) or No. 1 and No. 3 (E).

DISJUNCTION

Given the synapsis of four chromosomes as in Fig. 13-8, it is evident that there are four centromeres (W, X, Y, and Z) to go to two spindle poles at anaphase I. If W and X go to one pole and Y and Z to the other, each daughter nucleus will receive a complete set of genes (represented by A, B, C, and D) with no duplications or deletions. That is, *alternate* centromeres around the ring go to the same pole and such segregation is called *alternate disjunction* or merely disjunction, and the cell has *directed segregation*.

On the other hand, it is possible for centromeres W and Y to go to one pole and X and Z to go to the other or for W and Z to go to one pole and X and Y to go to the other. These conditions are called *adjacent disjunction* or nondisjunction because *adjacent* chromosomes go to the same pole. Both forms of adjacent disjunction produce daughter nuclei that are duplicate and deficient for chromosomal segments and genes. In plants adjacent disjunction is lethal and produces dead pollen grains or megaspores (Burnham, 1932; Pellew and Sansome, 1931). In animals viable gametes can be produced following adjacent disjunction because the genes in gametes do not seem to function (Sturtevant and Dobzhansky, 1930; Lindsley and Grell, 1969). Sturtevant and Dobzhansky found that compensating duplicate-deficient gametes of *Drosophila* when combined in the zygote produce viable individuals. But animal zygotes or early embryos will die if the duplications and deficiences are at all large (Novitski, 1951).

The two forms of adjacent disjunction are determined by which two centromeres go to the same pole. If the two centromeres going to one pole are nonhomologous (W and Y in Fig. 13-5), it is called adjacent-1. If the two centromeres going to one pole are homologous (W and Z in Fig. 13-5), it is called adjacent-2. In many species adjacent and alternate disjunction are about equally common, and the condition can sometimes be predicted if a hybrid has about 50% lethality in the F_2, which is sometimes called *semisterility*. Some plants with about 50% pollen and ovule abortion and about 50% alternate disjunction are maize, petunia, *Pisum sativum*, and *Sorghum versicolor*. On the other hand, a number of species and genera are known in which *alternate* disjunction ranges from about 70 to 95% and fertility is high. Such taxa are said to have *directed segregation*. Examples are *Oenothera, Datura, Triticum, Hordeum,* and others. The implication of the term directed segregation is that some condition prevents randomness of alternate and adjacent segregation. Nevertheless, of the two possible directed segregation conditions, (1) mostly alternate or (2) mostly adjacent, the latter is very rare (Haga, 1943, in *Lilium hansonii*); the former is very common.

It is often claimed that genes for alternate disjunction are necessary to raise the percentage of alternate disjunction substantially above 50%. The evidence for genetic control is the variation found among strains of a species or hybrid F_2 and F_3 (Thompson, 1956; Lawrence, 1963) and increases during generations when selected for fertility. But such claims do not specify how gene action can achieve nonrandom segregation nor is it indicated that a single gene accomplishes the increase in alternate segregation.

Alternate disjunction occurs when alternate chromosomes have their centromeres directed to the same spindle pole at late metaphase I (Fig. 13-8, *B*). This is a common and predictable arrangement in some species and is often called the *zigzag* arrangement. Adjacent disjunction occurs when adjacent chromosomes have their centromeres directed to the same pole. This is never a definite and determined condition but results only from the next condition. The so-called *open* arrangement of chromosomes in a ring means that the ring at metaphase is flat rather than zigzag and any two (or three, Hagberg, 1954) consecutive chromosomes may go to either spindle pole. The open arrangement results in statistically 50% alternate and 50% adjacent disjunction. But almost no species has more than 50% adjacent disjunction. In *Drosophila melanogaster* the male has from 50 to 60% and the female from 50 to 80% alternate disjunction (Brown, 1940). But gene action is hardly necessary or indicated, at least in some reported cases, since in many F_1 hybrids with a translocation derived from homozygous parents or in the first generation of spontaneous or X-ray formed heterozygosity the disjunction is *immediately* alternate. In rye Lawrence (1963) recorded alternate disjunction of from 68 to 82% immediately after irradiation. He assumed genetic preadaptation for alternate disjunction, whatever that means. Evans (1954) found 63% alternate disjunction in a trivalent of an F_1 hybrid of a grasshopper in the genus *Circotettix*. White (1957) found almost complete alternate disjunction in an F_1 hybrid between two geographical races of the grasshopper *Moraba scurra*. Yamashita (1950) produced translocations in hybrids of *Triticum monococcum* and *aegilopoides*, which are cytologically homologous. Hybridization produced rings of four with an average of 82.7% alternate disjunction. Similar results occur in hybrids of races of *Datura, Oenothera,* and *Gaura*. Either such species or races are genetically preadapted for complete alternate disjunction or there are other conditions that preadapt such hybrids to alternate disjunction. It must be emphasized that in these species alternate disjunction occurs regularly for all chromosomes. In many others with large rings alternate disjunction usually occurs, but just one chromosome in the ring making a mistake results in lethality.

Burnham (1956), in an extensive and intensive review of chromosome interchanges in plants, stated that "The species which show a high frequency of alternate segregation of the chromosomes at meiosis tend to have certain cytological features in common: the chromosomes are relatively uniform in length, the centromeres are median or nearly so, and the chiasmata are located at the ends or are terminalized completely." But he also concluded that "For each of the factors which have been suggested there is one or more species which does not behave in the predicted manner."

There have been a few cytological observations of some sort of mechanism during prometaphase and metaphase I that arranges the chromosomes in a ring so that alternate disjunction is greatly increased. Ostergren (1951) reported that in *Oenothera* chromosomes, all of which are in one ⊙14, he found a ratio of 2 open rings to 1 zigzag ring at early metaphase I which changes to a ratio of 1:1.2 by late metaphase, as though there were a mechanism arranging the chromosomes for more alternate disjunction. Hughes-Schrader (1943a, 1943b) has concluded in a mantid that alternate disjunction may not be a random or gene-controlled condition. Furthermore, whereas in diploid *Clarkia elegans* (Lewis, 1951) and barley (Tsuchiya, 1953) disjunction is almost exclusively alternate, in tetraploid meiocytes adjacent segregation is almost 50%. Burnham (1956) considered this "One bit of evidence against genetic factors affecting segregation in rings."

Gairdner and Darlington (1931) invoked two "forces" to "explain" alternate disjunction. One force is a spindle repulsion, occurring between the spindle poles and the chromosomes of the ring. The second force is a repulsion between the chromosomes, possibly localized at the centromeres, so that a ring is formed. The second force somehow causes alternate chromosomes to lie on opposite sides of the equator. This hypothesis is really not very helpful in explaining the phenomenon.

Thus, there seems to be some sort of mechanical (spindle) adjustment of the chromosomes in some species so that by early anaphase those in rings are arranged for alternate disjunction. It is possible that such a mechanism could occur in any meiocyte but is permitted to do so only when at least the conditions already cited from Burnham are all present. Certainly such a mechanism need not be specifically evolved to accomplish alternate disjunction, and it certainly is not in the cases already cited that immediately have alternate disjunction in the hybrid. To claim that the parents of such hybrids are preadapted is perhaps correct but does not explain much unless the factors are specified, and these factors are cytological rather than genetic. It is also interesting that almost never does adjacent disjunction amount to more than 50% although random segregation should equal one third adjacent-1, one third adjacent-2, and one third alternate. It seems that in all organisms having ⊙14 or larger circles the mechanism for alternate disjunction functions to some extent if it can, as much as it can.

If there is any crossing-over between a centromere and the break, as at T in Fig. 13-8, A, the results are changed. A crossover at T followed by alternate segregation would swap ends B and D and deletion-duplication would result in both daughter nuclei. Adjacent-1 would produce balanced ends, but adjacent-2 would be the same as without such interstitial crossing-over and would still have duplications and deletions, but ends would be swapped (Burnham, 1956). In maize if there is crossing-over in the interstitial region, the chromosomes that cross over pass regularly to opposite poles, there is no adjacent-2 disjunction, and alternate and adjacent-1 are equally common. Also in maize translocation heterozygotes that form chains of four rather than ⊙4 have very little adjacent-2 segregation at anaphase I.

Thus, it can be concluded that alternate disjunction is the "normal" form and it occurs if not prevented by certain chromosomal conditions. Of the adjacent segregations adjacent-1 is normal and usually occurs if alternate disjunction is inhibited. Adjacent-2 is most abnormal and occurs least.

Perhaps the best way to consider disjunction of chromosomes from a ring in meiocytes is to compare it to what seems to occur to normal bivalents during diakinesis, prometaphase, and metaphase I. At diakinesis the chiasmata of bivalents have more or less terminalized and the two homologous chromosomes are bowed away from each other, for unknown reasons. If the chiasmata are terminalized (as is necessary for alternate disjunction) the bivalent is a ring, and if the homologous centromeres are more or less medially located (as is usually true for alternate disjunction) they are remote from each other. During prometaphase the centromeres become attached to spindle fibers by the unilateral kinetochores. It is likely that one centromere, either one of the bivalents, attaches before the other. The first one attached then must turn its kinetochore toward the pole to which its spindle fibers extend if, as seems likely, the kinetochore of the centromere lies on one side of the centromere. This would turn the other chromosome and its kinetochore toward the other pole. Perhaps this constitutes the "dancing around" pointed out by Ostergren (1951). In this way, possibly, the bivalent becomes oriented to the two poles at metaphase I.

Hughes-Schrader (1943a, 1943b), on the other hand, concluded that spindle fibers to all three chromosomes of an X_1-Y-X_2 sex trivalent are formed at random (alternate or adjacent) but later, apparently, original spindle fibers are destroyed and new connections are made so that the trivalents always have alternate segregation at anaphase I. If this is a correct observation, there is no scheme to explain how the two X's "know" enough to go to one pole and the Y "knows" how to attach to the spindle so that it goes to the other pole.

It can be assumed that the same process should operate for a ring as for a bivalent but may be modified or restrained by the ring and chromosomal peculiarities. At diakinesis there is a ring of four chromosomes. One kinetochore of one chromosome acquires a spindle fiber and turns toward a spindle pole. This could bend the two adjacent chromosomes or turn their kinetochores toward the opposite pole into a zigzag, in which arrangement they would acquire spindle fibers toward the opposite pole. The fourth and most remote chromosome would be affected less by the twisting of the first chromosome and could have spindle fibers directed toward the same pole as the first chromosome. The result would be alternate disjunction, provided all chromosomal conditions are appropriate. This scheme could also explain alternate disjunction in sex trivalents. It is very similar to the scheme of Ostergren (1949, 1951).

The above scheme could hardly operate in a large ring of 10, 12, or 14 chromosomes as in *Paeonia, Datura, Rhoeo, Oenothera,* or *Triticum* because two or more centromeres at random would be apt to acquire simultaneously connections to poles at random so that rarely could regular alternate disjunction all around the circle be achieved, as is well known to occur. This scheme could explain, however, the condition in *Campanula* in which Darlington and LaCour (1950) found that the larger the ring, the greater the infertility. At this time it is possible to speculate only. Perhaps the arrangement of the two noncrossover and two crossover chromatids at the terminalized chiasmata that connect chromosome ends in the ring might have some influence in orienting the unilateral kinetochores in adjacent chromosomes toward opposite poles. Brown (1940) and Pipkin (1940) concluded that in *Drosophila,*

females at least, alternate disjunction from rings of four chromosomes is determined by the presence of chiasmata. If this assumption is correct, perhaps additional interstitial, unterminalized chiasmata would distort this relationship and produce the considerable adjacent disjunction that does seem to occur with unterminalized metaphase I chiasmata (Sax and Anderson, 1933). Sax and Anderson reported in *Tradescantia edwardsiana* a ratio of 2 alternate to 1 adjacent if all chiasmata were terminalized; but for one subterminal chiasma the ratio was 1:1.2, and for more subterminal chiasmata all segregation was adjacent. However it may be achieved, regular alternate disjunction *can* occur all around a circle of as many as 14 or more translocated and normal chromosomes.

LARGE RINGS

Rings of six chromosomes or more have been found or produced in many plant and some animal species (Burnham, 1956) often as a result of hybridization. Yamashita (1950) produced various translocations in two species of diploid *Triticum* by X-rays. After characterizing the different translocations as to the chromosomes involved, he was able to cross particular homozygous lines to produce individuals heterozygous for ⊙4, ⊙6, ⊙8, ⊙10, ⊙12, and finally ⊙14 wherein all chromosomes were involved. All of these had a high percentage of regular alternate disjunction. Similar methods have produced large rings in maize, barley, *Campanula, Datura,* etc. as discussed in detail by Burnham (1956).

Naturally occurring rings larger than ⊙4 occur in plants: *Paeonia* (Walters, 1941, 1942); *Oenothera* (Cleland, 1936); *Clarkia, Rhoeo* (Anderson and Sax, 1936); *Gaura* (Bhaduri, 1942); *Isotoma* (James, 1965); *Hypericum* (Hoar, 1931; Hoar and Haertl, 1932); *Chelidonium* (Nagao and Saki, 1939) and animals such as scorpions (Piza, 1950) and cockroaches (John and Lewis, 1958). These examples represent evolved systems, and the heterozygosity is maintained in the populations by the heterozygosity being better adapted, and/or by the population having lethal genes linked to translocations and close to the centromeres of normal chromosomes so that, when homozygous, as in chromosomal homozygotes, they actually cause the elimination or nonproduction of homozygotes. When such a cytogenetic system occurs, such that cytological and genetic heterozygotes persist, the condition is said to be a *balanced lethal* system.

Two examples from plants are instructive. Gairdner and Haldane (1933) in *Antirrhinum* and Rees (1961) in rye have found recessive genes for yellow plants such that only the heterozygotes for both ⊙4 and the genes (there is no homozygous normal) are normally green and vigorous. It is the chromosomal (translocation) heterozygosity that keeps the strains genetically heterozygous as one form of "permanent heterozygotes" (Carson, 1967a). When two such lethal recessive genes, one on the normal and the other on the translocated chromosome, are closely linked near the centromere, neither form of homozygous recessive can occur.

John and Lewis (1958) found in isolated populations of cockroaches in mines that males contained ⊙4 or ⊙6 with an average of 92% alternate disjunction. Of course, very regular alternate disjunction must be present for circles to persist. Piza (1950) reported ⊙7 and ⊙9 in scorpions of the genera *Tityus* and *Isometrus* accompanied by a variable chromosome number.

Paeonia californica ($2n = 10$) is an example of a species in which translocation circles seem to be evolving (Walters, 1942). All conditions of bivalents and of single rings have been reported: 5 bivalents; ⊙4 and 3 bivalents; ⊙6 and 2 bivalents; ⊙8 and 1 bivalent; and ⊙10 as well as some with two small rings. Mixtures occur within populations of this self-fertilizing species. It is assumed that there are lethals in some populations where plants with rings outnumber plants with five bivalents.

Isotoma petraea ($2n = 14$) of central and western Australia is a species that seems to present examples of the evolution of the final condition of ⊙14 and balanced lethals. In eastern Australia in closely related species 7 bivalents is by far the commonest meiotic condition reported but ⊙4 and ⊙6 or ⊙4 and ⊙4 have been found. All hybrids between parents with 7 bivalents also have 7 bivalents, indicating that only one genome occurs; they are all homokaryotypic. In the southwestern corner of Australia, southwest of Lake Barlee, the cytological condition is reversed. Only one of 14 populations examined had any 7-bivalent individuals. Each of the remaining 13 populations was uniform with respect to circle and number of bivalents. Some were ⊙6 and 4 bivalents; ⊙6 and ⊙6 and 1 bivalent; ⊙10 and 2 bivalents; ⊙12 and 1 bivalent; ⊙8 and ⊙6, and at the southwestern extreme of the range of the species, ⊙14 was found. Since each of these populations was uniform, it is evident that balanced lethals must be present to eliminate all homokaryotypic individuals.

COMPLEX HETEROZYGOSITY

Complex heterozygote is the term applied to an individual or species or population in which two sets of structurally different chromosomes are maintained by balanced lethals. In *Isotoma petraea* with 7 bivalents the chromosome ends can be designated as 1-2, 3-4, 5-6, 7-8, 9-10, 11-12, and 13-14. This can be considered the karyotype of the central and eastern Australian species and populations. A ⊙4 is produced by an exchange of either end between any two nonhomologous chromosomes, such as between 5-6 and 11-12 to give either 5-12 and 6-11 or 5-11 and 6-12. Since at meiosis homologous ends synapse, a circle of four is produced by alternation of the normal and the translocated chromosomes such as 5-6:6-11:11-12:12-5 (with the two 5's also paired). Then, if any one of these four chromosomes exchanges any end with any end of a previously uninvolved chromosome, such as 5-12 with 1-2, a ⊙6 will occur at metaphase I: 5-6:6-11:11-12:12-1:1-2:2-5. This can be repeated until all chromosomes form a single ring at metaphase I because there will be two sets, called *complexes*, of structurally different chromosomes such as complex 1: 1-2,3-4,5-6,7-8,9-10,11-12,13-14 and complex 2: 1-10,2-5,3-14,4-7,6-11,8-9,12-13, for example. When homologous ends are matched up, a ⊙14 is formed. There can be many different end-arrangement complexes that, when combined, can form one complete circle. There can also be many complexes that, when present together in a heterozygote, form various numbers of circles or circles and bivalents. For example, a complex 2 of 1-12,2-5,6-11,14-3,4-7, 8-9,10-13 synapsed with complex 1 above produces ⊙6 and ⊙8. Any pair of complexes can be perpetuated indefinitely through sexual reproduction if balanced lethals prevent homozygotes from occurring, if there are no further translocations, and if no outcrossing occurs. It is true that all true-breeding complex heterozygous plant species or populations are self-pollinating and that they have balanced lethals. However, new translocations can occur at any time that can modify the metaphase arrangement.

With a ring of any size having regular alternate disjunction the unmodified chromosomes involved in the ring (one of the complexes) all go to the same pole and all of the translocated chromosomes of the ring move as a group to the other spindle pole at anaphase I. Segregation of any bivalents is unimportant since they do not contain meaningful lethals. Genetically the two complexes are two linkage groups just as truly as are two homologous chromosomes, and this fact confused early geneticists of *Oenothera*. They knew that there were seven pairs of chromosomes but only one linkage group. Since the chromosomes of each complex stay together as a linkage group generation after generation, they can be given designations. Thus, such a heterozygote for two complexes is a true-breeding complex heterozygote, one complex passing through the pollen only, the other through the egg only.

Oenothera

Most of the present knowledge about true-breeding complex heterozygotes has derived from the study of the subgenus Euoenothera of the American genus *Oenothera* of the family Onagraceae during the past 70 years (Cleland, 1956, 1962; Carson, 1967a; many others). Just as the bivalent-pairing forms of *Isotoma petraea* occur in central Australia and ring-forming populations are limited to southwestern Australia with the ⊙14 populations found only at southwestern limit of the species range, so in *Oenothera* the California species of the subgenus Euoenothera, such as *Oe. hookeri*, have bivalent pairing whereas the central and eastern United States species, strains, or populations have large rings, especially rings of all 14 chromosomes, as in the *biennis, strigosa, parviflora* groups of species. In southwestern United States between California and the western edge of the ⊙14 species (Oklahoma, New Mexico, Utah, Nevada, and Arizona) there are intermediate *hookeri*-like forms with such configurations as ⊙4, ⊙6 and 2 bivalents or ⊙4 and 5 bivalents. These latter do not usually breed true because they lack balanced lethals.

The term balanced lethals refers to the condition in most ⊙14 oenotheras whereby each complex carries distinctive lethals so that it always gets into all progeny individuals and all progeny are "balanced" by having both complexes. This is accomplished by one complex having lethals so that it has to get into the progeny via the pollen whereas the other must get into the progeny via the egg. The above is expressed in very broad terms because these lethals act in many different ways and in various combinations. Some so-called lethals may merely slow down pollen tube growth or megaspore–embryo sac growth so that one complex always gets to the embryo sac via the fast-growing pollen tubes and the other always produces the functional embryo sacs. Other lethals may actually cause the deaths of the haploid pollen tube or embryo sac. If the two complexes are A and B, A may cause the death of all pollen grains or pollen tubes carrying it whereas B may cause the death of all embryo sacs having it in their cells. Thus, all male gametes that function will carry the B and all eggs will contain the A complex. These are called *gamete lethals*.

Zygote lethals function after fertilization, in the zygote or later, even as late as producing physiologically poor plants. Zygote lethality occurs in all homozygous zygotes or plants so that the only reproductive plants are the heterozygous ones. There are other lines in which only one complex carries appropriate lethals so that such strains do not breed true since half of the progeny from such a heterozygous plant will be homozygous and half heterozygous.

The condition of complex heterozygosity in

Oenothera is maintained also by self-pollination, which enforces inbreeding isolation, with only rarely any outcrossing. In general, such complex heterozygotes having balanced lethals as *Oenothera, Isotoma, Rhoeo*, etc. are self-fertilizing whereas their bivalent-forming ancestors were cross-pollinating.

Cleland (1956) has presented strong evidence that all *Oenothera* chromosomes are, as far as detailed study can determine, small and of the same length and all have median centromeres. Since nearly all chromosomes of both complexes of all $\odot 14$ races have undergone at least one exchange, it must mean that all surviving interchanges are for equal chromosome arm segments. Cleland proposed that unequal exchanges doubtless have occurred but have not survived because the unequal-armed condition would prevent regular alternate disjunction and they would be lost. He considers that all of the unique characteristics of *Oenothera* derive from the structural equality of the chromosomes in the original ancestral bivalent-forming oenotheras. This seems evident since not only all taxa of *Oenothera* but the related ring-forming genera *Clarkia* and *Gaura* have small and more or less metacentric chromosomes also.

The concept of complexes has already been discussed as sets of chromosomes having unique end arrangements. They are derived via the pollen or the egg only, all chromosomes of the complex go together to one spindle pole at anaphase I as a linkage group, and in the presence of balanced lethals, they get into the progeny only via the pollen or egg. The complex is a permanent association generation after generation. In *Oenothera*, following the working out of the end arrangements of some complexes, they were given Latin names. Thus the *hookeri* complex has the end arrangement 1-2,3-4,5-6,7-8,9-10,11-12,13-14 and is taken as the standard of comparison more or less arbitrarily but also as the probable ancestral end arrangement since it occurs in the homozygous, cross-pollinating, bivalent-pairing forms in California that lack balanced lethals etc. In contrast to that standard *hookeri* end arrangement of chromosome ends, 90 other arrangements of 14 different ends can occur on 7 chromosomes, and all have been found. There are 14 other end arrangments possible, such as isochromosomes: 1-1, 2-2, 13-13, 14-14, but they cannot occur with rings and are not found. A few examples of complexes and their end arrangements are the following (Cleland, 1936; Cleland et al., 1950):

Complex	End arrangement
acuens	1-4, 3-2, 5-6, 7-10, 9-8, 11-12, 13-14
velens	1-2, 3-4, 5-8, 7-6, 9-10, 11-12, 13-14
β Indian River	1-7, 3-12, 5-9, 2-10, 4-14, 11-8, 13-6
α Mountain Lake	1-4, 3-2, 5-14, 7-8, 9-10, 11-12, 13-6

Complexes of *Oenothera* are also designated as to the gametophyte by which they are transmitted, since one of the two in a complex heterozygote is transmitted via the pollen and the other via the embryo sac only. The complex transmitted through the embryo sac and egg is called the alpha (α) complex and the complex carried by the pollen and male gametes is called beta (β). Thus, a population of a species of *Oenothera* can be described by the two complexes present. Furthermore, specific phenotypic characters can be ascribed to particular complexes and they show up for that particular complex to some extent in the presence of various other complexes. In general, each complex also carries a specific lethal as for pollen, embryo sac, etc. For example, in the Chevy Chase collection of the *biennis* group 1 the complexes are the following (Cleland et al., 1950):

Complex	End arrangement
Alpha	1-2, 3-4, 5-14, 7-10, 9-8, 11-12, 13-6
Beta	1-14, 3-2, 5-9, 7-8, 6-12, 11-10, 13-4

These same complexes, together or separately, turn up in numerous other populations, yet adjacent populations may have more or less different complexes. Cleland et al. (1950) have concluded that there are certain recognizable taxa (not called species, however), the *hookeri, strigosa, biennis* 1, *biennis* 2, *biennis* 3, *parviflora,* and *grandiflora* groups. It is true that specific complexes and their slight modifications are variously characteristic of these taxa and imply phylogenetic relationships. But Cleland and associates long ago stopped using the term and concept *species* as meaningless in this and other subgenera of *Oenothera*. Rather they use the terms *group* and *race* for categories above the level of individual plant. A race includes all individuals containing the same two identical complexes. A group contains all those races having similar but not necessarily identical complexes.

Rhoeo discolor

Another well-known plant species having a ring of nearly metacentric chromosomes at metaphase I with all or most chiasmata terminalized and balanced lethals is the widely cultivated, monotypic, yucca-like member of the family Commelinaceae, *Rhoeo discolor* (Fig. 17-3). In most cytological features it is different from *Oenothera*. Sax (1931) made a detailed cytological study of this species and reviewed previous work. The chromosomes are not at all equal in length nor are many or any truly metacentric, as are all the chromosomes of *Oenothera* (Cleland, 1956). Sax (1931) and Flagg (1958) provided drawings of a somatic metaphase plate and an idiogram of metaphase I chromosomes, respectively. When arranged in descending order from most heterobrachial to most metacentric as a ratio of longer to shorter arms and with total relative length in parentheses, it is seen in Table 3 that, according to these data, there are no equal pairs, as there are in *Oenothera*, although there are no homologous chromosomes. Of these the data from somatic chromosomes are by far the more meaningful, but both show the same chromosomal conditions.

Another difference from *Oenothera* (that very regularly has a ring of all chromosomes) is that in *Rhoeo* one complete ring of 12 chromosomes occurs in only about one third of microsporocytes (Darlington, 1929). Walters and Gerstel (1948) in the diploid form studied found ⊙12 in 32%, chain of 12 in 38%, two separate chains in 15%, three chains in 14%, and four chains in 1% of microsporocytes. Regular disjunction to produce genetically balanced spores is certain only in ⊙12 and possibly in chain of 12. The 30% of two or more chains would only occasionally have regular complex disjunction. Even with ⊙12 or chain of 12 regular disjunction would occur if no adjacent segregation occurred, but it does.

All studies of meiosis in *Rhoeo* have reported considerable alternate disjunction. This means that alternate disjunction may occur for most chromosomes but somewhere in the ring two (or three) adjacent chromosomes pass to the same pole. That means that seven chromosomes go to one pole and five to the other. Such 7/5 distribution was reported in 33% of cells by Kato (1930), 30% by Walters and Gerstel (1948), and 7% by Darlington (1929). Two adjacent segregations in the same cell but directed toward opposite poles would give 6/6 distribution but sterility (Sax, 1931; Walters and Gerstel, 1948). Sax reported, "In about half of the first meiotic divisions observed the alternate chromosomes in the rings pass to opposite poles." In all cells with ⊙12 most chromosomes do have alternate disjunction, but it takes only one mistake in any cell to cause spore sterility. Even with chains and considerable adjacent disjunction the reported 80 to 90% of dead pollen grains is difficult to account for.

Sax (1931) was able to determine the regularity of the arrangement of chromosomes in rings and chains as had Kato (1930), but they disagreed to some extent. They both characterized the 12 chromosomes into two mor-

Table 3. Ratio of longer to shorter arms and total relative length

	Chromosomes No. 1 to No. 6					
Sax	3.4:1(22)	3.1:1(24)	2.6:1(18)	2.5:1(21)	2.4:1(17)	1.7:1(16)
Flagg	2:1(20)	1.9:1(25)	1.7:1(27)	1.7:1(19)	1.6:1(26)	1.5:1(23)
	Chromosomes No. 7 to No. 12					
Sax	1.6:1(23)	1.6:1(13)	1.5:1(20)	1.3:1(25)	1.3:1(16)	1.2:1(13)
Flagg	1.4:1(19)	1.08:1(27)	1.07:1(29)	1:1(24)	1:1(22)	1:1(10)

phological groups: (1) heterobrachial having definite short and long arms, and (2) isobrachial chromosomes with approximately equal arms. If H equals heterobrachial and I stands for isobrachial chromosomes:

Proposal	Arm-type of chromosomes
Kato	H-H-I-I-I-I-H-H-I-I-I-I
Sax	H-H-I-H-H-I-H-H-I-I-I-I

Sax further stated that the heterobrachial chromosomes of the pairs are regularly attached together in the ring by their short arms. He could also distinguish among the heterobrachial chromosomes. The order of chromosomes of different size and form which, nevertheless, have homologous ends thereby "proves" reciprocal translocation. Ovule abortion does occur for the same reason as pollen abortion, adjacent disjunction of some chromosomes of many rings (Carniel, 1960).

Thus *Rhoeo's* chromosomal variability in length and centromere location, possibly resulting from unequal translocations, does permit the continuity and partial fertility of the species but it is just barely competent. It contrasts instructively with *Oenothera*, which seems to have all the requirements as an ideal system.

Autotetraploid meiosis in *Rhoeo* (4n = 24) has also been studied (Walters and Gerstel, 1948). As in tetraploid *Oenothera* (Davis, 1943), numerous pairs and short chains are formed because each chromosome end can pair with as many as three homologous ends. In *Rhoeo*, ring bivalents (a pair of homologues with a chiasma at each end) were commonest, with open bivalents (a pair with one chiasma at one end) almost as common. Chains of 3 and of 4 chromosomes were also very frequent, but longer chains of from 5 to 12 chromosomes decreased in frequency of occurrence as the number of chromosomes in the chains increased. Rings of four chromosomes were fairly common, but larger rings were essentially absent. Occasional multivalents of odd shapes having three chromosomes united at their ends, two united to one end of another, were recorded such as to produce Y's or circles with a chromosome or chain extending outward.

Chromosome segregation was somewhat irregular so that about one half of the pollen grains at division had 12 chromosomes, about one quarter had 11, and about one quarter had 13. Progeny plants produced had chromosome numbers of 21, 23, 24, and 26. Pollen fertility in the diploid studied was about 33% whereas pollen fertility of the tetraploid was about twice that, about 73%.

Thus diploid, the common *Rhoeo discolor* has low pollen and probably ovule fertility so that seed set is low, probably from 5 to 10%; but because of balanced lethals all progeny that do result are complex heterozygotes.

Recently Wimber (1968) studied diploid plants from British Honduras that were called *Rhoeo discolor* variety *concolor* (2n = 12). It was shown to have only bivalent pairing and to contain in the homozygous condition and, of course without lethals, one complex of the usually cultivated heterozygous form. Wimber also accepted the nomenclatural change from *Rhoeo discolor* (L'Heritier) Hance to *R. spathacea* (Swartz) Stearn. Thus the usual cultivated complex heterozygous form with the underside of the leaves purple is *R. spathacea*, and the less robust bivalent-pairing form with all green leaves is called *R. spathacea* variety *concolor*.

Wimber found that reciprocal crosses between these forms could not be made and that the only successful hybridization was achieved when the ring-forming *R. spathacea* was the pistillate parent and variety *concolor* provided the pollen. There were two classes of progeny produced in equal numbers; one class had ⊙12, and the other had 12 bivalents. Therefore, the common ring-forming, complex heterozygous *Rhoeo* produces two sorts of eggs, containing either complex A or B. The pollen of *concolor* provided only one complex, complex A. Thus two kinds of zygotes and progeny were produced: AA that have bivalent pairing and AB that form ⊙12. Therefore, the lethality of *Rhoeo discolor* (that is, *R. spathacea*) is zygotic. The AA hybrids lack the lethality of the AA genomes of *R. spathacea*. The A genome provided by *concolor* lacks the lethal that is necessary to maintain the heterozygosity of *R. spathacea*.

Gossypium

In *Gossypium* the cultivated cottons have been shown to differ for very few translocations, considering the evolutionary differences. These translocations occur in the A genome (Gerstel, 1953; Menzel and Brown, 1954). The A genome of Old World diploids *G. anomalum* and *G. herbaceum* (A_1) is considered the most primitive. Genome A_2 of *G. arboreum* differs from A_1 by one translocation, and it is likely that the cultivated *arboreum* has been derived during cultivation from the also cultivated *G. herbaceum* (Hutchinson, 1954). It is not unlikely that the A genome in tetraploid *G. hirsutum* $(A_hD)_1$ is derived more directly from *herbaceum* than via *arboreum*. But the A_h genome of *hirsutum* differs from the A_1 genome of *herbaceum* by two distinct translocations so that in the hybrid at meiosis there are two ⊙4. When *hirsutum* (A_hD) is crossed with *arboreum* (A_2) a ⊙6, ⊙4 occurs at meiosis.

This can be summarized by end arrangements of five chromosomes as follows:

Genome	End arrangement
A_1	A-B, C-D, E-F, G-H, I-J
A_2	A-D, C-B, E-F, G-H, I-J
A_h	A-I, C-D, E-G, F-H, B-J

In the A_1A_2 hybrid there is a ⊙4 (A-B, B-C, C-D, D-A). In the A_1A_h hybrid there is a ⊙4 (A-B, B-J, J-I, I-A) and another ⊙4 (E-F, F-H, H-G, G-E). But the A_2A_h hybrid has the one ⊙4 as in the A_1A_h but also a ⊙6 (A-I, I-J, J-B, B-C, C-D, D-A). Thus the A-B chromosome of A_1 has been involved in two different exchanges, A-D, C-B, of A_2 and A-I, B-J of A_h. Therefore, it is more likely that A_h and A_2 have been derived separately from A_1.

DYSPLOID DECREASE

When all or most of the euchromatin that contains "active" genes is translocated from one chromosome to another nonhomologous chromosome and the latter reciprocally loses no essential genetic material, two new chromosomes are formed. One of them has all the important genes formerly present in two chromosomes, and the other has none. In subsequent generations the former may become homozygous, at least in some populations. If the latter is eventually lost, there has been a decrease in chromosome number by one pair. Thus a species (n = 4) with chromosomes AA, BB, CC, DD by transferring essential genetic material from, for example, C to B can produce a form with AA, B_cB_c, DD and have only three pairs. B_c represents the condition of one nearly intact chromosome, B, having a piece of another, C, translocated onto it. This sort of karyotype evolution is called *aneuploid decrease* or *dysploid decrease* and has been demonstrated to have occurred in a number of plants and animals where a sequence of haploid chromosome numbers occurs at the diploid level within a genus or family. In fact, whenever a chromosome number reduction is found, it is assumed that it has been accomplished by dysploid decrease as in *Drosophila*, *Crepis*, etc. There is no satisfactory alternative mechanism known. It does seem to be generally accepted that such decrease is common; but aneuploid increase, although theoretically possible (Darlington, 1937), is often considered to be uncommon (Swanson, 1957) but is probably achievable and not rare (see later).

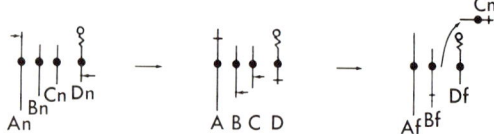

Fig. 13-9. Chromosome evolution by dysploid reduction achieved by translocation in *Crepis*. *C. neglecta* (n = 4) (left) has chromosomes designated $A_nB_nC_nD_n$ (subscript n for *neglecta*). *C. fuliginosa* (n = 3) (right) has chromosomes designated $A_fB_fD_f$ (subscript f for *fuliginosa*). The D chromosome in both has the nucleolus organizer, and the C_n is mostly heterochromatic. Study of meiotic pairing in the hybrid indicated that a small exchange (arrows, left) had occurred in *neglecta* between A_n and D_n because the A_n and D_f occasionally paired terminally. Another exchange (center) occurred between C_n and B_n (arrows) so that the C_n became almost completely heterochromatic and, at right, was lost. C_n and B_f paired commonly in the hybrid. By the loss of the remains of the C chromosome during evolution of *fuliginosa* from *neglecta*, the chromosome number was reduced from n = 4 to n = 3. (Data derived from Tobgy, H. A. 1943. J. Genet. 45:67-111.)

One of the first "proofs" of this mechanism of chromosome number reduction was that of Tobgy (1943) in which he demonstrated the reduction from $n = 4$ in *Crepis neglecta* to $n = 3$ in *Crepis fuliginosa* (Fig. 13-9). Following determination of good bivalent meiotic pairing in the parental species the hybrid was produced and its meiotic configurations analyzed. The mitotic chromosomes of these species are as illustrated in Fig. 13-9. Pairing at meiosis in the hybrid varied, but in essentially all cells there was at least one multivalent and often one or two univalents. The four chromosomes of *C. neglecta* were designated as A_n, B_n, C_n, and D_n (n for *neglecta*) and the three chromosomes of *C. fuliginosa* were called A_f, B_f, and D_f. Because of the way they paired, there is no C in *fuliginosa*. A_n and A_f were almost always united by a terminal chiasma at metaphase I as were also D_n and D_f. But A_n was very often paired terminally with D_f. B_f and C_n were also very often paired, with B_f often paired also with B_n. There is no C_f, but C_n is mostly heterochromatic.

Tobgy proposed that a translocation involving part of the short arms of A_n has transferred that piece to the long arm of D_f to produce the commonly seen A_f-A_n-D_f-D_n quadrivalent. The C_n centromere and most of its heterochromatin was lost after a translocation transferred its euchromatic tip to the end of B_f, since B_f and C_n are usually paired. Of course these translocations occurred separately in an $n = 4$ *C. neglecta* ancestral population to have given rise to an $n = 3$ "neglecta" and from it has subsequently evolved the $n = 3$ species *C. fuliginosa*.

It is not known how some D_n can be added to A_f, and A_n to D_f yet both chromosomes end up shorter in *fuliginosa* or how some C_n can be added to B_f and it becomes shorter than the original B_n. The only presently acceptable explanation is that the part of B_n translocated to C_n, before C_n was lost, was composed of heterochromatin. But too little is known about heterochromatin and its genetic content and about how chromosome mass can be decreased, as it often is, to give any really meaningful discussion.

Sherman (1946) has produced similar evidence from meiotic pairing between *C. foetida* ($n = 5$) and *C. kotschyana* ($n = 4$). It is likely that the original basic chromosome number in *Crepis* was $n = 5$ or 4. Perhaps 5 is more likely than 4; at least it requires only one increase, to $n = 6$, and dysploid increase has often been postulated but rarely demonstrated within genera.

Haplopappus gracilis, in the two forms "dibivalens" ($2n = 4$) and "tribivalens" ($2n = 6$), and the closely related *H. ravenii* ($2n = 8$) have been studied intensively in an effort to determine the chromosomal changes that have occurred during evolution and the probable direction of change (Jackson, 1962). Jackson has concluded (1965) that the most likely direction of change is from tribivalens to dibivalens (Fig. 13-10). The 5-chromosome hybrid between these two forms is highly fertile because of regular alternate disjunction from a trivalent (Jackson, 1964). At pachytene two short chromosomes of tribivalens are synapsed in tandem along almost all of the long submetacentric chromosome of dibivalens, clearly demonstrating that one of dibivalens is equivalent to the two of tribivalens.

Jackson has designated the shorter satellited chromosome of dibivalens as chromosome B and the long submetacentric as A. In tribivalens the B is identical to the B of dibivalens but there is no long submetacentric A. Rather, there are two short chromosomes: one, C, is acrocentric, and the other, D, is subacrocentric. The long arms of C and D have lengths very similar to the two arms of A. He proposes that a break in the long arm of either C or D close to the centromere and another in the short arm of D or C would, by reciprocal translocation, produce the A chromosome with two long arms and a very short chromosome, a supernumerary such as is often seen in *H. gracilis*.

Jackson favors such dysploid reduction rather than dysploid increase from dibivalens to tribivalens. He did propose a rather unlikely sequence of possible events for increase, by breakage and isochromosomes, and properly discards it as highly unlikely.

Jackson had earlier (1962) presented some evidence that *H. gracilis* (the $n = 2$ but not the $n = 3$ form) had originated probably from the

190 TEXTBOOK OF CYTOGENETICS

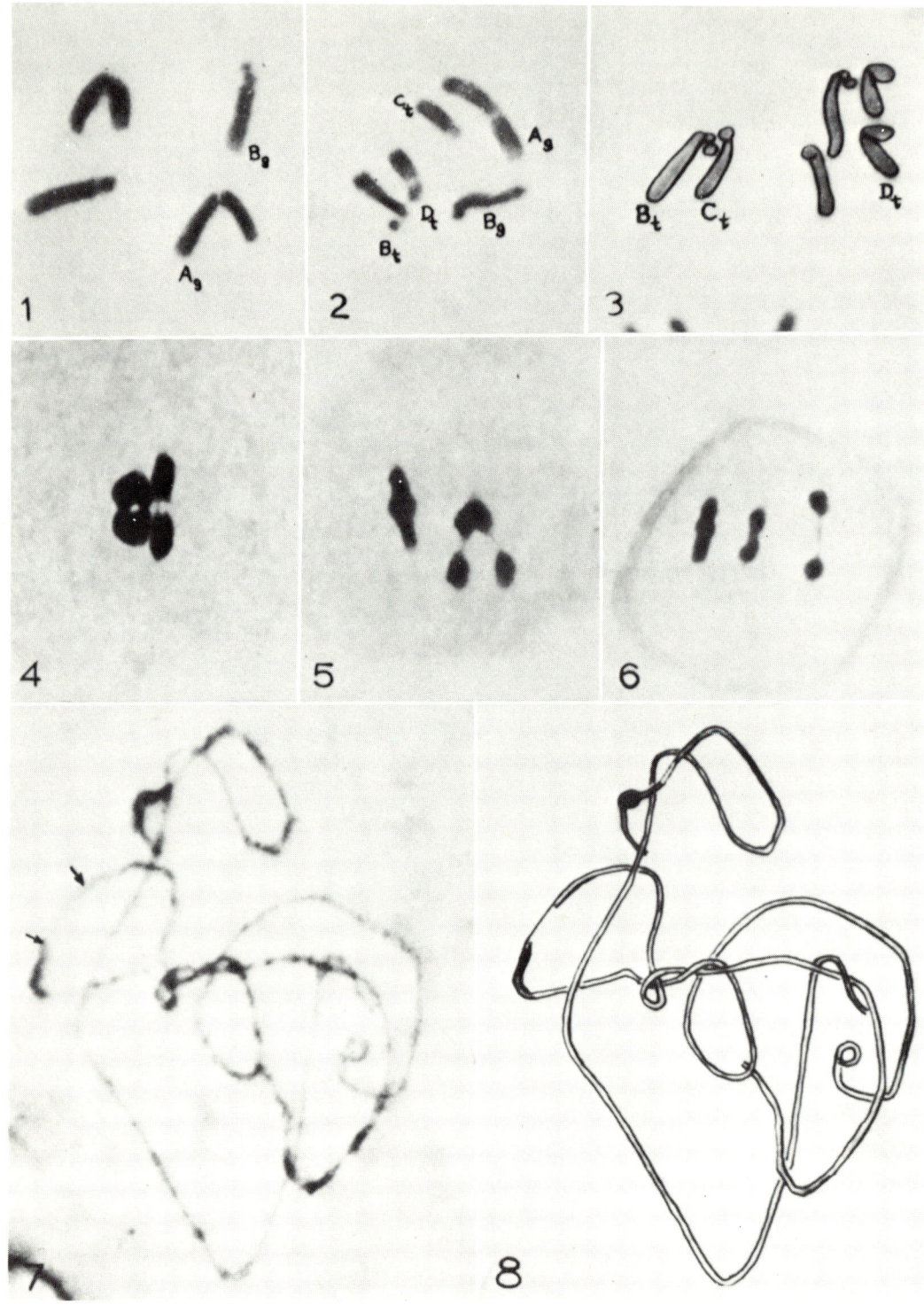

Fig. 13-10. For legend see opposite page.

Fig. 13-10. Chromosomes in *Haplopappus gracilis*. **1**, 2n = 2, the common (dibivalens) form; **3**, the 2n = 6 (tribivalens) form; and **2**, the 5-chromosome hybrid. **4, 5,** and **6,** The early anaphase I appearances in these three sorts of plants; two bivalents, **4,** in the n = 2 form; three bivalents, **6,** in the n = 3 form; and two associations, **5,** in the 5-chromosome hybrid. In **5** a bivalent (B_gB_t) and a trivalent ($A_gC_tD_t$) are evident. **7** and **8,** Pachytene in the hybrid demonstrating that two chromosomes of the tribivalens parent (C_t and D_t) are synapsed to one (A_g) of the dibivalens parent except for a short segment (arrows in **7**). Apparently a translocation between C_t and D_t transferred essentially all of C_t's euchromatin to D_t, thereby producing A_g of dibivalens, and C_t was subsequently lost. In the hybrid, C_t and D_t pair with different portions of A_g to produce the trivalent, **5, 7,** and **8.** All three forms occur in nature in a few restricted populations. It is possible, but not likely, that evolution has progressed from n = 2 to n = 3 by some unknown form of chromosomal evolution. (From Jackson, R. C. 1965. Amer. J. Bot. **52**:946-953.)

---

n = 4 *H. ravenii* by a series of unequal, reciprocal translocations and loss of centromeres as small supernumeraries. This was accompanied by a decrease in haploid chromosome set length from 17μ in *ravenii* to 13.5μ in n = 2 *gracilis*.

Chaenactis, another genus of Compositae, has yielded two demonstrated examples of dysploid decrease (Kyhos, 1965). The three species of the genus involved are closely related annuals of California, *C. glabriuscula* (n = 6), *C. fremontii* (n = 5), and *C. stevioides* (n = 5). Kyhos concluded that the two 5-paired species have been derived independently from the 6-paired *C. glabriuscula* (Fig. 13-11).

C. fremontii arose from *glabriuscula* following one translocation that eliminated the F chromosome centromere and a small portion of the E chromosome. Most of the long arm of F was added to the short arm of the E (Fig. 13-11, *I, II,* and *III*). The second translocation exchanged ends of the short arms of B and C, thereby transferring the nucleolus organizer to C. A similar exchange between B and D during the evolution of *stevioides* transferred the nucleolus organizer to D. At meiosis in both hybrids, *glabriuscula* × *fremontii* and *glabriuscula* × *stevioides*, instead of one nucleolus on the B bivalent, there is one nucleolus on one homologue of the B bivalent (the B from *glabriuscula*) and another nucleolus on one homologue of the C or D, as the case may be. That is the evidence for the exchange.

During the evolution of *C. stevioides* there were three simultaneous or two pairs of two different breaks (Fig. 13-11, *IV, V,* and *VI*). As a result, a portion of the short arm of F was translocated to the centromere and short arm of C, producing a small and probably heterochromatic chromosome which was lost. The remainder of the long arm of C was added to the long arm of B when a portion of the long arm of B was translocated to the short arm of F in place of its segment that had gone to C.

These translocations were deduced from the

Fig. 13-11. Translocations in *Chaenactis*. **I,** *C. glabriuscula* (n = 6) chromosomes A-F (left to right) with nucleolus organizer in short arm of chromosome B. A translocation (breaks at arrows in chromosomes E and F) leads toward *C. fremontii*. Following the translocation there is the loss of the short arm and centromere of F and part of the short arm of E. E has most of the long arm of F added to its short arm. A second exchange (arrows in **II**) interchanges the tips of B and C so that the nucleolus organizer in *fremontii* (n = 5) is on chromosome C. **III,** *C. fremontii*. **IV,** *C. glabriuscula* as in **I.** Three breaks (arrows) result in a piece of C on B (different from the exchange in **II**), a piece of B on F, and a piece of F on the centromere and short arm of C, which is lost. **V,** Two more breaks (arrows) exchange the ends of the short arms of B and D so that the nucleolus organizer is transferred to D. **VI,** The chromosomes of *C. stevioides*. Comparison of **III** and **VI** illustrates the differences between *fremontii* and *stevioides*. (Data from Kyhos, D. W. 1965. Evolution **19**:26-43.)

pairing at metaphase I in hybrids. In the *glabriuscula* × *fremontii* hybrid the E and F of *glabriuscula* were paired at opposite ends with the E + F chromosome of *fremontii* as a chain of three. In the hybrid *glabriuscula* × *stevioides* a chain of five was found consisting of the D, B, and C of *glabriuscula* and the B + C and D + B of *stevioides*.

Another example of chromosome number decrease was produced in *Godetia (Clarkia) whitneyi*, a species characterized by translocations (Hakansson, 1946). Semifertile F_1 plants with fairly regular meiosis were recovered from a cross of two monosomic plants that were structurally heterozygous for exchanges.

Probably the most "impossible" case of probable reduction in chromosome number has been reported in mammals, specifically the Asiatic deer genus *Muntiacus* (Chapter 10, Fig. 10-3). With one known intermediate the number seems to have decreased (in one step?) from about n = 23 of *M. reevesi* to n = 3 in *M. muntjak*, the two species being closely related enough to form hybrids!

These and other studies have demonstrated the mechanism of dysploid decrease in animal and plant evolution to the general satisfaction of most cytogeneticists and evolutionists. This method does lose some centromeres and some other chromosomal material usually said to be heterochromatin. In *Haplopappus* the small chromosomes that result from such unequal translocation are still "hanging around" in numerous populations, according to Jackson, as supernumerary chromosomes.

DYSPLOID INCREASE

One significant difference between the mechanisms of dysploid decrease and increase in diploid species is that the former is rigidly restricted to no loss of essential genes, therefore the rarity of monosomics, whereas increase of genes and chromosomes may produce duplication; but that is much more tolerable than deficiency. Furthermore, whereas there seems to be only one mechanism of chromosome number decrease, there are probably a number of methods by which increase is achieved.

Trisomy can be one possible mechanism that might lead through tetrasomy and diploidization to an increase of one pair of chromosomes. Certainly trisomy is not rare in experimental plant species and may be more common in nature than is presently known. At another level, when more than one pair of extra chromosomes occur, such species as *Claytonia* and *Draba (Erophila)* (Winge, 1940) demonstrate that a great deal of variation in chromosome number can be present without completely eliminating fertility. In the *Draba verna* complex Winge found populations having n = 7, 15, and 18 mostly but also 12, 16, 17, 20, 26, 27, 29, and 32. Furthermore, F_1 progeny of various intermediate chromosome numbers from crosses were often fertile and by the F_7 to F_9 generation were established as vigorous, true-breeding, and fully fertile new chromosome number lines. In fact, one line derived from a cross between n = 15 and n = 32 parents was established at n = 34, an increase over the chromosome number of the n = 32 parent.

Grant (1966a, 1966b) produced a "good" 38-chromosome synthetic "species" from two 36-chromosome parental species by the F_9 generation and had another line stabilized at 2n = 50. In both of these there was dysploid increase of one and seven pairs, respectively. The parental species were the 2n = 36 tetraploid self-fertile, desert annual species *Gilia maior* and *G. modocensis*. The F_1 hybrids were vigorous but almost completely sterile. The F_2 and F_3 were variable due to segregation, but they were of low vigor and low fertility. But vigor, fertility, elimination of aneuploids, bivalent chromosome pairing all increased and by F_9 were as "good" as in the parent species with little variability.

Gerstel (1945) derived lines with chromosome number increase from tetraploid *Nicotiana tabacum* (2n = 48) × *N. glutinosa* (n = 12) and the F_1 backcrossed with pollen of normal diploid (n = 24) *N. tabacum*. Among the various segregates were constant lines with n = 25 and n = 26. He concluded that these lines with increased chromosomes contained one or two pairs of *N. glutinosa* added to two complete sets of *N. tabacum*. These races he called alien addition races; and because backcrossing to *N. tabacum* eliminated the alien chromosomes, Gerstel concluded that inbreeding and self-fertility would be necessary to maintain such addition races. This conclusion is sound because almost all plants in which such experimental increase has been found are normally self-pollinating or they fail.

O'Mara (1951) reported an alien addition race of wheat with a pair of rye chromosomes, and there have been others.

Thus, one mechanism for increase in chromosome number is by the addition of various numbers of pairs of unmodified chromosomes following hybridization or, possibly, tetrasomy following nondisjunction.

The second mechanism for increase in chromosome number involves translocation and/or secondary or tertiary trisomics. Secondary trisomics have the extra chromosome as an isochromosome. An isochromosome has two structurally and genetically identical arms with centromeric and distal regions homologous. Tertiary trisomics have the two arms of the extra chromosome derived by translocation from two different nonhomologous chromosomes.

Rhoades (1955) reported the production of a line of n = 11 maize from the normal n = 10. Presumably a chromosome broke transversely across the centromere, thereby producing two nonhomologous telocentric chromosomes from one submetacentric. By breeding appropriately the line was made homozygous for the two telocentrics.

In *Drosophila*, Patterson and Stone (1952, p. 179) concluded that the only chromosome number increase in the genus has occurred from n = 6 to the n = 7 of *D. trispina*. Presumably in a species like *limpiensis* (n = 6) an unequal translocation occurred to produce a small and a larger chromosome, as is typical in dysploid reduction. But quite differently the small chromosome persisted and became a pair, and it was the large translocated chromosome that disappeared from the population. The two original untranslocated chromosomes also persisted. The net effect was like adding the small chromosome pair to *D. limpiensis*. *D. trispina*, therefore, has this extra small pair in addition to the "dot" pair so common in *Drosophila*.

An example, perhaps the only one clearly proved, of an increase of a pair of chromosomes in nature is that in *Clarkia* reported by Lewis and Roberts (1956). The species involved in the study were the self-compatible but usually cross-pollinated *C. ligulata* (n = 9) and three subspecies of the very closely related and reproductively similar (n = 8) *C. biloba* (Figs. 13-12 and 13-13). *C. ligulata* was known in 1956 from only two sites in the Merced River Canyon of California close to colonies of *C. biloba*. Cytological study of meiosis in interspecific hybrids between the three subspecies of *biloba* and *ligulata* revealed that *ligulata* differs for two translocations from *biloba* and that *biloba* subspecies *australis* differs from subspecies *brandegeae* by a small terminal exchange in the same chromosomes involved in the transloca-

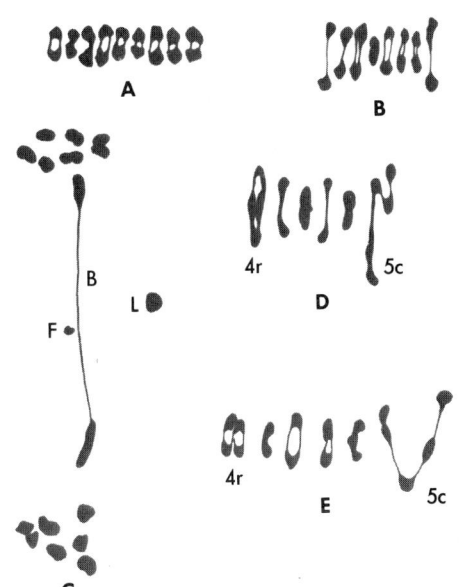

Fig. 13-12. Meiotic associations of chromosomes in: **A**, *Clarkia ligulata* (n = 9); **B**, *C. biloba* (n = 8); **C, D**, and **E**, interspecific hybrids. **A** and **B**, Good bivalent pairing in the parent species. **C**, A bridge: *B*, fragment; *F*, anaphase I association in the hybrid indicating specific difference for an inversion. **D** and **E**, The common metaphase I configuration for the 17 chromosomes in the hybrid was 4 bivalents, ⊙4, and chain of 5. About 65% of ⊙4's (*4r*) probably gave adjacent disjunction, as in **D**, rather than alternate disjunction, as in **E**. About 53% of PMC's of the hybrid had a chain of 5 (*5c*) with the chromosome variously oriented, two of the commonest being represented in **D** and **E**. (From Lewis, H., and M. R. Roberts. 1956. Evolution **10**:126-138.)

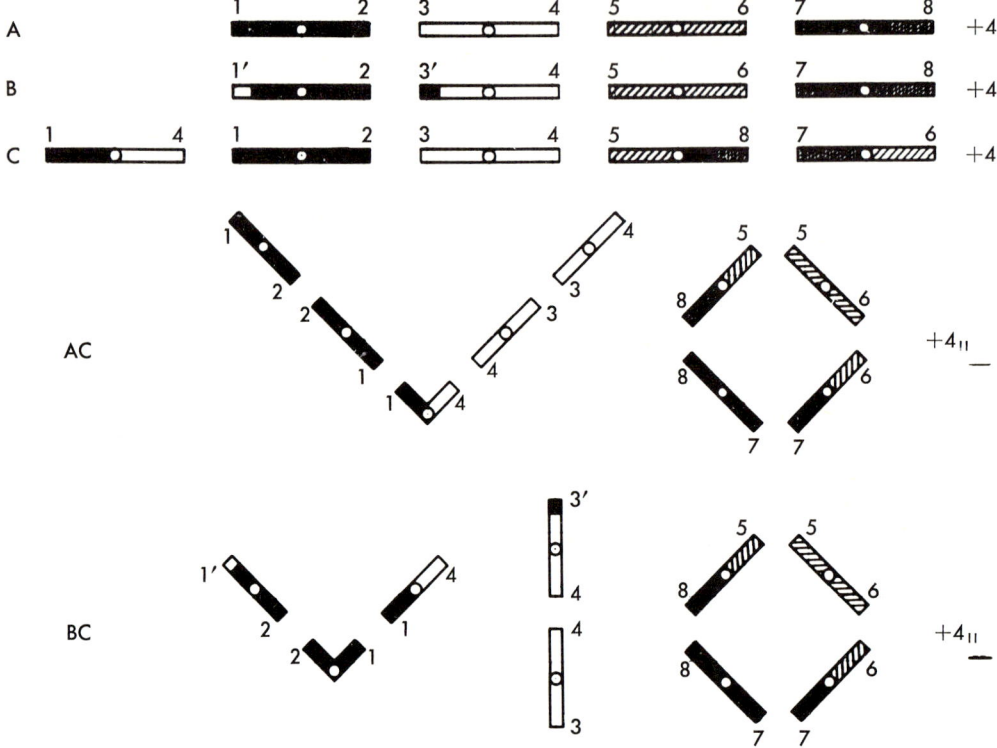

Fig. 13-13. Chromosomes of *Clarkia ligulata* and *C. biloba* derived in part from meiotic data in Fig. 13-12. **A** and **B,** Two subspecies of *C. biloba* (n = 8). **C,** *C. ligulata* (n = 9) AC and BC chromosome configurations at meiosis in interspecific hybrids. There are always four bivalents (far right) indicating considerable similarity. A quadrivalent was usually present, indicating that the species differed for a translocation among chromosome ends 5, 6, 7, and 8. The extra or ninth chromosome consists of arm 1 of chromosome 1-2 and arm 4 of chromosome 3-4 attached by a translocation to form chromosome 1-4. This gets involved in various pairing chains with pairs 1-2 and 3-4 that are present in both parents. The dysploid chromosomal increase came from a *C. biloba* (n = 8) plant that was a tertiary trisomic (2n = 17; one extra chromosome, 1-4) becoming a constant tetrasomic (2n = 18, n = 9). An additional alteration was an inversion. The result was *C. ligulata*. (From Lewis, H., and M. R. Roberts. 1956. Evolution 10:126-138.)

tion that produced the extra chromosome of *ligulata*. One translocation produced a ⊙4 in the hybrids but has nothing to do with the chromosome increase. Chromosomes 1-2 and 3-4 of *australis* are taken as the standard, and *ligulata* has them unmodified. But *ligulata's* ninth chromosome is a translocation product with its arms designated 1-4. Therefore, at meiosis in the hybrid there is a chain of five chromosomes with the extra ninth chromosome of *ligulata* in the middle: 1-2,2-1,1-4,4-3,3-4.

In the hybrid between *brandegeae* and *ligulata* a chain of three chromosomes rather than a chain of five occurs. This is explained by an exchange of small segments from the tips of arms 1 and 3 in *brandegeae*, giving 1'-2 and 3'-4.

That exchange prevents the 4-3,3-4 part of the chain of five from forming. Rather, the unchanged 3-4 pairs with the 4-3' as a bivalent. The remaining three chromosomes form a chain with the ninth chromosome at one end, like the first three chromosomes in the chain of five above: 1'-2,2-1,1-4.

This ninth-chromosome pair of *ligulata* makes *ligulata* a tertiary tetrasomic, the additional chromosome containing arms derived by translocation from two nonhomologous chromosomes. Thus it is duplicate for chromosome segments and the contained genes. Lewis and Roberts point out that a tertiary trisomic has been found in another species of *Clarkia*, *C. unguiculata*, and tertiary trisomics have been found in various other species of plants that have been intensively studied (see Burnham, 1956).

Thus, there are various mechanisms for the increase of chromosome numbers. The numerous assumptions of evolutionary chromosomal increase seem to be reasonable, such as that of Levan (1932, 1935), who considered the $n = 7$ North American species of *Allium* to be primitive and an an increase to $n = 8$ occurred during the evolution of the Eurasian species. But chromosomal number increase is not limited to a change of one pair at a time as is chromosomal decrease. Nevertheless, the present evidence is that chromosomal decrease is probably much more common than increase.

CHAPTER **14**

Inversions

> There can be no doubt, then, that formation of inversions is a very widespread method of evolution of the chromosomal apparatus, and in importance exceeds that of the translocations.
>
> Dobzhansky, 1941

Among organisms the major alterations of chromosome structure are translocations, which alter two nonhomologous chromosomes as one event, and inversions, each inversion being a reversal within the chromosome (see also Chapter 3). As the name implies, an inversion is the turning around within the chromosome of a long or short segment (Fig. 14-1). Since it is not possible for a normal end of a chromosome arm to attach to a broken surface of a chromosome arm, two breaks are required within one chromosome at the same time to form an inversion (Fig. 14-1, C). Apparently such double breaks are not uncommon, and it seems likely that such breaks can occur anywhere within the chromosome from the very end of the arm (Kaufmann, 1936) to the centromere. Thus, it would be rare or essentially impossible for a particular chromosome of a species to have two different inversion events with both breaks at identical loci. There is no evidence that such a coincidence has ever occurred. Dobzhansky (1941) stated that the chance is one in one million that two breaks in identical loci of the third chromosome of *Drosophila pseudoobscura* will occur, and the third chromosome of *D. pseudoobscura* is famous for inversions.

If one such inversion can occur in a chromosome, it is conceivable that two different inversions could occur at different times and at different break points in the same chromosome. Many such are known, especially in *Drosophila* (Bock, 1971). Dobzhansky (1941) reported 21 different inversions known in the third chromosome within the single species *pseudoobscura*. In Fig. 14-2 two breaks as in *A* (1 and 2) produce the inversion in *B*, and then two other breaks (3 and 4) as in *C* (the second overlapping the first) produce the sequence in *D*. Fig. 14-2, *E, F, G,* and *H* have two inversions, the second "included" within the first, also called *reinversions*. Combinations of three or two, overlapping and included inversions, are known. Kitzmiller et al. (1967) reported one

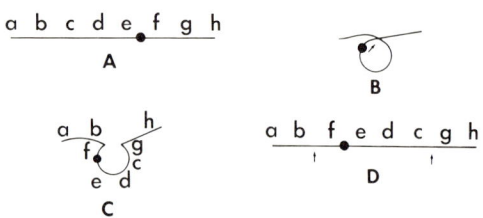

Fig. 14-1. Diagram of the formation of an inversion. **A,** The original chromosome and order of genes. **B,** Where two points of the chromosome are close together (arrow) a break occurs. **C,** Reunion of the broken ends produces an order of genes different from the original. **D,** The final result is that the segment c-f has been inverted to give a different order of genes.

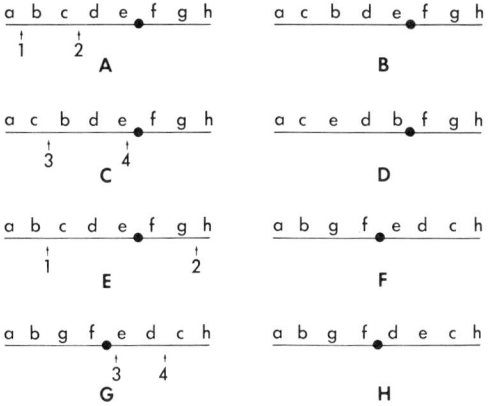

Fig. 14-2. A to D, Two overlapping inversions in one chromosome. A, Breaks at arrows 1 and 2 give the gene order in B. C, Two more breaks at arrows 3 and 4, 3 being within the first inversion, produce the gene order in D. The configuration at meiosis or in salivary gland cells (Fig. 3-10) with complete pairing of the genes within the two inversions is diagrammed in Fig. 3-11, 4. Two breaks in E followed by two breaks in G (within the already inverted segment) produce two included inversions (see Figs. 3-10 and 3-11, 3).

case of such a complex of inversions in the mosquito *Anopheles punctipennis*.

Inversions can be and are classified also into two different categories, depending upon whether the *centromere* is *not* included within the inverted segment (as in Fig. 14-2, A, where breaks at 1 and 2 give B); or, if the centromere *is* included within the inverted segment (breaks at 1 and 2 of E), the result is the sequence illustrated at F. A to B is called a *paracentric* inversion (*para-* meaning "beside"); that is, the inverted segment is, and both breaks are, beside or to one side of the centromere. On the other hand, if the centromere *is* included in the inverted segment, E to F, it is described as a *pericentric* inversion (*peri-* meaning "around"); that is, the inverted segment is around or on both sides of the centromere.

Inversions can be of almost any length (Fig. 14-3). There may be a minimum length determined by the minimum tightness that is physically possible in a loop of a chromosome, the

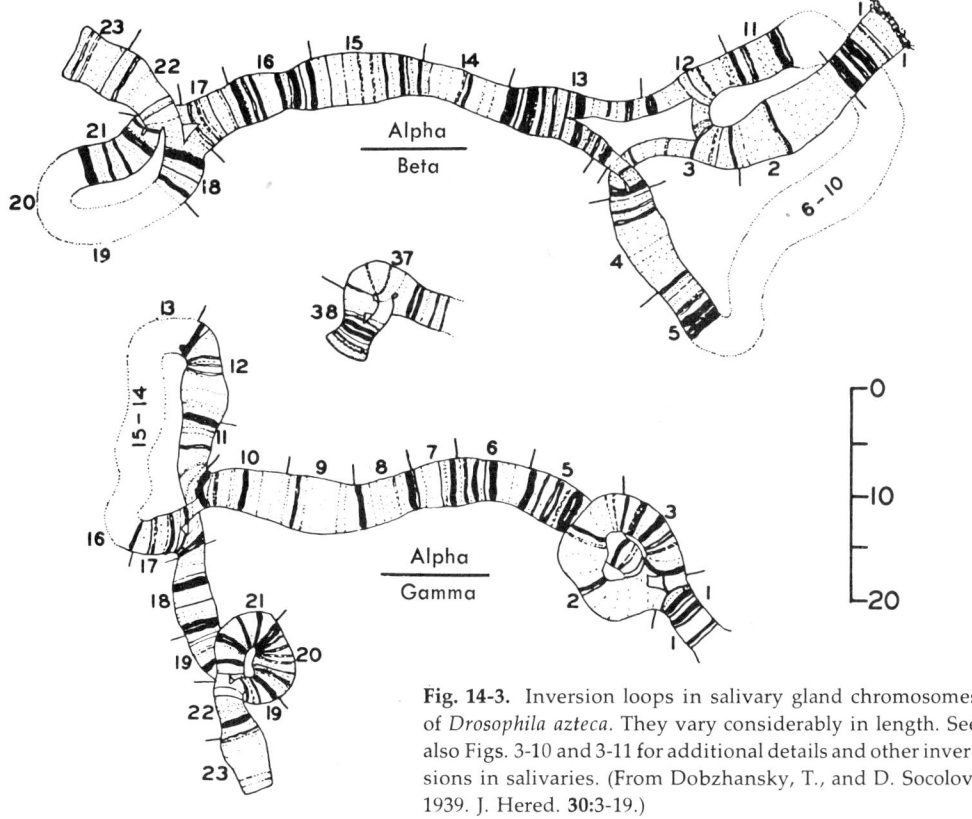

Fig. 14-3. Inversion loops in salivary gland chromosomes of *Drosophila azteca*. They vary considerably in length. See also Figs. 3-10 and 3-11 for additional details and other inversions in salivaries. (From Dobzhansky, T., and D. Socolov. 1939. J. Hered. **30**:3-19.)

loop being necessary for the inversion to be formed. There may be also maximum lengths determined by, perhaps, the way chromosomes are arranged within the nondividing nucleus when inversions are probably formed.

It is evident that paracentric inversions (Fig. 14-2, A to B) do not alter the general morphology of a chromosome with reference to the location of the centromere. The arm of the chromosome before and after the inversion is the same length. Thus paracentric inversions cannot be detected in mitotic metaphase chromosomes. Pericentric inversions, on the other hand, if they are long enough, can alter the observable form of the chromosome. In Fig. 14-2 the submetacentric chromosome E is converted to a metacentric at F by an unequal pericentric inversion. Such changes can be detected in metaphase chromosomes as in karyotype analysis (Fig. 13-7, B).

Neither type of inversion alters the genetic content of the chromosome, but both alter the linear arrangement of genetic sequence. Stebbins (1950) cites a number of plant species in which there are no evident morphological differences among the two homozygous and the inversion heterozygous conditions. If there is such a thing as so-called position effect, inversions would produce it since near the break point different genes are associated before and after the event. In Fig. 14-2 genes a and b are close together before the breaks, as are c and d. If gene a acts differently when close to c, than when close to b, there would be position effect.

Really detailed study of inversions can be made only of the giant interphase chromosomes of Diptera (Fig. 14-3) (see Chapter 3) and of pachytene chromosomes as of maize (Fig. 14-4) (see Rhoades, 1955, for an excellent coverage of this subject), although something can be learned of inversions from aberrations that are evident at anaphases I and II of meiosis. One aspect of salivary gland chromosomes, as of *Drosophila,* and of the pachytene chromosomes, as of maize, depends upon the paired condition of homologous chromosomes that makes visible the relatively inverted segments. Additionally, inversions can be studied in greatest detail and resolution in dipterous giant chromosomes because of the constant specific pattern of bands and interbands along chromosome arms. Thus, changes of positions of specific sequences of bands can be detected as inversions by comparison of such chromosomes in the homozygous condition in two species even when hybrids cannot be produced.

A cause-and-effect correlation has been es-

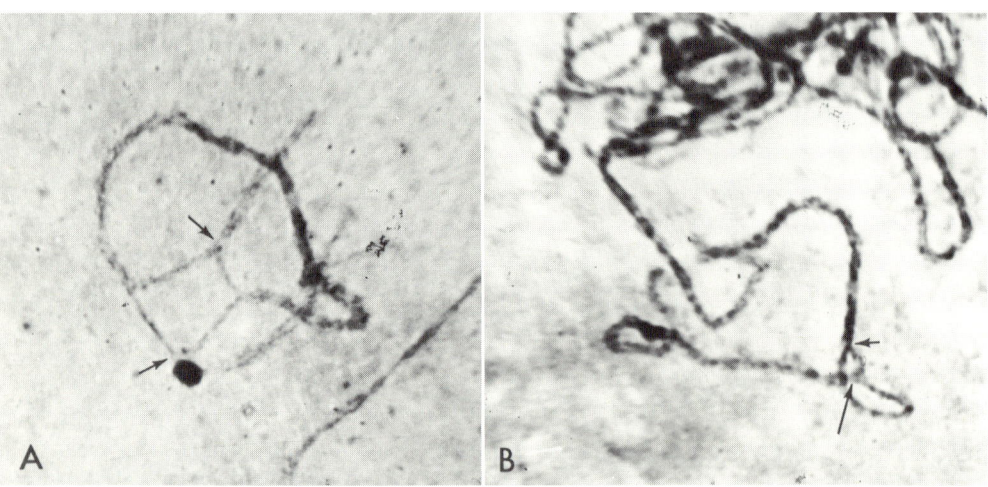

Fig. 14-4. Two early (1931) examples of inversion loops in pachytene chromosomes of maize. **A,** A large one (between arrows) with a large loop. **B,** A much smaller one (between arrows). See also Fig. 14-6. (From McClintock, B. 1931. Univ. Missouri Agric. Exp. Station Res. Bull. 163.)

tablished between heterozygosity for one or more inversions (and crossing-over within the inversion loop produced) and the occurrence of bridges with or without accompanying fragments at meiotic anaphases. This relationship is so well accepted that the presence of bridge-fragment aberrations at meiotic anaphase is adequate, though indirect, evidence for existence of inversions, heterozygosity for the inversions, and crossing-over within the inversion loops. It must be emphasized that such evidence is minimal because: most sites of inversion heterozygosity do not form inversion loops even if long enough to do so (Maguire, 1966; Nur, 1968) (Fig. 14-6); all types of crossing-over within a paracentric loop do not result in anaphase bridges with accompanying fragment; crossing-over within a pericentric loop does not form bridges (Fig. 14-5); environmental conditions such as temperature (Swanson, 1940) that affect the amount of crossing-over also affect the number of anaphase bridges; and many inversions may be located in chromosomal regions where chiasma frequency is very low. In other words, if bridge-fragment configurations are seen in *any* anaphase I cells, it can be assumed that the organism is heterozygous for at least one inversion.

Although crossing-over does often occur within both paracentric and pericentric inversion loops of heterozygotes, it is often stated that inversions suppress crossing-over. Actually, the genetic determination of suppressed crossing-over within a segment of a chromosome of heterozygous *Drosophila* (Sturtevant, 1931) was the first evidence that inversions occur. This was correlated with a change in genetic linkage sequence. Actual visualization of inversions in heterozygotes was possible with the cytological techniques for studying pachytene chromosomes of maize (McClintock, 1931, 1933) and salivary gland chromosomes of *Drosophila* (Painter, 1934a). What is meant by inversions suppressing crossing-over within the inverted segment is that in any meiocyte heterozygous for an inversion nearly all chromatids involved in exchange within the inversion loop (except for certain double exchanges) will produce chromosomal aberration at anaphase I (various sorts of bridges and fragments) such that the spore or gamete receiving one such chromatid will not function in producing progeny because of duplication and deficiency. Therefore, only those nuclei (often a sizable majority) containing chromatids that were *not* involved in crossing-over within the segment will be present in the progeny. Thus, most crossovers within inversion loops are not recovered, and so the segment tends to keep its genetic composition intact as long as heterozygosity persists. In maize Rhoades and Dempsey (1953) investigated an inverted segment 61 map units long and recovered between 1 and 2% crossover chromosomes only. All were the result of double exchanges.

It should also be emphasized that the extent of duplications, deficiencies, and acentric fragments resulting from crossovers within inversion loops is not determined by the size of the inverted segment but by its position in the chromosome arm. A crossover within a very small paracentric inversion loop remote from the end of the chromosome arm will produce a fragment twice as long as the distance from the crossover to the end of the chromosome arm. The duplication in the bridge and the bridge itself, however, are longer the closer the inversion is to the end of the chromosome arm.

INVERSIONS DURING MEIOSIS

Inversions are of little cytogenetic significance in somatic tissues except for the giant

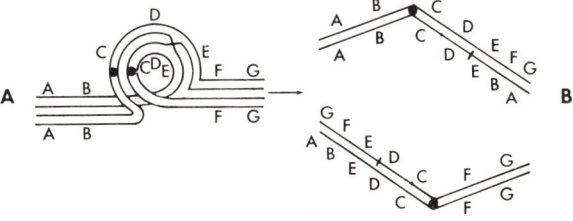

Fig. 14-5. Crossing-over in a *pericentric* inversion in **A** produces at anaphase I no bridge or fragment in **B** but does produce duplications and deficiencies. In the upper two-chromatid chromosome one chromatid is duplicate for the segment AB and deficient for the FG segment. In the lower chromosome one chromatid is also duplicate and deficient.

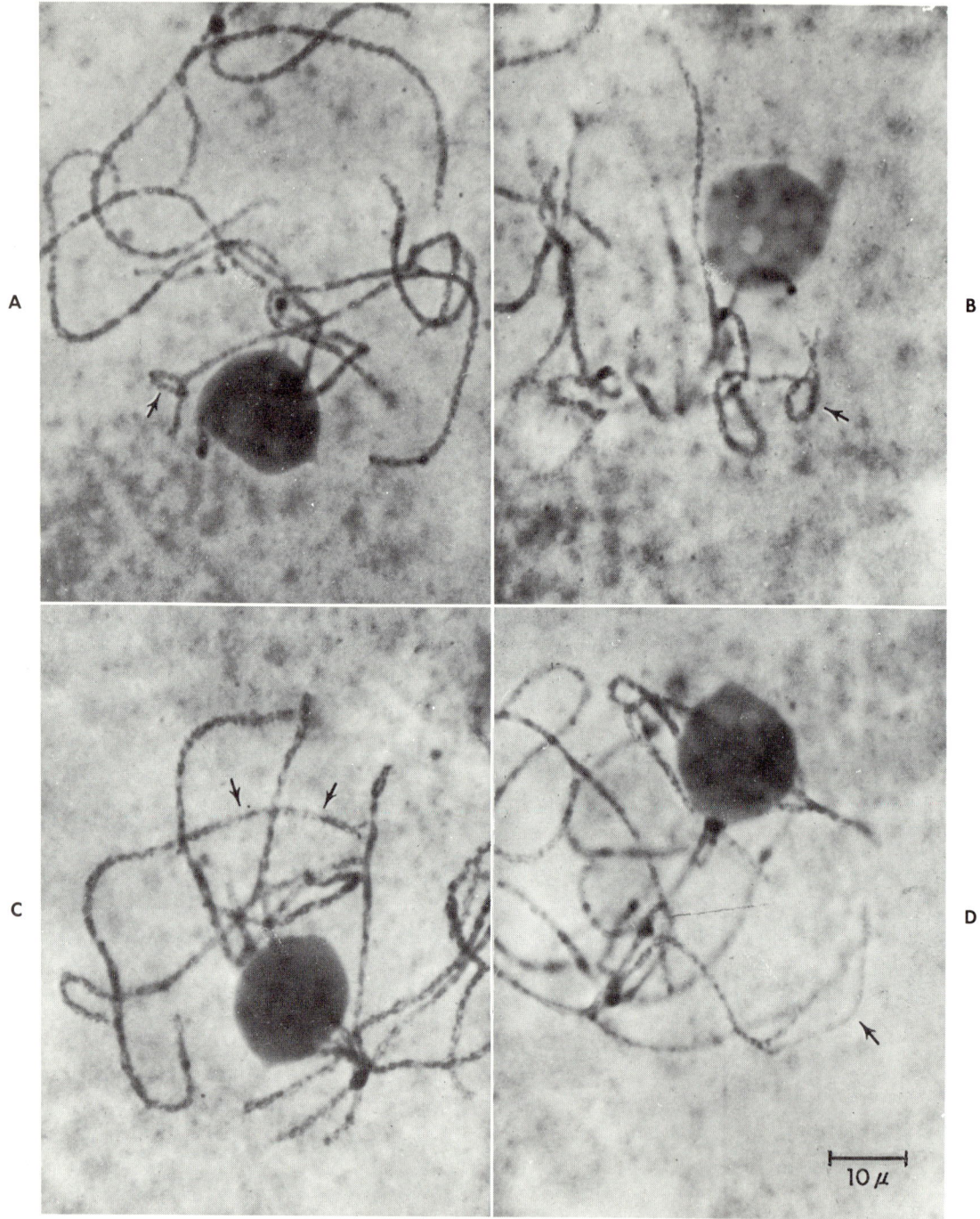

Fig. 14-6. What can occur in two homologous chromosomes that are heterozygous for a small inversion. **A** and **B**, The segments may pair *and* synapse to form a loop within which crossovers usually occur, at least in this case. This is called *reverse synapsis* and results in anaphase I bridge-fragment configurations, but it occurs in only about 34% of PMC's in this study. **C,** The inverted segments pair in reverse order, probably nonhomologously, synapsis probably does *not* occur, no bridge-fragments result. It was observed in 46% of PMC's and is called *rod pairing*. **D,** The relatively inverted segments neither pair nor synapse, called *pairing failure*, in 20% of PMC's in this case. The important conclusion is that inversion loops may occur in only a few cells, not in all. (From Maguire, M. P. 1966. Genetics **53:**1071-1077.)

polytene chromosomes of Diptera (Chapter 3) because they are generally undetectable. During meiosis, however, they can be studied cytologically and they produce genetic effects. These result from the meiotic phenomena of gene-by-gene synapsis, crossing-over, anaphase separation, and the lethality of duplications and deficiencies.

Synapsis brings homologous genes into side-by-side association. If two homologous chromosomes have the same order of genes, the two chromosomes will lie parallel throughout their length. However, if the two homologous chromosomes are heterozygous for an inversion, the following three conditions in the region of the inversion can occur (Maguire, 1966) (Fig. 14-6): (1) The two relatively inverted segments do not associate at all, reported in about 20% of maize PMC's by Maguire. (2) The two relatively inverted segments do appear paired (although not synapsed) and appear at pachytene no different from adjacent synapsed regions. She called this *rod pairing* and found in it about 45% of microsporocytes. (3) The two relatively inverted segments synapse gene by gene but "run" in reverse direction and form a loop. Maguire reported this condition in only about 33% of maize meiocytes and called it *reverse synapsis*. She further found the percentage of meiocytes having anaphase I bridges to be equal to the percentage of PMC's having reverse synapsis in the form of a loop, about 33%. This relates gene-by-gene pairing and inversion loops (reverse synapsis) directly to anaphase aberrations; that is, crossing-over can occur with reverse synapsis only. Rhoades and Dempsey (1953), also studied a paracentric inversion in maize and reported no anaphase I aberrations in 56% of PMC's. The 56% would include the percentage not forming a loop (no pairing and "rod pairing") plus double exchanges within the loop. Nur (1968) also found an exact relationship between homologous loop pairing and anaphase I bridges in a grasshopper.

Presumaby at homologous pairing and synapsis at meiosis the two homologous chromosomes pair gene by gene wherever they can, and if they are heterozygous for an inversion, they pair sometimes gene by gene between the relatively inverted segments. But when these segments do synapse, the synapsis must be gene for gene and a loop must result. By pachytene each homologue is double-stranded and crossovers have formed (Figs. 14-4 and 14-6). A crossover, at least a genetically detectable one, forms only between one chromatid of one homologue and one chromatid of the other (nonsister-strand crossing-over), but either chromatid of one homologue can exchange with either chromatid of the other (Fig. 14-7). Also, only one exchange can occur at a particular locus and for a distance on either side of that locus (positive interference).

In Fig. 14-7, A, a single crossover is represented in the paracentric inversion loop. The

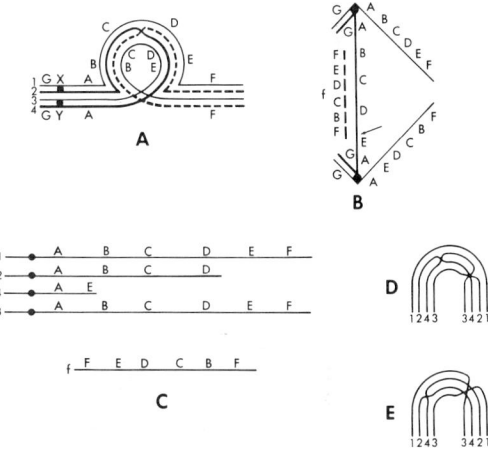

Fig. 14-7. Paracentric inversions and crossing-over. **A,** An inversion loop formed at meiosis by two homologous and synapsed chromosomes that are heterozygous for a paracentric inversion; x and y are the two centromeres; 1, 2, 3, and 4 are the four chromatids; A is the proximal region between the centromere and the loop; F is the distal region from the loop to the end of the arm; B, C, D, and E are segments of the loop. A crossover is indicated between chromatids 2 and 4 within the loop. **B,** The bridge-fragment configuration at anaphase I. Chromatids 2 and 4 are abnormal as a bridge and acentric fragment. The arrow indicates the point where the bridge finally breaks in this sequence of events. **C,** At the end of meiosis the four cells produced will receive one centromere and attached chromatids; 1 and 3 are normal, 2 and 4 are abnormal, and the fragment is lost in the cytoplasm. **D,** A three-strand double exchange within the loop. **E,** A four-strand double exchange within the loop.

four strands of the two homologues are numbered, 1 and 2 being sister strands as also are 3 and 4. Sister strands are united by centromeres X and Y. The exchange is represented between strands 2 and 4. If one starts tracing a chromatid strand at the centromere X and follows the heavy solid line of strand 2 to the exchange and there switches over to strand 4 and continues along the heavy solid line, one returns to the other centromere, Y. Thus, that one chromatid strand connects the two centromeres. Now if one starts at the other end of strand 2 (the heavy dashed line) and follows it into the loop and across the exchange, one returns to the homologous end of strand 4 without having encountered any centromere. This strand forms the acentric fragment. Strands 1 and 3, not being involved in a crossover, are normal.

At anaphase (Fig. 14-7, B) centromeres move toward opposite poles "dragging" the chromatids after them. The heavy solid-line chromatid that runs through both centromeres (chromatid 2-4) forms a bridge between the centromeres. The heavy dashed line (chromatid 4-2) being not attached to either centromere forms an acentric fragment and goes to neither pole. Chromatids 1 and 3 behave normally, going intact to opposite poles. Eventually the bridge breaks at random somewhere between the centromeres, for example between the gene segments D and E.

Following the second meiotic division the four nuclei formed will each contain one of the four centromeres and attached chromatids or fractions of chromatids. The two nuclei containing chromatids 1 and 3 of Fig. 14-7 will be genetically complete. But the nucleus contain-

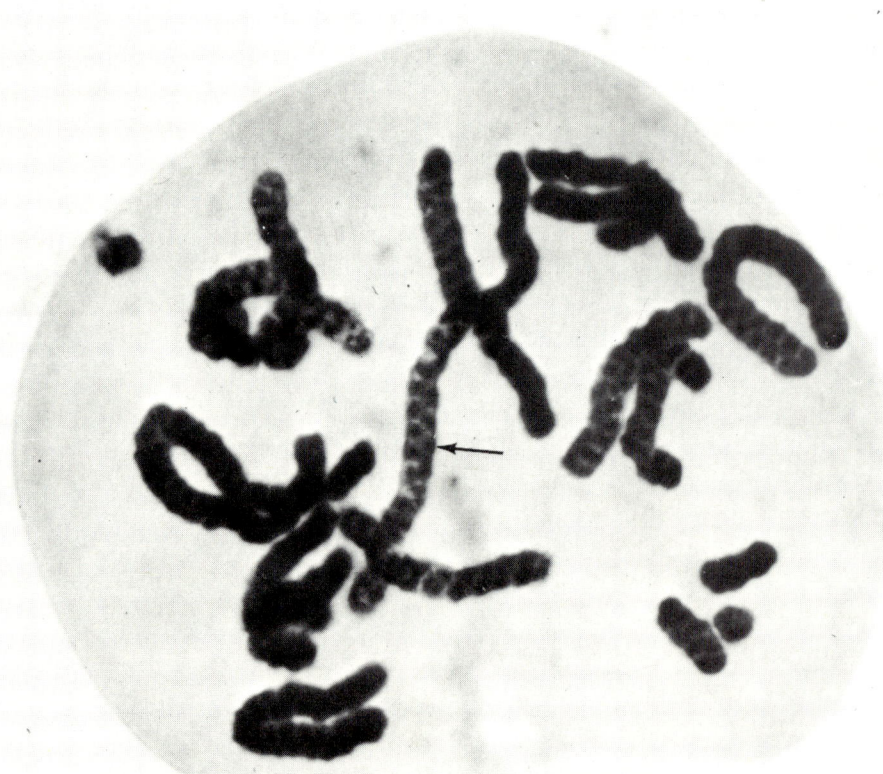

Fig. 14-8. A bridge between two anaphase I chromosomes of *Trillium erectum* following radiation treatment of 5 R per day for 30 days, from prepachytene through late pachytene. Fragments are also evident. (Courtesy A. H. Sparrow and R. F. Smith, Biology Department, Brookhaven National Laboratory.)

ing chromatid 2 will be deficient at least for gene segments E and F; and the nucleus containing chromatid 4 will be deficient at least for gene segments B, C, D, and F. These last two chromatids, the two originally involved in the crossover within the loop, will not appear in the progeny because of the lethal condition of their deficiencies.

Note that the acentric fragment is roughly twice as long as the distance from the right-hand end of the chromosome end to the crossover, and so its length is only slightly correlated to the size of the loop but is directly related to the distance from the end of the chromosome arm to the loop.

If the exchange occurs at other loci within the loop and if the bridge breaks at other positions, the resulting deficiencies will be different but all are lethal. An exchange outside of the paracentric loop as at A (in the "proximal" region) or at F (in the "distal" region), produces four normal chromatids. A second exchange between the same two chromatids (2 and 4 in this case) also within the loop, a two-strand double exchange, produces four normal chromatids cytologically as though no exchanges had occurred. Fig. 14-7, D, represents two crossovers within the loop, the first between chromatids 2 and 4, the second between 2 and 3. This is called a three-strand double exchange because one strand, No. 2, is involved in both and a total of three strands are involved. Fig. 14-7, E, represents a four-strand double exchange within the loop because there are two exchanges involving all four chromatids. There can also be an exchange within the loop and a second between the loop and the centromere in the "proximal" region (as at A) that produces distinctive anaphase results. A crossover within the loop and one distal to it, that is, as at F, is merely like a two-strand double. These different combinations of double exchanges produce distinctive results at anaphase I and in some at anaphase II, except that a two-strand double is cytologically about like no crossover at all although it is genetically detectable.

Following are the pachytene configurations, the resulting anaphase I and II conditions, and the number of viable spores resulting in plants, modified from Rhoades (1955):

	Anaphase I	
1. No exchange or two-strand double in loop	No bridge or fragment	4
2. Single exchange	One bridge and one fragment	2
Three-strand double in loop	One bridge and one fragment	2
Two- or four-strand doubles with one in loop and one in proximal region	One bridge and one fragment	2

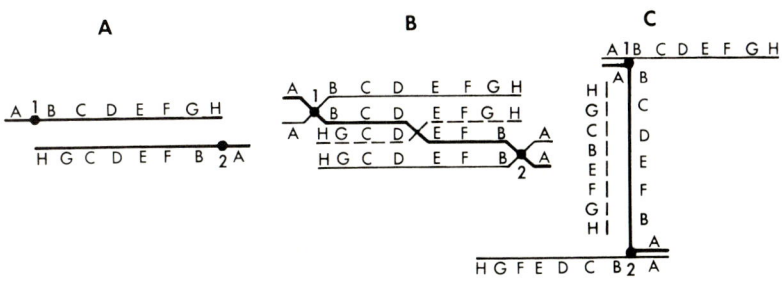

Fig. 14-9. **A**, An alternate method of representing two homologous chromosomes that are heterozygous for an inversion *(CDEF)*. In this method of representation the homologous regions AB and GH are not indicated as paired although in reality they are. **B**, The inverted segment is paired gene for gene with the noninverted segment, and a crossover between D and E is represented between two nonsister chromatids. The two chromatids involved are indicated by the dark line and the dashed line. The subterminal centromeres are labeled 1 and 2. **C**, At anaphase the centromeres pass toward the poles. One chromatid (dark line) is attached to two centromeres and becomes a bridge; the dashed chromatid is attached to no centromere, and at anaphase it becomes the acentric fragment.

Anaphase I—cont'd

3. Three-strand double with one in loop and one in distal region	One bridge and one attached fragment	2
4. Three-strand double with one in loop and one in proximal region	No bridge but one fragment	2
5. Four-strand double in loop	Two bridges and two fragments	0

Anaphase II

No bridges or fragments from types 1, 2, 3, or 5 above
One bridge, type 4 above, in one of the two cells

Thus, the study of meiotic anaphases can detect (indirectly by bridges and/or fragments) some of the combinations of crossovers between homologues that are heterozygous for an inversion, as well as detecting the existence of inversions. The types of anaphase configurations when the meiocyte is heterozygous for included, adjacent, or overlapping inversions can be quite complicated and can be determined diagrammatically, but no such organism (*Drosophila*) has had its meiosis studied for anaphase configurations.

McClintock (1939, 1941a) found in maize that following the break in the anaphase I bridge of a dicentric chromatid the cell containing it might be viable as a pollen grain or a megaspore. In either case, before the first postmeiotic mitotic division the chromosome with the broken end would double structurally so that there would be two side-by-side chromatids with broken ends at the same locus. These broken ends would then unite, as adjacent broken ends typically do. The result is a dicentric chromosome that forms another anaphase bridge at the first gametophyte division. By late anaphase this bridge also breaks at some random locus, and the cycle is repeated again in the developing embryo sac. McClintock called this the "bridge-breakage-fusion cycle." This cycle continues in the endosperm; but the end of the broken chromosome in the zygote, whether brought in by the male nucleus or already present in the egg nucleus, seems to "heal" and behave subsequently as a "normal" chromosome end, no longer forming bridges at anaphase.

As an example of frequencies of bridges in PMC's of a segregating progeny from a probable backcross of a hybrid between *Tradescantia humilis* × *canaliculata,* Swanson (1940) reported average percentages of anaphase I bridges that ranged from 2 to 32 among 28 plants. The continuous sequence in such a segregating progeny implies many small inversions and segregation for amount of crossing-over, the latter also varying from 1.8 to 2.45 chiasmata per bivalent. The types of bridges found were: one bridge and free fragment, one bridge and no fragment, one bridge and attached fragment, and two bridges in one chromosome arm. The fragments ranged from minute to longer than a chromosome arm, and a few circular fragments were observed. In general, the fragments were small but the bridges were long, both indicating that most inversions with crossovers were near the ends of arms. Such nearly terminal inversions with a contained crossover might not produce any fragment, a condition also reported by Stebbins and Elerton (1939) in American *Paeonia*. The most likely explanation of such a condition, as pointed out by Swanson, is that the fragment is covered by a chromosome arm. But an alternative is crossing-over within the loop and a crossover also in the distal region to form a three-strand double exchange. This condition produces a fragment attached or stuck to one of the arms of one of the separating anaphase chromosomes rather than lying free as is usual (Rhoades, 1955; Swanson, 1940). Swanson did observe such attached fragments but some might have been undetected.

INVERSIONS AND EVOLUTION

It is probably true that most individuals of most populations of most species are not heterozygous for any detectable inversions. In most interbreeding populations inversions are eliminated because of the numerous types of inviability they produce. But if enough individuals are examined, a few having inversions are usually found. Dubinin et al. (1936) found 525 individual chromosomes out of 34,500 examined of *Drosophila melanogaster* to have an inversion (1.5%), and Dobzhansky (1941) stated that inversions have been found in all species of *Drosophila* that have been examined for them. In contrast, some species are famous for large numbers and large percentages of individuals being heterozygous for inversions. Among animals, *Drosophila pseudoobscura* is unusual for the number of different inversions within especially the third chromosome of the species, for the distribution of them geographically and among populations, for the number of heterozygous individuals, and for the seasonal and annual variations within single populations (Dobzhansky, 1941; Koller, 1936; Strickberger and Wills, 1966; Pavlovsky and Dobzhansky, 1966; Crumpacker and Salceda, 1968). Kitzmiller et al. (1967) also found that the mosquito species *Anopheles punctipennis* has many inversions, eight having been found in one stock.

Drosophila has a pair of mechanisms by which heterozygosity for an inversion does not result in much sterility subsequent to meiosis. Some other animal species may have either or both of these mechanisms, but this is not known. The first mechanism, which is known in many Diptera (Brachycera), such as *Drosophila*, and some other insect species (possibly in females of Lepidoptera) is the *achiasmatic* condition in the male (or female of Lepidoptera). That is, the homologous chromosomes are paired (as in somatic pairing) during meiosis, but no synapsis or crossing-over occurs. Thus, heterozygosity for inversions in the male is transmitted without loops, or anaphase bridges and fragments, or any subsequent conditions that result in sterility.

In female of *Drosophila*, however, synapsis and crossing-over do occur. During meiosis in the egg the orientation of the two and four nuclei is such that one of the two noncrossover nuclei regularly become the nucleus of the haploid egg (Sturtevant and Beadle, 1936). The two nuclei resulting from the anaphase I bridge and the other noncrossover nucleus pass into the three polar bodies (Fig. 14-10). This is possible because the anaphase I bridge persists to the second division. That means that the two middle nuclei of the eventual row of four will be connected by the bridge whereas the two

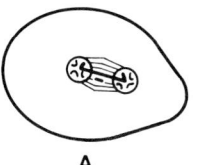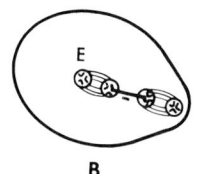

Fig. 14-10. First and second meiotic telophases in the egg of a *Drosophila* containing a bridge and fragment, **A**, resulting from a crossover within an inversion loop. The bridge persists to telophase II, **B**, connecting the two middle nuclei. The innermost nucleus becomes the egg nucleus **E**, and contains a noncrossover chromosome. The other three nuclei are extruded as polar bodies.

end nuclei will contain the two normal and complete chromosomes. The end nucleus toward the center of the egg becomes the haploid egg nucleus. Thus, most eggs do not contain chromatids that crossed over within the loop; they are fertile and produce normal progeny.

This is an unusual condition, but since angiosperms usually form a linear row of megaspores in the ovule, something like it might occur there, too, at least in those species in which walls do not form at interkinesis.

Among plants a few species seem to contain many inversions and even many individuals that are heterozygous for numerous inversions. *Paris quadrifolia* in the Tyrolean region of Europe was found to have one inversion or more in the heterozygous condition in every individual examined, and to have inversions in every arm of every chromosome of the set (Geitler, 1938). *Paeonia*, like *Paris*, has very large chromosomes and also like it is loaded with inversions, probably all rather small (Stebbins, 1938b; Walters, 1942). All species of *Paeonia* studied, both Old and New World species, have them. The maximum frequency found was 20% of microsporocytes having at least one anaphase bridge.

In contrast to individuals of good species, inversions are commonly found in interspecific hybrids, as is evident from anaphase bridges, or when band sequences of two closely related species or subspecies of *Drosophila* salivary gland chromosomes are compared (see Chapter 3). It seems evident that inversions play an important part in the origin of species because

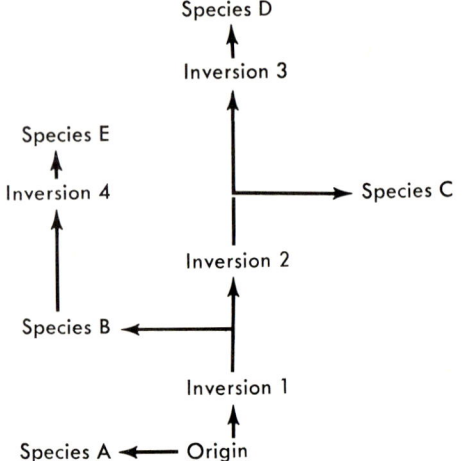

Fig. 14-11. Hypothetical scheme of inversions as they can and have been used in working out evolution in certain species and groups of *Drosophila*. It is assumed that any particular inversion has occurred only once and that all species or populations not carrying a particular inversion must have diverged before that inversion occurred. The direction of evolution is not always clear since the temporal sequence of inversion is not evident.

they (and translocations) "are the units from which are built up many of the isolating mechanisms separating plant [and animal] species" (Stebbins, 1950). When an inversion does occur within a species, it will usually be associated with the original condition in heterozygous individuals. There it will tend to be eliminated due to the percentage of lethality of cells following meiosis. But by chance, individuals homozygous for the inversion will occasionally be formed within the population. The heterozygous condition will keep the genetic nature of the segment intact because any crossing-over within the segment is lethal. Presumably, particularly "good" sets of genes and alleles within the heterozygous segments may and often do confer hybrid vigor on the heterozygotes, as in *Drosophila pseudoobscura*. Thus, there may be a net advantage to the heterozygous condition over either homozygous condition so that the inversion can spread and persist within the species. Sometimes, however, the homozygous condition for the inversion may become established in a large population of the species. Additional inversions, mutations, etc. in both the original and the inversion-containing populations can eventually produce a more or less complete reproductive isolation between the two forms as a step in species formation.

Since inversions are almost always produced sequentially, it can be assumed that inversion A preceded B or that B preceded A. It is impossible to determine the order of appearance of the various inversions without resorting to evidence derived from other sorts of evolutionary studies. However, the overall concepts of evolution within a species, as of *Drosophila*, do permit insertion of each inversion in a hypothetical sequence (Fig. 14-11). Some such studies have been made, as mentioned in Chapter 3.

CHAPTER 15

Karyotypes

The important fact to remember here is that chromosomes are not only structures which result as end products of a series of gene-controlled developmental processes; they are themselves the bearers of the genes or hereditary factors. This at once puts them into a different category from all other structures of the body. Chromosomes are not merely aggregates of discrete genetic units. To a certain extent they are units in themselves. We should expect changes in the chromosomes to bear a more direct relationship to genetic-evolutionary processes than do any other types of changes.

Stebbins, 1950

The concept of the karyotype was developed especially by Levitsky, as propounded in his famous 1931 publications. In contrast to genotype, the karyotype of an individual, species, etc. is the set of chromosomes, whether normal or abnormal (Figs. 11-2 and 13-7), as seen at mitotic metaphase. The set is made up of a certain number of chromosomes, whether diploid, aneuploid, or euploid, and the forms of those chromosomes both individually and as a set (Figs. 15-1 to 15-5, etc.). The diagrammatic arrangement in order of decreasing length is often designated as an *idiogram* (Fig. 15-1). The arrangement of cut-out photomicrographs of all the individual chromosomes of the somatic metaphase set arranged in pairs (when pairs exist) (Fig. 10-2), as in modern studies such as human cytogenetics (Figs. 19-1 and 19-2), is called a karyotype. As in the idiogram, the pairs are arranged in descending order of lengths and numbered in that order, chromosome No. 1 always being the longest. The karyotype of a triploid would show three of each kind of chromosome (Fig. 11-2), and in a trisomic the extra chromosome may be matched to the two homologues by similarity of physical features. The karyotype of a monosomic would have one single chromosome instead of a pair. Chromosomal polymorphism (Fig. 10-2) is often detected.

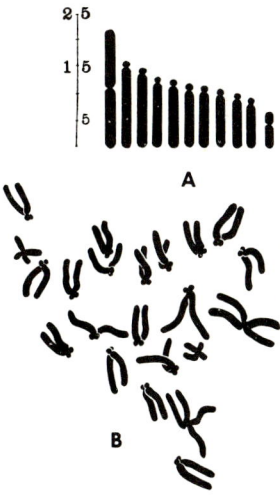

Fig. 15-1. A, Constructed idiogram or karyotype and B, a chromosome spread of the plant genus *Sagittaria*. In construction of the karyotype the chromosomes are arranged in sequence from longest to shortest and, if numbered, the longest is No. 1. In *Sagittaria* all species so far studied have the same karyotype. (From Brown, W. V. 1946. Bot. Gaz. **108**:262-267.)

The karyotype of a species is determined from karyotypes of a number of individuals, in part because of technical variation and in part because individuals vary. The karyotype as determined is the present condition or culmination of past evolutionary changes of various sorts. Karyotypes evolve because the individual chromosomes of the set undergo changes of form, are lost, new ones are added, etc. The main characteristics of karyotypes are discussed here.

CHROMOSOME NUMBER

The karyotype of a *species* consists of the usual number of chromosomes reported for that species; for example, in man the karyotype consists now of 46 chromosomes (Figs. 19-1, 19-2, etc.). Animal karyotypes of the two *sexes* are usually different, the male having XY, the female XX or some other combination (Fig. 10-3). *Individuals* may have exceptional karyotypes, such as human beings with 45, 47, 48, etc. chromosomes, trisomy, tetrasomy, monosomy, nullisomy, etc. Karyotypes of species vary with respect to numbers from n = 2 (Fig. 13-10) to a few hundred (Figs. 12-1 and 12-2). In a few species the karyotype of the germ cells is different from that of somatic cells, and in some individuals the karyotypes of cells of specific tissues may be different. In plants the karyo-

Fig. 15-2. Mitotic metaphase chromosomes of the plant *Haemanthus katharinae*. The variations in sizes and centromere positions are evident. (Courtesy Dr. A. H. Sparrow, Biology Department, Brookhaven National Laboratory.)

types of the gametophyte and sporophyte generations of the same life cycle are numerically different. In man and other species mosaics occur, having two or more different karyotypes in the same organism, organ, or tissue. There are exceptional species that regularly show polymorphism (Fig. 10-2), different individuals having quite different karyotypes but all being functionally normal. Polyploidy, of course, is not uncommon within plant and some animal species, producing different karyotypes among individuals, populations, or subspecies.

Variations among species of a genus (Fig. 15-3) or among genera of a family with respect to chromosome number are common in plants and animals. There are some genera in which all species have the same chromosome number, such as *Ceanothus*, 2n = 24 (Nobs, 1963), *Ribes*, 2n = 16 (Keep, 1962), *Pinus*, 2n = 24 (Mirov, 1967), *Sagittaria* (n = 11), although other genera of the family Alismaceae have other basic numbers and some polyploidy (Brown, 1946; Baldwin and Speese, 1955). In other whole families there is one common basic number, such as all Culicidae (mosquitos) have 2n = 6 (Baker and Aslamkhan, 1969; Aslamkhan and Baker, 1969), but some polyploidy occurs in some others as in the Asclepiadaceae, Caprifoliaceae, Fagaceae, and Rubiaceae. At the other extreme is the family Compositae with no clear basic

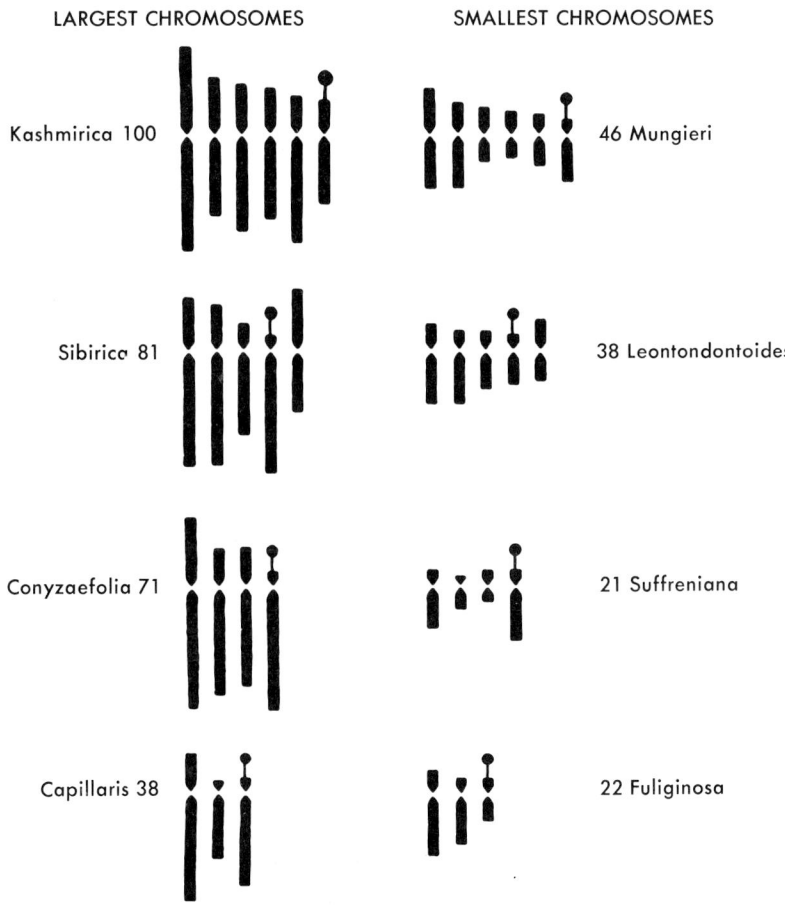

Fig. 15-3. Karyotypes of some species of the plant genus *Crepis*. The variation in chromosome numbers (from 3 to 6), in average lengths, regardless of number, in total length per cell, and centromere locations are obvious. Evolution within the genus is assumed to have started with n = 5 or n = 4. (From Babcock, E. B. 1947. The Genus *Crepis*. Part I. Calif. Pub. Bot. Vol. 21. [Originally published by the University of California Press, Berkeley; reprinted by permission of the Regents of the University of California].)

number and many genera having diploid species with three, four, five, or more different haploid numbers, such as *Haplopappus* with n = 2, 3, 4, 5, 6, 8, and higher, possibly polyploid, numbers; *Crepis*, n = 3, 4, 5, 6, 7; *Amsinckia*, n = 4, 5, 6, 7; *Brachycome*, n = 2, 4, 5, 6, 8, 9 (Smith-White et al., 1970), etc. There are even species composed of individuals, populations, or subspecies having various haploid numbers, such as the Australian *Brachycome lineariloba* with n = 2, 4, 5, 6 (Smith-White and Carter, 1970).

Mechanisms of chromosome number increase and decrease at the diploid level by translocations (including centric fusion and fission) have been examined in Chapter 13 and by polyploidy, in Chapter 12. These are the two main ways karyotypes change for number. These two mechanisms can, of course, be combined as in *Brassica* where three basic numbers 8, 9, and 10 originated by dysploid change, by increase or decrease or both. Such change was accompanied and/or followed by polyploidy among these three basic numbers to produce *secondary basic numbers* (called interchange polyploid-amphiploid numbers by Stebbins, 1950), in this case 17, 18, 19, 27, and 29.

Agmatoploidy (see Chapter 12) with or without polyploidy is effective in altering the number aspect of karyotypes in those groups of animals, such as Lepidoptera, Hemiptera, and Homoptera, and plants, such as Juncaceae and Cyperaceae, that have diffuse centromeres rather than the usual localized centromeres. Possibly the completely aneuploid genus *Carex*, which has all haploid chromosome numbers from 12 to 43, is an example of agmatoploidy *and* polyploidy. If so, it is far more extensive than the classical example *Luzula* (Nordenskiöld, 1956) since *Carex* has many times as many species as *Luzula*.

CHROMOSOME FORM

Alterations or differences in chromosome form (as distinct from mere size difference) within or between taxa are very important aspects of karyotype analysis (Fig. 15-4). These can concern either the whole set of chromosomes or individual chromosomes. Forms of sets are of two sorts. One sort has all chromo-

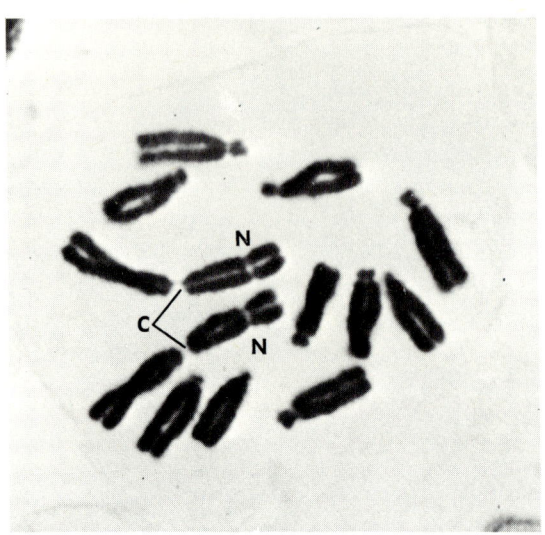

Fig. 15-4. A chromosome spread of the chromosomes of the plant *Vicia faba*. The two longest chromosomes seem almost divided into two by the long median primary (centromeric) constriction, C. In one arm of this chromosome is a secondary constriction, N, which is the location of the nucleolus organizer. The remaining chromosomes are acrocentric or subacrocentric. 2,000×. (Courtesy Dr. A. H. Sparrow, Biology Department, Brookhaven National Laboratory.)

somes of about the same size, called a *symmetrical karyotype,* such as all large (*Trillium, Tradescantia, Lilium,* salamanders) or all small (fungi, Juncaceae, Cyperaceae, Homoptera, Hemiptera). In contrast are groups having so-called *asymmetrical karyotypes,* which are characterized by the set containing some large and some very small chromosomes, as two nonintergrading classes. Such asymmetrical karyotypes are common among birds, lizards, and the plant family Agavaceae. As an example of the taxonomic use of whole karyotype forms is the classification of the xeric genera *Yucca, Agave,* and their relatives.

Yucca has a superior ovary and so used to be classified in the family Liliaceae. *Agave* and related genera have inferior ovaries and so were classified with the Amaryllidaceae. The difference between inferior and superior positions of ovaries is a major taxonomic difference. When these groups were studied cytologically, it was found that *Yucca, Agave,* etc. have very similar karyotypes with respect to both num-

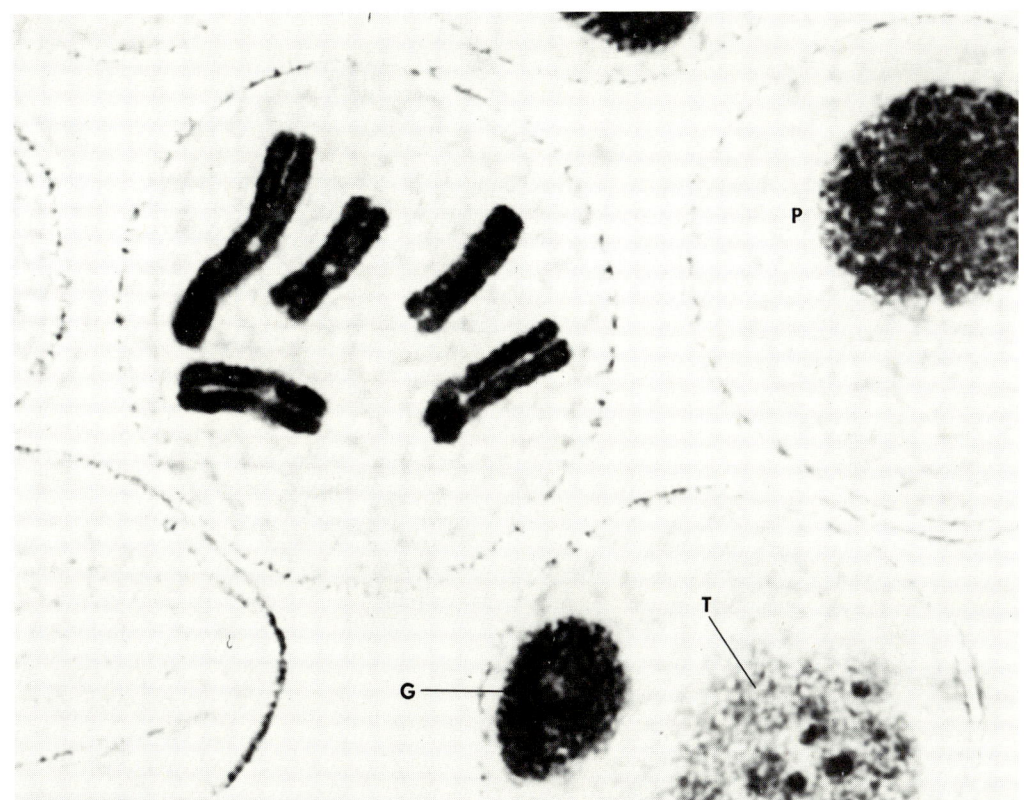

Fig. 15-5. Developing pollen grains of *Trillium*; a prophase nucleus, *P*; a metaphase showing the haploid number (n = 5) at the first pollen grain division. The loci of the centromeres are indicated by circular separations of the chromatids. The result of this division will be the two-celled condition, the generative, *G*, which will eventually divide to form two male gametes, and the tube nucleus, *T*, which is already changing, as its pale color indicates. (Courtesy Dr. A. H. Sparrow, Biology Department, Brookhaven National Laboratory.)

ber and overall form (Whitaker, 1934; Granick, 1944). There is polyploidy, but the haploid chromosome number is n = 30 of which 5 are large and 25 are small. Since such an evolutionary coincidence is considered to be essentially impossible and since the genera do have many real similarities, the cytological evidence of similar asymmetrical karyotypes compelled the union of these genera into one family, the Agavaceae.

The form of individual mitotic metaphase chromosomes is determined by the total length of the chromosome (chromosome width is not considered) and the length of the two arms. Total length varies from more than 30μ in *Trillium* (Fig. 15-5) to about 0.5μ in some fungi and some animals and plants (Chapter 8). The average chromosome length is about 5 or 6μ. Length of chromosome arms varies from almost zero (a truly telocentric chromosome would have no second arm) to the full length of a telocentric chromosome. Metacentric chromosomes have two equal or almost equal arms. Acrocentric chromosomes have one very short arm. The various dimensions are often combined into ratios.

The symbolism officially adopted in human cytogenetics (Chapter 19) is here used. That is, the long arm length is expressed as q, the short arm length as p, and the total length of the chromosome as $p + q$. The relative length (RL) = $\dfrac{p + q \times 1{,}000}{\text{length of haploid set}}$ that is, 1,000 times the length of a particular chromosome divided by the total length of the haploid set including the X chromosome. The arm ratio (AR) = $\dfrac{q}{p}$, the length of the long arm divided by the length of the short arm. The centromeric index (CI) = $\frac{p}{p+q}$, the short arm length divided by the length of the chromosome. Each chromosome can be characterized by either AR or CI or both. For example (Al Aish, 1969), human chromosomes No. 1 and No. 3 (see Chapter 19) have the same arm ratios and centromeric indices but different total (relative) lengths:

Chromosome	p + q	AR	CI
No. 1	81.84	1.07	48.3
No. 3	65.24	1.06	48.3

whereas chromosomes No. 10 and No. 11 have about the same lengths and centromeric indices but different arm ratios:

Chromosome	p + q	AR	CI
No. 10	43.38	1.33	34.8
No. 11	43.48	1.86	34.9

By such measurements and comparisons the various chromosomes of the set can be specified, and compared.

Another structure of the chromosome set useful in karyotype analysis, especially in marking certain chromosomes, but not all chromosomes, is the *satellite*. This is a long or short segment of one arm separated from the rest of the arm by a constriction. This thin connection may be as short as the centromere constriction, as in *Haplopappus gracilis*, or it may be very long and threadlike (Figs. 8-3 and 8-4). Very often it is near the end of the arm and the satellite is a small sphere. Satellites are usually on short arms, but cases are known of them on long arms as in the composites *Crepis pulchra, Kregia gracilis, Lygodesmia texana,* and *L. dianthopsis* as well as in *Aloe arborescens, Drosophyllum lusitanicum, Vicia lutea, V. pannonica,* etc. One pair of chromosomes usually bears the satellites, but there are numerous truly diploid species known that have satellites on two or more pairs of chromosomes such as *Lolium perenne* (Jain, 1966), four species of *Aegilops* (Riley et al., 1958), Old World cottons (Skovsted, 1933, 1934), *Plantago ovata-insularis* (Stebbins and Day, 1967), and *Elymus rechingeri* (tetraploid) has three pairs.

Polyploid species often have two or more pairs, but, as in hexaploid wheat, only one pair is usually evident. The satellite itself is not unusual, but the constriction is; it is the nucleolus organizer. Why that particular region of the chromosome arm that contains the redundant ribosomal DNA should form a constriction during mitosis is unknown since it apparently functions from telophase through interphase.

An excellent example of the use of satellites in karyotype study and of morphological and chromosomal incipient "speciation" is the study of *Elymus rechingeri* on Aegean islands (Heneen and Runemark, 1962). *E. rechingeri* is tetraploid and essentially self-pollinating. There is essentially no possibility of crossing between populations on any two islands. Thus, the small population on each island is reproductively isolated, self-pollinating, and inbred. Morphological differences have evolved on each island, accompanied by chromosomal changes, as represented by differences of the three pairs of satellited chromosomes, the only ones that could be distinguished among the 2n = 28 chromosomes of the set. There were five recognizable morphological types of the A pair, two forms of the B pair, and two forms of the C pair. These were observable changes in length of the satellites and long arms. They occurred in various combinations on different islands. On two islands only one satellited B was observed, the satellite constriction, the nucleolus organizer, not forming on that chromosome. The chromosome itself was present since all 28 chromosomes were counted. This condition on each island is an example of the *Sewell Wright effect* or of *genetic drift*, the rapid fixation (or loss) of alleles and chromosomal changes randomly in small, isolated populations.

Jayan (1970) reported even more extreme intraspecific variation in probably satellited chromosomes in 11 types of the tetraploid apomictic grass species *Cenchrus ciliaris* in India. The longest chromosomes had secondary constrictions in the short arms, but the number of such chromosomes varied from one to seven among the 11 genotypes.

The modern technique of making spreads of vertebrate mitotic chromosomes (see Chapter 19 on human cytogenetics) has resulted in a surge of karyotype studies during the past few years. This technique permits very exact study that in revealing details is at least equal to root tip squashed preparations of plants. It has opened up a new field of critical karyotype analyses that were previously impossible, human cytogenetics being merely the area of most intensive study.

Mitotic metaphase chromosomes can be altered in form by either translocations or pericentric inversions or both (see Chapters 13 and 14).

Meiotic chromosomes at pachytene can occasionally be used for karyotype study. The best example is maize and its relatives in which chromosome lengths, centromere positions, nucleolus organizer, nucleolus arm lengths and, rather uniquely, knobs are determinable (Rhoades and McClintock, 1935; Rhoades, 1955). These knobs are blobs of heterochromatin of various sizes that, in a particular strain, are permanent and of constant location. Within the species, however, they vary in number and location. In maize and its very close relative *Euchlaena* (which is now considered conspecific with *Zea mays*) the knobs are always interstitial, but in the distant relative *Tripsacum* they are terminal. Within a particular culture or cross being studied they are useful markers.

The only interphase chromosomes that can be used in karyotype study are the giant interphase polytene chromosomes of Diptera. As discussed in Chapter 3, these chromosomes approach the ideal for cytogenetic study and permit karyotype analysis almost to the location of individual genes.

EVOLUTIONARY ALTERATIONS

The evolutionary mechanisms that bring about observable alterations of karyotypes have been discussed in earlier chapters. Alterations of whole sets are produced by polyploidy, which is an irreversible evolutionary process (Stebbins, 1950, 1970). It is possible experimentally to produce polyhaploids that are capable of some reproduction (DeWet, 1965, 1968); but, as Stebbins stated (1970), there is no evidence that such a process, the reverse of polyploidy, has ever occurred in nature and probably cannot because the known polyhap-

loids are all poor things or are merely diploids of autopolyploids and so are nothing new in an evolutionary sense. They are merely reversions to the original diploid condition.

Given enough time, polyploids become amphiploids by diploidization. Such diploidized polyploids can subsequently produce polyploids themselves which may represent new lines of evolution with high basic numbers. Further numerical complication can evolve if dysploid increase or decrease occurred when they were amphiploids and if the original diploids disappeared. Thus, $n = 6$ species could establish amphiploids at $n = 12$. Dysploid change might give $n = 11$ species and $n = 13$ species. Subsequent additional polyploidy from such secondary diploids could give $n = 22, 23, 24, 25, 26$, and higher numbers. If the $n = 6$ species became extinct, the basic chromosome numbers for the group would be $x = 11, 12$, and 13. The haploid chromosome numbers 8, 9, 13, 15, 16, 21, and 26 suggested such an evolutionary change in ants (Crozier, 1970).

Dysploid karyotype changes of increase and decrease have been discussed in Chapter 13 on translocations. In many families and genera of plants and animals such changes are common. Whether intraspecific centric fusion and fission in certain animals are always produced by translocations has also been discussed.

Alteration by pericentric inversions seems to be fairly common, as in *Drosophila* (Patterson and Stone, 1952). This changes the form of a chromosome without modifying the genetic content. Aside from salivary gland or pachytene analysis such changes are difficult to prove (Fig. 13-7). However, as in human cytogenetics, when mitotic metaphase chromosomes are paired off in a karyotype and one member of one pair has the expected length but an altered arm ratio, it is reasonable to propose a pericentric inversion has occurred.

Karyotype analysis is quantitative, based upon measurements of chromosomes and chromosome arms. Martin and Hayman (1965) have devised a complicated analysis that they call *quantitative* in part because it adds complex statistical methods to chromosome arm measurements. Actually they add also the determination of the quantity of DNA per nucleus of each species involved. This quantity is added to the formulas for solution. It is difficult to see how this method is superior to simple measurements since amount of DNA has been shown quite adequately to be correlated to the length of metaphase chromosomal material in the cells, at least among closely related species.

Karyotype analysis is often a significant aspect of taxonomic studies that attempt to determine phylogenetic relationships, for example the study of the genus *Crepis* (Babcock, 1947). It is also useful in often indicating evolutionary changes of polymorphism within species. And in human beings it contributes to further understanding of that class of hereditary abnormalities caused by chromosomal aberrations.

CHAPTER **16**

Cytogenetics of hybrids

It has been assumed that the degree of chromosome pairing in species- or generic hybrids can be used as an index of chromosome homology and species relationships. But failure of chromosome pairing, or asynapsis, can be caused by genetic and environmental factors as well as by the lack of sufficient homology.

Sax, 1935

The term hybrid applies to results of a cross of two individuals that differ. The difference can range from a simple allelic heterozygosity for one gene to a cross between species of two genera. But it must be kept in mind that *any* cross that can be made must be between individuals that share overwhelmingly common genes and linkage arrangements. Crossable entities are far more alike than different, as is indicated by classification; and the more unlike they are, the more difficult becomes the cross, because one of the early differences involved as divergent evolution occurs is some partial internal isolation barrier of incompatibility. Sterility barriers usually start to form as soon as or even before two evolutionary lines start to diverge (Fig. 16-1), they become complete with greater separation, and eventually even sterile hybrids cannot form. If it were not so, there could hardly be species. Thus, at one end of the spectrum are simple genetic hybrids, but these are not usually amenable to cytogenetic study. Further along the spectrum cytological differences between the parental types may become apparent in the hybrids, and at the far end of the spectrum the cytological differences are extreme. But all of these differences are of the same sorts; they vary quantitatively only. It is those hybrids that show cytological differences that are the subject of this chapter. These can be either intraspecific or interspecific hybrids. They can also be either natural or experimentally produced.

INTROGRESSIVE HYBRIDIZATION

There is one form of noncytological genetic hybridization that warrants brief mention. That is a special form of evolution called introgressive hybridization (Anderson, 1949, 1953), which occurs between taxa that do not have complete sterility barriers between the parental species or subspecies or between the hybrid and at least one of the parental taxa. Following establishment of one hybrid or more where the ranges overlap or are close, the hybrids, because of location, abundance of one parental type, or greater ease of backcrossing to one parental taxon, repeatedly backcross to that particular parent. Eventually, forms occur that are almost like that backcross parental taxon but still have a few characters of the nonbackcross parent. That is, some characters and genes of the nonbackcross parent have "introgressed" into the other parental taxon. Repeated crossing and backcrossing may occur, accompanied by selection of adapted backcross types so that a continuous spectrum of types may occur between the F_1 hybrid and one pa-

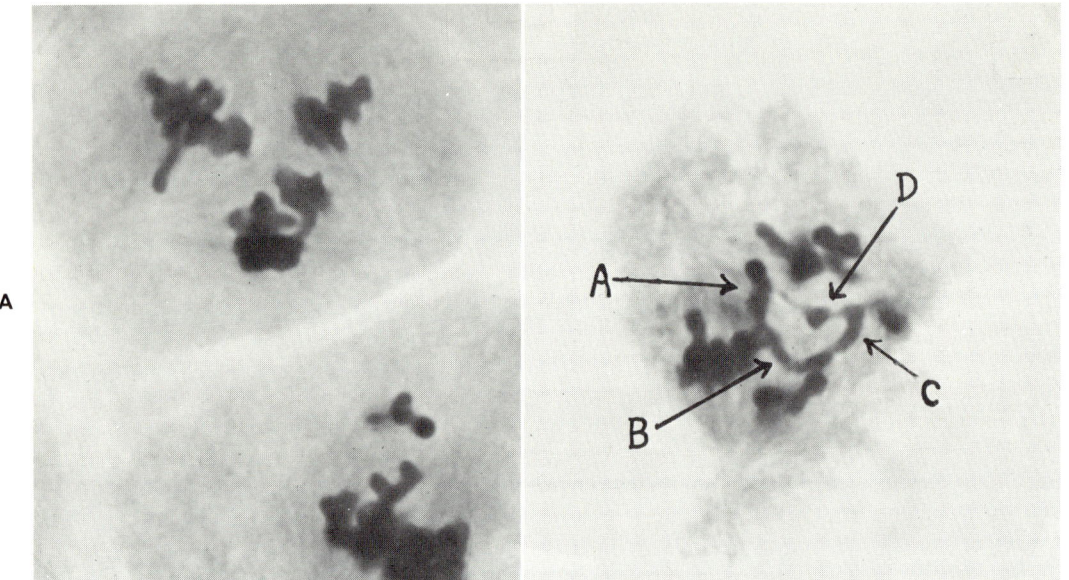

Fig. 16-1. Microsporocytes in a hybrid of *Elymus canadensis* (2n = 28) × *E. virginicus* (2n = 28) showing tripolar spindles and in **B** a quadrivalent *(A, B, C, D)* that is trying to go to three different poles. Such meiotic upsets are not uncommon in hybrid plants.

rental type. Under some conditions each parental taxon may introgress into the other. Usually the hybrid and introgressive types occur only in a "hybrid" or intermediate environment such as disturbed habitats in or near the overlap region. The result is that many individuals in the overlap region will be classified as being in the recurrent backcross taxon but having some characters of the other; many individuals will vary *in the direction of* the nonrecurrent backcross taxon. Introgression is rarely evident at the cytological level of chromosomes; it is of interest mostly to evolutionists, taxonomists, and geneticists.

INTRASPECIFIC HYBRIDS

Just as species hybrids range in irregularity of meiotic pairing and sterility from complete and normal meiosis and high fertility to very irregular meiosis and compete sterility, so do intraspecific hybrids. In general, however, intraspecific hybrids have at worst only slightly irregular meiosis and they are usually highly fertile. And, of course, whether a hybrid is really interspecific or intraspecific is often arbitrary, depending on the species concept or concepts of the taxonomists who are experts on the taxa involved.

Clarkia biloba (Roberts and Lewis, 1955) is an example of rather typical hybridization among subspecies. All crosses produced hybrids with eight bivalents, but progeny from subspecies *brandegeae* and either of the other two subspecies had reduced fertility of unknown origin.

Elymus glaucus (2n = 28, a tetraploid, x = 7), on the other hand, presents an unusual and extreme condition of meiotic irregularities and sterility in 22 F_1 intraspecific hybrids (Snyder, 1951). The parents, collected along a 75-mile transect from 250 feet to 8,400 feet in elevation in California, had regular meiosis and high fertility. Six of the hybrids had regular bivalent pairing, but all the remaining 16 had one or two quadrivalents and often a few univalents and trivalents. Nearly all had lagging chromosomes and bridges at both anaphase I and II. Chromosome fragmentation and heteromorphic pairs also occurred in a few hybrids. Three hybrids always had one quadrivalent and 12 bivalents, and eight hybrids had one quadrivalent in from few to many microsporocytes.

Fertility of the hybrids as indicated by good and empty pollen ranged from complete sterility in nine, less than 1% fertility in two, and 3% good pollen in three hybrids, about 25% in two plants, about 45% in two, and about 72% in two hybrid plants. Thus, 14 of the 20 hybrids were completely or almost completely unable to produce good pollen. The six hybrids that produced from 20 to 75% good pollen did not have the most regular meiosis but in general had less irregular meiosis than those hybrids that produced no good pollen.

Snyder concluded that this species is itself probably a species hybrid, probably an allotetraploid, one genome having been derived from the *Agropyron inerme – A. spicatum* group. The self-pollinating character of *E. glaucus* was inherited from that parental complex and would permit reproductive isolation of populations. If one of the parents had been an outcrossing species, the fixation of different gene combinations by enforced inbreeding could produce the morphological and physiological diversity that was found among the various populations. He also suggested the possibility of repeated backcrossing of different populations of *E. glaucus* to various species of *Agropyron, Sitaniom,* and perhaps *Hordeum*. Such introgession of different species would produce the populations isolated by sterility barriers that were found.

The results in *Elymus glaucus* are similar to what has been reported in the artifical hybrid "species" of wheat *(Triticum)* and rye *(Secale)*, called *Triticale*. Such hybrids have been synthesized a number of times by various breeders and from various strains of both wheat and rye and have produced constant but genetically different strains. Muntzing (1939) made various hybrids among six such constant strains. The various F_1 hybrids varied greatly because of the variability among the *Triticale* strains used, but the F_2 hybrids were of much lower fertility that the F_1's and had more irregular meiosis but varied greatly among themselves.

Galeopsis tetrahit (Labiatae) is a self-pollinating allotetraploid species ($2n = 32$). Muntzing (1938) made many hybrids among plants from various populations. He reported from 18 to 93% of microsporocytes in all such intraspecific hybrids had a few univalents at metaphase I, a few had one trivalent plus one bivalent, and rarely a quadrivalent was evident in some hybrids. About one half of the hybrids were partially but variously pollen- and seed-sterile, but almost half were fully fertile. He proposed some introgression from *G. biflora* into some populations as a source of variation within the species. He also reported similar variation among hybrids within the diploid species *G. speciosa*.

Clarkia dudleyana ($2n = 18$), like 14 of the 34 species of the genus, contains from few to numerous individuals that are heterozygous for translocations and so form rings or chains of various sizes at metaphase I of meiosis. Snow (1960) crossed individuals from various populations, and all hybrids produced contained such rings or chains even though the parents did not. The longest chain contained all 18 chromosomes. All natural or synthesized interchange heterozygotes with rings of four or six chromosomes were fully fertile because of regular alternate disjunction. Apparently, larger rings or chains of from 12 to 18 chromosomes could not maintain the regularity of disjunction and were highly sterile.

The intraspecific hybrids produced by Snow between populations demonstrated that different interchanges had appeared in different populations and had often become homozygous. Sterility was not genetic as in *Galeopsis* but chromosomal by irregular disjunction at meiosis.

Intraspecific evolution by interchanges of this sort have been revealed by intraspecific hybridization also in *Clarkia unguiculata, williamsonii,* and *amoena, Paeonia californica,* and *Datura metaloides* (Snow, 1960).

In general, hybrids within a species of normally outcrossing habit have normal meiosis unless the parents are recognizably very different subspecies. Intraspecific hybrids among populations of self-pollinating species may, on the other hand, have some meiotic irregularities. Hybrids between distinct subspecies sometimes have irregular meiosis that may approach that of species hybrids and will show a great deal of segregation in the F_2 generation (Clausen et al., 1947).

SPECIES HYBRIDS

The cytogenetics of a great many plant *species hybrids* has been studied since Sax (1935), for example, discussed the subject. A small sample only can be included in this chapter. The arrangement will be according to ascending ploidy, beginning with hybrids of diploid parents. Most animal hybrids fall within this first group. The subject of permanent hybridity occcupies Chapter 17 and will also intrude into this chapter inevitably.

Diploid × diploid

Since nearly all animals and most plants are diploid and since most allopolyploids originate as hybrids from crosses between diploids, it is obvious that this is a large and important class of hybrids. Diploid hybrids that persist as evolutionary lines are often called *allodiploids* or *amphihaploids*.

Animals. Natural hybridity among animal groups varies greatly. Among freshwater teleost fish, hybrids between any sympatric species of a genus are to be expected (Hubbs, 1955) and are easily produced (Hubbs, 1967). The latter reported that the ease of crossing and develop-

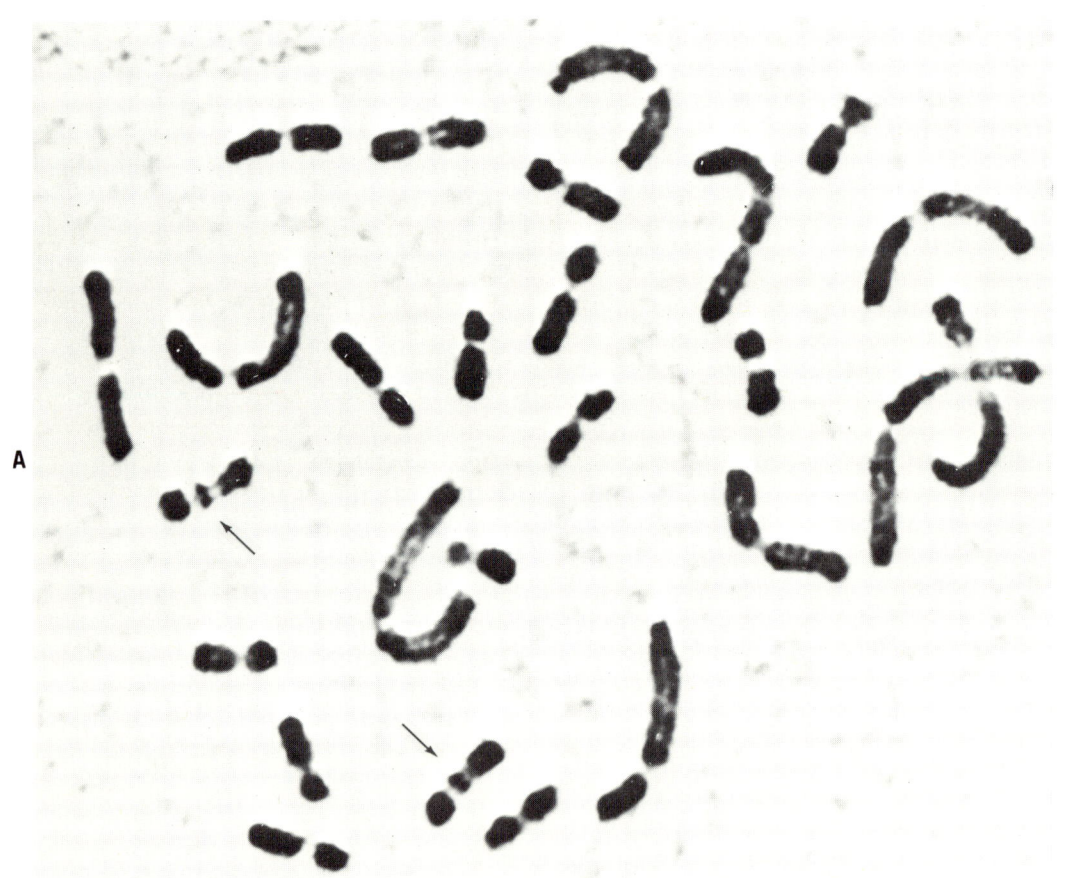

Fig. 16-2. Chromosome spreads of **A,** the frog, *Rana pipiens* (2n = 26), and **B,** the toad, *Bufo americanus* (2n = 22). The frog and toad karyotypes are different as to number, obvious secondary constrictions in one pair of frog chromosomes (**A,** arrows), and some acrocentric chromosomes (**B,** arrows) in the toad. The assumed frog-toad hybrid had no chromosomes with secondary constrictions and had more than the haploid number of acrocentric chromosomes. Thus it had no frog parent. (Courtesy Dr. J. P. Bogart, Louisiana Tech. University.)

mental stage reached by the hybrid (zygote, young embryo, late embryo, etc., to sexual adult) are closely correlated with taxonomic relationship. The subject of hybridity in animal groups has been discussed somewhat in preceding chapters. Thus, although there seems to be considerable hybridization going on among some animals, there has been very little cytogenetic study of such known hybrids, mainly because of the cytological difficulties of studying meiosis. Among insects, however, considerable cytological study of meiosis of hybrids has been done especially in the Orthoptera, such as by Evans (1954) on *Circotettix* and White (1957) on the Australian *Moraba scurra*. White (1970) presents evidence for believing that many or most parthenogenetic animals (reptiles, amphibians, fish, insects) are hybrid.

Examination of somatic chromosomes does sometimes provide some evidence on the subject, such as the resolution of the claimed frog-toad hybrid of *Bufo* × *Rana* (Kiley and Wohnus, 1968). Bogart et al. (1969) have subsequently demonstrated that the assumed hybrid was possibly a gynogenetic (parthenogenetic) diploid larva of *Bufo* containing no *Rana* chromosomes (Fig. 16-2). This conclusion was based upon broad hybridization experience with *Bufo* and from analysis of the somatic chromosome complements of *Rana pipiens, Bufo americanus,* and the assumed hybrid. Volpe et al. (1970) have given strong support to the conclusion of Bogart et al. but have concluded, probably correctly, that the assumed hybrid was the result of accidental fertilization of the *Bufo* egg by a *Bufo* sperm; the sperm of the frog cannot enter the toad egg.

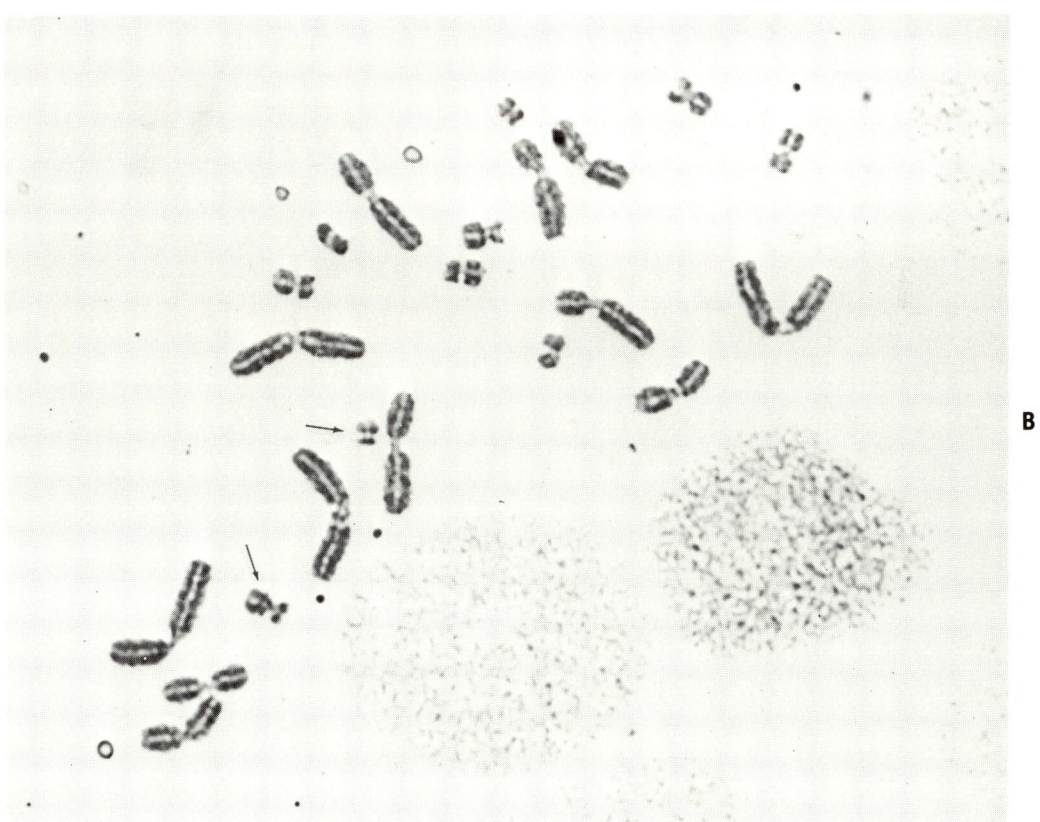

Fig. 16-2, cont'd. For legend see opposite page.

In Diptera the pairing of salivary gland chromosomes or the somatic pairing in neuron cells is somewhat comparable to meiotic synapsis. For example, Sears (1947) reported that the cross of *Drosophila quinaria* ($2n = 8$) × *D. subquinaria* ($2n = 12$) produced in neuron cells of the hybrid two associations of two acrocentrics with one submetacentric each. Similar associations occurred in *D. munda* ($2n = 8$) × *subquinaria, occidentalis,* and *suboccidentalis,* all $n = 12$. Natural and experimental intraspecific and interspecific hybridity for inversions has often been shown by salivary gland chromosome study in *Drosophila* (Patterson and Stone, 1952).

Experimental hybridizations have revealed that crosses within *Drosophila* species groups can often be made, but attempts between species of different groups almost always fail. Even within groups the developmental stages reached by the hybrids vary greatly, and in almost half of the cases success is achieved in only one direction in reciprocal crosses. Similar successful hybridization within groups or sections or subgenera, but mostly failure among species of different sections, has been reported in the plant genera *Nicotiana* (Goodspeed, 1945), *Ceanothus* (Nobs, 1963), *Quercus* (Grant, 1971), *Ribes* (Keep, 1962), *Pinus* (Grant, 1971), *Eucalyptus* (Pryor, 1959), and many others, and in the rodent genus *Peromyscus* (Dice, 1940). In *Crepis,* "hybrids between species which are less closely related, as judged from morphology, tend to be weak and sterile or, if vigorous, to be sterile or of very low fertility; whereas hybrids between more closely related species tend to be vigorous and more or less fertile" (Babcock, 1947). Thus, in general, the more different any taxa are morphologically, the more difficult is natural or experimental hybridization.

Cytogenetic study has been made in the so-called horse genus *Equus.* Makino (1951a) reported the chromosome numbers of the horse, *E. caballus,* and the donkey, *E. asinus,* incorrectly (as both $2n = 66$) and that meiosis in the hybrid, the mule, is irregular, the irregularities being sufficient to produce its hybrid sterility. Koulischer and Frechkop (1966), however, found by modern techniques of blood culture, spreading, and karyotype analysis that *E. caballus* has $2n = 64$, as earlier reported by Trujillo et al. (1962). *E. prjewalskii,* a mongolian horse, however, has $2n = 66$ and the hybrid has $2n = 65$. This hybrid, at least one of two produced, is fertile when backcrossed to *E. caballus. E. prjewalskii* has two more acrocentrics than *E. caballus;* otherwise the karyotypes are the same. Chromosome numbers of other species of *Equus* are (Hsu and Benirschke, 1967) *E. asinus* (donkey), $2n = 62$; *E. heminus,* $2n = 56$; and *E. zebra,* $2n = 32$. Thus the mule, a hybrid of horse and donkey, should have $2n = 63$.

Makino (1951a) listed meiosis as being regular in the F_1 hybrid between the mice *Mus musculus* and *M. molossinus,* both having $2n = 40$.

Meiosis of hybrid grasshoppers has been studied a number of times, of which a couple of recent examples are given. Evans (1954) reported some populations of *Circotettix undulatus* from the California-Nevada border as having the 20 autosomes typical of the genus but some other populations as having 22 autosomes and still others as being hybrid with 21 autosomes. Apparently, one metacentric of the 20-autosome type has undergone chromosome fission to produce two acrocentric autosomes. In the 21-autosome hybrid at meiosis one of each of the acrocentrics pairs (synapses) with the metacentric to form a trivalent. This trivalent is so arranged at metaphase I that the two acrocentrics go to the same pole by alternate disjunction about 63% of the time. This is an example of an intraspecific chromosomal hybrid.

White (1957) reported a somewhat similar case of intraspecific chromosomal natural and experimental hybridization in the Australian grasshopper *Moraba scurra.* The evidence is that a metacentric, the A-B chromosome, of the 14-autosome race has undergone breakage of some sort to produce two acrocentrics, the A and the B, of the 16-autosome race. At meiosis in hybrids a trivalent is formed by the two acrocentric autosomes (A and B) synapsing with the single A-B metacentric, and alternate disjunction occurs almost always. White reported that in some hybrids the A chromosome pairs by its proximal (centromeric) end to a different chromosome, an unusual acrocentric, C-D. He concluded that the so-called dissociation of the A-B chromosome was not a simple fission but was a translocation. Presumably the A-B broke in the A arm close to its centromere. Thus, the new acrocentric B chromosome kept the A-B centromere, but the acrocentric A acquired its centromere from the acrocentric C-D chromosome.

Plants. The cytogenetics of many diploid plant species hybrids has been studied. A few selected examples only can be given here. It should be noted at this point that in hybrids of all kinds that have only partial pairing at meiosis the variation is usually considerable among the various meiocytes examined. Therefore, averages and ranges of bivalent pairing or average numbers of univalents are usually expressed. Furthermore, the extent of chromosome pairing is strongly influenced by environmental conditions, especially in anthers (Sax, 1935; Douglas and Brown, 1971).

Probably the most sensitive cytologically observable aspect of meiosis in hybrids is the effect on chiasma frequency per bivalent or per cell. A reduction in chiasma frequency may be the only detectable effect of hybridization. Chiasma frequency is very sensitive to environmental factors also.

COTTON. The genus *Gossypium* is divisible into Old World (all diploid) and New World (diploid and tetraploid) groups. The diploid hybrids of greatest interest are those between Old World and New World diploids (n = 13) as well as between Old World (Asia, Australia, and Africa) diploids and New World (South, North, and Central America and Hawaii) tetraploids. Skovsted (1934) noticed that the Old World cottons (all are diploid) have n = 13, short (about 2.25μ) chromosomes whereas the diploid New World cottons have shorter (about 1.25μ) chromosomes.

Hybrids between the Old World diploid cultivated cottons *Gossypium arboreum* and *G. herbaceum* (Silow, 1944) have normal bivalent pairing but reduced number of chiasmata and are fully fertile. However, segregation occurs in the F_2 to produce mostly inferior types. These F_1's can be described as cryptic structural hybrids (Stephens, 1950). Hutchinson (1954) considered it likely that *arboreum* "arose in cultivation by differentiation from *G. herbaceum*." Such slight subspecific difference is indicated by the cytogenetics already discussed. The cross of either *arboreum* or *herbaceum* to the wild African *anomalum* produces 10% fertile hybrids with anaphase I bridges that indicate structural differences larger than those between the two cultivated species (Skovsted, 1937). Crosses between Old World diploids *arboreum* × *stocksii* and *herbaceum* × *stocksii* were as sterile as Old World × New World diploids (Beasley, 1942), indicating *stocksii* as a "good" species.

Skovsted also made crosses between Old World and New World diploids (as have others, Stephens, 1950; Phillips, 1966) such as *arboreum* × *thurberi* and *anomalum* × *thurberi*. All of these hybrids are sterile, and at meiosis not more than half of the chromosomes are paired. But even so, there were usually some anaphase I bridges, which indicate some crossing-over, and such structural differences as small inversions. Such diploid hybrids of specifically different diploid parents are often called *allodiploids*, and they are usually completely or almost completely sterile. From them allotetraploids can be formed, such as from the *arboreum* × *thurberi* diploid hybrid. Meiosis in tetraploids produced by chromosome doubling is characterized by mostly bivalent pairing and partial fertility (Stephens, 1950). The hybrid *G. arboreum* × *G. raimondii*, the latter another New World diploid, was sterile both when selfed or pollinated by others and also when backcrossed to either parent. At metaphase I the hybrid (2n = 26) had an average of about 14 univalents and 6 bivalents; trivalents and quadrivalents were very uncommon (Endrizzi and Phillips, 1960). The reason for the presence of such multivalent configuration in diploid hybrids is not understood but is not rare. Another case is cited in the *Aegilops* × *Secale* hybrids discussed later.

Hybrids between diploid species of *Gossypium* vary considerably, depending sometimes upon the closeness of relationships, but other crosses seem to fail for incompatibility reasons. The considerable cytogenetic work on

Table 4. Genomes of *Gossypium* species

Species	Genome	Origin
herbaceum and *arboreum*	A_1, A_2	Old World diploid cottons, Africa and India
anomalum and two others	B_1, B_2, B_3	South Africa and Cape Verde Islands
sturtianum and three others	C_1-C_4	All Australian
stocksii and one other	E_1, E_2	Northeastern Africa to Pakistan
thurberi, raimondii, and six others	D_1-D_7	New World diploids
hirsutum and *barbadense*	AD	New World tetraploid cottons

this genus has permitted rather exact characterization of genomes and the assignment of symbols to the various genomes (Beasley, 1942; Brown and Menzel, 1952) (Table 4). The diploid species have five basic genomes, from A to E, which are geographically restricted (Phillips, 1966).

Brown and Menzel (1952) stated that "Despite the cytological differentiation between the genomes A, B, C, D and E, members of each genome in *Gossypium* can be crossed with species in one or more other genomes. Conversely, species within a genome with little or no cytological differentiation may fail to cross. Each species shows a range of crossability with other species. Even though each species can be crossed with one or more other species, there is no diploid species which does not exhibit failure to cross with some species." The extreme of intergenomic sterility (genetic incompatibility) is shown by the New World diploids D_3 (*G. klotzschianum*) and D_6 (*G. gossypioides*). The latter crossed to other New World diploids produced no mature hybrids, and death occurred at latest by the cotyledon stage. A few hybrids with genomes A and B, however, were obtained, but they died before reaching maturity. A few mature hybrid plants were secured from D_6 crossed to tetraploids. Meiosis in the hybrid $D_6 \times$ AD had pairing between the small D chromosomes of the tetraploid and the D_6 chromosomes of *G. gossypioides* ranging from commonly 13 bivalents to 8 bivalents plus 2 or 3 trivalents uncommonly. The average of 11.23 bivalents was the lowest among all AD $\times$ D hybrids known (Brown and Menzel, 1952). There are no known chromosomal rearrangements among the various D genomes (Endrizzi and Phillips, 1960), so such incompatibility seems to be genetic.

Pairing between genomes in crosses between diploid species of *Gossypium* varies. The A and B genomes form an average of about 12 bivalents per meiocyte of a possible 13 (Endrizzi and Brown, 1968); whereas the D and A genomes, for example the A_2D_5 hybrid of *G. arboreum* $\times$ *raimondii*, had an average of only about 6 bivalents per meiocyte (Endrizzi and Brown, 1968). Thus, in *Gossypium*, genetic incompatibility prevents many hybrids and distorts or prevents genome analysis of hybrids among some diploids.

Phillips (1966) has gathered together data on all diploid $\times$ diploid species crosses in *Gossypium*, such as number of univalents, bivalents, trivalents, and quadrivalents. He considered the number of univalents, out of a possible 26, at meiosis in the hybrid as most instructive. In hybrids between species with the same genome, such as A $\times$ A, C $\times$ C, etc., the average number of univalents was almost always less than one. Between A and B genomes about 2.5 univalents per cell was average. Hybrids between C and E genome species had about 24 of 26 chromosomes as univalents at meiosis. He summarized the intergenomic hybrids as to *average univalent frequency* by a five-pointed star with lines drawn between genomes and the average univalent frequency stated (Fig. 16-3). He concluded from this analysis that genome E is the most remote from all the others and diverged in evolution earliest since it has univalent average frequencies with other genomes of 17, 22, 23, and 24. Genome D diverged shortly before genome C, having averages (except with E) of: D(11, 12,

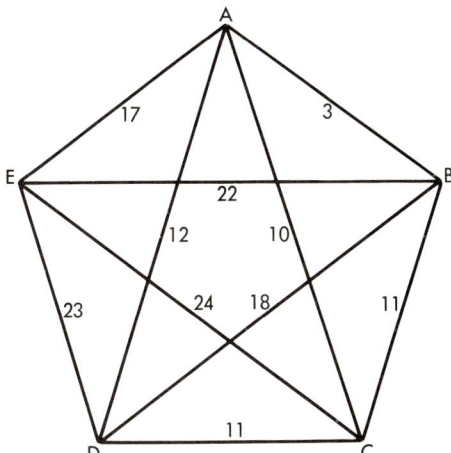

Fig. 16-3. A diagram indicating relationships among diploid species of *Gossypium* based upon average univalent frequency. The letters at the points of the diagram indicate the different genomes. The lower the number on a line connecting different genomes (the average univalent frequency), the more closely related are the two genomes. (From Phillips, L. L. 1966. Amer. J. Bot. **53**:328-335.)

and 18) but C(10, 11, and 11). Genomes A and B are very close, with the average univalent frequency of only 2.5.

Diploid × tetraploid and tetraploid × tetraploid cotton hybrids are discussed later.

AEGILOPS SQUARROSA × SECALE CEREALE. Wheat × rye and *Aegilops* × rye hybrids have been made and studied repeatedly since about 1930, and it has been generally concluded that there is essentially no homology between rye and either *Aegilops* or *Triticum* (but see Unrau, 1958, who considers that all genera of Hordeae may have homologous genomes to the extent of homeology). Meiosis in the *Aegilops squarrosa* × *Secale cereale* hybrid was subjected to detailed study (Melnyk and Unrau, 1959) and is used here as an example of diploid species hybrid, an allodiploid, or an amphihaploid. One significant advantage of this cross for meiotic study is that rye chromosomes are significantly larger than are those of *Ae. squarrosa* and the size difference remains at meiosis in the hybrid. Thus, the chromosomes of the two parents can be identified, and any pairing between rye and *Aegilops* can be determined with certainty.

The hybrid was produced by pollinating 20 florets of *Ae. squarrosa* with rye pollen. Eleven seeds were obtained, of which six provided embryos for artificial culture and one grew into the hybrid plant. This plant was completely sterile.

At meiosis 10% of cells had no bivalents. The average meiocyte of 100 examined contained 9.67 univalents of the total of 14 chromosomes. Almost none of the bivalents at metaphase-anaphase I had a chiasma in both ends, so they were described as *open* bivalents. Each cell had an average of 0.4 open bivalents consisting of one rye chromosome synapsed to an *Aegilops* chromosome, thereby forming a heteromorphic bivalent. It was considered proved, contrary to previous conclusions, that at least two chromosomes of each species are homologous enough to synapse. Since the D genome of hexaploid wheat was derived from *Ae. squarrosa* (McFadden and Sears, 1946), there must be some homology between at least hexaploid wheat and rye, but not much.

Some cells contained bivalents of *Aegilops* or rye or both. Trivalents, present in 25% of cells, consisted in some meiocytes of two rye and one *Aegilops*, and in others of two *Aegilops* and one rye chromosome. Two quadrivalents consisted of two chromosomes from each parent. And one cell contained a multiple association of three rye and six *Aegilops* chromosomes. At anaphase I the distribution of rye and *Aegilops* chromosomes seemed to be random to the two poles so the complete sterility of the hybrid was inevitable.

TRITICUM. A great deal of taxonomic, morphological, genetic, and genome analysis has been devoted to the cultivated wheats and their wild relatives in the genera *Triticum* and *Aegilops*. Hexaploid wheat was assigned the three genomes (each of which is a set of seven chromosomes) A, B, and D. The A genome was assigned to the only wild triticum, *Triticum aegilopoides*. Fairly early the D genome was shown to be derived from *Aegilops squarrosa* (McFadden and Sears, 1946), and eventually the B genome was considered to be most like that of the diploid *Ae. speltoides* by Sarkar and Stebbins (1956) on morphological grounds and by Riley et al. (1958) on certain cytological characteristics, even though Sears before 1946 had produced a hexaploid derived from a cross of the tetraploid *T. dicoccoides* × *Ae. speltoides* and found that it failed to have taxonomic or cytological similarities to hexaploid wheat (McFadden and Sears, 1946). Sharman (1943) had also concluded that *Ae. speltoides* does not have the proper number of nucleoli (part of the proof of Riley et al. in 1958) to be considered the source of the D genome. Apparently, *Ae. speltoides* is a complex of distinctive subspecies, and Sears and Sharman may not have used the best variety or "species" of *speltoides*. But the real reason for failure to relate *Ae. speltoides* with the B genomes was that these early workers were trying to relate it to the D genome and not the B. Failure was predictable.

One possible reason for difficulty in relating *Ae. speltoides* to the B genome may be the interactions of numerous meiotic genes that affect pairing. Hexaploid wheat has the strong 5B gene that inhibits pairing of homeologous chromosomes as well as the 5D and 5A genes, which promote pairing. *Ae. speltoides* has such a strong promoter of pairing that it is able to override the 5B inhibitor in a hybrid of *Ae. speltoides* × *T. aestivum* (Sears, 1969). Haploids of hexaploid wheat (polytriploids) consist of genomes A, B, and D and at meiosis have one or two loose pairs of homeologues. When the 5B chromosome is missing, the 21 homologues of the polytriploid do pair to the extent of 19, with an extreme of 15 involved in trivalents (Riley and Chapman, 1958). Morris and Sears (1967) and Sears (1969) among others are still not convinced that *Ae. speltoides* is the species that contributed the B genome to hexaploid wheat.

Much of the work proving this origin of tetraploid and hexaploid wheats was the analysis of meiosis in hybrids among the three polyploid levels of *Triticum* and *Aegilops*, especially tetraploids × diploids. Recently all species of *Aegilops* have been transferred to the genus *Triticum* by Bowden (1959). *Triticum aegilopoides*, the only wild diploid species of what used to be the genus *Triticum* before *Aegilops* was added to it, is now called *T. boeticum* and shares the A genome with the cultivated *T. monococcum*. Hybrids between these species have very regular meiosis and high fertility. But hybrids between either of these and any diploid of what used to be the genus *Aegilops* show low pairing and almost complete sterility. Pairing usually involves only about three or four of the seven chromosomes of each species, and most pollen aborts.

NICOTIANA. The examples just given are of true allodiploids showing little or no pairing at meiosis and complete sterility of the hybrid. However, many other hybrids between species having the same chromosome number have either normal bivalent pairing and high fertility or nearly normal pairing but sterility. Then there are hybrids between species of different chromosome numbers, and there are hybrids that produce multivalent configurations at meiosis. Even within the same genus hybrids between different species may be remarkably different for the above characters. For example, in the genus *Nicotiana* the hybrid between the two 12-chromosome species *N. paniculata* and *N. solanifolia* has nearly normal bivalent pairing whereas the hybrid between the two 12-chromosome species *N. paniculata* and *N. cavanillessi* produced very few and loosely paired bivalents (Goodspeed, 1934). Goodspeed concluded that the degree of pairing in hybrids reflects the genetic relationship between the species in the genus.

CLARKIA. Interspecific hybridization between diploid species in the genus *Clarkia* often shows specific differences for interchanges comparable to those intraspecific translocational differences revealed by crosses between populations (Snow, 1960). Lewis and Raven (1958), for example, found at meiosis in hybrids between *C. franciscana* and *C. rubicunda*, neither of which was heterozygous for exchanges, that the most common meiotic configuration in hybrids was a chain of five, a chain of three, and three bivalents. This common configuration indicates that the two species differ for at least three large translocations involving four pairs of chromosomes. A chain of chromosomes is not necessarily an unclosed ring. The ends of an unclosed ring are homologous; but a chain, especially of an odd number of chromosomes, may have ends that are not homologous and, therefore, could not form a circle. From one to four bridge-fragment configurations at first meiotic anaphase in the *Clarkia* hybrid indicated that the two species differ for at least four paracentric inversions. Since no double bridge at anaphase I nor bridges at anaphase II were seen, it was concluded that double crossing-over rarely occurs within the heterozygous inversions. Such results are not typical of hybrids between closely related species, but in *Clarkia* and some other genera of Onagraceae rapid evolution by translocations is common.

GRINDELIA. In the example just given hybridization produced chains, and in other examples rings occur in hybrids from parents that do not have chains or circles. In *Oenothera* and some other genera, parents that have rings can lose them or have them modified by hybridization. This latter is much rarer than the former since to lose a ring or chain the other parental species or variety must have chromosomal end arrangements present in the ring-forming heterozygous species. On the other hand, speciation is often accompanied by translocational difference between species so that species hybrids often have ⊙4 or ⊙6 or ⊙2, ⊙4. For example, *Grindelia hallii* (Compositae) × *G. camporum* or *hallii* × *procera* gives hybrids with 6 bivalents, whereas *havardii* × *oxylepsis* or *oxylepis* × *procera* produces 4 bivalents and ⊙4, and *havardii* × *procera* or *havardii* × *camporum* gives hybrids with 2 bivalents plus ⊙2, ⊙4 (Dunford, 1970). Other examples of intraspecific and interspecific hybrids with circles are discussed in Chapter 13.

Diploid × tetraploid

Genome analysis is largely achieved by making hybrids between allopolyploid species and diploid species thought to have contributed one or other of the genomes present in the polyploid. Ideally at meiosis in such a hybrid, if the diploid involved did indeed contribute a genome, its set of chromosomes should pair with one set of chromosomes of the polyploid, and the remaining chromosomes from the gamete of the polyploid should be unpaired. However, because chromosomal evolution continued in both the diploid and the allopolyploid species subsequent to the original hybridization that led to the origin of the allopolyploid, the ideal condition is rarely observed. Usually all the chromosomes from the two species that ideally should pair do not do so, and the number paired varies at meiosis among the meiocytes observed. If translocations, inversions, or certain meiotic gene mutations have occurred, as is often the case, further complications arise. And if the diploid species chosen as one presumed parent of the hybrid is only a close relative of but not the most exact descendent of the original diploid involved, only some pairing will occur.

DROSERA. Historically the most famous study of meiosis in a tetraploid × diploid species hybrid was in the genus of sundews, *Drosera*. It had long been known that natural hybrids occur between species of *Drosera* having different leaf shapes and other characters, such as between *D. intermedia* and *D. filiformis* (Macfarlane, 1898) and *D. rotundifolia* and *longifolia* (now *D. anglica*) in Europe (Rosenberg, 1903).

Rosenberg in his classical study (1904, 1909) reported that *D. rotundifolia* has n = 10 and *D. anglica* has n = 20. At meiosis in the natural hybrids there were 10 bivalents plus 10 univalents. Since that study, any triploid hybrid from a tetraploid × diploid natural or experimental cross that has at meiosis two sets of chromosomes paired as bivalent and one set unpaired as univalents is described as exemplifying the *Drosera scheme*. This scheme became the basis of genome analysis as discussed in the remainder of this chapter. It is apparent that *D. anglica* (2n = 40) contains two sets of chromosomes (genomes), one of which is essentially that of *D. rotundifolia* (2n = 20) since they pair at meiosis. The other set seems now to be that of the 2n = 20 American species *D. linearis* (Wood, 1955).

Wood found *D. rotundifolia* and *linearis* often growing together in Michigan with sterile morphological, 2n = 20 hybrids occasionally present. These hybrids, as do most species of the genus, reproduce asexually by dissemination of detached leaves that form buds. Wood found a very few fertile plants that were identical to the circumpolar tetraploid *D. anglica* (the *longifolia* of Rosenberg). Thus there are sterile allodiploid hybrids between *rotundifolia* and *linearis,* which, when the chromosome number is doubled, have produced the amphidiploid allotetraploid *D. anglica.* This species when crossed to *D. rotundifolia* (2n = 20) produces the triploid hybrid of Rosenberg.

Therefore, if the *rotundifolia* genome of 10 chromosomes is designated *R,* and that of *linearis* as *L, D. rotundifolia* (RR) × *linearis* (LL) gives the sterile hybrid RL. When this is doubled, the tetraploid *anglica* (RRLL) is formed. Rosenberg studied the tetraploid and diploid parents *anglica* (RRLL) × *rotundifolia* (RR) and the triploid hybrid (RRL) in which the two *rotundifolia* genomes pair, leaving the 10 *linearis* genome chromosomes as univalents.

TRITICUM. Another famous diploid × tetraploid hybrid is that produced by McFadden and Sears (1946) between the cultivated wheat tetraploid, *Triticum dicoccum,* or its wild close tetraploid relative, *T. dicoccoides,* and the wild diploid species *Aegilops squarrosa.* The researchers determined that *Ae. squarrosa* carried a group of major characters that distinguish the hexaploid wheats from the tetraploid. For this reason *Ae. squarrosa* was selected as the most likely contributor of the third, the D, genome of hexaploid wheat. The hybrid was triploid and completely sterile; but when the chromosome number was doubled by colchicine treatment, a hexaploid was formed which was very similar to the hexaploid *T. spelta.* It crossed easily with *T. spelta* and *T. vulgare,* and the hybrid had regular meiosis and was fully fertile. This demonstrated that the hexaploid wheats arose from a spontaneous hybridization between tetraploid wheat and *Ae. squarrosa,* according to archeological evidence, about 3,000 BC (Frankel and Munday, 1962).

McFadden and Sears (1946) also reviewed the subject of the origin of the B genome present in tetraploid wheats. They discussed the evidence of a number of cytogenetic analyses that indicated that *Ae. speltoides* had perhaps contributed a genome to tetraploid wheat. From analysis of meiosis in the hybrid of *Ae. speltoides* (2n = 14) × *T. turgidum* (2n = 28) they found 7 bivalents and 7 univalents in 71% of microsporocytes. Sharman (1943) also concluded that *Ae. speltoides* might be one diploid ancestor of hexaploid wheat but perhaps more likely of diploid wheat. His *Ae. speltoides* × *T. durum* (AABB, 2n = 28) triploid hybrid was doubled to form an allohexaploid, which he considered to be possibly AABBA'A'. He was close; it should have been AABBB'B' as we now know. Analyses to 1946 had indicated by irregular meiosis and sterility in the synthesized hexaploid that *Ae. speltoides* had not contributed the D genome. The workers had produced AABB × B'B'→ABB'→AABBB'B'. Such a hexaploid should be sterile, as it was.

Sharman did report that *Ae. speltoides* does have four nucleoli as do the tetraploid wheats. But his hexaploid wheat did not have eight nucleoli. Later the number of nucleolus organizers, two in the genomes of *Ae. speltoides,* and satellite size, the same as in hexaploid wheat (Riley et al., 1958), were used as evidence that *Ae. speltoides* had contributed the functional nucleolus organizers and the B genome.

At present it is considered proved that *Triticum boeticum* (formerly *T. aegilopoides*) contributed the A genome and *Ae. speltoides* probably the B genome to produce the tetraploid wheats. Later *Ae. squarrosa* contributed the D genome by hybridizing with the tetraploid wheat to form a sterile triploid. Somehow the triploid produced an allohexaploid wheat, which has subsequently been selected to produce the various 42-chromosome wheats. It is possible that hexaploid wheat arose directly by union of unreduced gametes of tetraploid wheat and unreduced gametes of *Ae. squarrosa.*

A genetic factor for bivalent pairing appeared in tetraploid wheat after its hybrid origin. The presence of this inhibitory gene, which prevents hemeologous pairing and limits pairing to homologous chromosomes, was responsible for much of the confusion resulting from genome analysis. Homeologous pairing did occur in any hybrids lacking the gene, but none occurred or was present when one parent was either diploid or hexaploid wheat, which have the 5B gene. How a gene can control pairing in this manner is unknown. It seems to affect premeiotic pairing (Feldman, 1966) rather than synapsis including the synaptinemal complex. But this is merely a finer resolution of the greater problem of why homologous or homeologous chromosomes in general can and do pair whereas chromosomes having less homology than is implied by homeology cannot and do not.

COTTON. Hybrids between diploid Old World cultivated cottons (2n = 26) and New World tetraploid cultivated cottons, *Gossypium hirsutum* and *G. barbadense*, both 2n = 52, have demonstrated that the New World allotetraploids arose from hybridization at some unknown time in the past between an Old World diploid and a New World diploid (Stephens, 1947). The questions of how *Gossypium* diploids can be native both to the Old World, from South Africa to Australia, and to the New World, from Peru to southwestern United States, and how and when some Old World diploid was able to hybridize in the New World with some New World diploid have no satisfactory answers. But genomes common to both worlds are not rare. Menzel and Martin (1971) have recently reported such in the cotton relative *Hibiscus*.

Early analysis of meiosis in Old World diploid × New World tetraploid cotton indicated 13 bivalents and 13 univalents; but Skovsted (1934) reported in addition to some bivalents and univalents, some trivalents and quadrivalents in most and occasional pentavalents and hexavalents in microsporocytes of such 39-chromosome triploids. Because reciprocal crosses always had at least 13 univalents he considered it proved that "the chromosomes of Asiatic cotton are homologous with half the number of chromosomes in New World cotton, while the remaining half are left out as univalents." The multivalents must represent some translocation difference between the Old World set and the more or less homologous one-half set of the New World tetraploids.

Some autosyndesis comparable to the multivalents found in triploid Old World cotton (Skovsted, 1934), in all Old World × New World diploid hybrids (Endrizzi and Phillips, 1960), and in all allopolyploids (Sarvella, 1958) may also be responsible for some multivalent formation. It is evident that there is considerable homology between the various Old World and New World genomes (Brown and Menzel, 1952) as well as homeology between chromosomes within each genome as evident in haploids of diploid species. But it is true in general that at meiosis in a triploid hybrid between an allotetraploid and a diploid in *Gossypium* there are approximately 13 bivalents and 13 univalents.

ORYZOPSIS. A somewhat different sort of evolution seems to have occurred in a group of three species of the grass genus *Oryzopsis* (Shechter and Johnson, 1968). *O. hymenoides* is tetraploid (2n = 48), *O. contracta* is also tetraploid, but *O. micrantha* is diploid and has 2n = 22. Morphologically *contracta* is intermediate between the other two species. The tetraploid hybrids, both natural and synthetic, between *hymenoides* (n = 24) and *contracta* (n = 24) have regularly 16 bivalents plus a few univalents and multivalents.

Hybridization attempts between *hymenoides* and *micrantha* have always failed, and *contracta* crosses with difficulty to the other species. It is obvious that *contracta* is not an amphidiploid derived from the other two species. Rather, Shechter and Johnson proposed that at some time in the past an essentially sterile, 2n = 35 hybrid resulted from a cross of *micrantha* and *hymenoides* with 11 *micrantha* and 24 *hymenoides* chromosomes. At some time a 24-chromosome gamete of this hybrid containing mostly *hymenoides* chromosomes but carrying segments translocated from the *micrantha* genome united with a normal 24-chromosome gamete of *hymenoides*. Subsequently, selfing produced the homozygous genome of *O. contracta*.

Tetraploid × tetraploid

Meiosis in tetraploid × tetraploid species hybrids varies greatly among such hybrids in the amount and types of pairing. That is, they vary in percentages of univalents, bivalents, and such multivalents as trivalents, quadrivalents, and possibly larger configurations. As in other categories of hybrids the percentages of bivalents present at meiosis imply the percentages of homologous and/or homeologous chromosomes, but the amount of cryptic structural, translocational, and inversional heterozygosity present modifies the pairing frequency.

The cytological literature, of course, contains reports of studies of pairing and genome analysis of many such hybrids. A few only can be discussed here as examples. Since most tetraploid species are allotetraploid, it is evident that each gamete will contain two genomes unless the hybridization that produced the tetraploid occurred long enough ago so that enough diploidization has taken place to convert two diploid genomes into one tetraploid genome. Thus there will be considerable variation at meiosis, depending upon the amount of cytological and genetic change that has occurred in either or both of the tetraploid parents of the species hybrid being studied.

TRITICUM. Earlier in this chapter diploid × diploid and diploid × tetraploid examples have been taken from published reports on genome analysis of *Triticum* (including *Aegilops*), in part because this group of diploid, tetraploid, and hexaploid species has been studied in more detail and by more investigators than any similar group of species.

Kihara was one of the leading students of genome analysis in this group and in 1929 published a report on the pairing of many tetraploid × tetraploid hybrids. Aase (1930) reviewed the analyses of such species hybrids from numerous reports and contributed original studies.

Hybrids between *Triticum* tetraploids (2n = 28) have a mode of 14 pairs with, in some microsporocytes, two univalents. Except for *T. timopheevi*, all tetraploid species are probably cultivated derivatives of the wild *T. dicoccoides* and can be considered as much varieties as species. There has been some diploidization in these species as would be expected, and the looser pairing with fewer chiasmata (at metaphase bivalents associated at one end only rather than at both ends) is typical of meiosis in intraspecific hybrids.

Pairing in tetraploid hybrids between tetraploid *Aegilops* (2n = 28) species is lower, from 3 to 10 bivalents, and some trivalents are typical. Most bivalents are of the open type (paired at only one end at metaphase I). Both Kihara and Aase reported that chromosomes of PMC of *Aegilops* species hybrids are very difficult to study since numerous "exotic" apparent multivalents involving four, five, six, or more chromosomes are common. *Aegilops* × *Triticum* tetraploid hybrids have even fewer bivalents, usually none, but in some hybrids as many as five or six. The numerous univalents are often scattered over the spindle at metaphase. Some meiocytes contain trivalents. It is evident that *Aegilops* and *Triticum* tetraploid species are less closely related than are the diploids, but most of this failure to pair is the presence of the 5B gene in tetraploid wheat that prevents pairing of homeologues.

Zohary and Feldman (1962) and Feldman (1965) concluded that in each section of the genus *Aegilops* the tetraploids have one genome in common but there is much variation in the second genome due to repeated hybridization between such tetraploids and diploids and tetraploids in other sections; thus each tetraploid contains one unaltered and one "modified" genome. The modified genomes result in blurred boundaries between species as well as pairing differences among hybrids. For example, the Pleionathera section of species of *Aegilops* is characterized by *all* tetraploid species having the C^u genome of the diploid *Ae. umbellulata* of the section, but associated in different tetraploid species with genomes S^v, C, M^e, M^b, M^t, and M^o. Other modifications of the M genome are associated with the D genome of *Ae. squarrosa*, which is common to all species of the section Pachystachys as well as hexaploid wheats.

AGROPYRON. A somewhat different pairing at meiosis was reported in some species hybrids in the grass genus *Agropyron* (Dewey, 1965). *A. spicatum* exists as diploid (SS) and as autotetraploid (SSSS) strains. *A. dasystachyum* and *A. riparium* are allotetraploids, each having 14 bivalents at meiosis. Meiosis in *A. dasystachyum* × diploid *spicatum* and *riparium* × tetraploid *spicatum* was characterized as to pairing configurations in the 2n = 21, 2n = 28 hybrids as follows (X represents the 14 chromosomes from either *dasystachyum* or *riparium*; S is the *spicatum* genome; Roman numerals indicate univalents, bivalents, etc.):

	I	II	III	IV	V
XS	6.8	6.9	0.1	0	0 = 2n = 21
XSS	7.05	4.7	3.0	0.6	0.01 = 2n = 28

The approximately 7 bivalents in the triploid hybrid from tetraploid *dasystachyum* × diploid *spicatum* indicate an S (or S_1) in the tetraploid species associated with another nonhomologous genome represented at meiosis in the triploid hybrid by 7 univalents. In the tetraploid hybrid the 7 univalents represent the non-S genome, but there must be three S genomes in the hybrid to produce the 4.7 bivalents, 3.0 trivalents, and 0.6 quadrivalents. Thus these two species contain one *spicatum* genome. *A. riparium* and *A. dasystachyum* are no more than two forms of one species, so the other genome is probably the same in both forms, according to Dewey. At meiosis in the autotetraploid *A. spicatum* an average of 8.28 bivalents plus 2.86 quadrivalents are formed.

Pentaploid × tetraploid

Solanum tuberosum, the potato, is tetraploid (2n = 48) which is probably more autoploid than alloploid. *S. curtilobum* is a closely related, seed-producing, tuberous pentaploid (2n = 60). Lamm (1941) produced hybrid seed from reciprocal crosses. Among 75 hybrids the chromosome numbers were usually from 50 to 56 (expected, 24 + 30 = 54), but the total range found was from 2n = 48 to 59. From these data it would be expected, and it was found, that the hybrids varied morphologically from *tuberosum*-like to *curtilobum*-like but most were intermediate. The hybrid plants varied greatly with respect to vigor and pollen fertility also. The *tuberosum* parent had been selected for high seed fertility but had only about 50% pollen fertility. The *curtilobum* parent had 43% pollen fertility, which is typical for this pentaploid species (Lamm, 1945). The cross *curtilobum* pistillate × *tuberosum* staminate was easily made, and the hybrids varied in pollen fertility from 14 to 90%. The cross with the *tuberosum* parent as pistillate was made with great difficulty, and all of its hybrids were completely sterile. At meiosis in these latter hybrids only univalents were observed at metaphase I. They passed irregularly to the poles at anaphase I, divided or undivided. But degeneration of meiotic cells began from metaphase I to interkinesis so that no microspores or pollen grains were formed.

Hybrids from the reciprocal cross having *curtilobum* as the pistillate parent had quite different meiosis; it was as normal as in the parents. Pairing and anaphase disjunction were as normal as in the parents in both anthers and ovules. Interestingly, meiosis in ovules of some *tuberosum* × *curtilobum* hybrids was normal and embryo sacs formed, but in others meiosis did not procede past interkinesis after irregular metaphase with chromosomes scattered all along the spindle. Such sterility differences and hybridization success between reciprocal crosses are not uncommon, and, as in this case, success is usually achieved with the higher chromosome number species as pistillate parent.

Hexaploid × diploid

Examples for this category are drawn from the extensive published reports on the tuber-bearing species of *Solanum* related to the potato of the section Tuberarium. The hexaploid most widely used is the Mexican species *S. demissum*. It and most of the diploids employed in the crosses are not closely related to the cultivated forms. In this section of from 150 to 200 species more than 60% are diploid (Hawkes, 1956); there are at least four triploid, many tetraploid, three or four pentaploid, and five hexaploid species known.

S. demissum (6n = 72, x = 12) is reported to be a typical allohexaploid (Howard and Swaminathan, 1952), although an average of one quadrivalent and occasional univalents and trivalents were reported by Schnell (1948) and Cooper and Howard (1952). Walker (1959) reported 36 bivalents in about 50% of PMC's.

In all diploids at meiosis 12 bivalents are usual but 11 bivalents and 2 univalents are occasionally seen (Walker, 1959), and in some species anaphase bridges and fragments have been reported. On the whole, however, these seem to be cytologically rather typical diploid species, some of which are, doubtless, heterozygous for inversions and other segmental differences.

The hybrid between a hexaploid (2n = 72,

6x) and a diploid (2n = 24, 2x) is a tetraploid (3x plus 1x) with a chromosome number of 2n = 48. As would be expected, there is variation in the meiotic irregularity among these hybrids, depending upon the diploid species employed. Nevertheless, there seems to be considerable meiotic irregularity at meiosis in the hybrids, according to Magoon et al. (1958) and Walker (1959), although some earlier reports (Howard and Swaminathan, 1952) cited 24 bivalents with different diploid species. In general (Magoon et al., 1958), there is a range of from 1 to 8 univalents, from 0 to 2 trivalents, and from 0 to 4 quadrivalents at metaphase in these hybrids. Usually, in a microsporocyte there are about four univalents and two trivalents and/or quadrivalents. That is, of the 48 chromosomes about 12 are not bivalents. At anaphase I the univalents behave in various ways, such as going to the poles early or late and lagging; bridges and/or fragments are occasionally produced. The result is that from 60 to 80% of microspores are unbalanced cytologically. One, two, or five rather than the expected four microspore nuclei are often produced in addition to micronuclei. Restitution nuclei often form at anaphase I or II. Nevertheless, from 20 to 50% of pollen is stainable and presumably "good."

Some such tetraploid hybrids of hexaploid *S. demissum* × diploid species do form a very few seeds whereas other hybrids are completely sterile. This sterility is not entirely the result of chromosomal imbalance but is the result of endosperm failure (Beamish, 1955). Beamish found that in some hybrids embryos did form but in others they did not. But in both, the endosperm developed to a point and then degenerated. Endosperm failure is a common cause of hybrid sterility (Brink and Cooper, 1947).

Hexaploid × tetraploid

Since the same principles of pairing apply to hexaploid × tetraploid species hybrids as to those already discussed, only one example, that of *Triticum* and *Aegilops*, will be discussed. Tetraploid wheats have the genomes A of *T. aegilopoides* and B of *Ae. speltoides*. The hexaploid wheats have in addition to A and B the D genome of *Ae. squarrosa*. Therefore, the original hybrid was between the two diploid species *T. aegilopoides* (AA) and *Ae. speltoides* (BB). This sterile AB hybrid had its chromosome number doubled spontaneously to produce the wild tetraploid *T. dicoccoides* (AABB) and the cultivated tetraploid wheat derivatives. At some later time some cultivated tetraploid wheat hybridized with *Ae. squarrosa* (DD) to produce a sterile hybrid ABD. This doubled spontaneously to form the first hexaploid (AABBDD) wheat, which fortunately was propagated. Similar hybrids probably formed repeatedly but to get both chromosome doubling and progeny may have occured only once.

Hybrids between tetraploid and hexaploid wheats regularly form 14 bivalents (AABB) plus 7 univalents (D) (Aase, 1930), but the pairing is a little looser than in either parent, there are fewer chiasmata, and occasional PMC's have 9 univalents when one pair does not form a bivalent at metaphase I. This is an example of the two parental species having two almost identical genomes in common.

Hexaploid wheat (ABD) × *Aegilops cylindrica* (DC), on the other hand, produces a pentaploid hybrid with a mode of 7 bivalents (DD) and 21 univalents (ABC) at metaphase I. But hexaploid wheat × *Ae. ovata* (C^uM^o) or *Ae. triuncialis* (C^uC) (Zohary and Feldman, 1962) produces pentaploid hybrids with a mode of from zero to two bivalents (Aase, 1930), as would be expected. Pairing in the hybrid of *Ae. ovata* (C^uM^o) × *cylindrica* (DC) has an average of 5.5 bivalents, indicating the justification for proposing that both have the C genome but that C and C^u are different enough to produce fewer than 7 bivalents in most PMC's.

Actually, the genome analysis proceeded historically from the observed pairing at meiosis in hybrids to designation of genomes, rather than as above, where genomes are given and then meiotic pairing is discussed.

Hexaploid × hexaploid

The commonest polyploid stage in plants is the tetraploid, but hexaploids are rather common in nature. At the hexaploid level there is enough genetic duplication among the genomes present that the gain or loss of one chromosome or a few does not significantly affect vigor or reproductive potential.

Love and Suneson (1945) studied meiosis in the hexaploid hybrid derived from a hexaploid wheat, *T. macha* (2n = 42) × *Agropyron trichophorum* (2n = 42). This hybrid should have had 2n = 42 but did have 2n = 41. At meiotic metaphase I it was quite variable, having from 13 to 31 (average 22.6) univalents, from 4 to 13 (average 7.7) nearly all open bivalents, chains of three from 0 to 3 (average 0.9), and chains of four from 0 to 1 (average 0.8). It is not surprising that this F_1 plant set no seed but a sister plant that may have had 2n = 42 did produce a few seeds. One F_2 plant was grown and was found to have 70 chromosomes; it was decaploid. The researchers assumed that a partially reduced (28-chromosome) male gamete from some wheat hybrid fertilized an unreduced (2n = 42) egg. This plant had an average of 18 bivalents and had multivalents to a chain of six in a very few cells, and there were usually more than 22 univalents.

SPARTINA TOWNSENDII. A famous hexaploid hybrid formed naturally on the coast of England between the native *Spartina maritima* (2n = 60, x = 10) and the American *S. alterniflora* (2n = 62). The hybrid, when synthesized, is completely sterile, with some irregular pairing. Apparently, doubling of the 2n = 62 chromosomes of the hybrid occurred to produce the fertile and very vigorous amphidiploid. This 2n = 120, 122, and 124 amphidiploid is a new species, called *Spartina townsendii* (Marchant, 1966).

Octoploid × diploid

FRAGARIA. The example for the octoploid × diploid class of hybridization will be the strawberries, *Fragaria*. The species involved are *F. vesca* and *F. viridis* diploids (2n = 14, x = 7) and *F. virginiana* and *F. chiloensis* octoploids (2n = 56). Senanayake and Bringhurst (1967) made crosses of these species in order to determine the genomic relations. Apparently, the two diploids have the same genome, A, and in the pentaploid hybrids (2n = 35) they gave similar results. In the pentaploid hybrids the average number of bivalents and univalents varied somewhat. The *viridis* × *chiloensis* hybrid had an average of 9.7 bivalents and 15.5 univalents; *vesca* × *chiloensis* had 10.4 bivalents and 14.2 univalents; and *vesca* × *virginiana* had 12.6 bivalents and 9.8 univalents. But the ranges of these were the same, from very few, 6 bivalents and 23 univalents, to many, 14 bivalents and 7 univalents. Multivalents were very rare, but there were never more than 14 bivalents.

Seven of the from 10 to 13 bivalents indicate that the octoploid has an A genome like that of the diploids or that it has four sets of one genome, BBBB, that form seven pairs in the hybrid. The occurrence of up to seven more pairs but usually fewer (from three to five) indicates two similar but not identical genomes either both in the octoploid or one there and one in the diploid.

Earlier workers had proposed the genome A for the diploid species and ABBC for the octoploid gamete. Thus the pentaploid hybrid would have AABBC and as many as but no more than 14 bivalents. Of course other possibilities can be imagined.

Senanayake and Bringhurst (1967) tested the AABBBBCC proposal by crossing the octoploid to an autotetraploid *F. vesca*, AAAA, to give a hexaploid hybrid. If the octoploid contributes ABBC and the autotetraploid contributes AA, the hexaploid hybrid should be AAABBC (2n = 42) and there should be 7 univalents, often about 7 bivalents but no more than 14, up to 7 trivalents, but *no* quadrivalents. But if the octoploid and diploid do not share the A genome, there should be many bivalents but no trivalents. What they found ranged from 12 bivalents and 18 univalents (2n = 42); to 20 bivalents and 2 univalents; to 1 quadrivalent, 16 bivalents, and 6 univalents; and many other combinations including 1 trivalent very often.

The occurrence of trivalents indicates that the diploid and octoploid species do share a genome, designated A. The presence very often of more than 14 bivalents (as many as 20) seems to indicate that AAABBC for the hexaploid hybrid is not correct. Changing C to A' would permit more than 14 bivalents. The occasional quadrivalent also indicates that perhaps C should be changed to A', as the workers proposed. Thus, the octoploid strawberries are now considered to be AAA'A'BBBB.

They further assumed that sterile hybrids followed by chromosome doubling was not the mechanism by which polyploid species arise in *Fragaria*. Rather, the occurrence of unreduced gametes is so very common (Bringhurst and Senanayake, 1966; Bringhurst and Gill, 1970) that they proposed combinations of unreduced gametes to give fertile polyploids immediately. Bringhurst and Gill produced many high polyploid hybrids very easily, 9x, 10x, 12x, and 14x. Ellis (1962) has also cited 16x and 18x plants of hybrid origin in his cultures. As would be anticipated, these high polyploids are partially fertile and also often produce aneuploids.

There are, of course, hybrids and allopolyploids at higher levels of ploidy than those so far discussed in this chapter. Presumably the same principles apply, and in theory genome analysis of high polyploids such as *Galium grande* (2n = 220) or *Hibiscus radiatus* × *H. diversifolius* (2n = 216) might be made.

In addition to the few examples discussed here of meiosis in species hybrids there are hundreds of others in the literature, of course mostly of plant species. Many are in cultivated genera, but many more are in wild forms. Discussions and data are often parts of cytotaxonomic and evolutionary studies (see Grant, 1971).

CHAPTER 17

Permanent heterozygosity

> When areas of the genome are shut off from recombination, either by inversions, chiasma localization or differential segments in translocations, heterozygosity can accumulate. Such regions also serve as traps for genes that may have lethal or semilethal effects in the homozygous condition. Accordingly, heterozygosity in such regions becomes increasingly obligatory, and return to a homozygous state becomes increasingly difficult.
>
> H. L. Carson, 1967a

Species vary greatly with respect to genetic heterozygosity. Many inbred diploid species of plants are highly homozygous. On the other hand, most animal and many plant species are and must be genetically heterozygous to survive, as inbreeding has often demonstrated. The same differences exist for chromosomal heterozygosity, which, of course, implies also genetic heterozygosity. In general, chromosomal heterozygosity tends to be eliminated during evolution; any inversions or translocations that may arise in a species either do not spread widely and are gradually reduced to homozygosity or increase and become fixed as new homozygous conditions in one or more subspecies. As Carson (1967a) expressed it, "In a cross-fertilizing, diploid organism, the heterozygous state appears to represent the essence of impermanence." But a number of sorts of mechanisms exist in nature that can perpetuate chromosomal and its accompanying genetic heterozygosity at a very high level, in some cases as true-breeding heterozygotes.

Evolution theory requires that any case of perpetuated heterozygosity must confer greater fitness (heterosis) on the heterozygotes in specific environments than the fitness possessed by either of the two kinds of homozygotes. The example of sickle cell trait in human beings in the presence of the malarial parasite is well known, where the heterozygote is better adapted than either homozygote. As a result of the superior fitness of the heterozygote, the condition called *balanced polymorphism* is achieved in the population. When this condition exists, both homozygotes and the heterozygotes exist in the population, and the proportion of heterozygotes depends upon time since the "mutation" occurred, the superiority of heterozygous fitness, etc. However, the condition of *obligate heterozygosity* is not such a balanced state necessarily but describes the situation in which the heterozygous condition is present in all or nearly all individuals, and there are few or no homozygotes. Such a state often requires the complete breakdown of the usual rules of mendelian inheritance (Darlington, 1956b). Carson (1967a) has discussed this topic of permanent heterozygosity in detail.

There are five mechanisms that can perpetuate this condition of obligate heterozygosity. Two of them, inversion heterozygosity and translocation heterozygosity, occur usually at the diploid level and in inbreeding systems; the other three allopolyploidy, apomixis or parthenogenesis, and diplopolyploidy, usually occur at polyploid levels. Some of these have been discussed to some extent in previous chapters but will be considered here together as cytogenetic mechanisms for permanent heterozygosity.

INVERSION HETEROZYGOSITY

As a known mechanism, inversion heterozygosity is limited to the Diptera, because analysis of salivary gland chromosomes is possible. Among the Diptera it is best known in certain species of *Drosophila*, especially *D. pseudoobscura*, and there especially in the third chromosome. As Crumpacker and Salceda (1968) stated, "Selective advantage of heterokaryotypes is known to be an important factor in the dynamics of third-chromosome polymorphisms in *Drosophila pseudoobscura*, and *persimilis*." That this was so was early suspected and later proved (Dobzhansky and Levene, 1948; Pavlovsky and Dobzhansky, 1966; Strickberger and Wills, 1966; Crumpacker and Salceda, 1968). In these species homozygotes do occur in large numbers, but many individuals in all populations and throughout the breeding seasons are heterozygous for one or more inversions on the third chromosome.

A system of *balanced lethals* within an inversion segment could convert such a balanced polymorphic system into obligate heterozygosity, as reported by Carson (1962) in *Drosophila mangabeirai* (see later). The term balanced lethals applies to two or a few, often closely linked genes such that when homozygous they cause the death of the homozygotes. In plants, one gene may cause death or inability to function of certain pollen grains whereas the other causes death or other form of elimination of those megaspores containing it, the "gamete lethals" of *Oenothera*. In this way heterozygotes only are formed or survive. In some species such balanced lethals occur with no known cytological modification of chromosomes, such as in the plants *Antirrhinum majus* (Gairdner and Haldane, 1933), in *Matthiola* (Frost, 1915), in maize (Bianchi and Morandi, 1962), etc. In each case a pair of lethal genes are heterozygous. If the lethals are linked then a mechanism to prevent or reduce crossing-over between them is essential to prevent their recombination. Inversion heterozygosity does seem to provide such a mechanism because any chromatids involved in exchange within the loop are themselves lethal so that the linkage of the lethal alleles remains permanently.

Drosophila mangabeirai is, in part, an example of such an obligate heterozygote, the heterozygosity being perpetuated by a system of three inversions (Carson, 1962). All individuals are heterozygous for these inversions, and crossovers either do not occur or produce a lethal condition. Associated with this condition is *selective autogamy* (Murdy and Carson, 1959). That is, following meiosis without fertilization in the egg the arrangement of spindles is such that two haploid nuclei containing reciprocally inverted chromatids fuse regularly to form a diploid heterozygous nucleus. Parthenogenesis follows. Males are rare, apparently sterile, and not needed.

Some other examples of probable obligate or near obligate heterozygosity probably correlated with balanced lethals in *Drosophila* are a population of *D. funebris* from England in which all larvae for a number of laboratory generations, the first from the wild, were heterozygous for three independent inversions on chromosome No. 5 (Berrie and Sansome, 1948). A population of *D. tropicalis* from Honduras had about 70% of individuals heterozygous for an inversion on chromosome No. 2. When inbred for 4 months more than 90% of individuals were heterozygous. Dobzhansky and Pavlovsky (1955) concluded that the homozygous condition is lethal by the pupal stage at the latest. Pavan et al. (1957) reported more than 50% inversion heterozygotes in a number of populations of *D. willistoni*.

Carson (1967a) cites similar cases reported for other insects that had inversion heterozygosity and parthenogenesis, as in *D. mangabeirai*. A probable case of permanent heterozygosity in white-throated sparrows that probably involves inversions is interesting cytologically.

White-throated sparrows

The white-throated sparrow seems to have a unique system of persistent heterozygosity (Thorneycroft, 1966). That is, although the total chromosome number is constant, the three longest chromosome types (except for the very longest pair) vary in form and number (Fig. 17-1). There are three forms for these three chromosomes and one, the M, is never paired. They can be designated as pair 2, pair 3, and M, but 2 and 3 are not always present as a pair and M or 3 may be missing. Thus five combinations have been found, but the sum of those chromosomes present is always four (Fig. 17-1). Thorneycroft reported 16 individuals having 2, 2, 3, 3 and no M; 15 birds with 2, 3, 3, M (only one No. 2); 3 birds with 2, 2, 3, M (monosomic for No. 3); 1 bird with 2, 2, 2, M (trisomic 2, nullisomic 3, monosomic M); and one individual with 2, 2, 2, 3 (trisomic 2, monosomic 3, and nullisomic M).

He found no relationship between any of these combinations and sex (sex chromosomes, as is typical of birds, are ZZ-ZW). He further assumed that really there are two pairs present, 2, 2 and 3, 3; but that rearrangements, especially pericentric inversions, have created not only the M that is genetically similar to the sub-acrocentric pair 2 or the acrocentric pair 3 but also 2's that can be morphologically like the 3, and 3's that are like the 2. The presence of only one M is determined by sexual selection. All birds of bright nuptial plumage have an M, and all individuals of dull nuptial plumage lack the M. Bright birds always mate with dull, so the progeny have only one M and heterozygosity is maintained.

TRANSLOCATION HETEROZYGOSITY

In Chapter 13 examples of true-breeding animal and plant species or populations that are heterozygous for reciprocal translocations have been discussed. In some species, exemplified by *Oenothera*, all the chromosomes occur in a ring at metaphase I of meiosis. In other species smaller rings of four or more chromosomes regularly appear in meiosis; homozygotes (also called homokaryotypes) are eliminated by systems of balanced lethals that remain together because no recombination occurs between them.

In *Oenothera* subgenus Euoenothera all crossing-over occurs in the distal regions near the ends of chromosome arms; there is none in the proximal regions on each side of the median centromere. Furthermore, in the ring-forming species alternate disjunction keeps whole sets of chromosomes together as effective linkage groups (Fig. 17-2). Mutations in the middle

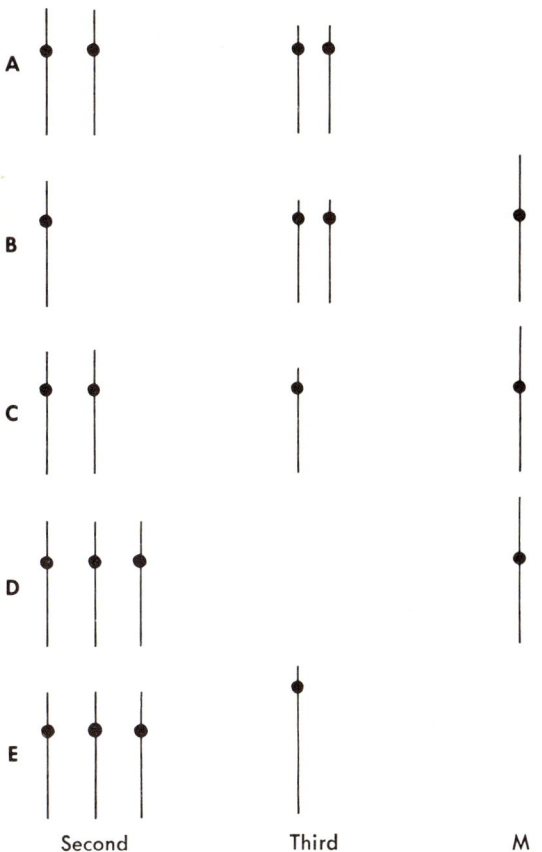

Fig. 17-1. In the white-throated sparrow the following combinations of the second, third, and M chromosomes have been reported. The sum of these is always four chromosomes, and they are distinctive enough for identification. (Modified from Thorneycroft, H. B. 1966. Science **154**:1571-1572.)

Fig. 17-2. Diagram of the 14 chromosomes of *Oenothera* in a ⊙14 at meiotic metaphase I. Centromeres are all median, and arms are all equal. Alternate disjunction is assured by alternate centromeres connecting to spindle fibers from the same pole. (From Brown, W. V., and E. M. Bertke. 1969. Textbook of Cytology. The C. V. Mosby Co., St. Louis.)

regions between the centromere and the break points in the arms of a complex, in addition to the lethals, can accumulate without any recombination until each complex is a distinctive genotype. If crossing-over does occur in the proximal or interstitial regions of arms, the result is inviable chromatids, which are eliminated. Apparently the ⊙14 condition has built up gradually by many random translocations from normally bivalent-pairing, open-pollinating, alethal species such as *Oe. hookeri* at the same time that various lethal mutations became organized into balanced systems.

Cleland (1968) has pointed out a very interesting condition in *Oenothera* subgenus Raimannia. Many species in this South American subgenus have ⊙14, as in Euoenothera, but numerous others have 7 bivalents. Very interesting is the finding that pairs of species or varieties very similar with respect to morphology and ability to cross possess the two extreme conditions, ⊙14 versus 7 bivalents. And fully as interesting is the fact that the ⊙14 species almost always have small flowers, are self-pollinating, and have balanced lethals; whereas the very similar 7-bivalent species have large flowers, are open-pollinating, and have no lethals. Furthermore, many of the 7-bivalent species have S-type genes for self-incompatibility, which, of course, the selfpollinating ⊙14 species cannot carry. Examples of these pairs of species and varieties are the following: ⊙14 *mollissima* versus 7-bivalent *affinis*; ⊙14 *humifusa* versus 7-bivalent *drummondii*; a ⊙14 unnamed variety versus 7-bivalent *grandis*. This is similar to the pair of *Rhoeo* species, *R. discolor*, ⊙12 (Fig. 17-3), versus *R. concolor*, 6 bivalents.

In the group 2 of Raimannia, the group that has been most studied, essentially all species have one complex, the A complex, or a slight modification, in common. This is the complex of two 7-bivalent species, *indecora* and *affinis*. Among the ⊙14 species that have in common the A *affinis* complex, the B complexes are very different and have nothing in common.

There is at present no indication of the evolutionary relationships of the pairs of species, which might have been derived from which. Cleland does consider that evolution of species

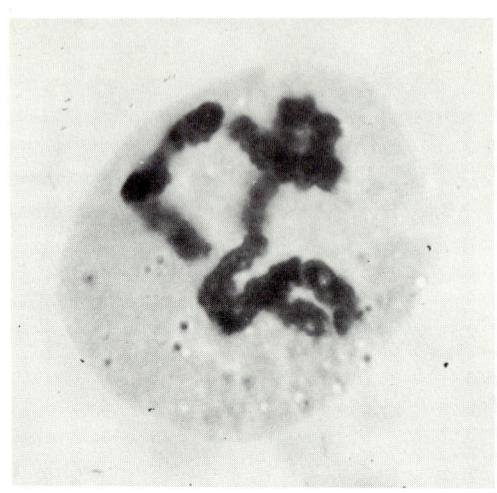

Fig. 17-3. Meiotic metaphase I in a PMC of *Rhoeo discolor* showing all 12 chromosomes united end to end in a ring. *R. discolor* chromosomes do not all have median centromeres, and they vary in length. The result is much less regularity than in *Oenothera*. (From Brown, W. V., and E. M. Bertke. 1969. Textbook of Cytology. The C. V. Mosby Co., St. Louis.)

in Raimannia has been at work for a longer period than in Euoenothera so that the species are more clear-cut morphologically and are often unable to cross. Therefore, it would seem that the pairs of species or varieties are of recent origin and must have evolved from a common ancestor that was either ⊙14 or 7-bivalent. Both systems, complex heterozygosity and homozygosity, work well enough for the species to persist.

The tendency to translocations seems to be common in the family Onagraceae, as is evident in the genera *Clarkia* and *Gaura* in addition to *Oenothera*.

Any species or population in which all or most individuals are regularly heterozygous for one or more translocations or inversions can be considered to be genetically heterozygous and heterotic and to have balanced lethals that permit it to be a permanent, true-breeding, heterozygous aggregate of individuals. In either case, the condition probably evolved from typical homozygous ancestors by steps.

AMPHIPLOID HETEROZYGOSITY

Amphiploid species are usually allopolyploids that have become diploidized to a condition of as good bivalent pairing and fertility as a good diploid species. As already discussed in Chapter 16, allopolyploids have hybrid origin and, with no segregation, become polyploid with two (or more) double sets (genomes) of homozygous chromosomes:

$$A_1A_1 \times B_1B_1 \rightarrow A_1B_1 \text{ (doubled)} \rightarrow A_1A_1B_1B_1.$$

Thus, homologous chromosomes, A_1 and A_1, are very homozygous, having been derived by doubling from the A_1 of the hybrid; but the more or less homeologous chromosomes of the two sets, A_1 and B_1, are heterozygous, as genetically different as were the two original species from which the amphiploid was derived. Therefore, the amphiploid is a true-breeding permanent heterozygote. Genetically, of course, it has to some extent four or more copies of each "unique" gene and so has segregation and three degrees of heterozygosity (Little, 1945), and recessive mutations are buried by the polyploidy. Nevertheless, for many unique genes the species act as diploids but require special cytogenetic techniques for study such as the use of monosomics, telocentric chromosomes, and isochromosomes, as in hexaploid wheats (Chapter 11).

ASEXUAL HETEROZYGOSITY

With any form of asexual reproduction extreme heterozygosity is possible, and segregation is essentially impossible. In some forms of parthenogenetic or apomictic reproduction normal meiosis may occur with recombination followed by autogamy (as in *Drosophila mangabeirai*) or diplospory in some plants, so that some segregation from one generation to another does occur. But in most asexually reproducing species there is no segregation.

It is also generally true that most parthenogenetic animals and apomictic plants are polyploid, triploid or higher, or euploid or aneuploid. Furthermore, most are also, probably, of hybrid origin. Therefore, they are apt to be as genetically heterozygous as the two species or subspecies from which they arose were genetically different.

White et al. (1963) found in one diploid species, a grasshopper of Australia, *Moraba virgo*, a combination of different sorts of inversion heterozygosity, "polyploidy," and parthenogenesis that together perpetuate its heterozygosity indefinitely. The "polyploidy" arises from a premeiotic doubling of the chromosomes in the egg to the tetraploid level. Meiotic pairing then occurs between the completely homozygous sister chromosomes produced by the premeiotic doubling. Thus the more or less heterozygous "homologous" chromosomes of the diploid condition do not synapse or exchange, and the numerous chromosomal alterations and any genetic heterozygosity are not affected by recombination. Parthenogenesis of the diploid egg produces progeny identical to the mother.

Heterozygosity is especially characteristic of facultatively apomictic species that form some sexual diploid and/or polyploid individuals in each generation. Usually, sexual forms of apomicts are very tolerant of pollen parents, and rather wide crosses often occur. This often produces agamic complexes. *Poa pratensis* (Kentucky bluegrass) is an example. Actually it and a number of related "species" form an interbreeding complex due to occasional hybridization that shows segregation of heterozygosity. Many of the highly segregating progeny from sexual reproduction in an apomictic species are themselves apomictic and if vigorous can start a new line with combination of characters (Clausen, 1954, 1961).

DIPLOPOLYPLOIDY

This form of permanent true-breeding heterozygosity, called permanent odd polyploidy by Grant (1971), is known to occur in a very few plants and animals. It is characterized by a basic diploidy, but with extra genomes that are transmitted regularly by one sex only. The number of extra genomes can vary to produce euploidy at various levels, but there is no recombination possible among these extra genome chromosomes. The dog roses of Eurasia, the Caninae, including the best studied, *Rosa canina* (Gustafsson, 1944), and the Australian species *Leucopogon juniperinus* (Smith-White, 1948, 1955) are the best known examples of diplopolyploidy. They will be discussed together because they have evolved essentially the same system. The dog roses have a basic

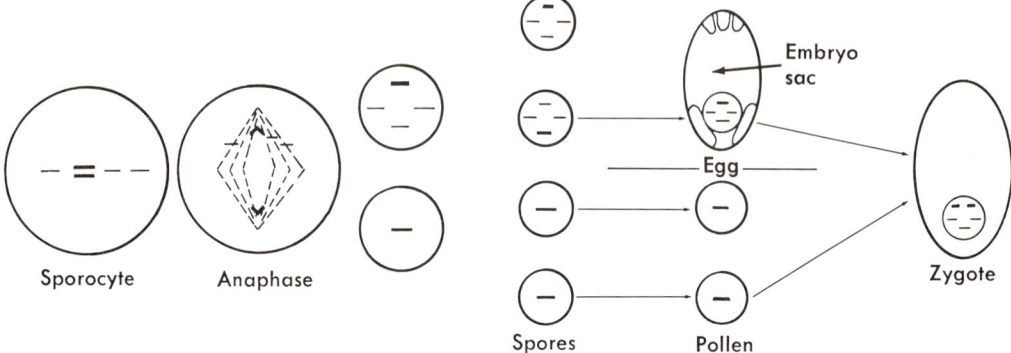

Fig. 17-4. Diagram of diplopolyploidy in a pentaploid species such as *Rosa canina*. Each short line represents one genome of seven chromosomes. The bivalent-pairing genomes are represented by heavy lines, the three nonhomologous genomes by faint lines. At meiosis I two genomes are paired as 7 bivalents whereas three genomes are present as 21 univalents. At anaphase I the bivalents disjoin regularly, but all 21 univalents go to one pole. Four spores are produced, two with 7 chromosomes each and two others with 21 chromosomes each. The embryo sac and egg are derived from a 21-chromosome megaspore, the pollen grains and male gametes from 7-chromosome microspores. The zygote formed has, again, the 35 chromosomes or five genomes of the parent sporophytes. The pollen of dog roses is not produced in as orderly a manner as this.

chromosome number of $x = 7$, and in chromosome number the various species are tetraploid ($2n = 28$), pentaploid ($2n = 35$), and hexaploid ($2n = 42$). *Leucopogon juniperinus* is triploid ($2n = 12$, $x = 4$).

During meiosis in both forms (Fig. 17-4) the haploid set of chromosomes provided by the pollen parent pairs with a homologous haploid set provided by the egg parent. *That* is the basic diploidy of such organisms. The remaining sets of chromosomes (genomes), one in *Leucopogon* and two, three, or four in the various dog rose species, are transmitted intact to the egg. They make the sporophytes polyploid. The chromosomes of these additional genomes do not pair at all, neither with the pairing genomes nor among themselves. So at metaphase I there is a set of bivalent chromosomes, four in *Leucopogon* and seven in dog roses, and from one to four genomes of univalents.

At anaphase I in both megasporocytes and microsporocytes of *Leucopogon* and in megasporocytes of dog roses (but *not* in microsporocytes of dog roses) one haploid set from the metaphase bivalents goes to each pole (as is typical of diploid meiosis), but *all* of the univalents pass undivided en masse to *one* pole. The second meiotic division is normal, with all chromosomes dividing and passing equally to the two poles. As a result, two types of spores are produced, half that are haploid and half that are polyploid with one genome less than the sporophyte.

At this point megaspores behave differently than do microspores—embryo sacs develop from polyploidy megaspores only, but pollen grains develop from haploid microspores only.

In *Rosa canina* the pollen grains are haploid also but are derived in a haphazard manner. At anaphase I all the univalents lag more or less on the metaphase plate whereas the seven bivalents segregate normally, as is typical of triploids, pentaploids, etc. (Chapter 12). As a result of this triploidlike division, numerous sorts of telophase I nuclei result. Some have only a haploid set of seven derived from the bivalents, and others have seven chromosomes derived from the bivalents plus various numbers of univalents. Meiosis II also has univalents lagging on the metaphase plates. Thus, from each microsporocyte numerous nuclei result, but some (about 15%) contain only a haploid set of seven chromosomes: these, and these only, form the functional pollen grains, and they are haploid ($n = 7$).

In both species the polyploid egg is fertilized by a haploid male gamete to restore the triploid (*Leucopogon*) and tetraploid, pentaploid, or hexaploid (*Rosa*) conditions in the embryos.

It is concluded that the two sets of chromosomes that pair are homologous, and for these the two species are regularly diploid. But *Leucopogon juniperinus* sporophytes contain also a set of completely nonhomologous chromosomes, and these are the univalents. *Rosa canina*, a pentaploid species, in addition to the diploid condition contains three sets of chromosomes that are not homologous among themselves but are not homologous with the diploid set either. Genomically, *Leucopogon* can be described as AAB (triploid) and *Rosa canina* as AABCD (pentaploid), although some evidence cited by Grant (1971) indicates $A_1A_1 A_2BC$ with A_2 chromosomes rarely pairing with A_1's. In addition to the pentaploid dog roses (the Caninae) there are tetraploid and hexaploid species. They all behave the same, being diploid for one genome but having two or four additional nonhomologous genomes. They are all true-breeding heterozygotes that perpetuate themselves by the same mechanism.

The pentaploid species of dog roses could have originated from a cross between a diploid and an octoploid (AA × AABBCCDD). Other pentaploid species could result from other diploids, such as BB or CC or DD crossing to the same octoploid. Of course, such tetraploid, pentaploid, and hexaploid dog rose species might have built up gradually from diploids and tetraploids (AA × AABB → AAB, × AACC → AABC, × AADD → AABCD, × AAEE → AABCDE), the triploid AAB having evolved the unusual meiotic segregation in the sporocytes. Actually the divisions in sporocytes of diplopolyploids are not very different from usual meiosis in hybrid triploids or certain hybrid pentaploids; it is merely genetic determination that all univalents pass to the same pole. In microsporocytes of the dog roses the meiotic divisions are rather typical of AAB triploids or AABCD pentaploid hybrids except that no spores function in dog roses unless they are purely haploid for the bivalent-pairing genome. Subsequently during evolution the diploids and triploids died out; all but the diploids and triploids are still extant.

Among animals diplopolyploidy seems to occur in the gall midges, family Cecidomyiidae, for example the genus *Miastor* (White, 1954). The germ line cells of males and females contain 48 chromosomes but the somatic cells of males, only 6 and of females, only 12. It seems likely that the "true" diploid chromosome number is $x = 6$. Therefore, the germ line can be thought of as octoploid with the male soma haploid ($n = 6$) and the female soma diploid ($2n = 12$). At meiosis in spermatocytes a first meiotic division typical of diplopolyploids occurs; 6 chromosomes go to one pole and 42 to the other at anaphase I. The secondary spermatocyte having six chromosomes divides by a typical second meiotic division, and the two resulting spermatids produce two sperm. The 42-chromosome secondary spermatocyte degenerates.

The details of meiosis in the female *Miastor* are obscure, but an egg that is haploid for 6 chromosomes plus 36 others seems to be produced, probably much as in ovules of the diplopolyploid plants. Fertilization of the 42-chromosome egg by the 6-chromosome sperm produces the 48-chromosome zygote and germ line. Some other gall midges have similar elimination of chromosomes from somatic cells and are probably also diplopolyploids, although chromosome numbers may vary.

A possible difference between the dog roses and *Leucopogon* on the one hand and *Miastor* on the other is that the plants are evidently allodiplopolyploids, consisting of two of one genome and from one to four additional different genomes; whereas the gall midge *may* be an autodiplopolyploid, having all genomes homologous.

BAND HETEROZYGOSITY

Permanent heterozygosity at a lower level of cytogenetics is revealed as asymmetrical bands and puffs of salivary gland chromosomes of various Diptera. Asymmetrical bands of paired homologous polytene chromosomes are evident as thin bands extending halfway across the chromosome but completed in the other half by much thicker half-bands. The thicker half-band represents *band enlargement,* a form of DNA amplification (Pavan and da Cunha, 1969a), and represents genetic heterozygosity of the individual. Asymmetrical bands have been found very frequently in natural and laboratory populations of *Sciara ocellaris* and other genera (see Pavan and da Cunha, 1969a, for references), and they are of the same type as Keyl (1965) reported in laboratory-produced, known hybrids of *Chironomus.*

Pavan and Perondini (1967) reported another type of asynchrony, called *asynchronous puffs,* in *Sciara ocellaris,* such that the puffing at the same locus is different for the two chromosomes (Fig. 3-7). Usually one side begins puffing before the other, and Perondini and Dessen (1969) found one asymmetrical puff that formed close to an asymmetrical band. The late-forming puff was spatially related to the enlarged half-band.

Pavan and Perondini (1967) concluded that "the frequency with which the asymmetric bands and the asynchronous puffs are found in individuals of natural populations, and their maintenance during long periods of time in natural populations as well as in laboratory cultures (over three years), suggests that those systems may have some heterotic effects on the viability and fitness of the flies." Thus, they may represent a cytologically visible form of permanent genetic heterozygosity.

It is evident that many systems have evolved among organisms to permit and maintain the inheritance of genetic and cytological heterozygosity, unaltered somewhat or completely by recombination. These systems all seem to perpetuate heterosis or fitness greater than that possessed by the homozygotes.

CHAPTER 18

Apomixis

> The apomictic terminology may seem unnecessarily complicated to an outsider.
> A. Gustafsson, 1946

In this discussion of apomixis no attempt at complete coverage of the subject is made. As Stebbins wrote in 1970, very little that is new in processes or concepts has been added since 1950, and there are numerous older and more recent detailed reviews generally available: Stebbins (1941, 1950); Gustafsson (1946-1947, the most extensive); Nygren (1954a); Fryxell (1957); Battaglia (1963); Grant (1971).

This chapter is limited to asexual reproduction in plants. The comparable condition in animals, parthenogenesis, has been treated in Chapter 12.

References to examples of types of apomictic processes will be limited to one or very few that are selected for familiarity of genera or for journal availability. The intent is to introduce the subject adequately for general comprehension without introducing the confusion of conflicting terminologies, long lists of examples or references, or the subtilties that occur in the literature of apomixis.

Furthermore, very little discussion of apomixis in plants other than the flowering plants, the angiosperms, will be included although asexual reproduction is common among lower plants and fungi. Actually, the term apomixis is generally restricted to asexual reproduction in angiosperms; other terms such as apospory, parthenogenesis, apogamy, etc. are used in discussions of asexual reproduction in other groups of organisms.

It must also be emphasized that apomictic processes have evolved independently in many families and genera of plants. According to the 1954 list of Nygren there were about 300 known apomictic (including viviparous) species in about 100 genera of 37 families. At least 60 species in at least 25 genera have been proved to be apomictic since 1954. Thus, apomixis must have originated independently at least 37 times (once at least in each family) and probably nearer 100 times, often in a number of genera or tribes of one family as in the Graminae and Compositae. In many cases acquisition of apomixis has been a recent development or is in the process of evolving or being tried. As a result, variation from sexuality to semiapomixis (facultative apomixis) often exists within the same genus, within the same species, and even within the same plant with sexuality. But even in species having apomixis of long standing the facultative condition can persist as a highly adaptive evolutionary compromise since it provides the variation and hybridity possibilities of sexuality and at the same time the high ecological adaptability of uniformity of the apomictic processes. A species or plant that is facultatively apomictic is both apomictic and sexual. In the grass complex called *Poa pratensis*, for example, not only is there often both asexual and sexual reproduction on the same plant, but both can occur in the same ovule. The same is true in many species of *Citrus*. Furthermore, two or more types of apomixis may occur in one genus, such as somatic and generative apospory in *Potentilla* and *Poa*.

CLASSIFICATION OF APOMIXIS

I. Vegetative reproduction — vivipary, etc.
II. Agamospermy — asexual reproduction by seeds
  A. Adventitious embryony — extra-embryo sac embryos
  B. Somatic apospory — supernumerary embryo sacs
  C. Diplospory — typical development but apomeiosis
  D. Generative apospory — no megasporocyte developed

Apomixis is the term applied to asexual reproduction in *plants* and includes a variety of subtypes. Vegetative reproduction, subtype I, includes such common and obvious forms of asexual reproduction as rhizomes, stolons, layers, vivipary, etc. In general usage, however, even among the students of apomixis, vegetative reproduction, even vivipary, is usually not included in the concept of apomixis. Subtype II includes asexual reproduction by seeds.

Technically this subtype is called agamospermy but is what most students of the subject usually mean by apomixis (Clausen, 1954). Agamospermy is not *obviously* asexual; it takes detailed study to determine whether a plant or a plant species is reproducing sexually or by agamospermy.

Some reviews of the subject also include the formation of haploid sporophytes (Chapter 11) in the overall concept of apomixis. Technically this may be correct, but in general it is not so considered, at least in angiosperms. When included, it is called nonrecurrent apomixis, nonrecurrent parthenogenesis, etc. Because the haploid sporophyte produced is sterile, that is the end of it.

The usual concept or connotation of the term apomixis is *recurrent agamospermy* by which a diploid or polyploid sporophyte reproduces by seed without chromosome number reduction (meiosis) or sexual union of gametes or gamete nuclei. The implication and result is that the progeny are usually genetically identical to the single parent, and the progeny are genetically identical among themselves. An indication of apomixis arises when two phenotypically distinctive plants are crossed so that the F_1 progeny that should be phenotypically like the pollen parent are, rather, all identical to

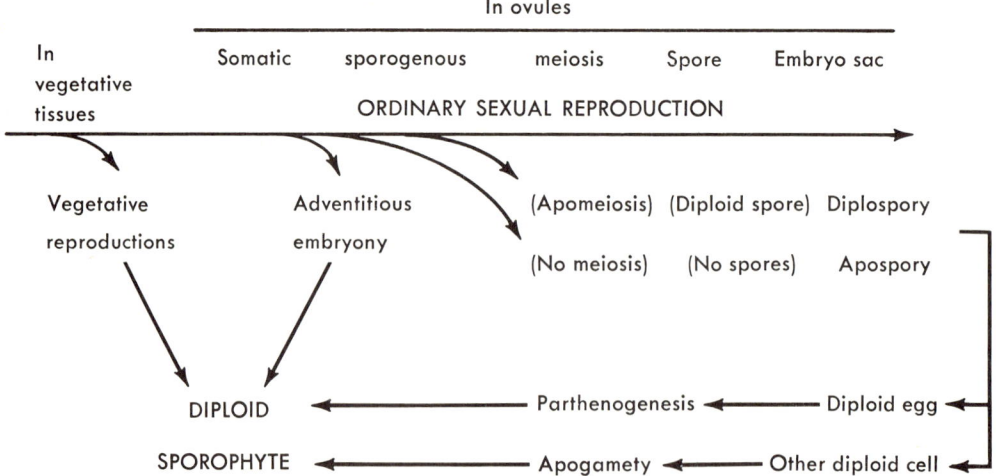

Fig. 18-1. Divergence of various types of apomixis in angiosperms from the ordinary sexual sequence of events. Sexual reproduction can accompany most forms of apomixis; it can accompany adventitious embryony or apospory in the same ovule, but diplospory and sexual reproduction must occur in different ovules.

the homozygous recessive pistillate parent. Apomictic reproduction can then be suspected. Actually, genetic determination of apomixis is usually made of species that are genetically unknown. Then strong maternal resemblance, such as maternal chromosome number (Einset, 1947b), especially following reciprocal crosses, indicates apomixis.

PSEUDOGAMY

Proof of apomixis (agamospermy) requires cytological examination of reproductive processes within the ovule. Meiosis (or breakdown of it) and pollen production (or failure) in the anther may have little or nothing to do with apomixis unless pollen is required, as it usually is, for processes other than syngamy. When pollination is required before seed production can occur in an apomict, as usually in *Poa*, the term *pseudogamy* is applied. Most apomictic species are pseudogamous because a male gamete is necessary for endosperm formation, and endosperm development is necessary for embryo and seed formation. That is usually not true with adventitious embryony. Most apomictic species are pseudogamous, although some species use pollen of related species (Simpson and Bashaw, 1969) or do not require pollination, as in *Poa nervosa* (Grun, 1955) and *Calamagrostis* (Nygren, 1946, 1954b). Therefore, development in the anther, although meiosis in the pollen mother cells (PMC's) may be very irregular, usually *does* lead to pollen production. But there can be considerable variation even within a genus. Among obligate apomictic species of *Crepis*, *C. occidentalis* and *C. intermedia* have essentially normal pollen but in *C. acuminata* degeneration occurs before meiosis can even begin (Stebbins and Jenkins, 1939). Other examples of irregular meiosis but pollen production are known in *Potentilla*, *Taraxacum*, *Hieracium*, and *Antennaria*.

AGAMOSPERMY

The term agamospermy applies to those forms of apomixis in which the asexual and ameiotic process occurs within ovules and the seed carries the apomictically produced embryo or embryos. Developments within the ovule are varied and complex. It is here that the terminology tends to become complex and confused because meiosis and female gametophyte (embryo sac) as well as failure of fertilization and embryo development are concerned.

Adventitious embryony

Adventitious embryony is the simplest condition of apomixis in ovules. This term connotes that one or more typical somatic cells of the interior of the ovule start to divide and eventually form embryos that are quite separate from the typically formed embryo sac. Adventitious embryony is little different from embryos forming in callus tissue on cut surfaces of stems (tomato), leaf (African violets, etc.), or even roots. Adventitious embryony is typical of the genus *Citrus* (Frost, 1938a, 1938b) and other genera of the family such as *Xanthoxylum* (Desai, 1962), Cactaceae of the genus *Opuntia* (Archebald, 1939), and numerous other genera in a number of mostly tropical (Grant, 1971) families.

Somatic apospory

Other forms of apomixis do require embryo sacs in ovules, but there are two sorts of embryo sacs that occur. In somatic apospory ordinary diploid somatic cells (rather than the usual sporogenous cells) within the ovule start to grow and become unreduced *embryo sacs*, each of which contains an unreduced egg, as in

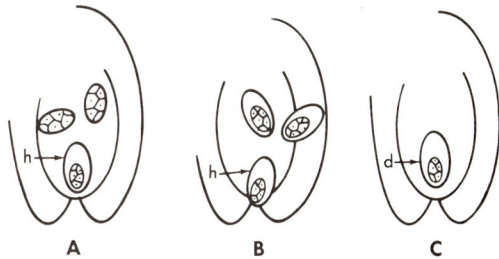

Fig. 18-2. **A,** Adventitious embryony; an ovule containing a normal haploid embryo sac, *h*, with a sexually produced embryo, as well as two adventitious embryos. **B,** Apospory; an ovule containing a normal haploid embryo sac, *h*, with a sexually produced embryo, as well as two diploid aposporous embryo sacs, each containing an apomictic embryo. **C,** An ovule containing a typical but diploid embryo sac, *d*, with an apomictic embryo, as in diplospory.

Crepis (Babcock and Stebbins, 1938), panicoid grasses (Brown and Emery, 1958; Simpson and Bashaw, 1969; Bashaw et al., 1970). There may be from one to a few such aposporous embryo sacs in one ovule, usually in addition to a normal or degenerated haploid embryo sac derived from a sporogenous cell. The term apospory means that the embryo sac (female gametophyte) that functions does not arise from a spore, as is usual in sexual reproduction of plants, but from an ordinary somatic cell.

Diplospory

Diplospory is the remaining form of apomixis, and it, too, requires an embryo sac. But in contrast to apospory, diplospory (diploid spore) does require that a spore of sorts become the embryo sac. The spore, however, is one of the products of the first apomeiotic division; there is no division comparable to the second meiotic division. Diplospory is essentially identical with sexual reproduction in angiosperms except that meiosis does not occur (ameiotic division) (Fig. 18-3) or occurs incompletely (semiheterotypic or pseudohomeotypic divisions) and syngamy does not occur either. With diplospory there is usually only one embryo sac and one embryo per ovule, as in sexual reproduction.

Generative apospory

Generative apospory is merely the extreme modification of meiosis and diplospory, where a typical mitotic division has replaced meiosis; the mitoticized megasporocyte (egg mother cell, EMC) has become a somatic cell, and nothing like megaspores are formed. The megaspore mother cell develops by mitosis directly into the embryo sac. It differs from somatic apospory by having only the one aposporous or diplosporous embryo sac in the usual position of sexual and diplosporous embryo sacs. Apomictic species of *Antennaria* (Stebbins, 1932) and *Zephyranthes texana* (Pace, 1912; Brown, 1941) are examples of generative apospory. That generative apospory is really merely an extreme form of diplospory is evident in the grass genus *Calamagrostis*, such as *C. canadensis*, in which some populations are normally sexual with meiosis and megaspore production in ovules whereas other populations have a mitotic division in the EMC, which becomes directly the embryo sac (Nygren, 1954b).

Agropyron scabrum. A recently described example of diplospory is the New Zealand species *Agropyron scabrum* (Hair, 1956). Four populations were studied. All plants of the completely sexual population, A, were hexaploid (2n = 42, x = 7) with regular meiosis in anthers and ovules, and the plants were morphologically quite uniform.

Population B consisted of facultative apomicts of which four plants were hexaploid (2n = 42), one was aneuploid (2n = 43), and one was 9-ploid with 2n = 63. In this group meiosis or fertilization or both were apt to fail with production of polyhaploid (2n = 21) and 9-ploid individuals as well as aneuploids. Failure of meiosis produced diplospory. In all four populations, sexual or apomictic, pollination and union of a male gamete with the polar nuclei was required to form the necessary endosperm. In population B the individual plants varied in fertility and morphology.

Population C was essentially diplosporous and pseudogamous although meiosis in anthers was somewhat abnormal, varying from all bivalents to some univalents, trivalents, quadrivalents, and, rarely, pentavalents. Some individuals were aneuploid (2n = 57) or 9-ploid, but half (two) were hexaploid.

Population D was uniformly hexaploid. The division in the megaspore mother cell was mitosis, as was true in some plants of popu-

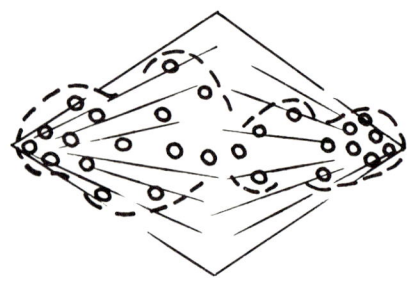

Fig. 18-3. Diagram of the formation of a restitution nucleus at telophase. The nuclear envelope (dashed lines) is forming adjacent to the scattered chromosomes and will eventually enclose all (the unreduced number) within one restitution nucleus (see Fig. 4-5 for comparison).

lation C. Usually with a mitotic division in EMC's the result of the division is the two-celled embryo sac. But in *Agropyron scabrum* a dyad of "megaspores" was produced, as in typical diplospory, one of which became the embryo sac. Therefore, this is not quite an example of generative apospory.

In this species the whole range from strictly sexual (population A) to obligate diplospory with mitosis (population D) is evident with facultative intermediates (populations B and C) still in existence. The series indicates the direction of evolution.

PARTHENOGENESIS

Diplospory and apospory are usually pseudogamous (require pollination). In either type, if the diploid egg develops without fertilization, the process is called, properly, *parthenogenesis*, for example in *Crepis* (Stebbins and Jenkins, 1939) and *Antennaria* (Stebbins, 1932). But some few cases are known of some other unreduced nuclei of the embryo sac developing into an embryo. That condition is called *apogamety*, meaning a nongamete cell of the embryo sac occasionally produces the embryo. Apogamety is not common in angiosperms but is common in ferns (Manton, 1950; Walker, 1962).

SEMIGAMY

Semigamy describes the condition of a male gamete actually entering the egg but the two nuclei do not fuse (Battaglia, 1946). Battaglia reported in *Rudbeckia laciniata* and *R. speciosa* that the male nucleus divided at one side of the base of the egg at the same time that the unreduced parthenogenetic egg nucleus divided. Some cells of the embryo were derived from the male nucleus to form somewhat of a mosaic. Coe (1953) found in *Cooperia* that the male nucleus divided once to form two cells at the base of the two-celled embryo but contributed nothing to the embryo proper as Battaglia (1955b) reported for *Rudbeckia sullivantii*. In these species some factor inhibits or does not stimulate nuclear fusion in the egg. Male gamete penetration may be required to initiate division of the egg, but in *Rudbeckia sullivantii* 50% of eggs develop parthenogenetically without sperm penetration. Semigamy is, therefore, questionably parthenogenesis.

FACULTATIVE APOMIXIS

There are two possible conditions of apomixis in a species or individual. If a whole individual or a whole species is always apomictic, it is described as an *obligate apomict*. If, on the other hand, some individuals of a species or some flowers on a plant that is basically apomictic are sexual, then it is said to be *facultatively apomictic*. Of course it is essentially impossible to characterize any species as an obligate apomict from any sort of sample because somewhere there might be individuals, populations, or subspecies that are sexual or facultatively apomictic. It is possible, however, to characterize a species as a facultative apomict. Examples of facultatively apomictic species are *Hieracium* of the subgenus *Pilosella* (Rosenberg, 1917), *Poa, Potentilla, Rubus, Citrus,* and others; examples of obligate apomicts are the American polyploid species of *Crepis* (Stebbins and Jenkins, 1939) and *Cooperia* (Coe, 1953).

APOSPORY AND APOGAMY

Among lower plants, such as mosses and ferns (Steil, 1951), the term apospory refers to the development of gametophytes (diploid) from sporophyte cells rather than from spores. In the same groups the terms apogamy or apogamety are used to connote the derivation of a sporophyte from a gametophyte cell rather than from a gamete. Such conditions have given rise to haploid sporophytes and/or diploid gametophytes as well as polyploids. Such a cycle can be continued experimentally. The existence of haploid sporophytes and diploid gametophytes has demonstrated that these stages of the plant life cycle are not dependent upon but are merely correlated with the chromosome conditions of diploidy and haploidy. Apogamy occurs in the ferns *Adiantum, Asplenium, Dryopteris, Polystichum, Pteris,* and other genera (Grant, 1971).

APOMEIOSIS

Since diplospory uses the normal plant life cycle but substitutes some sorts of modified meiotic divisions, called *apomeiosis*, for meiosis in the megasporocyte (the EMC), the characterization of the various types of apomeiotic division have received considerable attention. Such divisions range from almost meiotic to very typical mitotic. The most meiotic of such apomeiotic divisions is a restitution anaphase I, such as in *Erigeron karwinskianus* variety *mucronatus* (Battaglia 1950) where 80% of megasporocytes form restitution nuclei but 20% form reduced embryo sacs via normal meiosis. That is, a restitution nucleus is formed when movement of chromosomes is so erratic that chromosomes are spread all along the spindle at telophase I so that one nuclear envelope encloses all (the diploid number) of the chromosomes into one restitution nucleus. Such a genetically determined apomeiotic division would have pairing and crossing-over. More extreme apomeiotic divisions are asynapsis, followed by creation of a restitution nucleus or division of all the asynapsed chromosomes in a sort of mitotic division. And finally a very typical mitotic division may occur in the EMC. In this latter type, if there were megaspores produced, diplospory would result. But if the EMC itself acts like a megaspore (as is typical) so that the products of this mitotic division are in fact the two nuclei of the binucleate stage of embryo sac development (so that megaspores are really not formed), the process can be called generative apospory. But since many completely sexual species "skip" producing megaspores (only four haploid nuclei) in this same way, this term may be superfluous.

Apomeiosis in anthers and in ovules of the same apomictic plant or even flower does not always correspond. Meiosis in the anther may be quite normal in such obligate apomicts as *Crepis occidentalis* and *intermedia* (Stebbins and Jenkins, 1939) or *Potentilla argentea* (Muntzing, 1931), but with apomictic development of some sort in the ovules. *Poa nervosa* anthers either do not even form or, if they do, they have no sporogenous tissue or PMC's (Grun, 1955). In *Calamagrostis* anthers there is variable breakdown; in *C. chalybaea* and *purpurascens* no pollen is produced, *purpurea* rarely does so, *lapponica* does occasionally, and *canadensis* is variable among populations from full pollen formation to none at all (Nygren, 1946, 1954b).

POLYPLOIDY

Nearly all apomictic species are polyploid. From Nygren's (1954a) and Stebbins's (1950) lists of apomictic species the only diploids are *Citrus* subspecies, *Nothoscordum bivalve* ($2n = 16$), *Sarcococca pruniformis* ($2n = 28$), *Arabis holboellii* ($2n = 14, 21$), some species of *Eugenia*, some forms of *Hieracium umbellatum*, *Alnus rugosa* ($2n = 28$, $x = 14$), *Calycanthus fertilis* and *C. floridus* ($2n = 22$, $x = 11$), *Ranunculus auricomus* ($2n = 16, 32, 40, 48$), possibly *Spathiphyllum patinii* ($2n = 18$, $x = 9$), and possibly *Euphorbia dulcis* if the $2n = 12$ form rather than the $2n = 28$ form was studied. Recent additions to the list of apomictic species are also essentially all polyploid. As in *Crepis* (Stebbins and Jenkins 1939) and *Townsendia* (Beaman, 1957) the diploids are sexual, and the polyploids, including triploids often, are apomictic. In *Calamagrostis* (Nygren 1946, 1954b) and *Antennaria* (Stebbins, 1932) all species seem to be polyploid, but the apomictic species tend to be higher polyploids than the sexuals. In *Crepis* and *Townsendia* the polyploid apomicts are euploid whereas in *Calamagrostis* and *Poa* each apomictic species is characterized by many aneuploid chromosome numbers. But also in *Calamagrostis* and *Poa* some of the apomictic species are facultatively apomictic.

An interesting result of the polyploidy of apomicts is that they can produce viable "haploids" (called "polyhaploids") (DeWet, 1965, 1968). Because they have the genes for parthenogenesis, haploid eggs that are sometimes formed by the polyploids can produce polyhaploid progeny. And because the polyhaploids are still at the diploid level at least, they can have a full set of genes and be viable.

The polyploidy of apomicts may be either allopoidy, as in *Crepis*, or autoploidy, as in *Townsendia*. Although genes seem to be necessary for apomixis, there seems to be something about polyploidy per se that tends to bring out the apomixis as it also seems to bring out parthenogenesis in animals.

Polyploidy is not absolutely essential for apomictic reproduction, but diploidy is rarely correlated with any sort of agamospermy except adventitious embryony, the simplest type. Polyploidy seems to be necessary for polygenic inheritance but also for the balance "of oppositional, accumulative, subtracting and complementary effects" (Clausen, 1954) that characterize the inheritance of apomicts, especially the permanently facultative apomicts that have a balance or equilibrium between the asexual and the sexual forms of reproduction in the same genotype. Clausen further stated that "recombination in facultative apomicts takes place at the level of the genome or of the chromosome rather than at the level of the gene." Polyploidy is necessary to buffer such a form of recombination that requires toleration of wide hybridization. It is generally concluded that polyploidy is more essential to apomixis than is hybridization.

HYBRIDIZATION

A great many apomictic species are also partly sexual, which permits hybridization. The polyploid condition also permits wide hybridization because "polyploid plants possess highly buffered genotypes" (Clausen, 1961) that permit hybridization, and introgression. In genera like *Poa* the high polyploidy and apomixis permit the persistence of aneuploidy types. Clausen cited the results of planting seeds of *Poa ampla* ($2n = 63 \pm$, 9-ploid, $x = 7$). From 90 to 95% of seed are produced apomictically and give 63-chromosome progeny, but from 5 to 10% have chromosome numbers from $2n = 60$ to 147. These latter are produced sexually and are generally weak and are eliminated by the 63-chromosome apomicts in competition. They are viable, however, in spaced plantings. He further stated that "among *Poa* hybrids, the chromosome number frequently shifts as much as 21 chromosomes from one sexual generation to the next and it may jump either up or down along the scale. Sexual types have been known to lose as many as 36 chromosomes in one generation and become apomictic. Presumably it is the balance between the genes within the chromosomes, rather than the number of chromosomes, that determines whether or not a plant is able to organize embryos without fertilization." Natural hybrids in *Poa* are frequently found, and hybrids have often been produced artificially. In fact the *Poa pratensis* group of facultative apomicts is one grand "hybrid swarm" with chromosome numbers from 33 to about 150, with chromosomes or whole genomes from such species as *Poa alpina, scabrella, ampla,* etc. (Clausen, 1961).

Stebbins concluded that the apomictic and polyploid species of *Crepis* are all allopolyploids of a rather few endemic sexual diploids.

But Stebbins, Gustafsson, and Clausen all agree that hybridization is not necessary for apomixis and it often restores sexuality. Such hybridization-restored sexuality can be very important in the start of new lines that later reachieve apomictic reproduction by segregation. Clausen (1954) cited work of Rutishauser on *Potentilla*, that hybrids between two apomicts are sexual, but his own work in *Poa* usually produced a progeny of both apomicts and sexuals. In subsequent F_1 to F_4 generations lines are established that breed true for percentage of apomixis, but the lines vary from essentially sexual to as apomictic as the original parents. Segregation for apomixis (or sexuality) is very much like segregation for fertility or vigor, that is, polygenic.

AGAMIC COMPLEX

The term agamic complex was coined by Babcock and Stebbins (1938) to describe the taxonomically very difficult species of *Crepis*. Stebbins (1941) explained the agamic complex as follows: "The best way to analyze a species complex in which polyploidy and apomixis are prevalent is on the basis of its diploid sexual members. Among the apomictic forms an enormous number of 'micro-species' could be recognized, all forming a more or less continuous series of intergrading forms, but that not a single one of these possessed any new morphological characteristics. Every feature of the apomicts could be explained either as a result of the recombination of characteristics found in two or more of the sexual diploids, or as the direct result of an increase in the chromosome number. The diploids are distinct from each other. The difficulties which botanists have had in distinguishing between these species are due entirely to the presence of auto- and allopolyploid apomicts. Such a species complex is called an agamic complex." He then cited the following 12 genera containing such complexes: *Festuca, Poa, Potentilla, Rubus, Citrus, Wikstroemia, Erigeron, Antennaria, Chondrilla, Taraxacum, Crepis,* and *Hieracium.* Some others that can now be added are *Calamagrostis, Townsendia, Parthenium, Dichanthium, Bothriochloa, Crataegus, Youngia,* and *Ixeris.*

Within an agamic complex the usual concepts of species do not hold except for the diploid progenitor species. By allopolyploidy and/or autopolyploidy from these diploid species numerous polyploid apomictic populations have arisen as microspecies or agamospecies. Since each agamospecies of the complex has arisen separately, it is slightly different from every other agamospecies, and they form continuous spectra between and among the various diploid species, mostly within the usual variability of the species involved.

An example of a widespread agamic complex is the Old World *Bothriochloa-Dichanthium* complex (Harlan and DeWet, 1963). Each "genus" consists of a number of agamospecies such as *D. annulatum, D. aristatum,* and *D. caricosum* (DeWet 1965, 1968). Each of these agamospecies consists of a few populations of sexual 2n = 20 diploids, some sexual tetraploids, and many allotetraploids and autotetraploids. The apomicts range from obligate to facultative. The sexual diploids produce about 1% of unreduced male and female gametes that produce sexual autotetraploids, which, in turn, can produce hybrids when pollinated by apomictic tetraploids. Some autotetraploids occasionally produce haploids (polyhaploids) and, rarely, such haploids are vigorous and fertile. They are essentially like the sexual diploids found in nature but nothing new. They can, in turn, produce tetraploids again. This tetraploid-diploid-tetraploid cycle is interesting but seems to have no evolutionary significance (Stebbins, 1970) and does not represent an evolutionary reversal of polyploidy.

The agamic complex of European blackberries of the genus *Rubus* consists of "innumerable microspecies," or 691 microspecies grouped around 109 taxonomic species (Gustafsson, 1943, 1946-1947). Grant (1971) has recently discussed the agamic complex in detail.

GENETICS OF APOMIXIS

Numerous breeding programs of apomictic forms have been performed by a number of investigators in a number of genera. Such programs involve crosses between sexual and obligate or facultative apomicts of various chromosome numbers or between facultative and obligate apomicts. It should be recalled that apomicts usually produce pollen and many are pseudogamous, actually requiring pollination. The genetic results are, to say the least, vague. There is no clear case of single or even two-gene inheritance. The complex inheritance results from possible hybridity, polyploidy certainly, the multiple sequence of conditions that must occur to achieve apomixis, and, often, the occurence of both complete sexuality and complete apomixis in the same organism. Nevertheless, it is possible to consider two categories of genetic determination of apomixis.

Vivipary and adventitious embryony

These two forms of rather simple apomixis seem to be determined rather simply by a few genes, and they often occur in diploid forms. Vivipary, the conversion of a flower before it is formed into a little plant or bulbil, has been studied in the genus of onions, *Allium,* where it is common. In a single flower head, from few to all flowers may be converted to bulbils, so sexuality and apomixis can occur in the same plant. Levan (1937) made such a genetic study with *Allium pulchellum* (2n = 16), a purely sexual diploid form and a diploid sexual derivative of the triploid obligate apomict, *A. carinatum* (2n = 24). By constant removal of bulbils of a plant of *A. carinatum* some flowers were produced that formed a few seeds sexually. Among them were some diploid sexuals, diploid apomicts, and triploid apomicts, thus demonstrating the heterozygosity for a vivipary gene of the triploid species. One of the diploid flowering plants derived from *A. carinatum* was crossed to the sexual diploid *A. pulchellum* and 152 F_1 plants were grown. Of these, 66 were variously apomictic (viviparous) and 86 produced only flowers. The F_1 ratio indicated a single gene for vivipary, but the wide variability of expression of vivipary among the 66 apomictic plants indicated segregation for a number of modifying genes.

Adventitious embryony inheritance has been studied in *Citrus* where it is very common. This form of apomixis is unrelated to anthers and embryo sacs, the from zero to ten embryos merely forming directly from "stimulated" somatic cells anywhere in the ovule. It is conceivable that this simple condition could be controlled by one gene or a very few. At the same time the plant is quite regularly sexual with normal meiosis in anthers and ovules, reduced embryo sac, fertilization, and a sexually produced embryo. That is, many such species are normally sexual but also produce adventitious embryos as a superimposed rather than an alternative condition. There does seem to be genetic control of sexual reproduction by abortion at some stage (Desai, 1962) or inability of sexually produced embryos to compete with the perhaps heterotic apomictic embryos.

Various results have been achieved from different crosses (Frost, 1926, 1938a, 1938b). Some crosses give progeny that are all highly apomictic. Other crosses give purely sexual progeny with great segregation in the F_2 generation. In some cases adventitious embryony is lost, which would indicate a recessive nature of the few alleles that produce it.

The relatively few genes seemingly exercising genetic control of vivipary and adventitious embryony are not necessarily genes controlling sexual reproduction. In vivipary flower primordia, long before genes for sexual reproduction have a chance to act, are diverted into the formation of little plants or bulbils. In some onions the bulbil generation will grow and produce a third generation of bulbils while still on the original parent. In adventitious embryony also the genes for this form of apomixis are not mutations of sexual reproduction genes although some alleles may occur to inhibit more or less the sexual reproduction.

Diplospory and apospory

Only a few generalizations have come from attempts to work out the genetics of the more complicated forms of apomixis, diplospory and apospory. In these forms most of the genetic changes have been mutations of meiotic and other sexual reproductive genes such as occur in many regularly sexual species, such as parthenogenesis, asynapsis, pollen sterility, various forms of degeneration, and heterosis. For these forms to be complete in apomixis meiosis must not occur, and parthenogenesis must follow. But some aspects of sexual reproduction must be preserved in most species, such as pollen formation, pollen tube growth, fertilization of the polar nuclei, endosperm development, and often the premeiotic development of the megaspore mother cells.

It is also true that in most of these types of apomicts there is occasional failure to inhibit one or more of the usually inhibited sexual phenomena. For example, meiosis may occur without fertilization to produce haploid (really polyhaploid) progeny. Fertilization of unreduced eggs is common by reduced, unreduced, or aneuploid male gametes. Failure of meiosis varies considerably among individuals of diplosporous forms, from semiheterotypic, through pseudohomeotypic to mitotic divisions and often includes asynapsis and restitution nuclei. The final step is failure to form two "spores," the megasporocyte developing directly into the embryo sac in generative apospory.

A significant factor producing almost insurmountable difficulty in genetic analysis is polyploidy, which almost always accompanies or makes possible these forms of apomixis. Certainly polyploidy per se does not produce apomixis since there are many highly polyploid sexual species, and even within apomictic genera there are some sexual forms at higher levels of polyploidy than some apomictic forms.

Stebbins, Gustafsson, and others who have attempted genetic study of such apomicts have concluded that, in general, apomixis is recessive to sexuality and depends upon a delicate genic balance that is easily modified by selfing, crossing, or changes in chromosome number. Some examples that have provided evidence for this conclusion will be given.

Muntzing (1940) found a $2n = 36$ polyhaploid of a $2n = 72$ apomict of *Poa*. This polyhaploid was still at a polyploid level in the genus that has $x = 7$. Muntzing reported that it had regular meiosis (18 bivalents) and when selfed was sexual, forming both reduced and unreduced eggs. This seems to indicate that sexuality is dominant to apomixis.

In *Potentilla*, diploid sexual × diploid sexual gave all sexual progeny, 8n sexual × 4n apomictic (sexuals at a higher polyploid level than the apomict) gave mostly sexual progeny; but 4n sexual × either 4n or 6n apomicts gave mostly or all apomicts. But sexual diploid × apomictic diploid produced sexual progeny, indicating dominance of sexuality (Gustafsson, 1947).

Noack (1939) made reciprocal crosses between sexual diploid and apomictic tetraploid *Hypericum perforatum*. When the sexual diploid was the pistillate parent, all progeny were triploid. With the apomictic tetraploid as pistillate parent, triploid and pentaploid hybrids were formed, indicating that both reduced and unreduced eggs are formed and can be fertilized. All progeny were apomictic, but, since in triploids there are two sets of the apomictic genome to one of the sexual and the pentaploid hybrid has four apomictic genomes to one sexual, these results hardly prove that apomixis is dominant to sexuality.

Parthenium has been studied by a number of people, and Powers (1945) derived a genetic scheme to explain meiosis in this and at least some other apomicts. Some sexual diploid plants of the apomictic species *P. argentatum* produce polyploids that are apomictic, but some polyhaploids from polyploid apomicts are apomictic, and some polyploids are sexual. Thus, polyploidy per se cannot produce apomicts unless the diploid has the proper genes already. What the polyploidy does to bring out the apomixis is unknown. It is likely that doubling the chromosome number and genes produces a different genetic balance of some sort.

Powers proposed the following genetic scheme:

A, a are a pair of alleles for reduction versus failure of meiotic reduction.
B, b are a pair for fertilization versus failure of fertilization.
C, c a pair for nondevelopment without fertilization versus parthenogenesis.
AABBCC is the normal meiotic and sexual type.
aaBBCC produces unreduced embryo sacs and eggs, but eggs can be and must be fertilized to develop as triploids.
AAbbCC produces haploid embryo sacs and eggs, but these cannot be fertilized nor can they develop parthenogenetically.
AABBcc produces haploid eggs that are fertilized and develop; the parthenogenetic ability is hidden and not required.
aabbcc produces apomicts; they would be obligate apomicts.

Such a simple scheme cannot apply to species like *Poa* and other apomicts that have facultative apomixis in single plants. In such a plant sexual reproduction, *AABBCC*, and apomixis, *aabbcc*, would both have to be present, an obvious impossibility. Also, as Gustafsson states, in *Poa* hybrids, crosses of two apomictic species often produce sexuals that need fertilization, or self-pollination of a facultative apomict produces mostly sexual progeny. This scheme cannot apply to all apomictic species but might apply to some. Power's scheme also indicates that the origin of apomixis is in the diploids, which is hardly likely generally.

Genetic control of meiosis in anthers and in ovules of apomicts may be provided by different genes, or the same genes may effect different results in the two different environments. It is true that in the same plant the extent of meiotic irregularity may be very different, ranging from no difference to greater irregularity in anthers or greater irregularity in ovules.

It is now known that there are many apomictic "species" of angiosperms and that apomixis has evolved independently in each family or subfamily or genus. It is evident also that there are a rather small number of methods by which it can be achieved. It is for that reason that the same or almost the same scheme occurs in numerous genera of different families, such as adventitious embryony in the Rutaceae, but also in some other families.

In many species of apomicts the change from sexuality to apomixis seems to be evolving because in the same genus or even species a graded series of changes is found. In *Agropyron scabrum* (Hair, 1956) populations ranging from completely sexual to obligately apomictic were studied. The sexual have regular meiosis. Some facultatively apomictic populations have almost normal meiosis except that at telophase a restitution nucleus forms. This is called the *semiheterotypic* form of division. In other populations the apomeiotic division is more mitotic, the chromosomes through prophase in the megasporocyte appear meiotic, but by metaphase I they are like mitotic chromosomes and divide at anaphase like a mitosis. This is called the *pseudohomeotypic* form of apomeiotic division. And finally, in the obligate populations, the apomeiotic division is a typical mitotic division. All of these apomeiotic divisions produce a dyad of unreduced spores. The final step, to generative apospory, has not occurred or been found in *Agropyron scabrum*. Similar sequences of evolutionary stages from meiosis to or toward generative apospory occur, as in species of *Hieracium, Antennaria,* and *Calamagrostis*. Yet all of these stages are able to achieve the same result, at least some unreduced embryo sacs and eggs. Nevertheless, plants with the intermediate stages of apomeiosis are also transitional toward obligate apomixis, being facultatively apomictic. These transition forms are also upset by environmental conditions such as having meiosis in the EMC's under certain conditions whereas other conditions permit mitotic divisions, as in *Calamagrostis* (Nygren, 1946). However, the generally greater success of facultative apomicts over obligate apomicts in survival and evolution means that evolution as far as the obligate condition does not usually occur.

There are two conclusions about the evolutionary potential of apomicts: (1) that agamospermy is a "blind alley of evolution" (Darlington, 1939; Stebbins, 1950) or (2) that they may have as much evolutionary potential as most sexual species (Gustafsson, 1946-1947). Grant (1971) wisely refused to lump all apomicts into one category and then propound a single generalization about its evolutionary potential. If evolutionary potential of sexual species is based upon mutations, recombination, polyploidy, hybridization, vigor, fertility, and production of sibling species or microspecies acted upon by natural selection, then there seems to be no evolutionary inhibition for many facultative apomicts such as *Rubus, Poa, Calamagrostis, Taraxacum, Hieracium*. It is likely, though, that obligate apomicts may well be in an evolutionary blind alley as to giving rise to new genera (Stebbins, 1950), but the odds are in favor of the obligate polyploid apomictic "species" of American *Crepis* outlasting the American diploid sexual species of that genus.

CHAPTER **19**

Human cytogenetics

For more than three decades efforts have been directed toward the solution of the problem of the chromosomal picture in man but, owing to the very great difficulty in obtaining normal material properly fixed, the results have been contradictory to an unusual degree.

H. M. Evans and O. Swezy, 1929

Until the late 1950's only an occasional cytologist would attempt to determine the correct chromosome number for the species *Homo sapiens*. The field of human cytogenetics, and that of vertebrates in general, has developed only recently and depended for its origin on the development of a satisfactory technique. During the past decade the field has attracted many workers and has begun to relate cytology with human heredity and certain inherited abnormalities such as mongolism (for a short review see German, 1970).

One generalization already derived from the study is that only very trivial cytological deviations from the normal are to be expected in postnatal individuals. That is, a mongoloid idiot, whose condition is now generally called Down's syndrome, is considered a very abnormal human being but the cause is merely an extra chromosome of the very smallest size, and the only known monosomics (lacking one whole chromosome) lack only an X chromosome or one of the smallest of the autosome chromosomes (Al Aish et al., 1967). In other words, trisomy or monosomy of the large or medium-sized autosomes larger than No. 13 is not found in adults because, it is assumed, such an anomaly would be lethal before birth. The outstanding exceptions to this generalization are rare mosaics. Two other generalizations are that all individuals must have at least one X chromosome and that any organism having a Y chromosome (a few exceptions are known) is more or less male.

CHROMOSOME NUMBER IN MAN

The quotation at the head of this chapter was written in 1929. It could have been written and been equally correct as late as 1955. The correct normal diploid chromosome number in average man and woman was strongly indicated only in 1956 (Tjio and Levan; Ford and Hamerton) and definitely established by 1960 as evidenced by the Denver classification of that year. It took about 80 years after Flemming in 1882 made the first attempt to determine the characteristic number to correctly determine it.

There seem to have been three periods in this study. From 1882 to about 1920 most workers reported a low somatic chromosome number, mostly $2n = 24$ but ranging from $2n = 16$ to 36 and with extremes of $2n = 8$ and 73. The second period began about 1920 when $2n = 48$ became by far the commonest reported number (see for example Painter, 1923, 1924), although $2n = 47$ was occasionally claimed. The third period began in 1956 when Tjio and Levan, using new techniques, reported $2n = 46$ for the first time, except the expressed possibility by Painter (1923). That number was quickly established as the correct chromosome number of normal men and women (Ford et al., 1958; Tjio and Puck, 1958; Levan and Hsu, 1959; Chu and Giles, 1959), and the abnormal cytological correlation with mongolism was made (Lejeune et al., 1959; Böök et al., 1959).

The chromosome number of 2n = 47 occasionally reported in some men led some cytologists to propose an XO sex chromosome scheme, but the popularity of 2n = 48 in males and females seemed to fit the XY scheme. Subsequently the X and Y sex chromosomes were identified considerably more exactly than Painter (1923) or Hsu (1952) was able to do.

TECHNIQUES

The techniques that have permitted the rapid development of a field of human cytogenetics provide (1) many cells at metaphase having (2) the chromosomes well spread with few overlaps, (3) the techniques themselves not producing any artifacts, and (4) a simple, relatively painless method of getting test material from the person being studied. Although bone marrow can be examined at once (Tjio and Wang, 1962) because of the many normally dividing cells, it is not easy to get at. Therefore, most study is made of a sample of peripheral blood (Moorhead et al., 1960), which is easily obtained. This method removes the erythrocytes

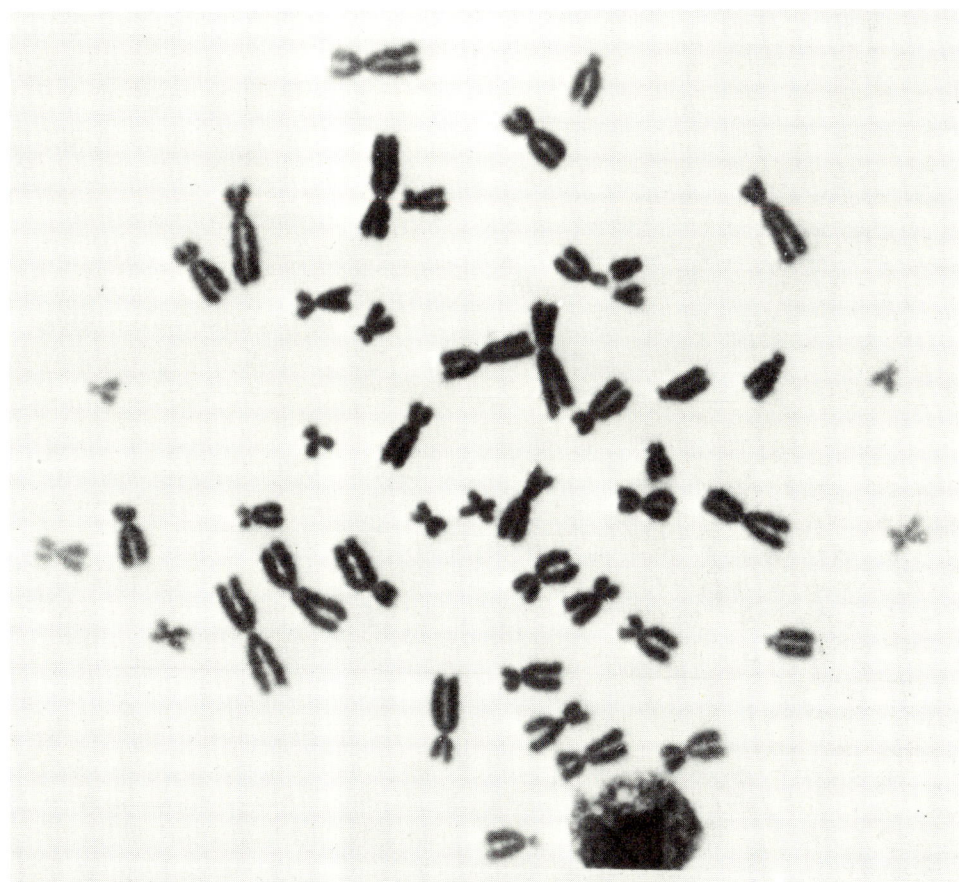

Fig. 19-1. A spread of human somatic metaphase chromosomes derived from a blood leukocyte by the standard technique. Each chromosome consists of two sister chromatids united at the centromeres, which appear as constrictions. The variation in total chromosome lengths and the relative lengths of arms of each chromosome are evident. (Courtesy Dr. M. S. Al-Aish.)

and requires a short-term in vitro culture. During the sterile, 3-day culture period the white bood cells in the plasma are exposed to phytohemagglutinin (Nowell, 1960), which very quickly converts the postmitotic G_1 nuclei to interphasic nuclei (Tokuyasu et al., 1968), stimulates phosphorylation and dephosphorylation of nuclear protein (Kleinsmith et al., 1966), and thereby stimulates RNA synthesis, protein synthesis, enzyme activity, and morphological change, called *transformation*. That is, the phytohemagglutinin treatment somehow sends many cells into division. After 3 days of culture, colchicine (or a compound of similar effect) is added to stop all mitotic divisions as they reach metaphase. As a result, during the from 3 to 5 hours in colchicine many dividing cells accumulate at metaphase. After various manipulations including centrifuging three or four times, the cells are given the very important hypotonic treatment that is necessary for good chromosome spreading, they are fixed, put on slides, dried, and stained. The procedure is long and involved and varies slightly among laboratories. For a detailed discussion of most techniques, see Priest (1969). The desired result is a slide with many well-spread metaphase figures such as Fig. 19-1. Examination of numerous chromosome groups by an experienced observer and analysis (often of photographs) with preparation of a karyotype is required to determine abnormalities. This same method can be used for cytogenetic study of other vertebrates. Technique is very important in this field because fine details of length, form, etc. must be observed.

HUMAN KARYOTYPE

As a result of many studies of human chromosomes by many people by these new techniques, the normal human, somatic, metaphase karyotype has been established internationally. The first such agreement was the 1960 Denver classification, which has been revised slightly at the London (1963) and the Chicago (1966) Conferences. Although eventually all 22 autosomes and the sex chromosomes may be identified (some experts believe they can do so now),

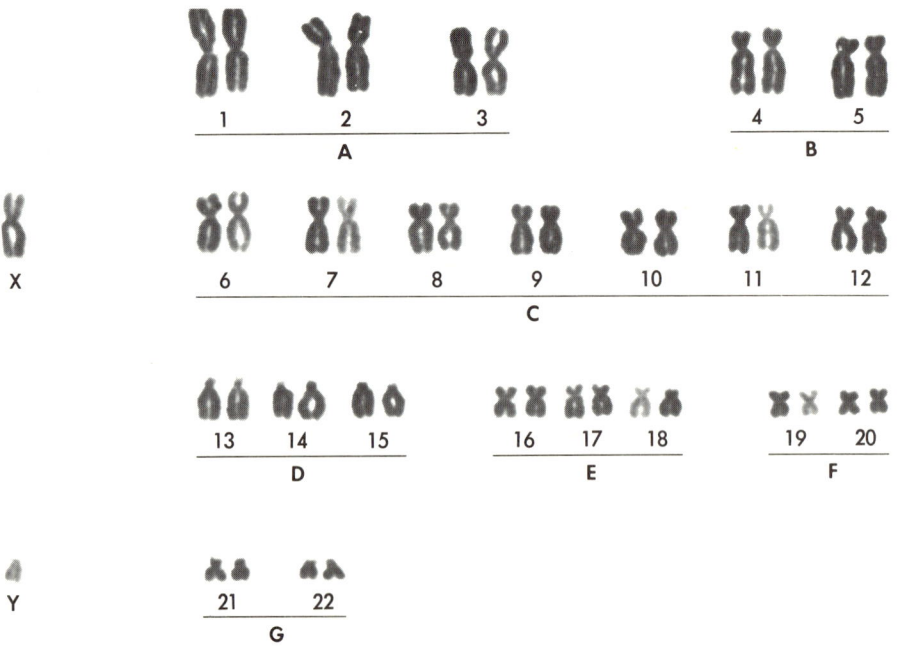

Fig. 19-2. Denver classification of the human karyotype. The 22 pairs of autosomes (1 through 22) are arranged in seven groups (A through G) according to total lengths. The chromosomes within each group are separable with more or less difficulty. (Courtesy Dr. M. S. Al-Aish.)

the standard, or Denver, classification arranges them according to decreasing total lengths (Fig. 19-2). Pairs 1, 2, 3, 16, 17 and 18, it is generally agreed, can be determined with certainty. Because so many pairs cannot, however, the whole set is arranged in groups of similar sizes and forms designated as A through G plus X and Y (Fig. 19-2). Chromosomes No. 1, No. 2, and No. 3, the A group, are the longest and are approximately metecentric; No. 4 and No. 5, the B group, are almost indistinguishable in total length, arm ratio (length of longer arm relative to the shorter arm), centromeric index (ratio of length of shorter arm to length of the whole chromosome), absence of satellite, etc.; the X chromosome, which is isocyclic in the male, is just shorter than the B-group chromosomes and slightly longer than the seven pairs of C-group chromosomes; the six acrocentric chromosomes of the D group have very short arms, all of which may bear satellites or at least satellite stalks (Al Aish, 1970) (Fig. 19-4); the three pairs of E-group chromosomes are shorter than those of the D group but have relatively and actually longer short arms; pairs 19 and 20 constitute the F group and are short metacentrics; and the two pairs of G-group chromosomes are nearly like the Y, which is often included in that group, and they are all very small and have tiny short arms. All G-group chromosomes except the Y may have satellites. The use of letter designations of groups is gradually being discontinued for numbers only, 1 through 22.

At metaphase each chromosome is double, consisting of two chromatids that are held together by or at the centromere. Apparently none of the three sorts of chromosomes (22 autosomes plus X plus X or Y) is actually telocentric, and it is only the shortest arms, the short arms of D- and G-group chromosomes, that have the satellites. The Y chromosome can usually be distinguished from the 21 and 22 pairs by somewhat greater length, less divergence of or closed long arms, poorly defined matrix, lack of satellites, and obliqueness of the terminal ends of the long arms. Furthermore, the length of the Y chromosome varies among males but seems to be the same in families (Cohen et al., 1966; Unnerus et al., 1967).

Secondary constrictions (Fig. 19-3), in contrast to primary constrictions, have been reported in the long arms of at least chromosomes No. 1, No. 9, No. 10, No. 16, and the long arm of the Y. Al Aish (1970) has shown differences on the short arms of the three D-group chromosomes, No. 13, No. 14, and No. 15 (Fig. 19-4). Chromosome No. 13 has two tandem satellites, No. 15 has one satellite, and No. 14 has only a satellite stalk. He also presented evidence that permitted him to distinguish among these chromosomes on general morphological characters.

Another technique for distinguishing otherwise indistinguishable chromosomes is autoradiography (Schmid, 1963; German, 1967; Gianelli and Howlett, 1966, 1967). It is known that heterochromatin replicates its DNA later than euchromatin during the synthetic period of interphase (Takagi and Sandberg, 1968). It is also evident that different chromosomes have the heterochromatic segments differently distributed along their arms; that is, some chromosomes may have a block of heterochromatin

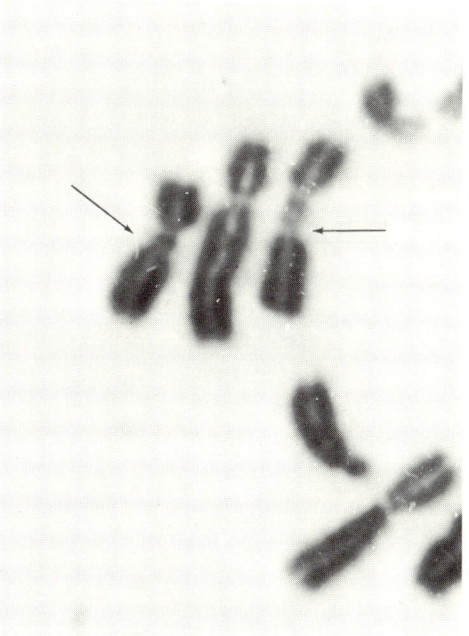

Fig. 19-3. Secondary constriction in the long arms of the two chromosomes No. 9. Between them lies a B group chromosome, probably chromosome No. 5. (Courtesy Dr. M. S. Al-Aish.)

256 TEXTBOOK OF CYTOGENETICS

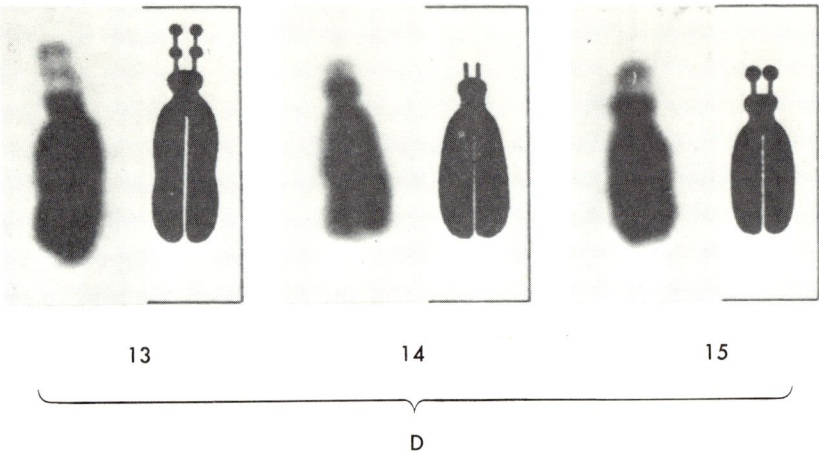

Fig. 19-4. High magnification micrographs taken by differential interference technique and interpretive drawing of the D group, satellite-bearing chromosomes No. 13, No. 14, and No. 15. The differences in satellites of the three chromosomes are evident. (From Al-Aish, M. S. 1970. Canad. J. Genet. Cytol. **12**:160-163.)

at or near the end of the long arm whereas a morphologically similar chromosome may have the whole short arm and the region of the long arm near the centromere (the proximal region) heterochromatic. Therefore, if the chromosomes are given radioactive thymidine during the early S period only, or only during the late S period, otherwise indistinguishable chromosomes may be distinguished by where along a chromosome the radioactivity is evident and where it is not (Fig. 19-5). Other newer techniques that reveal the distribution of heterochromatin in chromosomes are discussed later.

There is some variation of "normal" karyotypes that is obviously due to technique. But there seem to be variations of the normal that are real. That is to be expected since in biology there never is an absolute, only a statistical "normal." Some of this variation may actually represent abnormal conditions, but having effects within the normal variation of human phenotypes. For example, the short arms of G-group chromosomes vary considerably, although complete deletion is not found (Court Brown et al., 1965). Ten chromosomes, the D and G groups, have satellites on their short arms (Ferguson-Smith and Handmaker, 1961); but almost never can satellites be seen on all 10 in one cell. Often, too, so-called enlarged satellites are seen on some of these chromosomes and, although seemingly inherited (Cooper and Hirschhorn, 1962), do not seem to be related to abnormal phenotypes.

Apparently, normal variation in chromosome length, other than artifact, is common. This is especially true of the sex chromosomes, both the Y (Makino et al., 1963; Cohen et al., 1966; Unnerus et al., 1967) and the X (Bishop et al., 1965). Some variation in chromosome length may be normal but indistinguishable from artifact at this time. Structural anomalies of the Y chromosome are numerous and distinctive, such as isochromosomes, dicentrics, deletions, pericentric inversions, and translocations (Jacobs, 1969, for review). The long arm may become satellited by translocation (Sinha and Bejar, 1968; Schmid, 1969). Attempts have been made to relate these to clinically abnormal phenotypes, but the genetic and cytological variability of human beings makes such conflicting claims unreliable. However, abnormally long Y's are heritable (Nuzzo et al., 1966; Schmid, 1969).

Although the standard Denver classification of human chromosomes is generally followed, nevertheless new proposals are occasionally put forward with detailed justifications (Levan et al., 1964; Al Aish, 1969). Following excellent techniques and independent exact measurements of nonoverlapping metaphase plates from eight individuals, both males and fe-

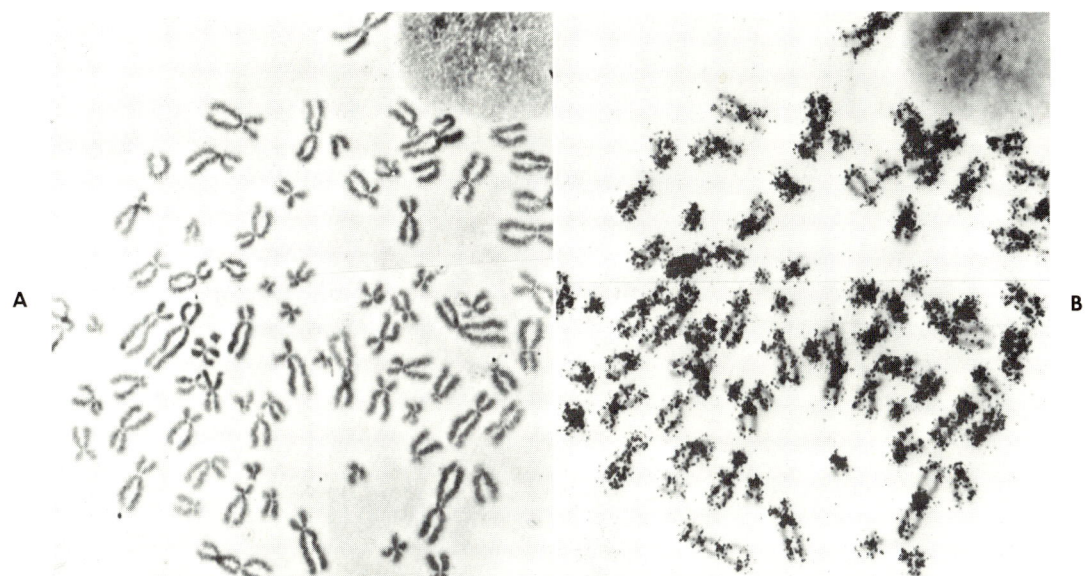

Fig. 19-5. Autoradiogram of human chromosomes treated with tritiated thymidine late in the S period so that late labeling regions of constitutive heterochromatin are labeled. (Courtesy Dr. A. Zweidler.)

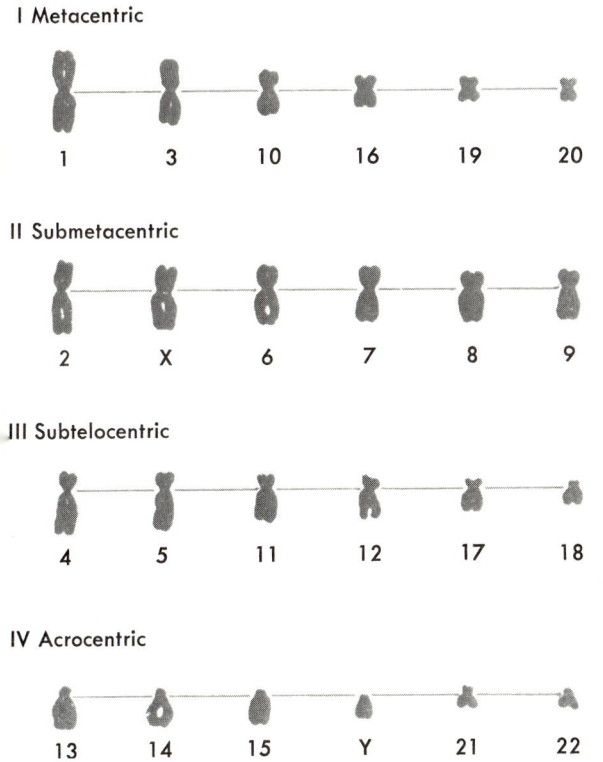

males, by two competent investigators, the data were compiled and analyzed by Al Aish. The results are summarized in Table 5. He then proposed a tentative grouping according to centromere location. Four groups result (Fig. 19-6): I, metacentrics (1, 3, 10, 16, 19, and 20); II, submetacentrics (2, X, 6, 7, 8, and 9); III, subtelocentrics (4, 5, 11, 12, 17, and 18); and IV, acrocentrics (13, 14, 15, Y, 21, and 22). Within groups the chromosomes are arranged according to decreasing total lengths. He then proposed a modified karyotype consisting of isolated chromosomes and groups of two or three: 1-2-3, 4-5, X-6, 7-8-9, 10, 11-12, 13-14-15, 16, 17-18, 19-20, Y, 21-22 (Fig. 19-7). This grouping indicates that cytogenetics of human mitotic chromosomes are approaching the original hope of being able to identify with confidence each and every chromosome of the set.

Fig. 19-6. The four basic forms of human chromosomes: metacentric, submetacentric, subtelocentric, and acrocentric. Even among the chromosomes of one group there is relative variation of arm lengths, and such groups must be characterized arbitrarily. (From Al-Aish, M. S. 1969. Canad. J. Genet. Cytol. 11:370-381.)

Table 5. Data on normal human chromosomes derived from measurements of short and long arms*

No.	AR†	CI‡	No.	AR	CI
3	1.06	48.3	9	1.79	35.9
1	1.07	48.3	11	1.86	34.9
20	1.15	46.4	12	1.91	34.4
19	1.20	45.4	18	2.02	33.2
16	1.33	43.0	4	2.38	29.6
10	1.33	42.8	5	2.43	29.2
X	1.45	40.7	21	2.77	26.5
2	1.53	39.5	22	3.19	23.9
7	1.55	39.1	Y	3.67	21.3
6	1.59	38.6	13	5.00	16.7
8	1.61	38.3	15	5.65	15.0
17	1.75	36.3	14	5.93	16.4

*The chromosomes are arranged from most metacentric to most acrocentric.
†AR, the arm ratio, the ratio of the length of the long arm over the length of the short arm. An exactly metacentric chromosome has an AR of 1.00 whereas a truly telocentric chromosome would have an AR of infinity.
‡CI the centromeric index, the ratio of the length of the short arm over the length of the whole chromosome. An exactly metacentric chromosome has a CI of 50 whereas a truly telocentric has a CI of zero.
Data from Al-Aish, M. 1969. Canad. J. Genet. Cytol. **11**:370-381 with permission.

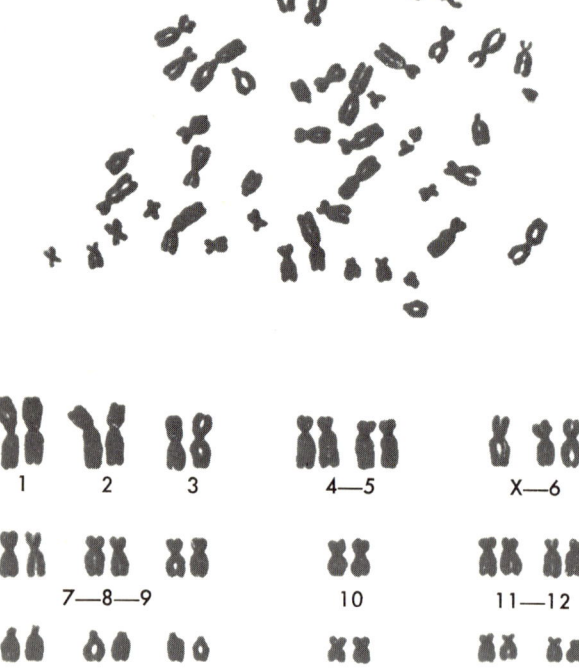

Fig. 19-7. Normal human chromosomes. Top, A spread of chromosomes of a normal person. Bottom, An arrangement of the karyotype somewhat different from the usual Denver classification. The numbers of the chromosomes are the same, but letter grouping (A through G) is discontinued. Instead of the six lettered groups there are twelve recognizable groups based upon relative total length, arm ratio, and centromeric index of each chromosome. (From Al-Aish, M. S. 1969. Canad. J. Genet. Cytol. **11**: 370-381.)

MEIOSIS

Meiosis in human beings is the same as in other mammals (Chen and Falek, 1969). Painter (1923, 1924) and Evans and Swezy (1929) gave early, detailed accounts of spermatogonial and meiotic divisions in the male and von Winiwarter (1901) in the female foetus of from 5 to 7 months. It is known (Ohno et al., 1962) that in the ovary the oogonia cease division by the fifth foetal month and the primary oocytes have completed early prophase I (through diplotene) before birth. For example, pachytene and other meiotic phases occur in the 5½-month female foetus (Ohno, 1965). The ova remain arrested, "frozen" in the unusual diplotene, called the *dictyotene*, for from about 12 to 50 years. As an ovum matures, alone or with one or a few others, during a particular menstrual cycle it resumes meiosis, reaching metaphase I on the day of ovulation. Only if fertilized does it complete the second meiotic division; if unfertilized it reaches metaphase II only. The superficial appearance of dictyotene chromosomes is much like interphase (Ohno et al., 1962).

In the tubules of the testes the spermatogonia divide a few times apparently in "spiral waves running the length of the tubule." Both Painter (1923) and Evans and Swezy (1929) reported three types of dividing cells among the Sertoli, or nurse, cells in the tubules: (1) typical spermatogonial divisions, (2) cells with balled or rounded chromosomes, and (3) giant cells with the tetraploid number of chromosomes, the last being infrequent. The second type of cells occurs in groups of six or seven cells. Among typical spermatogonia prophases are much less frequent than in somatic tissues. Evans and Swezy reported chromosome lengths in spermatogonial metaphases as ranging from 6.2μ to 0.78μ, the smallest for the tiny Y chromosome. These lengths are considerably shorter than lengths determined by recent techniques of from 2 to 10μ (Osgood et al., 1964). Oddly, the earlier researchers figured no other chromosomes comparable in size to the Y, and the next smallest were three to four times the length of the Y. That is, they reported that the Y is very distinct from the probable G-group chromosomes. The same was true of somatic cells of various tissues. The consistent counts of $2n = 48$ and $n = 24$ by Painter, Evans and Swezy, and others such as Hsu (1952) rather than the $2n = 46$ by recent techniques is interesting. Chen and Falek (1969) were able to see centromeres at late diplotene in spermatocytes and to recognize the various groups of the human karyotype.

During the interphase preceding meiosis there is a noticeable difference from all other interphases: the X and Y appear already paired and remain condensed as pycnotic, pear-shaped heterochromosomes. They remain in that condition until metaphase I without ever becoming euchromatic, according to Evans and Swezy, Painter, and others.

Subsequent meiotic events in spermatogenesis are typical of mammalian meiosis in general. At pachytene the chromosomes form a bouquet configuration, the ends against the nuclear envelope opposite to the centrosome that lies just outside the nuclear envelope. The leptotene is reported to consist of single chromosomes that pair during zygotene (or synaptitene) followed by the development of the synizetic knot. Following pachytene there is a diffuse stage during which the chromosomes lose much of their staining capacity, are dotted with minute granules, become shorter and broader, and have jagged, diffuse outlines. That stage may be comparable to diplotene of the standard scheme of meiosis and centromeres may be seen (Chen and Falek, 1969). According to Evans and Swezy in describing this stage and diakinesis, "The spiral twisting of the chromosomes is, in most of the larger ones, rather elaborate and affords abundant opportunity for 'crossing over'—the well-known genetic hypothesis." Recent counts of chiasmata indicate about 54 in the 22 autosomal pairs. During metaphase I the longest chromosomes still show a number of apparent chiasmata, and only the three very smallest seem to have only one (Ohno, 1965). The sex bivalent is different, however, in that the X and Y are in a drawn-out and end-to-end arrangement.

Recently the electron microscope has added the knowledge and hypothesis of the synaptinemal complex to our knowledge of meiosis (see Moses, 1968, for a recent review). The presence of the synaptinemal complex in man was established the first year of its discovery and conception (Fawcett, 1956) and has been reported and studied in detail subsequently (Woollam and Ford, 1964; Woollam et al., 1966; Solari and Tres, 1967; Comings and Okada, 1971a).

It has now been established that the synaptinemal complex (see Chapter 6) is formed only during early prophase of the first meiotic division in probably all sexually reproducing (and, therefore, meiotic) organisms. It is a specially formed structure, probably necessary for ge-

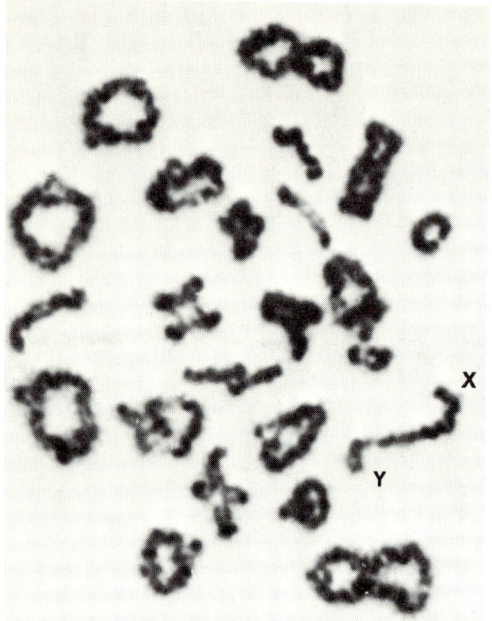

Fig. 19-8. Meiotic chromosomes of human male at pachytene. The XY bivalent is marked. Large chromosomes have two or three chiasmata whereas small chromosomes have one or two. The tiny Y is attached typically end to end with the much larger X. (Courtesy Dr. J. Melnyk, City of Hope Medical Center, Duarte, Calif.)

netic crossing-over since it is rarely present in achiasmatic species or sexes, such as male *Drosophila*. There must be genes for its formation, and mutations of one or more of such genes can probably produce asynaptic forms, strains, sexes, and species.

The details of the synaptinemal complex cannot be discussed here, but enough studies of it in spermatogenesis of man have been made to demonstrate that it is typical in man (see Chapter 6). It is a flat structure, largely of protein, formed between the two synapsed homologues, and contains short segments of the DNA strands of the homologues. It seems as though it is a structure that puts a large number (but small percentage) of homologous genes very close together at the center of the structure or elsewhere (Comings and Okada, 1971a) so that crossing-over can occur. It, like the parallel paired chromosomes between which it lies, makes contact at the ends only

with the nuclear envelope (at telomeres) where the chromosomes, synaptinemal complexes, and the associated portions of the nuclear envelope are all modified.

The sex bivalent or sex vesicle, in man and mouse at least, is somewhat different (Solari and Tres, 1967). Although the X and Y chromosomes at pachytene are paired in short regions, especially where they contact the nuclear envelope, no synaptinemal complex has been seen; they are paired but not synapsed.

One current concept of meiotic recombination proposes that already paired homologues (Brown and Stack, 1968) with ends attached to the nuclear envelope form the synaptinemal complex between them which brings hundreds or thousands of homologous genes into juxtaposition but only a few actual crossovers are formed randomly by some unknown mechanism.

SEX CHROMATIN BODY

In 1949 Barr and Bertram discovered a sex difference in the neurons of the cat. Very rapidly thereafter the same difference was found in various other mammals and man, and in other types of cells including the cells of the inside of the cheek and in certain white blood cells (Davidson and Smith, 1954). This difference occurs in interphase nuclei of some cells of the normal female where there is a mass of stainable chromatin against the nuclear envelope called the sex chromatin body or Barr body (Fig. 19-9, *A*). In mature polymorphonuclear leukocytes some nuclei (about 5%) have a thin extension of the nucleus as a stalk with a swollen end, called the "drumstick" (Fig. 19-9, *B*). Actually the swollen end of the drumstick is a sex chromatin body. It is believed that most cells of the female body have a Barr body at least at some time but all at any particular time do not show it. In the late postmitotic epithelial cells from the inside of the cheek one sex chromosome body is seen in up to 35% or even 60% of cells of a normal female. That is, it does not usually occur in more than 50% of such cells and is even less common in cells of some other tissues, although Schmid (1967) reported as high as 65 to 75% in hair root cells.

It has now been established that, in general,

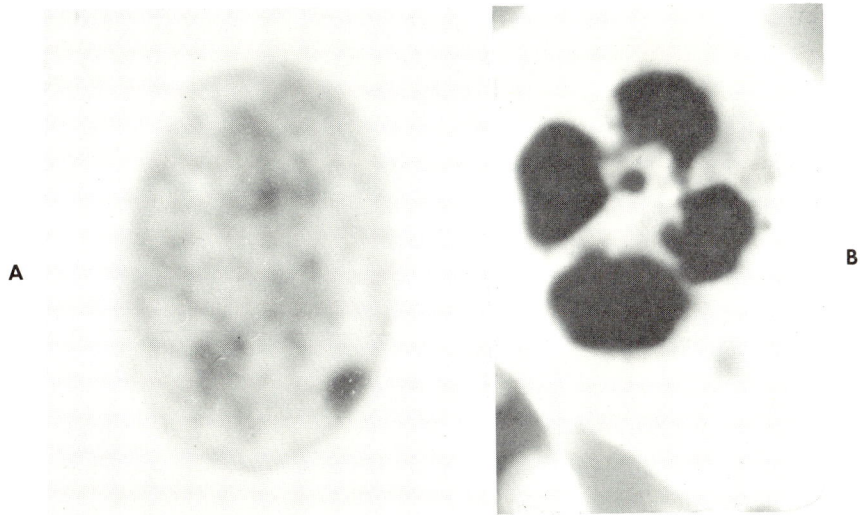

Fig. 19-9. Two forms of the sex chromatin body in two different types of cells. **A,** An example of a Barr body in a cell from the inner surface of the cheek. **B,** A white blood cell with a 4-lobed nucleus (black) and a "drumstick" extending into the center from one of the lobes. Both are from a normal female. (Courtesy Dr. M. S. Al-Aish.)

human nondividing nuclei tend to have a maximum of one less sex chromatin body than they have of X chromosomes. A normal female (XX) has one. An abnormal person, "male" or "female," having three X's will have some cells with a maximum of two Barr bodies. A female with Turner's syndrome (XO) has none, etc.

The sex chromatin body is a mass of heterochromatin composing most of one X chromosome. Thus, in a female interphase cell one X is present as euchromatin and so is invisible, the other is sometimes condensed as heterochromatin and so is visible as a Barr body. Furthermore, the X chromosome that enters a mitotic division from a Barr body will have its daughter chromosomes after telophase also form Barr bodies, and so on indefinitely. A post-telophase Barr body X chromosome will replicate its DNA late in the S period along with other late-replicating heterochromatin. Since the heterochromatic Barr body is largely euchromatin made temporarily heterochromatic, it is called *facultative heterochromatin*.

Apparently during early stages of mammalian embryo (sixteenth week in human beings) there are no Barr bodies, but eventually in some or most cells deactivation occurs so that a Barr body will form (Hill and Yunis, 1967) from *either* X chromosome, probably at random (Lyon, 1961). Thus, different cell lines and tissues will differ as to which X (maternal or paternal) will form the Barr body, and so each normal female is a mosaic for her functional X's. For example, Sinha and Nora (1969) found a 13-year-old girl who, in addition to being an XO/XX mosaic, had an X/X translocation so that one X was much larger than the other. In buccal smears two sex chromatin bodies were seen, one average in size, the other larger. Since it is likely that heterochromatin is genetically inactive, each cell will be hemizygous, a sort of genetic haploid or monosomic for those genes and alleles contained within the heterochromatin of its Barr body, but the whole multicellular organism will be a genetic mosaic having one X functioning in some tissues but the other X functioning in the remaining tissues. Nuclei having two X's but no Barr body may have genes of both X's functional.

Geneticists assume that this removal of one X in each cell from transcriptional activity is an example of so-called dosage compensation. That is, in any female cell only one X is to function, and so one must be removed from activity. Thus, cell for cell, the male and female are equivalent in being genetically monosomic for X chromosome activity. These ideas constitute the Lyon hypothesis, and the heterochromatic X is sometimes referred to as being Lyonized, after Lyon (1961; see also Russell, 1964). Somehow this hypothesis does not seem to apply to all postmitotic cells, and the genetics of sex-linked genes in heterozygous females seems no different from that of autosomal genes.

Material for study of whether sex chromatin bodies are present or not and, if present, how many is acquired merely by drawing a little blood or gently scraping some loose cells (gunk) off the inside of the cheeks of the mouth. Thus anybody can easily be checked as "female" or "male": athletes, inmates of penal institutions or institutions for the feebleminded, or other individuals suspected of sex chromosome abnormalities as a first screening for cytogenic study. Moore (1966) contains twenty-six articles by various authors on many aspects of sex chromatin.

CYTOGENETICS OF ABNORMALITIES

Cytological abnormalities are frequent, about 1% of births (Court Brown, 1967; Ferguson-Smith, 1967) or 0.5% (Sergovich et al., 1969; Ratcliffe et al., 1970; Lubs and Ruddle, 1970), and are often related to abnormal phenotypes, to congenital defects, many of which are called syndromes. Such aberrations are almost always cytologically minor but produce almost always at least some mental retardation and various other physical abnormalities. Such cytological aberrations can be profitably subdivided into those of sex chromosomes (heterosomes) and those of autosomes. Of course there are individuals showing both in the same cell or in cells of different tissues, as in mosaics. The probable reason for so many more individuals having sex chromosomal anomalies than autosomal is that the X and Y chromosomes carry rather few of the more important genes, so that extra chromosomes do not create too abnormal a phenotype.

It is generally believed that a large proportion of all chromosomal aberrations in man are produced in the oocytes, although mosaics must result from abnormal chromosomal distributions at anaphase of early foetuses. Aneuploidy probably always results from such abnormal anaphase distributions at meiosis or mitosis, called nondisjunction, when two chromosomes stick together and go to the same pole.

In large surveys of consecutive births it is generally found that chromosomal aberrations occur in from 0.5 to 1.0% (1.5% from mothers more than 34 years of age) and that sex chromosome and autosome aberrations are equally common (Lubs and Ruddle, 1970).

Sex chromosome abnormalities

All viable individuals must have at least one X chromosome and usually at least another X or Y in addition to the 22 pairs of autosomes. Nevertheless, occasionally individuals with only an X are found, the XO genotype. XO's are often found in some tissues of an individual associated with other tissues which have at least two sex chromosomes, in one sort of so-called mosaic. Almost always a Y chromosome determines a generally male phenotype, although the acutal male-determining factors constitute a small fraction of sex chromosome anomalies. Nuclear sexing of buccal mucosa cells of newborn children reveals about 1.5 individuals per thousand (that is, per mil, or %o), of whom most are chromatin-positive males (van den Berghe, 1970).

Klinefelter's syndrome. Klinefelter's syndrome is produced by a variety of exceptional sex chromosome numbers. They all have at least two X's and one Y and include such observed combinations as XXY, XXXY, XXXXY, XXYY, XXXYY, and various mosaics such as XXY/XX, XXY/XY, XXY/XXXY, and XXXY/XXXXY. Since all of these individuals have one or more chromosomes more than the usual diploid number of 46, they are described cytologically as aneuploid or as chromatin-positive males. Klienfelter's syndrome is a term used to specify the condition of a group of phenotypic males. They must be and are male because each has at least one Y chromosome. They are often sterile, having small testes and various other abnormal characters (stigmata) in common which constitute the syndrome. About 1% of males in institutions for the men-

Table 6. Commonest human chromosomal aberrations

Chromosomes	Sex	Result
44A* + X†	Female	Turner's syndrome, 1:3,500 of females
44A + XXY†	Male	Klinefelter's syndrome, 1:500 of males and 1% of males in institutions
44A + XYY	Male	XYY syndrome, 1:850 of males
44A + XXXY	Male	Rare
44A + XXX,XXXX	Female	Multiple X; XXX 1:850 of females
44A + 21‡	Either	Down's syndrome, 21 trisomy, 1:600 births
44A + 21 + XXY	Male	Mongolism and Klinefelter's syndrome
43A (−21)	Either	21 monosomic, very rare
44A + 18	70% female	E or 18 trisomy, rarely viable, 1:2,200
44A + 13-15	60% female	D or 13-15 trisomy, uncommon, 1:4,300
B (No. 5) deletion	Either	Cri du chat syndrome, lethal
G-group deletion	Either	Ph¹ chromosome, a form of leukemia
Translocations	Either	Such as 21/21, Down's syndrome
Inversions	Either	A few tentatively reported

*A, autosomes or nonsex chromosomes
†X and Y, sex chromosomes
‡21, 18, 13, etc., particular autosomes

tally subnormal are sex chromosome-positive, having one or more Barr bodies (Close et al., 1968). Klinefelter's syndrome is fairly common in the human population, occurring in about one in 500 males.

XYY condition. The XYY condition (Lubs and Ruddle, 1970), sometimes called XYY syndrome, occurs in about one in 550 male births (Lubs and Ruddle, 1970) and has been correlated two or three times with tall, aggressive males who are often enough criminals to be quite commonly found among inmates of prisons (Telfer et al., 1968; Close et al., 1968). But studies by Kessler and Moos (1969) and numerous others have failed to detect any such correlation; in fact, such males may be less aggressive than normal! It has been argued that XY individuals (normal males) are taller and more aggressive than XX individuals, normal females; therefore, by dosage effect it would be expected that XYY males would be taller and more aggressive than XY males. In most XYY individuals intelligence ranges from normal to somewhat deficient. Such men are fertile but do not produce XYY sons. In fact, meiosis in XYY males shows only the XY condition (Melnyk et al., 1969) as though "selection toward chromosomally normal spermatocytes occurs before meiosis in XYY males" (Thompson et al., 1967).

Turner's syndrome. Turner's syndrome applies to a type of female having only one X (XO) or a more or less equivalent condition in which an abnormal second X is present. Examples of such abnormal second X's are these: It is an isochromosome, it is an X with part deleted (deleted X), or it is a small, abnormal X. Of course, some such individuals are mosaics such as XO/XX, XO/XXX, XO/XY, etc. (McKusick, 1968).

Turner's syndrome, or gonadal dysgenesis, occurs about once in 3,500 "female" births. This "low" figure results probably from most or many XO conceptions ending in spontaneous abortion. About 6% of spontaneous abortions are found to be of the XO type. It has now been established that about 12% of conceptions end in spontaneous abortions, and about 20% of these are found to possess detectable chromosomal aberrations such as XO, 6%; trisomy for a D chromosome, 3%; trisomy for an E chromosome, No. 18 usually, 3%; trisomy for a G chromosome, or Down's syndrome, 2%; triploidy, 4%. D-trisomy infants live for a few weeks only, 18-trisomy infants survive a few months to a few years, and triploidy is almost never found in live-born human beings.

At least some persons with Turner's syndrome are lacking for the Y or X from the father (the X present has an allele present only in the mother such as color blindness), in which case the X or Y either was not present in the sperm or was lost very early in embryogenesis. Similar tests have also shown that in other individuals the missing X is the maternal one. The phenotype of Turner's syndrome ranges from male to female.

Multiple X females. Females having three, four, or five X chromosomes have been found by cytogeneticists. Females with three X's occur at the rate of about one in 850 females. The effect is quite mild, however, only sometimes, apparently, producing sterility. Their progeny are chromosomally normal. They can be detected by having two or more sex chromatin bodies in interphase cells. Individuals with XXXX are very defective mentally.

XY females. Although it is generally true that any individual possessing a Y chromosome is male, XY females do occur. By definition, any Y-containing individual is male; but there are two known types of individuals that are superficially XY females. The first group of that sort is said to have the "testicular feminization syndrome" or TFS (Boczkowski, 1968). Such individuals are legally and rather typically females. "The habitus is completely and often attractively feminine with well-developed breasts" (Boczkowski). On the other hand, they have no ovaries, fallopian tubes, or uterus, and the vagina is short. Histologically they are male since they have testes, but these are more or less undescended and produce no sperm. And, of course, they are cytologically male. Of 50 cases examined by Boczkowski 45 were XY, but 4 were various sorts of mosaics: XO/XY/XX; XY/XYY/XXY; 46/47 with XXY; and XO/XY/XXY; and one was possibly XXXY. This condition is evidently determined by a maternally transmitted gene that blocks male sex differentiation subsequent to the early critical differentiation of testes from primordial tissue (Court Brown, 1967). At present there is no evidence that this syndrome is the result of chromosomal aberration, nor have genetic linkage studies related it to any known gene or chromosome, although Boczkowski considered the syndrome to be sex-modified rather than male-limited. A second group of XY females occurs within those having pure gonadal dysgenesis, many of whom are XX. This form of dysgenesis produces individuals having no gonads at all, often not even rudiments but sometimes as slightly evident as in Turner's (XO) syndrome. This XY female condition too is probably genetically determined. They are apparent females, evidently because *any* individual will develop into a female unless made male by the presence of embryonic testes (Court Brown, 1967).

XX males. XX males are different in testes histology from individuals with Klinefelter's syndrome and are probably not XX/XXY mosaics. It has been proposed that they started out as XXY but after adequate testes development to establish maleness the Y was lost, or that the male-determining genes of the Y are present on other chromosomes (Court Brown, 1967).

Intersexes. Intersexes and so-called hermaphrodites, having phenotypes characterized by varying development of both male and female reproductive organs, occur in human population. Many or most of these show no detectable sex chromosome anomalies, but some are mosaics, such as XX/XY. They are determined genetically rather than cytologically.

Autosomal abnormalities

Abnormalities of autosomal chromosomes seem to be more severe than most anomalies of sex chromosomes except, perhaps, the multiple X condition. Most autosomal aberrations are trisomics but translocations, monosomics, inversions, and deletions are known among postnatal human beings at the present cytogenetic level of detectability. Autosomal aberrations of all kinds are cytologically trivial in postnatal people, involving trisomy of the small (D, E, and G groups) chromosomes or small deletions, duplications, or translocations. Green et al., (1968) stated that at least some chromosomal aberrations are not random and that gene mutations affecting DNA replication of particular chromosomes, which are dominant but heterozygous, can explain some reported cases. More severe aberrations, such as triploidy, tetraploidy, trisomy, and monosomy, are found in spontaneously aborted embryos and foetuses. Court Brown (1967) cites findings of two out of three very young abortions and one out of five first trimester abortions as having such severe aberrations. Less severe aberrations occur in from one in 100 to one in 250 live births.

Trisomy. Trisomy for large autosomes creates genetic imbalance involving many more genes than trisomy for small autosomes. Since trisomics for small chromosomes are semilethal, trisomy for large autosomes is lethal, usually before birth.

The commonest of the autosomal aberrations is trisomy of chromosome No. 21 or No. 22 (No. 21 by convention or definitely proved), which produces the condition known as mongolism, G trisomy, or Down's syndrome. The same condition can be produced by a translocation and is discussed later. Mongolism occurs once in about every 600 births. The frequency is directly correlated to maternal age, rising from one in 1,000 births between ages 20 and 30, to about 4 per thousand at age 37, to about 10 to 15 per thousand at age 42, to a maximum of about 15 to 18 at about age 45 (Turpin and Lejeune, 1965; Collman and Stoller, 1962; Wahrman and Fried, 1970). Trisomy of a small acrocentric autosomal chromosome in a chimpanzee produced conditions very similar to those of human Down's syndrome (McClure et al., 1969).

This is the best authenticated correlation between a chromosomal aberration and phenotypic (syndrome) effect. The phenotype is common, familiar, and obvious. It includes very severe mental retardation, heart and other malformations, peculiar eyelids, short stature, etc.

Although Down's syndrome is well correlated with maternal age, a number of workers (Mella and Lang, 1967; Fialkow, 1967) suggest that causes of aberrations such as viral infection and immunological differences between mother and foetus may at least predispose the child to Down's syndrome.

More severe than 21 trisomy, perhaps because it involves a larger chromosome, is E or 18 trisomy (E syndrome), which also has a recognized syndrome of effects. It occurs in one of 2,200 births. Survival beyond early infancy is uncommon, but those who do are mentally retarded and physically defective and unusual. This syndrome is sometimes found without the individual having any chromosomal aberration (Ferguson-Smith, 1967).

D-group trisomy (13-15 trisomy or D syndrome) is less commonly found (one of 4,300 births) and causes a syndrome more severe than E trisomy. The frequency of D plus E trisomy in newborn infants is about 7 in 10,000 (Marden et al., 1964), and, as with 21 trisomy, frequency rises with maternal age. Unlike 21 trisomy, which is equally common in both sexes, D and E trisomies are more common in females, 60% with D and 70% with E (Priest, 1969), probably because of greater lethality in males before birth. Sinha (1968) and a few others have reported probable trisomy of large C or B chromosomes in very young children but the conditions were lethal at a very early age.

Among spontaneous abortions additional obviously lethal trisomies have been found (Carr, 1965) such as pair 3, pair 16, B, or C. Trisomy of such large A, B, and C chromosomes is not found in postnatal individuals because it is lethal.

Partial trisomy. Partial trisomy occurs when an individual has in addition to the two normal chromosomes of a pair a piece of that chromosome on another chromosome also. An example is the insertion of a piece of some chromosome into chromosome No. 1 to produce the oral-facial-digital syndrome (Ruess et al., 1962) that occurs in females and affects the parts indicated in the name as well as, often, some mental retardation. Such conditions arise by exchange or translocation.

Translocation. In translocation, also called exchange, two homologous or, usually, nonhomologous chromosomes exchange pieces of variable sizes. Thus most of a long arm of chromosomes No. 21 may be added to most of the short arm of a D-group chromosome (No. 14 or No. 15) (Fig. 19-10). The usually viable con-

266 TEXTBOOK OF CYTOGENETICS

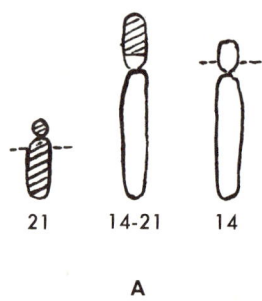

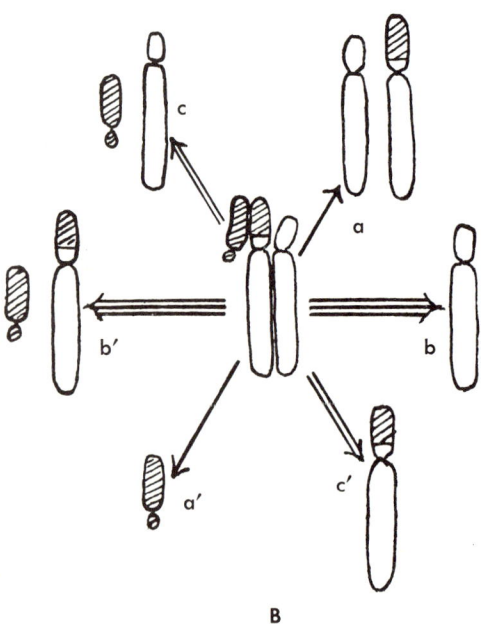

Fig. 19-10. A, Diagram of an exchange between a chromosome No. 21 long arm and a No. 14 short arm so that the translocated 14/21 chromosome has most of both chromosomes. B, The six possible gametes for these chromosomes following meiosis (center) and random disjunction of the three chromosomes. Gametes formed and effect in the zygote or later: a, duplicate for the long arm of No. 14, lethal; a', deficient for chromosome No. 14, lethal; b, deficient for chromosome No. 21, lethal; b', duplicate for chromosome No. 21, viable; c, normal haploid gamete, viable; c', abnormal but viable. Zygotes formed with normal gametes: b', gives one No. 14, two No. 21, one translocation; Down's syndrome. c gives two No. 14, two No. 21; normal. c' gives one No. 14, one No. 21, one translocation; normal, a carrier.

dition consists of one normal No. 14 or No. 15, one normal No. 21, and the translocation chromosome that is genetically about equal to a second No. 14 or No. 15 plus a second No. 21. Thus, the possessor of such a chromosome is phenotypically normal but has only 45 chromosomes and is called a "carrier." Translocations between other chromosomes have been reported such as D/D, D/G, G/G, and even a D/E translocation (Hauchteck et al., 1966).

Other examples of translocation that have been reported involving chromosome No. 21 are the addition of most of a long arm of No. 21 onto the short arm of No. 22 (Soudek et al., 1968; Cohen and Davidson, 1967) or short arm of a B-group chromosome (Aarskog, 1966), or the addition of the long arm of one chromosome No. 21 onto the short arm of another No. 21. This last produces a chromosome with two No. 21 long arms (see later for a very similar but different condition known as a 21 isochromosome).

Translocation between a No. 21 and either a D-group or No. 22 chromosome produces one type of translocation mongolism. Such an individual has only one of the normal D-group or No. 22 chromosomes, two normal chromosomes No. 21, and the translocation chromosome. Such a person is essentially a 21 trisomic but with 46 rather than 47 chromosomes and is, phenotypically, mongoloid. Persons who are phenotypically normal but have only 45 chromosomes may also possess such a translocation chromosome. Such a person is called a carrier of translocation mongolism and, when mated to a normal person, can produce a mongoloid child (Soudek et al., 1968); but such carriers have been found to produce much below the expected percentage of Down's syndrome children. Carriers are occasionally found when parents of mongoloid children are examined cytologically, but most translocation mongoloids seem to have arisen de novo. It is now possible to detect such individuals before birth, during the twelfth to sixteenth week of pregnancy by amniocentesis (Jacobson and Barter, 1967; Nadler, 1969). This method of detecting chromosomally aberrant foetuses involves the preparation of cells for karyotype analysis extracted from a sample of amniotic fluid.

Fig. 19-10, A diagrams the constitution of the chromosomes involved in a carrier. When meiosis occurs in such a person the three chromosomes, one normal No. 21, one normal D-group (No. 14 indicated here), and the translocation chromosome, synapse to form a trivalent (Fig. 19-10, B). At anaphase these three chromosomes can segregate in three different combinations to form six chromosomally different haploid cells, A and A' or B and B' or C and C'. A, A', and B are lethal because of excessive duplication or deficiency. Of those chromosomes of concern here, B' is viable with No. 21 duplication, C is a perfectly normal haploid cell, and C' is genetically balanced enough to be viable with the translocation chromosomes only. These three different viable gametes, when combined with normal gametes from the chromosomally normal parent, produce three types of zygotes and progeny: normal plus C gives a perfectly normal child; normal plus C' produces another carrier; and normal plus B' produces a mongoloid child since it is essentially trisomic for No. 21 although having 46 chromosomes.

It would seem, then, that a carrier should produce about one mongoloid child in three. Actually the ratio is about one in five. It is assumed that either the C-C' type of segregation is more frequent than the B-B' or that the B' gametes or mongoloid zygotes are semilethal.

A carrier with a 21/21 translocation chromosome or a 21 isochromosome to be normal would have only 45 chromosomes and no normal chromosome No. 21. Such a carrier would produce viable gametes with the abnormal chromosome and all progeny would be mongoloid.

Monosomics. Aside from the 45-chromosome XO condition, which is monosomic for the X chromosome and causes Turner's syndrome, autosomal monosomics are extremely rare. The only ones reported are *mosaics* lacking one G-group chromosome in some tissues, except for one reported case of *complete* G-group monosomy (Al Aish et al., 1967, who also cited a similar second case of a female child reported by T. Ikeuchi et al., 1967, Proc. Jap. Acad. **4310**:986-990). The young girl is mentally retarded but trainable, with some poor motor involvement and various superficial characteristics of ears, eyes, mouth, etc. It is generally considered that all monosomics except rarely 21 monosomics are eliminated very early by spontaneous abortion. Such 21 monosomics deviate from the normal in a direction and manner opposite to the deviation of 21 trisomy, and, therefore, the condition has been described as "antimongolism."

Deletion. Deletion occurs when a portion, always a very small portion, of one chromosome is missing, such as about half of the short arm of a B-group chromosome (Aarskog, 1966; Warburton et al., 1969) or about half of the long arm of a G-group chromosome. Such individuals are described also as "partial monosomics" and are always heterozygous for the condition; only one chromosome of the pair may be deleted in a viable individual.

The B chromosome deletion (probably chromosome No. 5) produces a phenotype called either the B chromosome deletion syndrome or the *cri du chat* syndrome (Fig. 19-11). This condition is lethal, death occurring in infancy. The syndrome includes growth deficiency, mental retardation, etc., and a weak, high-pitched cry that sounds like the cry of a cat, hence its name of cat-cry syndrome. Apparently deleted No. 4 *or* No. 5 gives the syndrome but only deleted No. 5 causes the cat cry (Ferguson-Smith, 1967). This may imply partial homeology of chromosomes No. 4 and No. 5 if true.

The G-group chromosome deletion, of part of a long arm, has been correlated with chronic myelogenous leukemia. The deleted short chromosome is often called the "Philadelphia" or Ph^1 chromosome. The Ph^1 chromosome is found not throughout the body but only in the bone marrow of the leukemic individual (but see Wahrman et al., 1967, for the Ph^1 chromosome in peripheral blood). For this reason it is not known whether the deletion is cause or effect of the bone marrow condition.

In some cases (Aarskog, 1966) it is claimed that the deletion is associated with translocation. A deficiency in the short arm of No. 18 has been repeatedly noted, but in only half of

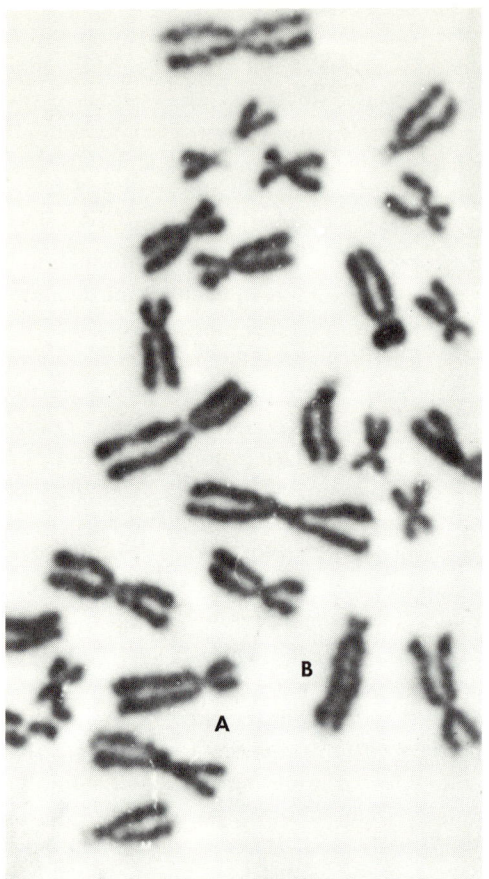

Fig. 19-11. Normal chromosome 5(A) and a deleted short arm chromosome 5(B) that causes the cat-cry or *cri du chat* syndrome. (Courtesy Dr. M. S. Al-Aish.)

the cases has it been associated with a syndrome (Ferguson-Smith, 1967).

Deletion of most of the short arm of the X chromosome is discussed in the next section.

Isochromosomes. An isochromosome is assumed to arise when a normal centromere breaks, is broken, or divides crosswise rather than longitudinally. As a result, each of the two daughter chromosomes consist of two arms that are genetically identical from centromere to distal end, such as two long arms or two short arms of the original chromosome (Fig. 19-12, *A*) The only likely isochromosomes found in human beings are of the long arm of a G-group chromosome, probably No. 21, and of the X chromosome. Such an isochromosome is indistinguishable from a No. 21 long arm translocated onto another long arm of another chromosome No. 21 (Fig. 19-12, *B*) and like it produces mongolism if a normal No. 21 is present also.

The case for an isochromosome of the X chromosome is very strong since the unusual chromosome, in females only, is much like chromosome No. 3, but detailed study shows it to consist of two long arms of the X, and the short arms, of course, are missing. Furthermore, that it is indeed a modified X chromosome is, in a sense, proved because such individuals have in some interphase nuclei one unusually large Barr body, a condition possible only when two X chromosomes are present and one is double. Such individuals are monosomic for the short arm of the X chromosome but trisomic for the long arm. Nevertheless, they are phenotypically like XO individuals, they have Turner's syndrome.

That deletion of most of the short arm of one of the X chromosomes can produce, essentially, Turner's syndrome is demonstrated by some individuals with that syndrome (female, of course) having one normal X and one X chromosome possessing an abbreviated short arm. Although they are "normal" females, since they have two X chromosomes, they are XO (with Turner's syndrome) for the short arm. Thus, in all cases, Turner's syndrome is determined by the hemizygous condition for the short arm of the X. Such individuals possessing a normal and a deleted X are symbolized as Xx, and they have usually a very small Barr body.

Inversion. Inversion, the turning around of a portion of a chromosome, is usually internal and either includes (pericentric) or does not include (paracentric) the centromere. Pericentric inversion often alters the arm ratio of the chromosome (Soudek et al., 1968) and so is often detectable as a chromosome changed in form (Fig. 19-13). Crossing-over often occurs within the inversion. In a person heterozygous for such an inversion it is assumed that a loop is formed and crossing-over is random within the loop. The random crossing-over within a pericentric loop produces a variety of chromosomes and, therefore, genetic types of gametes and zygotes. Commonly it produces chromo-

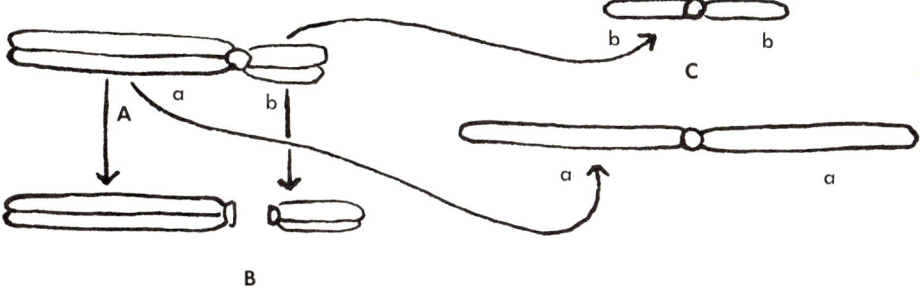

Fig. 19-12. The origin of two isochromosomes from a normal chromosome by a transverse break through the centromere. If the arms do not straighten out, *B*, telocentric chromosomes result. If the arms do straighten out, two absolutely metacentric chromosomes are formed, *C*.

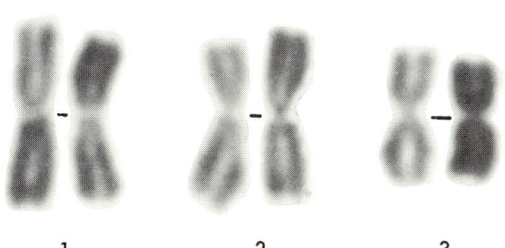

Fig. 19-13. Human chromosomes, pairs, 1, 2, and 3. The left chromosome No. 2 is normal with a longer arm downward. The right chromosome No. 2 has two equal arms as a result of an unequal pericentric inversion. Presumably, the effect has been to move the lower end of the long arm of normal No. 2 onto the upper end of the short arm. This condition was found in a mentally retarded girl and also in her normal father. (Courtesy Dr. M. S. Al-Aish.)

somes having duplicate segments and missing segments, both of which produce abnormal phenotypes. A few cases of pericentric inversion in human beings have been tentatively documented (Fig. 19-13).

Triploidy. Triploidy, although fairly common in spontaneous abortions, is very rare in postnatal human beings. One triploid male individual (Böök and Santesson, 1960) had 69 chromosomes including XXY but was more "male" than are Klinefelter syndrome XXY trisomics. Butler et al. (1969) also reported a 69,XXX triploid infant but female. Three triploids were reported by Edwards et al. (1967), and Schmid and Vischer (1967) reported a 48,XXY/XXXY mosaic boy described as showing "double aneuploidy and double mosaicism." They cited three other cases also.

Miscellaneous aberrations. Aamendares et al. (1969) described a family in whom there was, in addition to the normal chromosome set, a tiny centric fragment with a satellite on each end. They proposed that it consisted of parts of two satellited short arms of two D- and/or G-group acrocentric chromosomes. One member of the family was, in addition, a G trisomic.

Various types of chromosomal aberrations such as chromatid and/or chromosome breaks, fragments, dicentrics, etc. (Fig. 19-14) are known to result from radiations, chemicals, possibly LSD (for example, Cohen et al., 1967a; Irwin and Egozcue, 1967; but see Loughman et al., 1967, for negative results), or possibly cyclamates (in vitro by Stone et al., 1969; Stoltz et al., 1970), extracts of allogenic lymphocytes (Fialkow, 1967), viral infections (Lubs and Samuelson, 1967; Mella and Lang, 1967), even aspirin (Loughman et al., 1971), and genes, such as the rare recessive mutant causing Bloom's syndrome (German, 1969; 1970) or the also rare mutation causing Fanconi's anemia (German, 1970). The gene for Bloom's syndrome produces breaks across the centromere or in arms, sometimes producing interchanges (Fig. 19-14).

Quadriradial chromosomes, as in Fig. 19-14, indicate previous somatic crossing-over (German, 1964), which in turn presupposes probable somatic pairing. Fig. 19-15 indicates the generally accepted cause of quadriradials: somatic pairing of homologues, somatic crossingover, and a subsequent metaphase quadriradial configuration. German reports quadriradials only rarely in tissue culture and in normal and some abnormal people, but much more commonly in an individual having Bloom's syndrome, a genetic condition producing high frequencies of chromosomal breaks. Similar quadriradials have been reported in plant and animal cells following chromosome-breaking treatments such as with radiations or nitrogen mustard. They are never

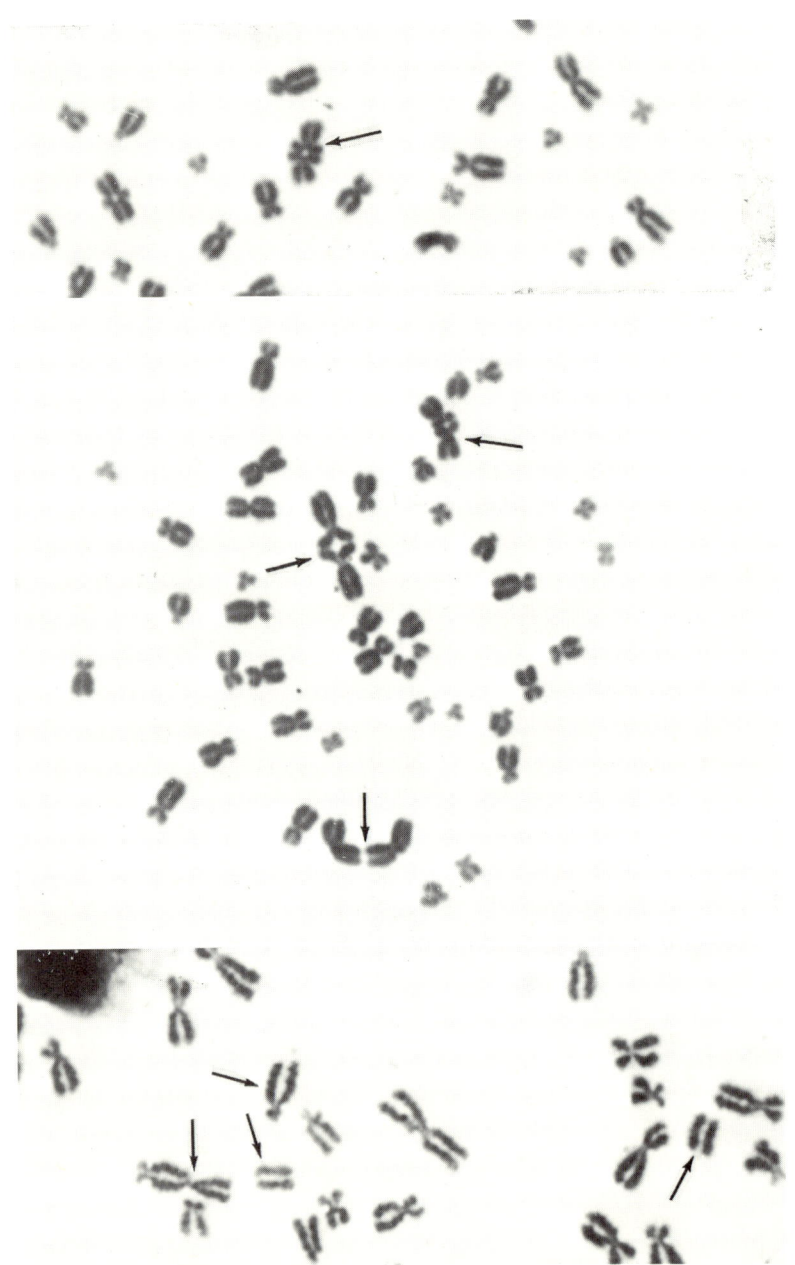

Fig. 19-14. Chromosome aberrations in dividing lymphocytes from a patient with Bloom's syndrome. Among normal chromosomes are dicentric, acentric, quadriradial, and other abnormal chromosomes, such as a quadriradial forming, with telomeric association of two homologues.

common, implying the rarity of somatic crossing-over in normal cells, but they do indicate that somatic exchange can occur.

Steele et al., (1966) reported a B-group ring chromosome with mosaicism in a newborn child with the *cri du chat* syndrome. The latter is usually correlated with a B-group deletion. Ring D chromosomes (probably No. 13) have been reported by Gerald et al. (1967) and by Reisman et al. (1965). In both individuals both

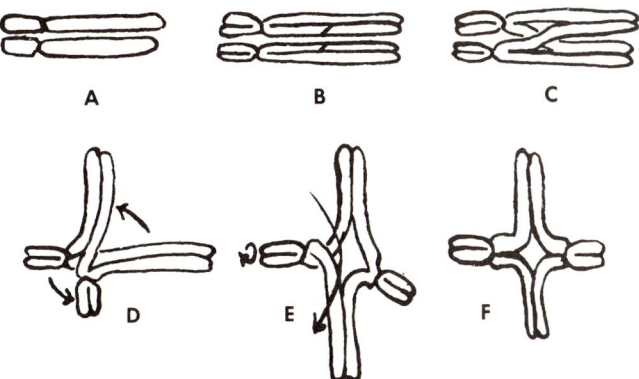

Fig. 19-15. Diagram of sequence of events in the presumed formation of a quadriradial in mitotic metaphase chromosomes. *A*, Homologous chromosomes somatically paired, probably at anaphase. *B* and *C*, During interphase each chromosome doubles, and an exchange occurs between two nonsister chromatids. *D* and *E*, During prophase the arms twist about, and the chromosomes rotate until centromeres (constrictions) come to lie opposite each other and long (or short) arms opposite each other, as in *F*, by metaphase (see Fig. 19-14). Anaphase separation and subsequent divisions should be quite normal.

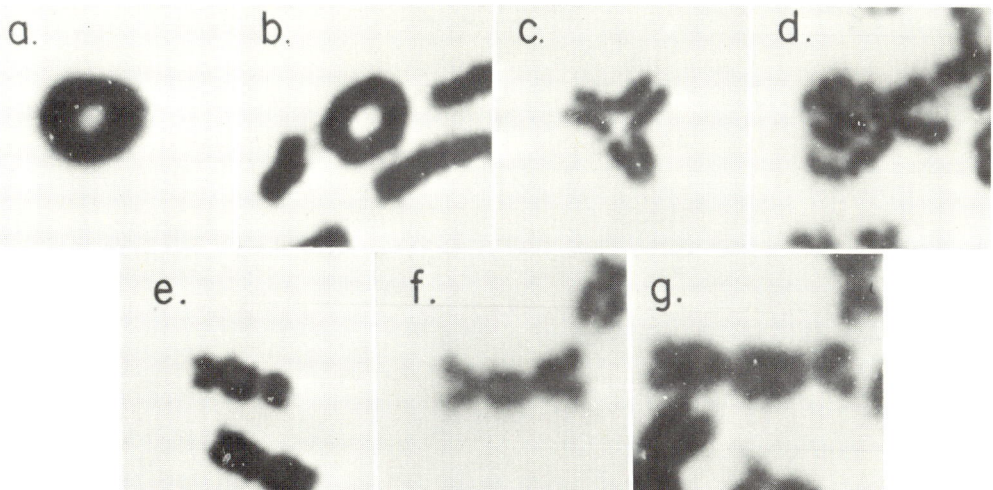

Fig. 19-16. Some examples of structural rearrangements and alterations of human chromosomes. **a** and **b**, ring C group chromosomes; **c**, a triradial chromosome consisting of three chromatids; **d**, an unusual secondary association, a so-called pregnant chromosome. **e**, **f**, and **g**, dicentric chromosomes. (Courtesy J. Schulman and Dr. M. S. Al-Aish.)

ends of chromosome No. 13 were lost after the double break, and apparently one end contains the gene for the haptoglobin alpha-chain (Bloom et al., 1967) (Fig. 19-16).

Cohen et al. (1967b) discuss centric fragments and rings of X and Y chromosomes. Centric fragments consist of a centromere and two very short arms. They can be described also as doubly deleted chromosomes. Since they are centric, they persist during cell divisions as well as any "normal" chromosome. There are a number of reports of such small, deleted X and Y chromosomes. Rings, on the other hand, have, in a sense, only one arm, which is attached to the centromere at both ends. Separation of sister chromatids at anaphase is often difficult. If the anaphase rings are interlocked or are dicentric, breakage at various places and reunion must occur if the ring persists. Rings of various sizes containing duplications and deficiencies result. Most reports of ring chromosomes are of ring X's. Rings may not be passed on at mitosis in leukocytes and may represent a case of unstable chromosomal conditions.

Magenis et al. (1970) reported a heritable, so-called fragile site on chromosome No. 16 that indicated the probable localization of the α-haptoglobin locus. Since this fragile site is observable in the long arm of chromosome No. 16 about one third of its length from the tip and is closely linked to the α-haptoglobin

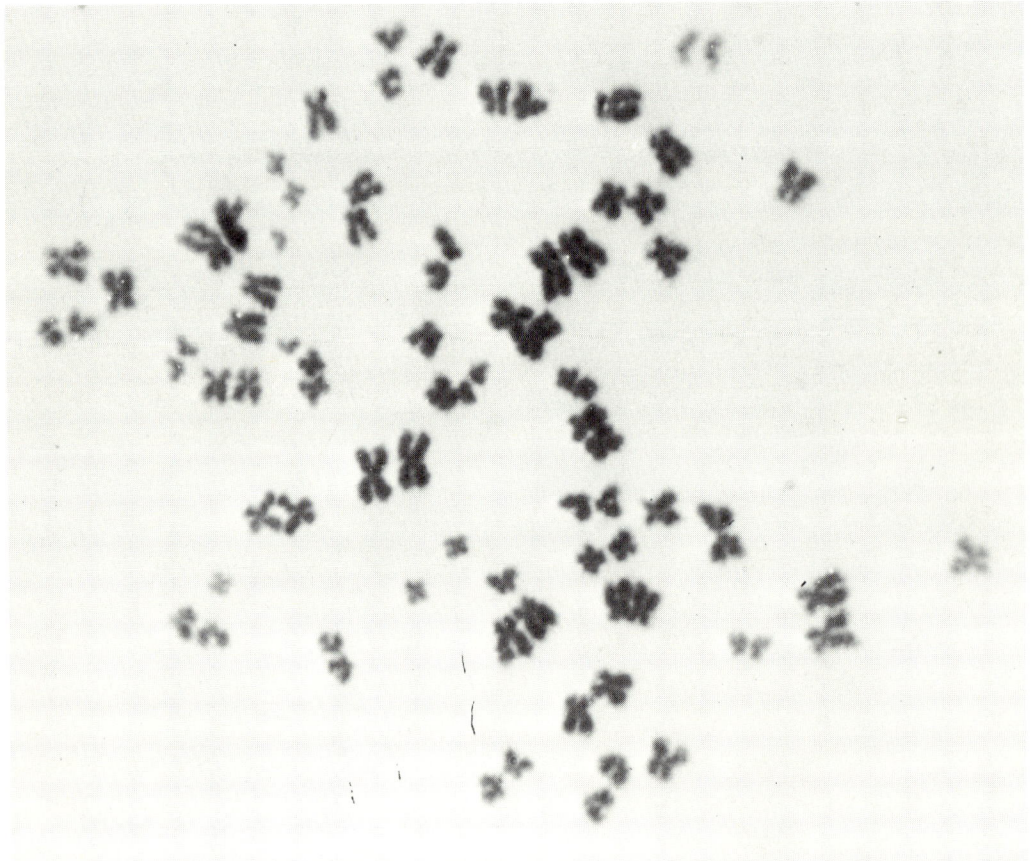

Fig. 19-17. Metaphase of human chromosomes that had remained in colchicine so that a second (endoreplicative) division had occurred. Each two-chromatid chromosome lies adjacent to its sister. Each chromatid of the previous metaphase (in colchicine) separated slightly from its sister and then doubled (in colchicine), with little or no movement, to produce this arrangement. The technique has caused some pairs to move slightly. (Courtesy Dr. M. S. Al-Aish.)

locus, the latter must be similarly located. The break is inherited in a simple mendelian manner from father or mother to son or daughter. Interestingly, the deleted fragment seems to endoreplicate itself since some lymphocytes had as many as 12 such fragments in one metaphase spread. The researchers postulated "that the control of replication for the distal portion of this chromosome is not the same as for the remainder of the chromosome." Since Lejeune et al. (1969) and Shaw (1968) have reported similar fragile sites on chromosome No. 2 that also showed selective endoreplication of fragments, it may be that detached pieces in general might have multiple replications during one interphase since such acentric fragments should be lost at anaphase and not persist from one division to another. Such fragments could have repeatedly entered the S period while chromosomes remained in the long G_1 period of leukocytes. Excessively long periods in colchicine can also cause endoreplication (Fig. 19-17) and must be avoided in the preparatory technique.

Effects of radiation on chromosomes persist for many years in an individual, probably for life. Examination of lymphocytes of people exposed in Japan in 1945 and the Marshall Islands in 1954 (Lisco and Conrad, 1967) have demonstrated the persistence of aberrations, especially of translocations, for many years. This is possible in dividing cells only if the aberration is stable. If such aberrations are unstable, the lymphocytes must exist undivided in the G_1 phase for years. The latter seems likely (Court Brown, 1967).

Marfan's syndrome has been somewhat questionably related to large satellites (Ferguson-Smith, 1967).

There are innumerable reports of various aberrations, such as ring chromosomes, translocations, deletions, etc., that are occasional or have not been related to known phenotypes. They are beyond the scope of this treatment.

Effect of aging. A number of cytogeneticists have noted a rise in aneuploidy (45 chromosomes) in old people (Court Brown et al., 1966; Court Brown, 1967), at least in their lymphocytes. This loss starts in women at about age 55 and in men at age 65. Apparently, women tend to lose an X and men a Y. In either case only a small percentage of lymphocytes show these losses.

The correlation of maternal age and 21 trisomy has been mentioned, but other trisomies also increase with maternal age.

CYTOGENIC MAPPING

At the present time genetic study of human beings has revealed about 800 known genes that range in effects from lethality to triviality. Furthermore, it has been established that man is highly heterozygous for a great many genes. Lewontin (1967) estimated that the heterozygosity for blood cell antigens is 0.16, which agrees with earlier estimates for human enzymes. The goal of human cytogenetics would be to map all known genes on specific chromosomes of the karyotype.

Cytogenic mapping is the determination of the loci of particular genes within the karyotype. The location may be merely to a particular chromosome, or to a specific arm of a particular chromosome, or more exactly within an arm such as proximal or distal. To do this, a particular mutation must be associated with a specific chromosomal aberration, usually a small deletion. Progress is being made in this field of cytogenetics (Gerald et al., 1967; Renwick, 1969) but little has been achieved as yet (German, 1970). An example is the linkage of the α-haptoglobin locus to the fragile site in the long arm of chromosome No. 16 already discussed. Robson et al. (1969) using translocations also located the α-haptoglobin locus in the long arm of No. 16.

Another example is the association of the gene for the haptoglobin alpha-chain with two cases of ring D chromosome (Bloom et al., 1967). By autoradiographic study the D chromosome is identified as No. 13. The assumption is that the ring is a result of an exchange within chromosome No. 13 with the breaks close to both ends. The two ends have been lost, thereby producing the genetic mutation. Therefore, the gene must be located close to one or the other end of chromosome No. 13. Thus the α-haptoglobin locus might be on chromosome No. 13 or No. 16.

Less exact deletion mapping is illustrated

by Nance and Engel (1967). They could only specify that a syndrome and its specific characteristics were related to deletion of the short arm and satellite region of a G-group chromosome.

The human X chromosome is the only one mapped for a number of genes (Berg and Bearn, 1968) such as those for hemophilia and ichthyosis. At present more than 85 genes are known linked to the X chromosome, and about 10 of these have been mapped relative to one another in one or other arm.

Martin and Sprague (1969) have proposed a novel method for specifying linkage groups. It is known that in long-continued tissue culture that occasionally tetraploid cells arise and multiply, constituting from 1 to 15% of mass cultures. In tetraploid cultures derived from single tetraploid cells they found from 4 to 17% diploid cells, which evidently arose suddenly by some sort of "reduction division." Such doubling and reduction is known in fungi as constituting a parasexual cycle. Since the original tissues were heterozygous for observable aberrations (a long submetacentric No. 16, partial short-arm deletion of No. 18, and partial deletion of short arm of a D chromosome), the diploid cells produced by reduction from tetraploid could be scored for homozygosity or heterozygosity, and any homozygous diploid cells would be the result of whole chromosome recombination. Such diploid recombinant cells were found in appreciable numbers by karyotype analysis. It was determined also that no such somatic recombination (XX and YY) of sex chromosomes occurred from an original male (XY) culture.

Tetraploid cells could arise from cell and nuclear fusion, formation of a restitution nucleus, as from endoreduplication. Reduction from tetraploid to diploid is most likely achieved by formation of a tetrapolar spindle rather than a bipolar and an exact diploid set of chromosomes somehow moving to each. The researchers suggest that this technique of a parasexual cycle with somatic recombination of entire linkage groups could form the basis of a formal genetic analysis in man. The possibility also exists that spontaneous or induced somatic crossing-over and genetic recombination at the diploid level will be discovered, thus completing a truly parasexual cycle and permitting higher resolution human genetics.

At the present time human genetics and cytology are making rapid progress, but true human cytogenetics is in the early formative period. It is still too inexact by current methods to locate many known genes with great accuracy or certainty; for example, the disagreement of Smith et al. (1968) and Danes and Bearn (1968) for the location in the human genome of the cystic-fibrosis gene, the location of the MN blood group locus on chromosome No. 2 (German et al., 1968) or on chromosome No. 4 (Weitkamp, 1969), or the "probable assignment of the Duffy Blood Group locus to chromosome No. 1" (Donahue et al., 1968).

In recent years fluorescent DNA-binding agents, especially quinacrine mustard (Caspersson et al., 1970b), have been used to distinguish human as well as other animal and plant chromosomes (Caspersson et al., 1968, 1969, 1970a) at metaphase and interphase or during meiosis (Pearson et al., 1970; Pearson and Bobrow, 1970). Apparently this type of dye binds preferentially to the DNA of constitutive heterochromatin. Since some chromosomes are nearly all heterochromatin, some have very little, and other have numerous loci along arms but in different patterns, it is possible to distinguish among morphologically similar chromosomes by their different fluorescent patterns (Caspersson, et al., 1970a).

A third technique, devised by Pardue and Gall (1970), for showing characteristic differences among chromosomes and along their arms was adapted to human chromosomes of cultured cells in 1971 by Arrighi and Hsu, Gagné et al., Sumner et al., and Patil et al. The technique uses the well-known chromosomal and blood staining mixture called Giemsa. Various treatments were proposed, but all produce light and dark stained regions along chromosome arms that facilitate chromosome identification. The stained regions correspond in most loci with the fluorescent regions produced by quinacrine mustard; the most evident difference between these two techniques is the bright fluorescent region at the distal end of the long arm of the Y chromosome, which does not stain with Giemsa. It has been proposed that the bands, and especially the centromeric regions stained by the Giemsa and quinacrine mustard treatments, represent regions of heterochromatin that contain highly redundant satellite DNA that has, presumably, been denatured and then annealed by the Giemsa treatment. The more unique DNA, the euchromatin, does not reanneal and so does not stain. This Giemsa method should make positive identification and detection of aberrations of chromosomes relatively easy (Craig-Holmes and Shaw, 1971; Yunis and Yesmineh, 1971).

In fact, Craig-Holmes and Shaw (1971) in Hsu's laboratory have detected by this denaturation-renaturation procedure seven variants among chromosomes 1, 9, 16, D group, F group, and G group in only four individuals of such constitutive heterochromatin. Most polymorphism was in the amount of chromatin adjacent to the centromere. They concluded from this small and preliminary study that there is "a very high frequency of variability of heterochromatin in the population," and this variability constitutes "a new class of chromosomal polymorphism in humans."

Another technique that gives somewhat the same evidence on distribution of heterochromatin along the various chromosomes is autoradiography. If a pulse of radioactive thymidine is given early during the synthetic period, the early-replicating euchromatin is labeled; if the pulse is given late in the S period, the late-replicating heterochromatin is labeled. Such studies can distinguish otherwise indistinguishable chromosomes (Fig. 19-5).

A technique for DNA/RNA hybridization to reveal constitutive heterochromatin used as a stain has recently been introduced (Pardue et al., 1970; Hsu, 1970). In this technique cells or organisms are made to produce radioactive RNA's. These are extracted and separated into various sorts by gradient centrifugation. Spreads or squashes of chromosomes are made on slides and variously treated. Then the chosen radioactive RNA in solution is put on the chromosomes. After a reasonable period of exposure the slide is washed and an autoradiograph made. It is found that the radioactivity (as a stain) is located in specific regions of the chromosomes and permits specific designation of chromosomes of the set.

It is still too early to predict the cytogenetic value of cell hybridization, as between human and mouse by Weiss and Green (1967) or Migeon and Miller (1968) possessing all mouse chromosomes but only one human chromosome of group E; the value of transformation (Aaronson and Todaro, 1969); accumulation of large masses of human metaphase chromosomes followed by fractionation into mixtures of different "Denver" size groups (Schneider and Salzman, 1970); the detection of "mutants" by use of antiserums in an isoantigenic selective system (Adman and Pious, 1970); or the value of pachytene preparation (Yerganian, 1957; Hungerford, 1971) as in *Zea mays*. It is inevitable that genetic and cytogenetic precision will become greater in the future.

Recently the method of transduction has been successful in introducing bacterial, viral, and mammalian genes into mammalian cells in culture via viruses (Merril et al., 1971; Qasba and Aposhian, 1971). Since the method of transduction has contributed greatly to the genetic knowledge of viruses and bacteria (Chapter 20), it is anticipated that its application to the genetically and cytologically far more complex human mammalian cells will add greatly to knowledge of human genetics as well as lead toward achievement of "genetic engineering," the introduction into individuals of desired genes at will.

CHAPTER 20

Cytogenetics of prokaryotic types

> It can be concluded that the method described, whereby both the single stranded and double-stranded regions of various heteroduplexes of viral DNA can be accurately measured, permits the construction of precise molecular maps, including the assignment of both position and size to various genes.
>
> B. C. Westmoreland et al., 1969

By prokaryotic types is meant those organisms (bacteria, glue-green algae, actinomycetes, etc.) or quasi-organisms (viruses, mitochondria, and plastids) that have a single naked nucleic acid polymer as a "chromosome" or "genophore" rather than having the complex chromosome of the Eukaryota. All of these are asexual and have no true mitosis, nothing like meiosis, no nucleus, and in many other ways, too, they differ from the Eukaryota. Although they have neither sexual reproduction nor meiosis, not only are genetic studies possible with some of them but great advances in genetic knowledge in the broadest sense have derived from research on them during the past 25 years. Such research has been made possible, in part, by technical advances in biochemistry and the use of the electron microscope (Miller et al., 1970a; Miller et al., 1970b) and earlier studies of viruses, bacteria, and techniques for detecting recombination of biochemical mutants (Beadle and Tatum, 1941). Some of these characteristics of bacteria and viruses that make genetics of these organisms and quasi-organisms possible are called lysogeny (Lwoff and Gutmann, 1950) in which, as Bordet (1925) wrote, "The power of reproducing the bacteriophage is woven into the hereditary web of the bacterium"; bacterial transformation (Griffith, 1928; Avery et al., 1944); transduction (Zinder and Lederberg, 1952); and bacterial conjugation (Lederberg and Tatum, 1946). These have been discussed in detail by Jacob and Wollman (1961) in a general treatment of bacterial genetics, by Stent (1963) on the molecular biology of viruses, or by Campbell (1969) on episomes.

It is especially in this area of investigation that the concept of cytogenetics is broadest. This is well expressed in the words of Thomas et al. (1968):

> The development of practical genetics came about by the continuous interplay between those doing recombination experiments and those studying stained chromosomes in the light microscope. In retrospect those two fundamentally different kinds of study merge with each other so that when we say the word "chromosome" it takes some patience to decide whether this is something that you see in the microscope or whether it is a graph drawn on the basis of observed recombination frequencies. With the advent of microbial genetics it became apparent that there was no analogous cytogenetics to supply a physical model upon which to interpret the results of recombination experiments. The discovery of the basic structure of DNA in 1953 did much to relieve the vacuum, yet as time went on, it became clear that this was not enough . . . . In the future it should be no great surprise if we use interchangeably, say, "genetic map" or "DNA molecule" just as we loosely say "chromosome" although we all know that we

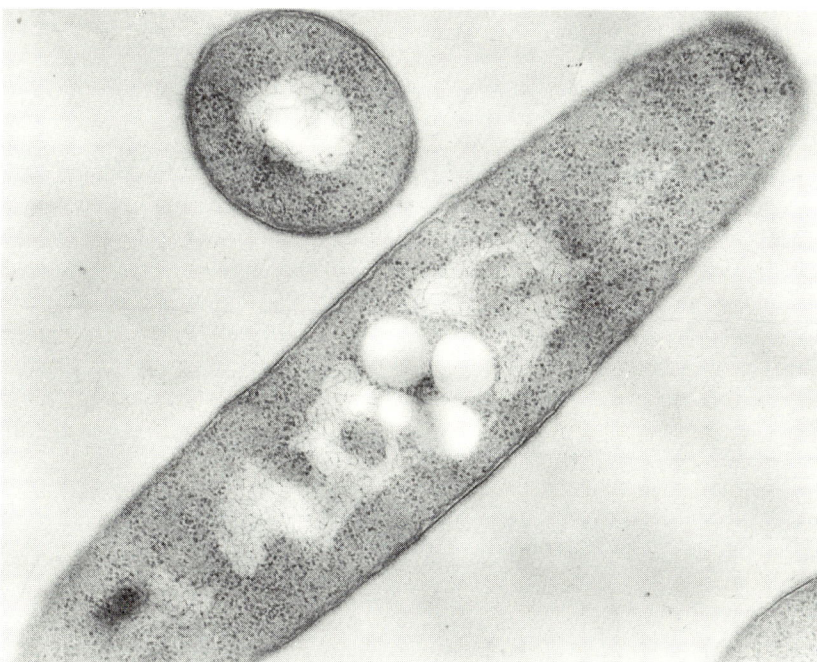

Fig. 20-1. Electron micrograph of two bacterial "cells" (*Bacillus aneurinolyticus*) showing filaments of the genophore in the irregular, lightly stained prokaryons. Presumably, there is as much genophore in such a bacterial cell as is indicated in Fig. 20-8. It is not unlikely, however, that a great deal of genophore, at least that portion or segments containing the codes for messenger RNA (Figs. 20-5 and 20-6), is external to the prokaryon, out in the darkly stained riboplasm where the ribosomes are located. (Courtesy Dr. J. L. Witliff, University of Texas Cell Research Institute.)

mean two different things. . . . Although it is now possible to visualize DNA molecules it is not possible to visualize the sequence of nucleotides . . . . Thus molecular geneticists have nothing comparable to the salivary chromosomes of flies. Instead they have resorted to ingenious density labeling procedures (perfected mainly by Matt Meselson and his collaborators) or to autoradiography (as exemplified by the works of Taylor and Cairns).

That is, much of the "seeing" of prokaryotic genophores, the cytogenetics of them, consists of graphs and numbers. But in this discussion seeing genophores by electron microscopy is emphasized whenever possible. The genetics of bacteria and viruses (bacteriophages, mostly) are so mutually involved that it is impossible to discuss the genetics of one without devoting equal attention to the other. For example, the general discussions of viruses (Stent, 1963), or genetics of bacteria (Jacob and Wollman, 1961), or of episomes (Campbell, 1969) show this.

GENOPHORE AND REPLICATION

Viruses, plastids, mitochondria, and Prokaryota have in common a single nucleic acid strand or polymer as the ultimate hereditary material. In all except the RNA or single-stranded DNA viruses, this polymer is a double DNA helix, although Miller et al. (1970b) reported genophores of *Escherichia coli* to be 40 Å in diameter, which is twice the diameter of a Watson-Crick double helix. They considered it to be coated with specific protein, and that it may be so coated in vivo. It is so very different from the usual eukaryotic chromosome that it should have a different name. "Genonema" (gene thread) and "genophore" (gene bearer) have been proposed. Here the term genophore will be used.

Evidence from molecular biology and electron microscopy (Cairns and Davern, 1968; Kuempel, 1970) indicates that the genophore in vitro and probably in vivo, when not replicating, is usually a single Watson-Crick double helix. It is known to exist also in linear form as in the lambda virion and the yeast mitochondrion. It ranges in length from 1μ in some RNA and single-stranded DNA viruses, to 250μ in *Mycoplasma*, to $1,500\mu$ in some chloroplasts, to more than $2,000\mu$ (2mm) in some bacteria. The double-stranded DNA genophores start replicating at one specific point (Lark et al., 1963), and the single replicating point forms a fork that proceeds around the circular genophore and can be observed by electron microscopy (Bonhoeffer and Gierer, 1963; Bode and Morowitz, 1967). Actually there are two forks, one at the site where replication began and the other at the moving replication point (Fig. 20-2). There are two strands connecting the two forks, and the whole thing is called the *initiation plus replication loop*. Replication is also indicated by an increased rate of DNA synthesis, in *E. coli* about halfway through the cycle or 20 or 25 minutes before division (Clark and Maaloe, 1967; Helmstetter, 1967), although some DNA synthesis occurs all the time. In rapidly dividing cells multiple replicating forks indicate that a second round of multiplication can start before the first has been completed (Yoshikawa and Sueoka, 1963; Helmstetter and Cooper, 1968).

Somehow or other (Sueoka, 1968) the two circular genophores produced, although greatly twisted and compacted, are able to separate cleanly from one another, aided, perhaps, by the membrane (Chapter 4) (Fig. 20-3). It seems evident that the double helix must unwind as it replicates into two separate strands. In the bacterium *E. coli* the replication rate is uniformly about 30μ per minute or 10^5 bases per minute. Therefore, the unwinding rate of the double helix must be about 10,000 revolutions per minute (Cairns, 1963; Cairns and Davern, 1968). Supercoiling of the genophore has been repeatedly reported in viruses, chloroplasts, mitochondria, and bacteria. If the supercoiled condition exists in the nonviral cells, the mechanism of genophore separation is even more difficult to conceive.

Multiple genophores certainly occur in many bacteria, mitochondria, and, possibly, chloroplasts (Boasson and Laetsch, 1969). Usually these are separate in distinct prokaryons somewhat like a multinucleate eukaryotic cell, but two or more per prokaryon are possible and are certain just after replication is complete, unless separation occurs during replication.

Shatkin (1971) has reported up to ten or more separate segments in each influenza reovirus virion which have "segmented ribonucleic acid genomes." Numerous other plant and animal viruses have incomplete genomes

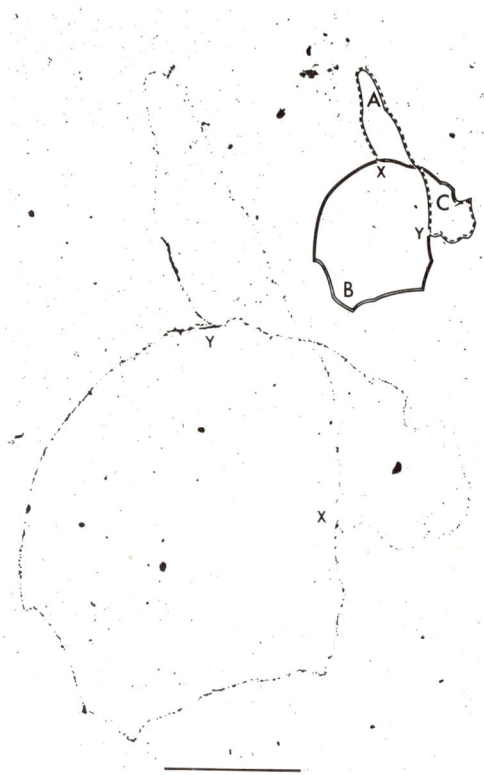

Fig. 20-2. Autoradiogram of the genophore of *E. coli* K-12, Hfr, labeled with tritiated thymidine for two generations and extracted with lysozyme; exposed for 2 months. The series of dots in the emulsion indicate the presence of the invisible genophore. This genophore is replicating and has two forks (X and Y in diagram). From Y, through B, to X it consists of two strands of the single double helix that are "hot," and so there are many dots. From X, through A or C, to Y there is one old hot strand and one newly replicated cold strand each, so there are only half as many dots along these filaments, between the two forks X and Y. The scale shows 100μ, and the whole circular genophore is about $1,100\mu$ around. (From Cairns, J. 1963, Cold Spring Harbor Symp. Quant. Biol. **28**:43-46.)

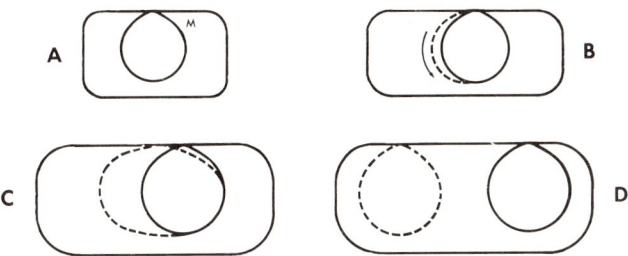

Fig. 20-3. Diagram of the common models of separation of two bacterial genophores following or during replication. **A,** A single genophore attached somehow to membrane, *m*, before replication. **B** and **C,** Replication has begun and two forks are evident, one (arrow) is progressing around the genophore, the other is stationary (see Fig. 20-2). **D,** Replication is complete and the two genophores are separated by movement at the membrane, or by mesosomes, or otherwise. Actually, this representation is far too simple because the genophore really (see Fig. 20-8) has a circumference about 200 times that of the whole cell. Such trivial movement as indicated here (from about 1 to 2μ) would not separate two replicated genophores more than 1%.

(satellite viruses) that require the presence of another virus in the same cell in order to produce virions (Kassanis, 1962, 1968).

Another difference between eukaryotic chromosomes and DNA genophores is that the latter are able to replicate and transcribe at the same time, although in different places. Presumably, transcription of RNA's occurs in front of and behind the single replicating fork. In eukaryotic chromosomes replication starts in many places along a chromosome at the same time, and during the period of replication there is no transcription.

Although the term genophore is used here, it must be understood that the term chromosome is often used in reference to this nucleic acid strand of bacteria and genome of viruses, as is also the term nucleus for the prokaryon.

VIRAL CYTOGENETICS

It is inappropriate here to discuss in detail the nature of viruses, how they increase in numbers, the molecular (or biochemical) genetics of viruses, etc.; but some facts, however derived, about the genetic information and electron microscopy of viral genophores *are* cytogenetics.

The extracellular, infective, free viral particle is called a *virion*, which consists of one or more species of protein and a genophore consisting of either RNA or DNA that is either single- or double-stranded and that has a molecular weight between about 1 and 130 million daltons (Table 7).*

The genophores of viruses, therefore, range in length from about 0.5 to 70μ and they can be circular (Streisinger et al., 1964) or not in different viruses or at different phases of the cycle. Table 7 contains some of these data for a few viruses, especially for bacteriophages, which are the main sources of genetic and cytogenetic data of viruses. Cohen (1967) has recently discussed briefly the chemistry and structure of bacteriophage nucleic acids.

Following adsorption of bacteriophage particles (for a recent illustration of a complex phage particle see Eklund, et al., 1971) onto the surface of a bacterial cell, such as *E. coli*, the DNA genophore is introduced into the bacterial protoplasm. It then produces messenger RNA for the formation of enzymes, some of which break down the host DNA for viral use (Garen and Skoar, 1958) and for formation of protein and viral DNA that are assembled somehow into virions within the host prior

---

*One dalton is the approximate mass of one hydrogen atom or one sixteenth of the mass of one oxygen atom of mass 16. Because the determination of molecular weights of macromolecules has an error of from 5 to 15%, the very small difference between calling the mass of a hydrogen atom 1 or 1.0078 is so trivial that it can be ignored. Therefore, for large polymers the molecular weight can be expressed equally well as so many daltons.

Table 7. Data on some genophores

Genophore	Molecular weight (daltons)	Length (microns)	Shape
Viruses			
DNA, double-stranded			
T2, T4, T6	120×10^6	60	Permuted, not circular
Lambda (λ)	33×10^6	17	Not permuted, circular
T1	42×10^6		
T3	54×10^6		
T5	85×10^6	About 44	Circular
T7	25×10^6	About 13	Circular
P1	About 58×10^6		
P2	About 24×10^6		
P22	About 27×10^6	About 13.5	Circular
H39	About 400×10^6	About 200	
Mammalian polyoma virus, SV40, rabbit and human papillomas			Circular
Herpesvirus			Circular
Some poxviruses, adenovirus			Not circular
DNA, single-stranded			
ϕX174	1.7×10^6		Circular
M13, fd	About 10^6		Circular
RNA, double-stranded			
Reovirus, wound tumor virus			
RNA, single-stranded			
MS2, f_2, R17, M12, Q	About 10^6		
Bacteria			
Haemophilus influenzae	2.2×10^9	1,000	
E. coli	3.3×10^9 (5.4×10^{-15} gm)	1,100	
B. subtilis (DNA/spore)		1,700	
PPLO	500×10^6	250	
Mitochondria			
Protozoan	620×10^6, 200×10^6 (*Tetrahymena*, 222×10^6)		Linear
Fungal (except yeast)	20×10^6 to 100×10^6		
Yeast	10×10^6		
Animal, sea urchin	14×10^6		Linear
Angiosperms	300×10^6		Circular
Chloroplasts			
Acetabularia	805×10^6		
Tobacco	40×10^6		

to host destruction, called lysis (Stent, 1963).

If two (or more, Hershey and Chase, 1951) genetically different viral genophores, such as r^+h and rh^+, are simultaneously introduced into the same bacterium, their product genophores can be recombinant, produced somehow and probably by crossing-over involving breakage and reunion (Hershey and Burgi, 1956; Meselson and Weigle, 1961; Tomizawa and Anraku, 1964; Meselson, 1964; Simon, 1965). At least recombinant particles are recovered, such as rh and r^+h^+, "in equal frequencies in a mass lysate; this is not the case when the yield from a single cell is observed" (Simon, 1965). This is equivalent to genetic recombination in eukaryotes (Hershey and Rotman, 1948; Delbruck and Bailey, 1946; Hershey, 1946). The proportions of such recombinant types in the progeny permit calculations of the distances that separate the genetic determinants in question, and genetic maps can be constructed.

Another capability of viral genophores is that one can combine with the bacterial host genophore without killing the bacterium. This integrated segment, often referred to as the *prophage*, then functions as part (about 1 to 2%) of the bacterial genophore indefinitely and through divisions. Such prophages can modify the genetic activity of the host bacteria such as infected *Corynebacterium diphtheriae* causing diphtheria or infected *Clostridium botulinum* producing the toxin (Eklund et al., 1971). Certain treatments can later separate the viral genophore, often with some of the bacterial genophore attached to the viral genophore so that a virion can carry that portion of one genetic type of bacterial genophore (an *episome*) into another genetic type. This process is called *transduction*. The recipient bacterial cell is then "diploid" for that particular piece of genophore, and recombination can occur. Phages P1 and P22 are so-called generalized transducers able to transduce genes from any part of the bacterial genophore, in contrast to *lambda*, which can occasionally transduce only a certain region of the bacterial genophore that is always closely linked to certain of the host genes (Jacob and Wollman, 1961). Transduction is one mechanism used in bacterial genetics. It is also possible to have two strands of viral genophore in a bacterial cell, one integrated into the bacterial genophore and the other attached to an episome carrying part of a bacterial genophore. Recombination can occur between the two homologous viral genophores (Meselson, 1968), and the exchange is reciprocal. Campbell (1969) has discussed episomes in detail.

Cytogenetics of virus genophores (called molecular cytogenetics by Campbell, 1969) can be achieved by DNA-DNA hybridization of two structurally different genophores following strand separation. When the two complementary strands come together (anneal) they form a doubled strand. Such a strand is called a *heteroduplex* if the two strands differ genetically or structurally (Thomas et al., 1968; Westmoreland et al., 1969).

Thomas et al. (1968) summarized some of their work on genophores of various phages. They stated that annealing strands of DNA in vitro must be about the same as what occurs in vivo. Phages T2, T4, P22, etc., all of the larger bacteriophages, have linear duplex DNA genophores in the virion but circular form in the host cell. Thomas et al. were able to convert linear genophores into circular ones in vitro by use of the enzyme exonuclease III or alkali, which denature the 3' ends of the double DNA strand but not the 5' ends. This exposed two intact 5' ends as single strands having the same nucleotide sequence. When this had been done, the two 5' ends could base-pair, since they are complementary, and form circles. This proved the duplicate sequences at both ends. The circular form could be seen by electron microscopy. Their work also supported the hypothesis that some of these viruses, such as T4, have *permuted* genophores. That is, in a population of such a virus the linear genophores can end with any locus, but always both ends must have the same sequence of nucleotides to permit them to form circles, as they do.

Westmoreland et al. (1969) worked with wild-type as well as genetically known deletion and substitution forms of the *lambda* bacteriophage. The double-stranded DNA genophores were extracted and purified. By standard or modified methods the double strands were separated into left and right strands by equilibrium centrifugation in cesium chloride gradient containing polymerized uridine-guanine. They were then mixed, wild-type left or right strand with deletion forms right or left strands, as well as wild-type with substitution forms. As a result, hybrid left-right double strands were reformed, one strand left or right of wild-type and one right or left of mutant. These were then prepared for and examined by electron microscopy. There was a remarkable similarity between these heteroduplex genophores and maize pachytene chromosomes. Substitution forms contain somewhere along the DNA strand a piece of nonhomologous DNA de-

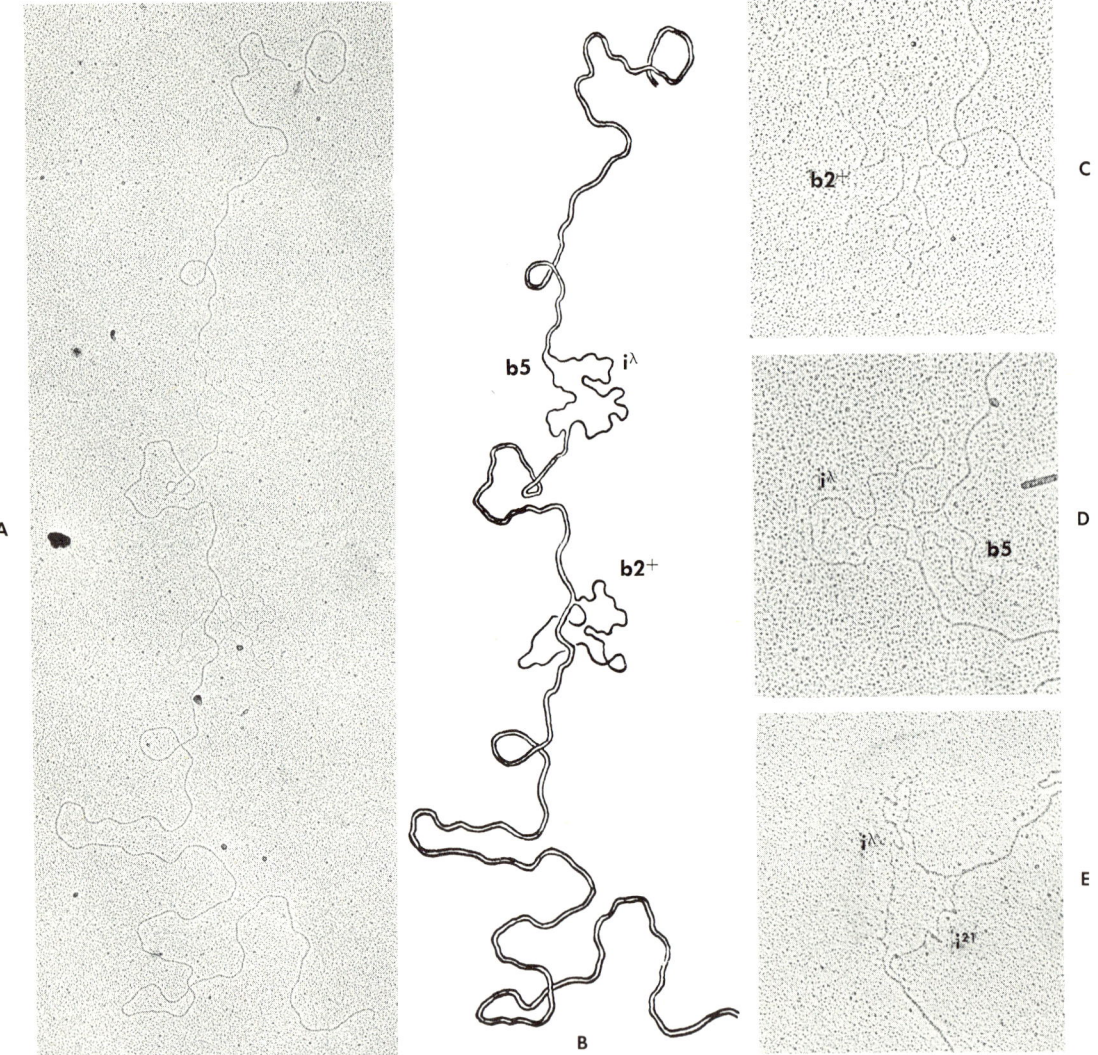

Fig. 20-4. Electron micrograph of viral genophore aberrations produced by heteroduplex formation, and an interpretive sketch. The two different strands, previously separate, were brought together by DNA-DNA hybridization. **A,** Heteroduplex formed of strand l of λ^+ and strand r of λ b2b5. **B,** Drawing of **A** showing a relative "deletion" at b2$^+$ and an unpaired (substitution) region where base sequences are different in b5 and i$^\lambda$. **C,** Another "deletion." **D** and **E,** Two other unpaired regions of substitution. (From Westmoreland, B. C., W. Szybalski, and H. Ris. 1969. Science **163:**1343-1348.)

rived from the bacterial host or other viral genophores.

Under the electron microscope a loop was formed in the longer wild type strand opposite a deletion in a mutant strand, and failure to pair occurred at a substitution (Fig. 20-4). These deletions and substitutions were measured for length and position within the genophore and compared to genetic analysis in the standard cytogenetic method so that cytological genetic maps can be constructed. They did not report any inversions or translocations, and genetic analysis has not reported any either, so perhaps these forms of chromosomal aberrations cannot occur at the level of the genophore.

Inman (1966) examined by electron microscopy the genophores of the lambda phage after treatment by mild heat and formaldehyde. The treatment produced partial denaturation at three rather definite sites (0.5_2, 0.7_3, and 0.9_8), which appeared as locally separated but not broken strands.

Deletions reduce the length of the genophore and so modify the rate of movement of strands through a gradient. As a result, deletions can be detected and percentage of length deleted determined. This and the microscopic measurement methods check each other.

Recently Gillies, et al. (1971) have published electron micrographs showing in vitro "reaction cores" of *Reovirus* producing simultaneously a number (up to 8) of mRNA strands. The significant aspects of this work are that mRNA is being transcribed directly from the double-stranded viral RNA, and, *uniquely*, the viral genophore is able to transcribe somehow in a condensed condition. There is no other known case of condensed nucleic acid being able to transcribe or replicate.

This section closes with the last short paragraph of the paper by Thomas et al. (1968), the opening of which was cited earlier: "As for the future it may be soon possible to study those structures involved in genetic recombination. If this proves possible the parallel between cytogenetics and molecular genetics will be complete. Finally the procedures are now available to make a renewed effort at dissecting the anatomy of vertebrate chromosomes."

BACTERIAL CYTOGENETICS

Most of what is known about cytogenetics of Prokaryota is the result of genetic and biochemical studies. Just as knowledge of genophores and genetics of viruses required equal knowledge of bacteria (bacteriophages) (Stent, 1963), so similar knowledge of bacteria very often requires use and knowledge of viruses (Jacob and Wollman, 1961) and episomes (Campbell, 1969). The quotations from Thomas et al. (1968) already cited indicate that information about chromosomes and genetics of viruses and bacteria, no matter how acquired, is still cytogenetics if it relates genes and genophores. One can "see" these relations not only with various sorts of microscopes but with centrifuges, cesium chloride gradients, annealing of nucleic acids, genetic recombination, and other microbiological and molecular biological techniques. These can all produce "pictures" of chromosomes, partial chromosomes, linkage maps, etc.

Nevertheless, true cytogenetics of bacteria is likely now that actual RNA transcription from the genophore is visible by electron microscopy (Miller et al., 1970b) (Figs. 20-5 and 20-6). Miller et al. were able to actually photograph the developmental sequence of growth of the RNA polymers along the DNA structural genes in *E. coli* and ribosomes on them probably translating the messenger RNA into protein. The genophore is probably DNA plus a protein coat, the whole being about 40 Å thick. Granules of 75 Å that are probably RNA polymerase molecules "move" along the genophore, transcribing m-RNA fibrils as they go. Immediately ribosomes become associated with each m-RNA fibril that extends outward from each polymerase molecule so that the longer the m-RNA fibril, the more ribosomes are associated along it; a maximum of about 20 is indicated. It is assumed that the ribosomes are translating protein, as a polyribosome that is attached at its growing end to the polymerase molecule and through it to the genophore. Thus a sequence of increasingly long RNA polymers is attached according to length along an active region of the genophore (an operon, up to 3μ long), ribosomes are attached along each RNA polymer, and, pre-

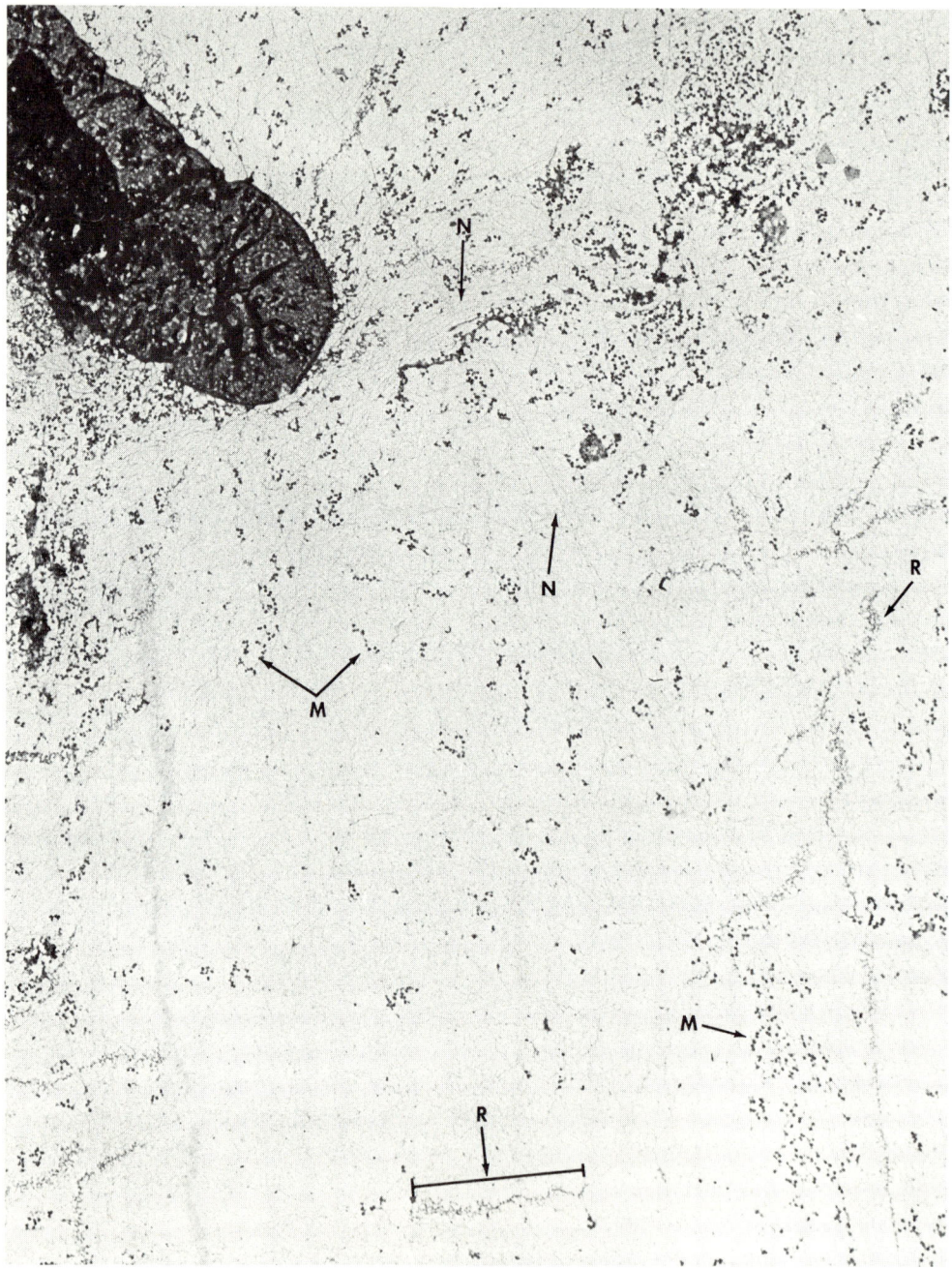

Fig. 20-5. Electron micrograph of two sorts of genes in action. Many structural genes are producing messenger RNA, which, in turn, is being translated by polyribosomes, *M* (see Fig. 20-6,*A* and *B*). Few ribosomal DNA cistrons are producing sequentially the 16s and 23s ribosomal RNA, *R* (see Fig. 20-6,*C*). The micrograph also shows many nontranscribing segments of genophore, *N*, which constitute a large percentage (most) of the total observable segments. 21,500×. (Courtesy Miller, O. L., B. A. Hamkalo, and C. A. Thomas, 1970. Science **169**: 392-395. Copyright 1970 by the American Association for the Advancement of Science.)

sumably, polypeptide strands of growing protein are associated with each translating ribosome.

If this picture represents the in vivo condition, and if all or nearly all ribosomes are permanently attached to m-RNA fibrils, as Miller et al. believe, then the m-RNA fibrils (which may be about 0.5μ long) must extend outside the region of the bacterial prokaryon where ribosomes are located in vivo. This is a picture of continuous activity in bacteria; DNA replication continuous with transcription, and RNA transcription continuous with protein translation. In eukaryotic cells these activities are separated in time and place.

It is obvious that the study of genetics of asexual, haploid, ameiotic organisms, such as bacteria, is impossible by the classical methods developed for Eukaryota. Fortunately, bacteria occasionally become "partial diploids" in

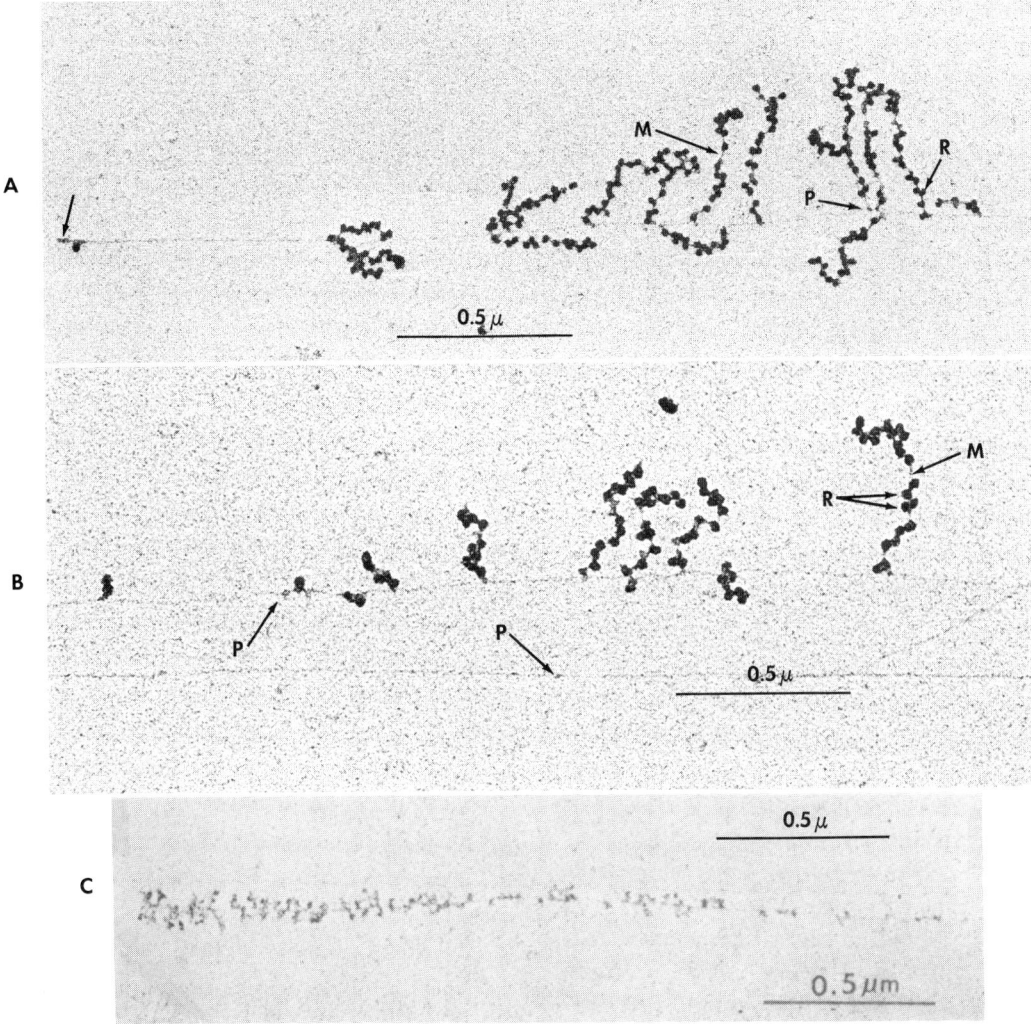

Fig. 20-6. Bacterial genes in action. Each of these is one linear segment of a genophore interpreted as one gene or one cistron. Attached along the gene are a graded series of trancribed RNA filaments, M, still attached to the genophore at a molecule of RNA polymerase, P. In **A** and **B** the ribosomes, R, are translating proteins as polyribosomes of as many as 40 individual ribosomes. Presumptive polymerase molecules seem to occur on the nontranscribing strands also (P, lower strand). (From Miller, O. L., B. A. Hamkalo, and C. A. Thomas. 1970. Science **169**:392-395. Copyright 1970 by the American Association for the Advancement of Science.)

which genetic recombination occurs, and the mechanism by which the partial diploidy is achieved lends itself by manipulation to providing data for determination of genetic linkage. As a result, the genetics and cytogenetics of bacteria not only can be and have been studied in great detail but have contributed vast amounts of knowledge of bacterial "chromosomes" (which will be called *genophores* hereafter), gene recombination, DNA replication, etc.

Transformation

There are a number of ways by which bacteria that are normally haploid can become temporarily partially diploid, at least enough for recombination to occur. The first method discovered, called transformation, has not only been a useful technique but it first strongly indicated, in 1944, that DNA and not protein is the ultimate hereditary material.

In 1928 Griffith reported that the simultaneous injection of mice with (1) viable, noncapsulated, and avirulent pneumococci of type II and (2) encapsulated but heat-killed virulent pneumococci of type I caused the death of the mice. He also recovered from them living encapsulated virulent type I bacteria. That is, dead type I bacteria in a mouse somehow changed living avirulent type II forms into living virulent type I forms. In 1931 Dawson and Sia achieved the same transformation in vitro, and Alloway in 1932 produced the transformation of type II to type I from extract of type I pneumococcus. The culmination of this line of study was performed by Avery, MacLeod, and McCarty in 1944 when purified DNA from type I was able to transform type II bacteria into type I and the transformed bacteria remained type I for generations. Subsequently, similar transformation of one bacterial type into a genetically different form has been achieved in a number of bacteria. But if the DNA is extracted form a form that differs genetically for two or more genes that are widely separated on the genetic map, almost never is the recipient transformed for more than one gene. It is concluded that the recipient receives only a very short fragment of the donor genophore. The bacterium is a partial diploid for that particular segment and the whole genophore. Some other organisms in which transformation has been reported are *Haemophilus influenzae* (Alexander and Leidy, 1951), *Bacillus subtilis* (Spizizen, 1958; Bodmer and Laird, 1968), and *Xanthomonas* (Corey and Starr, 1957). Considerable study of recombination by transformation has been done during the past 10 years (Bodmer and Laird, 1968).

Conjugation

Historically, the second mechanism by which bacteria can become partial diploids by acquisition of part of a genophore from a different genetic form is conjugation (Lederberg and Tatum, 1946).

What seems to happen is that two bacterial cells (most of this work has been done with *E. coli*) come together by random motion (illustrated by Anderson et al., 1957, with electron microscopy) and form a connection between any part of either bacterium (Fig. 20-7). The genophore of the donor starts to enter the

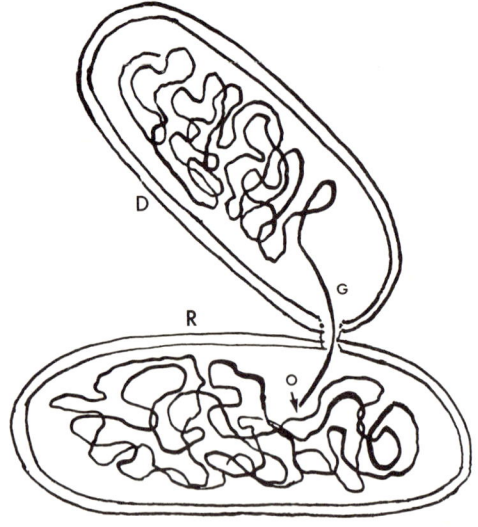

Fig. 20-7. Bacterial conjugation. The two cells must be different strains. The recipient, R, is normal; the donor, D, is unusual in having an episome called the sex factor, either F or Hfr. The genophore, G, of the donor always enters the recipient with the origin, O, leading. All of the donor genophore does not enter the recipient but enough does for genetic recombination and genetic study.

recipient via the connection, the origin (O) being the tip of the genophore first entering the recipient. This is one-way transfer (Hayes, 1953), and only certain strains can act as donor and others as recipients. Very rarely does the whole genophore of the donor enter the recipient. Almost always only a short portion enters to make a partial diploid or "zygote" or "merozygote" of the recipient (Jacob and Wollman, 1961). The formation of a merozygote by transformation, conjugation, sexduction, or transduction is called *meromixis*. In the partially diploid merozygote the partial genophore *apparently* more or less pairs with the homologous portion of the complete genophore of the recipient so that recombination can occur and genetic maps can be constructed (Lederberg, 1947). Why such intimate association of homologous regions of genophores and fragments should occur in merozygotes is unknown. It can be stated as being due to "the nature of homologous DNA," yet replicated homologous genophores separate from one another during bacterial division. Rarely does the partially diploid condition last very long, probably only until genophore replication or until division of the recipient cell (Lederberg, 1949).

Since the donor genophore always enters the recipient origin (O) first, it is possible to determine the order of genes behind the origin by breaking the entering genophore at various times and determining which genes have entered by each specific time (Wollman and Jacob, 1955). Thus a time-of-entrance linkage map can be computed.

Problems with producing linkage maps for all the known genes of the *E. coli* genophore eventually led to the proposal of a circular genophore (Jacob and Wollman, 1958). Jacob and Wollman also proposed that the origin of the genophore is not always at the same point in the circular linkage map. That fact permits the introduction of various portions of the genophore into various recipients and permits piecemeal building up and, eventually, production of the complete circular linkage map.

Quite early (Hayes, 1952) it became evident that certain strains of *E. coli* could act as donors and other strains as recipients. Thus the donor strain was considered to have a *sex factor* or F factor. Donors were considered to be F^+ and recipients F^-. F^+ mixed with F^- bacteria would produce conjugation between about two bacteria in a million. The F factor is not attached to the genophore (Lederberg et al., 1952), and usually only the F factor enters the recipient cell. Sometimes a portion of the genophore becomes attached to the F factor and can be introduced into a recipient to form a merozygote. This method of merozygote formation is called *sexduction*. The F factor is now known to contain or consist of DNA (Falkow and Citarella, 1965) of about 35×10^6 daltons and to contain 50% DNA hybridizable with the *E. coli* genophore (Falkow et al., 1966). Thus, the F factor is much like a nonlethal virus which can mutate (Campbell, 1969) and which contains from about 1 to 2% as much DNA as the bacterial genophore.

Jacob and Wollman (1961) stated that Cavalli-Sforza in 1950 found a sex factor (called Hfr, for high frequency of recombination) that achieved conjugation as donor 1,000 times as often as F^+. Hfr can "mutate" to F^+ and it is rarely transmitted by conjugation or otherwise to recipient strains. Thus, it behaves quite differently from F^+. It has since been determined that Hfr is an F factor integrated into the genophore of the donor strain where it replicates at the same rate as the genophore and occupies the opposite end of the genophore from the origin, the distal end. It thereby determines the particular origin for that strain of Hfr. When the F factor is integrated into the genophore, it opens the ring into a linear form, in which form part of the genophore can be introduced into a recipient cell. Wherever the F factor is inserted into the genophore it determines the genetic location of the origin in that particular strain and, therefore, it can be mapped. Different strains of Hfr exist and have different series of genes following the origin into the recipient. For the Cavalli strain the order is: origin, lactose, proline, leucine, threonine, arginine, and streptomycin; for the Hayes strain it is: origin, threonine, leucine, galactose, tryptophan, histidine, and streptomycin (Wood, 1967). The Hfr condition also represses the ability of the F factor to be freely trans-

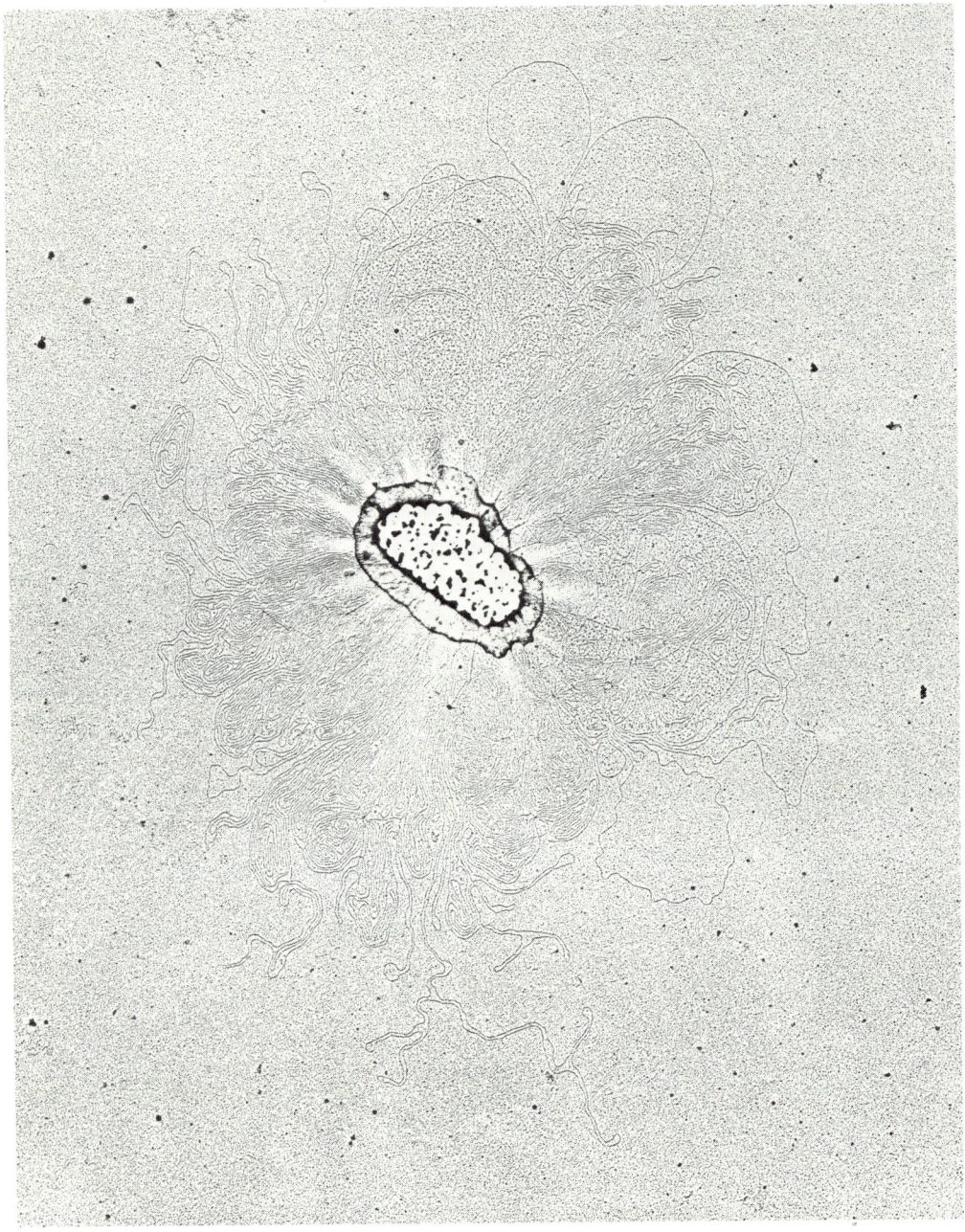

Fig. 20-8. An "exploded" bacterial cell of *Haemophilus influenzae* with the genophore spread out in a plane around it, shadowed, and photographed by electron microscopy. The remains of the bacterial cell are about 2 or 3μ long. Presumably, the genophore is a single linear or circular strand. In the living bacterium all of it was confined within probably one prokaryon such as in Fig. 20-1. The great extent of the genophore should be kept in mind during any discussion of genophores. 23,900×. (Courtesy Dr. L. A. MacHattie, Department of Medical Cell Biology, University of Toronto.)

mitted during conjugation (Jacob and Wollman, 1961).

There are several Hfr strains: Hayes, Cavalli, AB735, Hfr1, and Ra1. All of these except the Hayes seem to produce only one recombinant type per merozygote (Lederberg, 1957). There is probably no DNA replication of the donor DNA in the recipient cell previous to division and segregation, or at least not more than one round of replication of recipient male DNA (Tomizawa, 1960). Homogeneous colonies are achieved after two divisions (Lederberg, 1957). In the Hayes Hfr strain, however, segregation is delayed to between three and nine divisions; several recombinant classes are produced (Anderson, 1958), and there is probably DNA replication in the recipient and probably some descendent cells (Wood, 1967).

It is evident that just what the DNA's (genophore and fragment) do in the recipient and descendent cells following bacterial conjugation to produce the evident recombinants is highly conjectural. It seems that they must lie side by side, at least at one point, for crossing-over to occur, but this may not be so. Wood (1967) stated that the *simplest* model for genetic recombination involving the break and reunion method of crossing-over is based upon about five assumptions as follow: (1) Both donor DNA and recipient DNA are double-stranded and hemizygous. (2) The donor DNA does not replicate within the recipient before integration. (3) Integration is achieved by double-strand interaction between recipient and donor DNA. (4) Integration is fairly rapid (within the recipient before division). (5) There is no crossing-over between recombinant and recipient DNA genophores thereafter. This seems to be true except when the donor is the Hayes strain of Hfr which gives multiple recombinant classes. Copy-choice recombination is not completely ruled out as a possibility; but, in general, breakage and reunion seems to produce crossing-over in merozygotes (Siddiqi, 1963; Oppenheim and Riley, 1966). If the donor (male) DNA is integrated into the recipient genophore (Tomizawa, 1960), what happens to the "replaced" fragment of the recipient genophore, since hemizygosity results? The genetics of such mass matings is fairly well understood, but the physical basis is a number of unclear hypotheses. More analyses of colonies derived from single bacterial merozygotes isolated by micromanipulation, such as those of Lederberg (1957) and Anderson (1958), seem to be needed. Actually seeing what the DNA is doing in merozygotes would be desirable and may eventually be possible.

Gross and Caro (1965) discussed the relationship of such genetic transfer from donor to recipient to DNA replication. Of the following three alternatives that have been proposed, the first two relate transfer to DNA replication as essential, the third does not: (1) The model of Jacob and and Brenner (1963) proposes that the genophore in the Hfr donor replicates *as* genetic transfer occurs so that recipient and donor would each end up with a complete or partial donor genophore, one strand of which would have been produced after mixing of the positive and negative strains. Gross and Caro consider that the experimental evidence they cite somewhat favors this model. (2) The model of Bouck and Adelberg (1963) proposes that DNA replication is complete *before* transfer begins and the double strand (or one double strand) passes into the recipient with all or most DNA synthesized before mixing of the Hfr and F^- strains. (3) Gross and Caro state that a third possibility exists, "that there is no connection between transfer and replication: contact with the recipient would somehow proceed without regard to the state of replication."

Nagata (1963) found that when the F factor is integrated into the bacterial genophore DNA replication starts at or near the F factor and progresses toward the origin. This is the opposite direction to transfer at conjugation.

Thus the F or Hfr factor is the size of a virus (35×10^6 daltons), it is DNA, and it can mutate, become integrated into the genophore, assume control of the replication of the entire genophore, and accomplish sexduction. It is very much like a nonlethal virus that makes its host conjugate (Campbell, 1969).

Transduction

Transduction is the third mechanism by which partial diploid bacteria can be formed.

It was first reported by Zinder and Lederberg in 1952. This mechanism requires the aid of a bacterial virus of the bacterium, and the condition is called the *prophage.*

Usually the entrance of an infective viral particle into the bacterium results in the formation of numerous virions and the death by lysis of the bacterium. On the other hand, certain strains of bacteria, such as of *E. coli, Salmonella typhimurium,* or *Shigella,* when infected by certain temperate strains of virus do not result in formation of many virions or death of the bacterium by lysis. Rather, the resulting strain of bacteria, slightly altered by the presence of the virus, lives and reproduces fairly normally. Occasionally a bacterial cell will form virions and lyse, and the characteristic of doing so may be the only evidence that the bacterial strain contains the virus. This condition of harmless infection with occasional or rare lysis is called *lysogeny;* lysogenic bacteria and lysogenic strains of bacteria are referred to. "The noninfectious form in which lysogenic bacteria perpetuate the ability to produce the phage has been called *prophage"* (Jacob and Wollman, 1961).

Apparently what happens when a temperate phage particle enters a bacterium and produces a lysogenic strain is that the viral genophore is inserted (integrated) into the bacterial genophore and the two become, essentially, one (Franklin et al., 1965; Gratia, 1966). They reproduce as one, and at least some of the genes of the viral segment function. Circular viral genophores, such as that of P22 (Gough and Levine, 1968), assume a linear map configuration as prophage (Smith and Levine, 1965), as would be expected. Usually for every temperate phage there is a specific locus of the host genophore where the viral genophore can become attached (Jacob and Wollman, 1961), although the virus P2 can insert in various sites of strain C of the *E. coli* genophore (Six, 1966). The insertion of the prophage naturally increases the map distance between genes on each side of the insertion (Signer, 1966).

Transduction, then, is achieved when a temperate virus that had been integrated into

Fig. 20-9. An example of superhelical coiling of a genophore of a virus. Superhelical coiling is often reported among genophores of viruses, mitochondria, plastids, and bacteria. It may or may not occur in vivo. (Courtesy Dr. L. A. MacHattie, Department of Medical Cell Biology, University of Toronto.)

a bacterial genophore separates along with part of the host genophore and forms a merozygote; and usually, but not always (Zinder, 1953), the viral genophore produces lysogeny of the host. The frequency of transduction of a given character from a donor strain of bacteria to the recipient strain is about one virion in a million, and usually different characters are transmitted by separate acts of transduction. Apparently very small segments of host genophore are picked up at random by the phage genophores (Jacob and Wollman, 1961).

As in other-formed merozygotes, the short introduced segment of bacterial genophore associates with its homologous region of the whole genophore of the recipient, the association being intimate enough for recombination by breakage and reunion to occur. For closely linked genes it is possible to determine if both can be "cotransduced" by one viral prophage, and, if so, by what percentage of transducing acts. Jacob and Wollman (1961) cited the cotransduction of the M and *Arg* markers in 30% of all selections, which gives some additional information to the experts of the closeness of linkage of these two genes.

MITOCHONDRIAL AND PLASTID CYTOGENETICS

The popular symbiosis hypothesis of the origin of mitochondria (and plastids) proposes (Ris and Plaut, 1962; Sagen, 1967; Ellis, 1969) that they arose from the entrance of prokaryotes into the cytoplasm of either prokaryotes or eukaryotes between about one and two billion (1 to 2×10^9) years ago. One type evolved into the photosynthetic organelles of plants, called plastids; the other into the aerobic respiratory organelles of all eukaryotic cells, called mitochondria. Regardless of whether this unprovable origin is historically true or not, the similarities between them and prokaryotes are so uniformly consistent that at present the symbiotic origin is scientifically true, as the best scheme known at this time. These quasi-symbiotic organisms share with the Prokaryota essentially all of the differences that distinguish between the Eukaryota and Prokaryota. The latter, as well as mitochondria and plastids, have: no Golgi apparatus, no endoplasmic reticulum, no nucleolus, no nuclear envelope, genophores rather than chromosomes, no mitosis, no meiosis, no sexuality, no centrioles or basal bodies, no mitochondria or plastids, one genophore rather than two or more chromosomes, 70s rather than 80s ribosomes, 4s, 12s, and 21s rather than 5s, 18s, and 28s ribosomal RNA (plastids: Stutz and Noll, 1967; mitochondria: Fan and Penman, 1970); and mitochondria, at least, are affected like bacteria but not like cytoplasm of eukaryotes by inhibitors of protein synthesis, such as cycloheximide, erythromycin, chloramphenicol, and streptomycin (Scragg et al., 1971). Furthermore, mitochondria and plastids are quasi-organisms because they reproduce only by division, have genes that can and do mutate, have DNA and various types of RNA's, the complete system for protein synthesis (plastids: Smillie et al., 1967; mitochondria: Roodyn and Wilkie, 1968, and Scragg et al., 1971; polymerases: Bosmann and Martin, 1969). That is, they are partially autonomous intracellular symbionts that do, however, rely upon and are dependent on the nuclear genes and cytoplasm to some extent. They vary considerably in amount of DNA. Chloroplasts (Fig. 20-10) seem to have about as much DNA as bacteria (Ellis, 1969), enough for several thousand genes (Whitfield and Spencer, 1968); that is, as in Table 7 from about 805×10^6 daltons or 10^{-16} gm in the alga *Acetabularia* (Green and Burton, 1970: Kirk, 1963) to about 40×10^6 daltons in tobacco (Tewari and Wildman, 1966; Kirk, 1963). DNA amounts for *Chlorella, Euglena, Chlamydomonas,* and lettuce also fall within this range, up to 10^{-14} gm per plastid. Presumably, there is the same amount of DNA in leukoplasts or chromoplasts as in the green chloroplasts of the same land plant, but in these variously differentiated plastids different genes are active.

Mitochondria also vary in DNA content. Nass (1969) states that mitochondria of protozoa and most fungi have from 20×10^6 to 200×10^6 daltons of DNA; of angiosperm plants, about 300×10^6; but of most animals, especially vertebrates, only about 10×10^6 daltons of DNA (see Table 7). But even within a group of vertebrates (urodeles in contrast to anuran amphibians) and between liver and oocytes

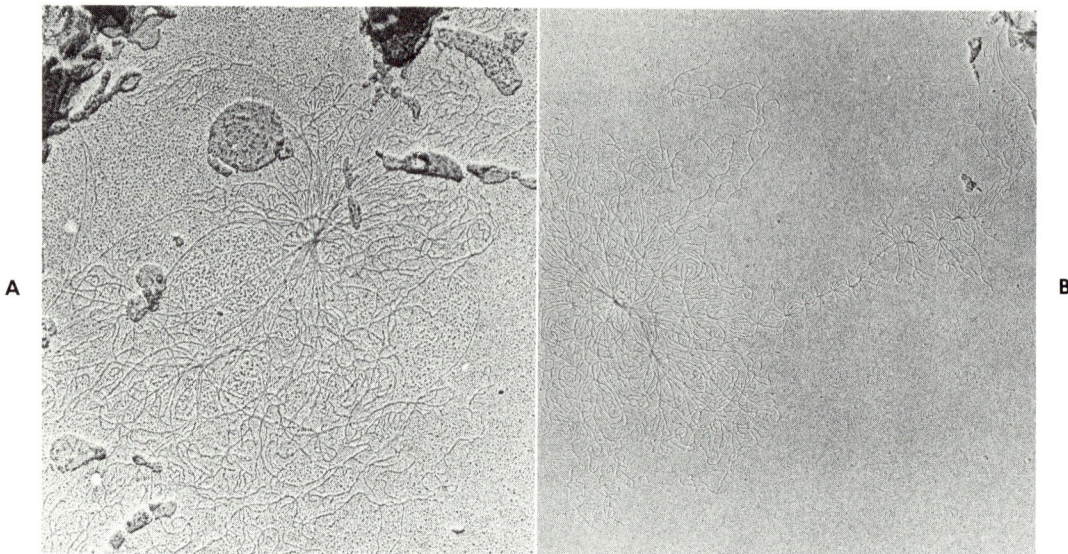

Fig. 20-10. Genophores of chloroplasts of the alga *Acetabularia*. The interesting arrangement in **B** extending outward from the general mass of the genophore may or may not be an artifact of preparation. (**A**, From Green, B. R., and H. Burton. 1970. Science **168**:981-982. **B**, Courtesy Dr. B. R. Green.) **A**, 31,250×; **B**, 19,500×.

of the same individual (Wolstenholme and Dawid, 1968) the circumferences vary, from 4.77 to 4.96μ in urodele oocytes versus 5.79 to 5.82μ in anuran oocytes, and about 4.68μ for liver of urodeles.

Very unusual "prokaryons" of one particular mitochondrion in the cell of certain flagellate protozoa have been described as the so-called kinetoplast (Burton and Dusanic, 1968). This is a large, transversely lying mitochondrion just interior to the basal body. Other mitochondria of the cell are normal. The prokaryon nearly fills the mitochondrion and stains for DNA so intensely and it is so large that it is clearly visible in the light microscope. It has its filaments running transversely to its length, and some sort of transverse alternation of masses of such filaments and blank regions produces a distinctive organization. It incorporates radioactive thymidine as does the nucleus. It divides transversely into two or more end-to-end prokaryons. Analogously, it seems to be to a typical prokaryon as a ciliate macronucleus is to a ciliate micronucleus or as a giant polytene chromosome is to a typical chromosome. It is likely to have a great deal of DNA and to represent an evolved type of mitochondrion just as green algal chloroplasts have evolved in vascular plants into leukoplasts, chromoplasts, and nongranal chloroplasts.

Wagner (1969) has discussed the genetics of mitochondria among eukaryotic organisms in general but especially of *Neurospora* and yeast. Most of these are biochemical mutants such as the so-called poky type of *Neurospora* and the so-called petite type of *Saccharomyces cerevisiae* which have deficiencies in respiratory capacity due to mutant inability of the mitochondria to synthesize certain enzymes. The mutations of the mitochondria affect the phenotypes of the organisms. Nevertheless, the nucleus of the cell does affect the mitochondria and plastids. There are many nuclear mutants known of many plants, many in the same species, that prevent obvious chlorophyll formation in plastids (Jinks, 1964); and there are plastid mutations that also affect chlorophyll formation.

Garnjobst et al. (1965) and Diacumakos et al. (1965) have demonstrated that certain mutant mitochondria can operate independently of nuclear controls. Isolated mitochondria of poky or *abn*-1 *Neurospora* injected into wild-type cells result in the cells taking on the mutant characteristics of growth rate, morphology, reproductive characteristics, and cytochrome patterns. In these examples the mutant mitochondria either drove out or at least dominated the wild-type mitochondria already in the cells (Wagner, 1969).

In maize, however, a peculiar result has been noticed when two inbred strains that differed for mitochondrial density (as measured by centrifugation in a linear sucrose gradient) were crossed and the density of the mitochondria of the hybrid determined (Sarkissian and McDaniel, 1967; Sarkissian and Srivastava, 1967). The mitochondria of strain *wf*45 sedimented as one band; those of *Oh*9 sedimented as two bands, one of which corresponded to that of *wf*45; and mitochondria of the hybrid sedimented as three bands, the upper and lower corresponding to the two bands of *Oh*9. The third or middle band of the hybrid was new and possibly corresponded to mitochondria of the upper band and lower band types that had fused. But, if that is the correct explanation of the intermediate band, why they should not fuse in the *Oh*9 strain is unknown.

Another example of possible allelic genetic strains in the mitochondria of maize has recently been reported by Miller and Koeppe (1971). One, the susceptible strain, occurs in the *cms*-T (Texas) type cytoplasm; the resistant strain of mitochondria studied occurs in the cytoplasm of maize strain PAG 15029. The toxin is produced by a new race of the southern corn leaf blight fungus, *Helminthosporium maydis*, which was so destructive of corn in 1970. The toxin does not affect the isolated resistant mitochondria, but isolated susceptible mitochondria are caused to swell irreversibly by effect on the mitochondrial membrane system, which in turn stops ATP production. Deprivation of ATP kills the cells and plants of the susceptible strains.

Whether the "gene" that produces susceptible or resistant mitochondria as a "mutation" is mitochondrial or nuclear has not been fully determined; but because these alternative conditions are cytoplasmically inherited Miller and Koeppe consider that they are "probably hereditarily transferred in an extranuclear and extrachromosomal manner," that is, within mitochondrial genophores.

The researchers refer to essentially similar strains of mitochondria in oats that are susceptible or resistant to the toxin, victorin, produced by the related fungus *H. victoriae*. Cytogenetic study by heteroduplex formation of such genophores might yield significant results.

The genetics of plastids and other forms of extrachromosomal inheritance have been discussed by Jinks (1964). Ris and Plaut (1962) first adequately demonstrated that DNA is present in plastids. Unique enzymes of various sorts including polymerases, unique ribosomes, and ribosomal RNA are also found in plastids as in mitochondria (Nass, 1969).

The cytogenetics of mitochondria has been treated recently and briefly by Nass (1969). As is true for viruses and bacteria, much of our knowledge of the mitochondrial genophore has been derived from molecular biological studies, but there are also numerous electron microscopic studies of the isolated genophores of numerous organisms.

The DNA of mitochondria has a density different from that of the nuclei of the same cells and separates out in density gradients as a satellite band (Luck and Reich, 1964). The genophores of mitochondria of numerous animals and yeast are short, having a length of only from 4.7 to 5.5μ (Nass et al., 1965), and have molecular weights from 9 to 10×10^6 daltons. The indicated differences are in part due to technical factors, but real length differences do exist, such as from 4.7 to 4.8μ for mouse fibroblast and from 5.1 to 5.2μ for chicken liver mitochondria treated together as well as separately. The genophores are usually circular (Sinclair and Stevens, 1966), and when extracted and spread on a protein film they range from quite open to greatly twisted and contorted. In this respect they resemble genophores extracted from virions. In both cases, and in bacteria even more so, the genophores in vivo must be highly compacted to fit within the small volume of the capsid or prokaryon, or that specific region of a mitochondrion or plastid occupied by the DNA. The problem of genophore separation at mitochondrial division is much simpler with its 5μ genophore than for an equal-sized bacterium with a genophore of from 1,000 to $2,000\mu$. But details of such division are completely unknown.

Isolated genophores also appear occasionally with perimeters twice (dimers) or more the expected 5μ, as two or more circles interlocked, as linear strands, and as "nicked" single or interlocked rings. Nicked rings have one or more single-strand *scissions* (regions of single strands) rather than the continuous double helix. It is probable that dimers and probably interlocked rings exist as such within the living mitochondria. Yeast (Sinclair et al., 1967) and protozoan (Nass, 1969) mitochondria, on the other hand, seem to be linear in vivo. Because each end has a protruding single strand, the ends can be annealed to form circles, as is true of the lambda virus. Nass also reports that tumor cells contain fewer than normal numbers of mitochondria but high numbers of dimers, oligomers, and interlocked mitochondrial genophores. It is proposed that inability of such forms to separate results in fewer mitochondria or that arrest of mitochondrial division may result in formation of the odd genophores. The assumption is that they are abnormal.

Genophores of mitochondria seem to be attached to the mitochondrial membrane, especially in actively growing cells. Similar attachment has been reported in bacteria, and it has been proposed that such attachment may be involved in the mechanics of genophore separation. The twistings or coilings of the pulled-out coil, the so-called supercoils, that are often or usually seen of mitochondrial and viral genophores are assumed, though definitely not proved, to reflect the condition in the confined particle, but such complexity of form does complicate all models of mechanisms of separation. It is also assumed that a nick, a short region of single-strandedness, could permit the supercoiled condition to untwist. But, as Nass stated, "The mechanical aspects of DNA replication in mitochondria (initiation points) may remain obscure for some time. Even in studies with bacteria, which are more amenable to experimentation than mitochondria are, the number of working models available is roughly proportional to the number of investigators making the studies."

Electron microscopy and biochemical techniques of so-called petite mutations of yeast mitochondria have failed to detect (Moustacchi and Williamson, 1966) or have shown reduced amount (Tewari et al., 1966) or significantly different base composition (Mounolou et al., 1966) of DNA from wild type. The result is lack of respiratory enzymes, cytochromes, and some dehydrogenases and abnormal, membrane-deficient mitochondria (Yotsuyangi, 1962).

Mitochondrial genophores seem to be self-replicating in the same way as circular viral genophores, showing in the electron microscope the double-forked initiation plus replication loop.

It seems that the cytogenetics of prokaryotes and quasi-organisms such as viruses, mitochondria, and plastids is and will be based more upon molecular biological studies including molecular genetics than upon observation of genophores in the electron microscope. But the latter is now significant and will continue to give information on the physical basis of molecular biological hypotheses. Together they represent cytogenetics in the broadest sense.

Literature cited

Aamendares, S., L. Buentello, A. Cuevas-Sosa, and J. M. Cantu-Garza. 1969. Familial extra centric bisatellited chromosome. Cytogenetics 8:177-185.

Aaronson, S. A., and G. J. Todaro. 1969. Human diploid cell transformation by DNA extracted from the tumor virus SV40. Science 166:390-391.

Aarskog, D. 1966. A new cytogenetic variant of translocation Down's syndrome. Cytogenetics 5:82-87.

Aase, H. 1930. Cytology of *Triticum, Secale,* and *Aegilops* hybrids with reference to phylogeny. Res. Stud. St. Coll. Washington 2:5-60.

Abraham, S., I. H. Ames, and H. H. Smith. 1968. Autoradiographic studies of DNA synthesis in the B chromosomes of *Crepis capillaris.* J. Hered. 59:297-299.

Adman, R., and D. A. Pious. 1970. Isoantigenic variants; isolation from human diploid cells in culture. Science 168:370-372.

Aksel, R. 1967. Quantitative genetic analysis of characters in wheat using chromosome substitution lines (theoretical considerations). Genetics 57:195-211.

Al-Aish, M. S. 1969. Human chromosome morphology. I. Studies on normal chromosome characterization, classification and karyotyping. Canad. J. Genet. Cytol. 11:370-381.

Al-Aish, M. S. 1970. Human chromosome morphology. II. Studies on the satellites and identification of the large acrocentric chromosomes. Canad. J. Genet. Cytol. 12:160-163.

Al-Aish, M. S., F. de la Cruz, L. A. Goldsmith, J. Volpe, G. Mella, and J. C. Robinson. 1967. Autosomal monosomy in man. Complete monosomy G (21-22) in a four-and-one-half-year-old mentally retarded girl. New Eng. J. Med. 277:777-784.

Alexander, H., and G. Leidy. 1951. Determination of inherited traits of *H. influenzae* by desoxyribonucleic acid fractions isolated from type specific cells. J. Exp. Med. 93:345-359.

Allen, C. E. 1935. The genetics of bryophytes. Bot. Rev. 1:269-291.

Allen, C. E. 1940. The genotypic basis of sex-expression in Angiosperms. Bot. Rev. 6:227-300.

Allen, C. E. 1945. The genetics of bryophytes. II. Bot. Rev. 11:260-287.

Allen, N. S., G. B. Wilson, and S. Powell. 1950. Comparative effects of colchicine and sodium nucleate. J. Hered. 41:159-163.

Alloway, J. L. 1932. The transformation in vitro of R pneumococci into S forms of different specific types by the use of filtered pneumococcus extracts. J. Exp. Med. 55:91-99.

Alverdes, Fr. 1912. Der Kern in den Speicheldrusen der *Chironomus*-Larve. Arch. Zellforsch. 9:168-204.

Amabis, J. M., and D. Cabral. 1970. RNA and DNA puffs in polytene chromosomes of *Rhynchosciara*: inhibitions by extirpation of prothorax. Science 169:692-694.

Ambühl, H. 1953. Uber die Mitoseverhältnisse im Felchenei (*Coregonus lavaretus* L. Steinmann). Schweiz. Z. Hydrobiol. 15:221-234.

Ammermann, D. 1969. Release of DNA breakdown products into the culture medium of *Stylonychia mytilus* ex-conjugants (protozoa ciliata) during the destruction of the polytene chromosomes. J. Cell Biol. 40:576-577.

Anderson, E. 1949. Introgressive Hybridization. John Wiley and Sons, Inc., New York.

Anderson, E. 1953. Introgressive hybridization. Biol. Rev. 28:280-307.

Anderson, E., and K. Sax. 1936. A cytological monograph of the American species of *Tradescantia*. Bot. Gaz. 97:433-476.

Anderson, E. G. 1935. Chromosomal interchanges in maize. Genetics **20**:70-83.

Anderson, L. M., and C. J. Driscoll. 1967. The production and breeding behaviour of a monosomic alien substitution line. Canad. J. Genet. Cytol. **9**:399-403.

Anderson, T. F. 1958. Recombination and segregation in *Escherichia coli*. Cold Spring Harbor Symp. Quant. Biol. **23**:47-58.

Anderson, T. F., E. L. Wollman, and F. Jacob. 1957. Sur les processus de conjugaison et de recombinaison chez *E. coli*. III. Aspects morphologiques en microscopie électronique. Ann. Inst. Pasteur **93**:450-455.

Antropova, E. N., and Y. F. Bogdanov. 1970. Cytophotometry of DNA and histone in meiosis of *Pyrrhocoris apterus*. Exp. Cell Res. **60**:40-44.

Archebald, E. E. A. 1939. The development of the ovule and seed of jointed cactus, *Opuntia aurantiaca* (Lindley). S. Afr. J. Sci. **36**:195-211.

Arrighi, F., and T. C. Hsu. 1971. Localization of heterochromatin in human chromosomes. Cytogenetics **10**:81-88.

Aslamkhan, M., and R. H. Baker. 1969. Karyotypes of some *Anopheles, Ficalbia*, and *Culex* mosquitoes of Asia. Pakistan J. Zool. **1**:1-7.

Atwood, S. 1937. The nature of the last premeiotic mitosis and its relation to meiosis in *Gaillardia*. Cellule **46**:391-409; also Proc. Nat. Acad. Sci. U. S. A. **23**:1-5.

Avdulow, N. P. 1931. Karyo-systematic studies in the grass family. (In Russian, German summary.) Bull. Appl. Bot. Suppl. 44.

Avers, C. J. 1958. Histochemical localization of enzyme activity in the root epidermis of *Phleum pratense*. Amer. J. Bot. **45**:609-613.

Avers, C. J., and R. B. Grimm. 1959. Comparative enzyme differentiation in grass roots. I. Acid phosphatase. Amer. J. Bot. **46**:190-193.

Avery, A. G., S. Satina, and J. Rietsema. 1959. Blakeslee: the Genus *Datura*. The Ronald Press Co., New York.

Avery, O. T., C. M. MacLeod, and M. McCarty. 1944. Studies on the chemical nature of the substance inducing transformation of pneumococcal types. Induction of transformation by a desoxyribonucleic acid fraction isolated from *Pneumococcus* type III. J. Exp. Med. **79**:137-158.

Babcock, E. B. 1947. The Genus *Crepis*. I. The Taxonomy, Phylogeny, Distribution, and Evolution of *Crepis*. University of California Press, Berkeley, Calif.

Babcock, E. B. 1949. Supplementary notes on *Crepis*. II. Phylogeny, distribution, and Matthew's principle. Evolution **3**:374-376.

Babcock, E. B. 1950. Supplementary notes on *Crepis*. III. Taproot versus rhizome in phylogeny. Evolution **4**:358-359.

Babcock, E. B., and G. L. Stebbins. 1938. The American species of *Crepis*: their relationships and distribution as affected by polyploidy and apomixis. Carnegie Inst. Wash. Publ. No. 504.

Bachmann, K., O. B. Goin, and C. J. Goin. 1966. Hylid frogs: polyploid classes of DNA in liver nuclei. Science **154**:650-651.

Badr, F. M., and R. S. Badr. 1970. The somatic chromosomes of a wild population of rats: numerical polymorphism. Chromosoma **30**:465-475.

Bajer, A. 1965. Behavior of chromosomal spindle fibers in living cells. Chromosoma **16**:381-390.

Baker, R. H. 1968. The genetics of "golden," a new sex-linked color mutant of the mosquito *Culex tritaeniorhynchus* Giles. Ann. Trop. Med. Parasit. **62**:193-199.

Baker, R. H., and M. Aslamkhan. 1969. Karyotypes of some Asian mosquitoes of the subfamily Culicinae (Diptera: Culicidae). J. Med. Entom. **6**:44-52.

Baker, R. H., A. S. Nasir, and M. Aslamkhan. 1968. The salivary gland chromosomes of *Anopheles pulcherrimus* Theobald. Parasitologia **10**:167-177.

Baker, R. H., and M. G. Rabbani. 1970. Complete linkage in females of *Culex tritaeniorhynchus*. J. Hered. **61**:59-61.

Baker, R. J., and T. C. Hsu. 1970. Further studies on the sex-chromosome system of the American leaf-nosed bats (Chiroptera, Phyllostomatidae). Cytogenetics **9**:138-141.

Baker, W. K. 1968. Position effect variegation. Advances Genet. **14**:171-223.

Bakerspigel, A. 1959. The structure and manner of division of the nuclei in the vegetative mycelium of *Neurospora crassa*. Amer. J. Bot. **46**:180-189.

Baldwin, J. T. 1938. *Kalanchoe:* the genus and its chromosomes. Amer. J. Bot. **25**:572-579.

Baldwin, J. T. 1941. *Galax:* the genus and its chromosomes. J. Hered. **32**:249-254.

Baldwin, J. T., and B. M. Speese. 1955. Chromosomes of taxa of the Alismataceae in the range of Gray's Manual. Amer. J. Bot. **42**:406-611.

Baldwin, J. T. 1942. Polyploidy in *Sedum ternatum* Michx. II. Cytogeography. Amer. J. Bot. **29**:283-286.

Barr, H. J., and R. Ellison. 1971. Quinacrine staining of chromosomes and evolutionary studies in *Drosophila*. Nature **233**:190-191.

Barr, M. L., and E. G. Bertram. 1949. A morphological distinction between the neurones of the male and female, and the behaviour of the nucleolar satellite during accelerated nucleo-protein synthesis. Nature **163**:676-677.

Barry, E. G. 1969. The diffuse diplotene stage of meiotic prophase in *Neurospora*. Chromosoma **26**:119-129.

Bashaw, E. C., A. W. Hain, and E. C. Holt. 1970. Apomixis, its evolutionary significance and utilization of plant breeding. Proceedings of Eleventh International Grassland Congress, p. 245-248.

Basile, R. 1969. Nucleic acid synthesis in nurse cells of *Rhynchosciara angelae* Nonatl. and Pavan, 1951. Genetics (Suppl. 61) **1**, 1969:261-273.

Bateson, W. 1909. Mendel's Principles of Heredity. London.

Bateson, W., E. R. Saunders, and R. C. Punnet. 1906. Experimental studies in the physiology of heredity. Report 3, Evaluation Comm., Roy. Soc.

Battaglia, E. 1946. Ricerche cariologiche ed embriologiche sul genere *Rudbeckia*—VIII: Semigamia in *Rudbeckia laciniata*. Nuovo G. Bot. Ital. **53**:483-511.

Battaglia, E. 1947. Divisione eterotypica in cellule somatiche di *Sambucus ebulus* L. Nuovo G. Bot. Ital. **54**:724-733.

Battaglia, E. 1950. L'alterazione della meiosi nella reproduzione apomittica di *Erigeron karwinskianus* var. *mucronatus*. Caryologia **11**:165-204.

Battaglia, E. 1954. Assenza di centromero localizzato in *Heleocharis uniglumis* (Link) Schult. Caryologia **6**:319-332.

Battaglia, E. 1955a. A consideration of a new type of meiosis (mis-meiosis) in Juncaceae *(Luzula)* and Hemiptera. Bull. Torrey Bot. Club **82**:383-396.

Battaglia, E. 1955b. Unusual cytological features in the apomictic *Rudbeckia sullivantii* Boynton et Beadle. Caryologia **8**:1-32.

Battaglia, E. 1963. Apomixis. In P. Maheshwari (editor). Recent Advances in the Embryology of Angiosperms. International Society of Plant Morphology, Delhi.

Battaglia, E. 1964. Cytogenetics of B-chromosomes. Caryologia **17**:245-299.

Bauer, H. 1936. Structure and arrangement of salivary gland chromosomes in *Drosophila* species. Proc. Nat. Acad. Sci. U. S. A. **22**:216-222.

Bauer, H. 1967. Die kinetische Organisation der Lepidopteren-Chromosomen. Chromosoma **22**: 101-125.

Bauer, H., and T. Dobzhansky. 1937. A comparison of gene arrangements in *Drosophila azteca* and *D. athabasca*. Genetics **22**:185-197.

Beadle, G. W. 1930. Genetical and cytological studies of mendelian asynapsis in *Zea mays*. Cornell Univ. Agr. Exp. Sta. Mem. No. 129, pp. 1-23.

Beadle, G. W. 1931. A gene in maize for supernumerary cell division following meiosis. Cornell Univ. Agr. Exp. Sta. Mem. No. 135.

Beadle, G. W. 1932a. A gene in *Zea mays* for failure of cytokinesis during meiosis. Cytologia **3**:142-155.

Beadle, G. W, 1932b. A gene for sticky chromosomes in *Zea mays*. Ind. Abst. Vererb. **63**:195-217.

Beadle, G. W. 1935. Crossing over near the spindle attachment of the X chromosomes in attached X triploids of *Drosophila melanogaster*. Genetics **20**: 179-191.

Beadle, G. W., and E. L. Tatum. 1941. Genetic control of biochemical reactions in *Neurospora*. Proc. Nat. Acad. Sci. U. S. A. **27**:499-506.

Beaman, J. H. 1957. The systematics and evolution of *Townsendia* (Compositae). Contrib. Gray Herb. No. 183:1-151.

Beamish, K. I. 1955. Seed failure following hybridization between hexaploid *Solanum demissum* and four diploid *Solanum* species. Amer. J. Bot. **42**:297-304.

Beasley, J. O. 1938. Nuclear size in relation to meiosis. Bot. Gaz. **99**:865-871.

Beasley, J. O. 1940. The production of polyploids in *Gossypium*. J. Hered. **31**:39-48.

Beasley, J. O. 1942. Meiotic chromosome behavior in species, species hybrids, haploids, and induced polyploids of *Gossypium*. Genetics **27**:25-54.

Beatty, R. A. 1967. Parthenogenesis in Vertebrates. In C. B. Metz and A. Monroy (editors). Fertilization. Academic Press Inc., New York.

Beçak, M. L., W. Beçak, and M. N. Rabello. 1966. Cytological evidence of constant tetraploidy in the bisexual South American frog *Odontophrynus americanus*. Chromosoma **19**:188-193.

Beçak, M., L. Denaro, and W. Beçak, 1970. Polyploidy and mechanisms of karyotype diversification in Amphibia. Cytogenetics **9**:225-238.

Beçak, W. 1969. Gene action and polymorphism in polyploid species of amphibians. Genetics (Suppl. 61) **1**, 1969:183-190.

Becker, H. J. 1957. Uber Rontgenmosaikflecken und Defektmutationen am Auge von *Drosophila* und die Entwicklungsphysiologie des Auges. Z. Ind. Abst. Vereb. **88**:333-373.

Bedichek, S. 1938. Sex balance in the progeny of triploid *Drosophila*. Rec. Genet. Soc, Amer. **7**:94-95. (Abstr.)

Beermann, W. 1952. Chromosomenkonstanz und spezifische modifikationen der chromosomenstruktur in der enturcklung und organdifferenzierung von *Chironomus tentans*. Chromosoma **5**: 139-198.

Beermann, W. 1960. Der Nukleolus als lebenswichtiger Bestandteil des Zellkerns. Chromosoma **11**: 263-296.

Beermann, W. 1961. Ein Balbiani-Ring als Locus einer Speicheldrüsen-Mutation. Chromosoma **12**:1-25.

Beermann, W. 1965. Cytological aspects of information transfer in cellular differentiation. In E. Bell (editor). Molecular and Cellular Aspects of Development. Harper & Row, Publishers, New York.

Beermann, W., and G. F. Bahr. 1954. The submicroscopic structure of the Balbiani-ring. Exp. Cell Res. **6**:195-201.

Beermann, W., and U. Clever. 1964. Chromosome puffs. Scientific American, **210**:50-58.

Beermann, W., and C. Pelling. 1965. H^3-thymidinmarkierung einzelner chromatiden in Riesenchromosomen. Chromosoma **16**:1-21.

Bell, E. 1969. I-DNA: its packaging into I-somes and its relation to protein synthesis during differentiation. Nature **224**:326-328.

Bell, E. 1971. Informational DNA synthesis distinguished from that of nuclear DNA by inhibitors of DNA synthesis. Science **174**:603-606.

Bellamy, A. W. 1936. Interspecific hybrids in *Platypoecilus*: one species ZZ-WZ the other XY-XX. Proc. Nat. Acad. Sci. U. S. A. **22**:531-536.

Belling, J., and A. F. Blakeslee. 1922. The assortment of chromosomes in triploid daturas. Amer. Nat. **56**:339-346.

Belling, J., and A. F. Blakeslee. 1924. The configurations and sizes of the chromosomes in the trivalents of 25-chromosome daturas. Proc. Nat. Acad. Sci. U. S. A. **10**:116-120.

Belling, J., and A. F. Blakeslee. 1927. The assortment of chromosomes in haploid daturas. Cellule **37**:355-361.

Berendes, H. D., and G. F. Meyer. 1968. A specific chromosome element, the telomere of *Drosophila* polytene chromosomes. Chromosoma **25**:184-197.

Berg, K., and A. G. Bearn. 1968. Human serum protein polymorphisms. Ann. Rev. Genet. **2**:341-362.

Berger, C. A. 1938. Multiplication and reduction of somatic chromosome groups as a regular developmental process in the mosquito, *Culex pipiens*. Carnegie Inst. Wash. Contrib. Embryol. No. 167, p. 210-232.

Berger, C. A. 1940. The uniformity of the gene complex in the nuclei of different tissues. J. Hered. **31**:3-4.

Berger, C. A., R. M. McMahon, and E. R. Witkus. 1955. The cytology of *Xanthisma texanum* D. C. III. Differential somatic reduction. Bull. Torrey Bot. Club **82**:377-382.

Berger, C. A., and E. R. Witkus. 1943. A cytological study of C-mitosis in the polysomatic plant *Spinacia oleracea*, with comparative observations on *Allium cepa*. Bull. Torrey Bot. Club **70**:457-467.

Berghe, H. van den. 1970. Nuclear sexing in a population of Congolese metropolitan newborns. Science **169**:1318-1320.

Berrie, G. K., and F. W. Sansome. 1948. Wild population studies; *Drosophila funebris* near Manchester. J. Genet. **48**:151-152.

Bhaduri, P. N. 1942. Further cytological investigations in the genus *Gaura*. Ann. Bot. **6**:229-244.

Bianchi, A., G. Bellini, M. Contin, and E. Ottaviono. 1961. Non-disjunction in the presence of interchanges involving B-type chromosomes in maize and some phenotypical consequences of meaning in maize breeding. Z. Ind. Abst. Vererb. **92**:213-232.

Bianchi, A., and A. Morandi. 1962. A case of balanced lethal factors in maize. Heredity **17**:409-414.

Bianchi, N. O., and M. S. A. Bianchi. 1965. DNA replication sequence of human chromosomes in blood cultures. Chromosoma **17**:273-290.

Bianchi, N. O., and M. S. A. Bianchi. 1969. Origin of the pattern and chronology of chromosome replication in vertebrates. Genetics (Suppl. 61) **1**, 1969: 233-242.

Bianchi, N. O., J. Contreras, and N. Dulout. 1969. Intraspecies autosomal polymorphism and chromosome replication in *Akodon molinae* (Rodentia, Cricetidae). Canad. J. Genet. Cytol. **11**:233-242.

Bishop, A., M. Leese, and C. E. Blank. 1965. The relative length and arm ratio of the human late-replicating X chromosome. J. Med. Genet. **2**:107-111.

Bisset, K. A. 1952. The evidence for mitotic spindles in bacteria. Science **116**:154-155.

Blackwood, M. 1956. The inheritance of B chromosomes in *Zea mays*. Heredity **10**:353-366.

Blair, A. P. 1941. Variation, isolating mechanisms and hybridization in certain toads. Genetics **26**: 398-417.

Blakeslee, A. F. 1921. The "Globe," a simple trisomic mutant in *Datura*. Proc. Nat. Acad. Sci. U. S. A. **7**:148-152.

Blakeslee, A. F., and B. T. Avery. 1919. Mutations in the Jimson weed. J. Hered. **10**:111-120.

Blakeslee, A. F., and J. Belling. 1924. Chromosomal mutations in the Jimson weed, *Datura stramonium*. J. Hered. **15**:195-206.

Blakeslee, A. F., J. Belling, M. E. Farnham, and A. D. Bergner. 1922. A haploid mutant in the jimson weed, "*Datura stramonium*." Science **55**:1-3.

Bloom, A. D., S. Neriishi, N. Kamada, T. Iseki, and R. J. Keehn. 1966. Cytogenetic investigation of survivors of the atomic bombings of Hiroshima and Nagaskaki. Lancet **2**:672-674.

Bloom, S. E., P. S. Gerald, and L. E. Reisman. 1967. Ring D chromosome: a second case associated with anomalous haptoglobin inheritance. Science **156**:1746-1748.

Bloom, S. E. 1970. Trisomy-3,4 and triploidy (sA-ZZZW) in chick embryos: autosomal and sex chromosomal nondisjunction in meiosis. Science **170**: 457-458.

Bloom, S. E., and E. G. Buss. 1966. Triploid-diploid mosaic chicken embryo. Science **153**:759-760.

Boasson, R., and W. M. Laetsch. 1969. Chloroplast replication and growth in tobacco. Science **166**: 749-751.

Bock, I. R. 1971. Intra- and interspecific chromosomal inversions in the *Drosophila bipectinata* species complex. Chromosoma **34**:206-229.

Boczkowski, K. 1968. Genetical studies in testicular feminization syndrome. J. Med. Genet. **5**:181-188.

Bode, H. R., and H. J. Morowitz. 1957. Size and structure of the *Mycoplasma hominis* H39 chromosome. J. Molec. Biol. **23**:191-199.

Bodmer, W. F., and C. D. Laird. 1968. Molecular mechanism of recombination in *Bacillus subtilis* transformation. In W. J. Peacock and R. D. Brock (editors). Replication and Recombination in Genetic Material. Australian Academy of Science, Canberra.

Boettger, C. 1922. Uber freilebende Hybriden der Landschnecken *Cepaea nemoralis* L. und *Cepaea hortensis* Müll. Zool. Jahresber. (Syst.) **44**:297-336.

Bogart, J. P. 1967. Chromosomes of the South American amphibian family Ceratophridae with reconsideration of the taxonomic status of *Odontophrymus americanus*. Canad. J. Genet. Cytol. **9**:531-542.

Bogart, J. P., W. F. Blair, and C. Hubbs. 1969. Comments on the "cytologically verified" hybrids

between *Bufo* and *Rana,* two genera of anuran amphibians. Cytogenetics **8**:30-34.
Bonhoeffer, F., and A. Gierer. 1963. On the growth mechanism of the bacterial chromosome. J. Molec. Biol. **7**:534-540.
Bonner, J., M. Dahmus, D. Fambrough, R. C. Huang, K. Marushige, and D. Tuan. 1968. The biology of isolated chromatin. Science **159**:47-56.
Bonnevie, K. 1911. Chromosomenstudien. III. Chromatinreifung in *Allium cepa.* Arch. Zellforsch. **6**:190-253.
Bonnier, G. 1933. Crossing-over in triploids of *Drosophila melanogaster* with attached X-chromosomes. Hereditas **17**:342-362.
Böök, J. A., M. Fraccaro, and J. Lindsten. 1959. Cytogenetical observations in mongolism. Acta Paediat. Scand. **48**:453-458.
Böök, J. A., and B. Santesson. 1960. Malformation syndrome in man associated with triploidy (69 chromosomes). Lancet **1**:858-859.
Bordet, J. 1925. Le problème de l'autolyse microbienne transmissable on du bacteriophage. Ann. Inst. Pasteur **39**:717-763.
Bosemark, N. O. 1956a. On accessory chromosomes in *Festuca pratensis.* IV. Cytology and inheritance of small and large accessory chromosomes. Hereditas **42**:235-260.
Bosemark, N. O. 1956b. Cytogenetics of accessory chromosomes in *Phleum phleoides.* Hereditas **42**: 443-466.
Bosemark, N. O. 1957. On accessory chromosomes in *Festuca pratensis.* V. Influence of accessory chromosomes on fertility and vegetative development. Hereditas **43**:211-235.
Bosmann, H. B., and S. S. Martin. 1969. Mitochondrial autonomy: incorporation of monosaccharides into glycoprotein by isolated mitochondria. Science **164**:190-192.
Boss, J. 1954. Mitosis in cultures of newt tissues. II. Chromosome pairing in anaphase. Exp. Cell Res. **7**:225-234.
Boss, J. 1955. The pairing of somatic chromosomes: a survey. Texas Rep. Biol. Med. **13**:213-227.
Bouck, N., and A. E. Adelberg. 1963. The relationship between DNA synthesis and conjugation in *Escherichia coli.* Biochem. Biophys. Res. Commun. **11**:24-27.
Boveri, Th. 1902. Ueber Mehrpolige Mitosen als Mittel zur Analyse des Zellkerns. Vererb. Phys. Med. Ges. N. F. **35**:65-88.
Boveri, Th. 1904. Ergebnisse uber die Konstitution der Chromatischen Substanz des Zellkernes. Jena.
Bowden, W. M. 1959. The taxonomy and nomenclature of the wheats, barleys and ryes and their wild relatives. Canad. J. Bot. **37**:657-684.
Boyer, S. H., et al. 1971. Silent hemoglobin alpha genes in apes: potential source of Thalassemia. Science **171**:182-184.
Bradley, M. V., and J. C. Crane. 1955. The effect of 2, 4, 5-trichlorophenoxyacetic acid on cell and nuclear size and endopolyploidy in parenchyma of apricot fruits. Amer. J. Bot. **42**:273-281.
Braun, R., C. Mittermayer, and H. P. Rusch. 1965. Sequential temporal replication of DNA in *Physarum polycephalum.* Proc. Nat. Acad. Sci. U. S. A. **53**:924-931.
Breland, O. P., and G. Gassner. 1961. Notes on chromosome complement of the mosquito, *Aedes aegypti* (L.). J. Texas Acad. Sci. **13**:391-397.
Bremer, G. 1959. Increase of chromosome number in species-hybrids of *Saccharum* in relation to the embryo-sac development. Bibl. Genet. **18**:1-99.
Breuer, M. E., and C. Pavan. 1955. Behavior of polytene chromosomes of *Rynchosciara angelae* at different stages of larval development. Chromosoma **7**:371-386.
Bridges, C. B. 1914. Direct proof through non-disjunction that the sex-linked genes of *Drosophila* are borne by the X-chromosome. Science **40**:107-109.
Bridges, C. B. 1916. Non-disjunction as proof of the chromosome theory of heredity. Genetics **1**:1-52, 107-163.
Bridges, C. B. 1922. The origin of variation in sexual and sex-limited characters. Amer. Nat. **56**:51-63.
Bridges, C. B. 1923. Aberrations in chromosomal material. Eng. Genet. and Family **1**:76-81.
Bridges, C. B. 1925a. Haploidy in *Drosophila melanogaster.* Proc. Nat. Acad. Sci. U. S. A. **11**:706-710.
Bridges, C. B. 1925b. Sex in relation to chromosomes and genes. Amer. Nat. **59**:127-137.
Bridges, C. B. 1928. Chromosome aberrations and the improvement of animal forms. J. Hered. **19**: 349-354.
Bridges, C. B. 1929. Variation in crossing over in relation to age of female in *Drosophila melanogaster.* Carnegie Inst. Wash. Publ. No. 399, p. 63-89.
Bridges, C. B. 1935. Salivary chromosome maps. J. Hered. **26**:60-64.
Bridges, C. B. 1938. A revised map of the salivary gland X-chromosome. J. Hered. **29**:11-13.
Bridges, C. B. 1939. Cytological and genetic basis of sex. Sex and Internal Secretions. Ed. 2. Bailliere, Tindall and Cox, London.
Bridges, C. B., and P. N. Bridges. 1939. A new map of the second chromosome. J. Hered. **30**:475-476.
Bridges, C. B., E. N. Skoog, and J. C. Li. 1935. Genetical and cytological studies of a deficiency (Notopleural) in the second chromosome of *Drosophila melanogaster.* Genetics **21**:788-795.
Bringhurst, R. S., and T. Gill. 1970. Origin of *Fragaria* polyploids. II. In reduced and doubled-unreduced gametes. Amer. J. Bot. **57**:969-976.
Bringhurst, R. S., and Y. D. A. Senanayake. 1966. The evolutionary significance of natural *Fragaria chiloensis* × *F. vesca* hybrids resulting from unreduced gametes. Amer. J. Bot. **53**:1000-1006.
Brink, R. A., and D. C. Cooper. 1947. The endosperm in seed development. Bot. Rev. **13**:423-541.
Brinkley, B. R., and E. Stubblefield. 1966. The fine

structure of the kinetochore of a mammalian cell in vitro. Chromosoma **19**:28-43.

Britten, D. M., and J. W. Hull. 1948. Mitotic instability in *Rubus*. J. Hered. **48**:11-20.

Britten, R. J., and E. H. Davidson. 1969. Gene regulation for higher cells: a theory. Science **165**:349-357.

Britten, R. J., and D. E. Kohne. 1968. Repeated sequences in DNA. Science **161**:529-540.

Brosseau, G. E. 1960. Genetic analysis of the male fertility factors on the Y chromosome of *Drosophila melanogaster*. Genetics **45**:257-274.

Brown, D. D. 1966. The nucleolus and synthesis of ribosomal RNA during oogenesis and embryogenesis of *Xenopus laevis*. Nat. Cancer Inst. Monogr. **23**:297-309.

Brown, D. D., and I. B. Dawid. 1968. Specific gene amplification in oocytes. Science **160**:272-280.

Brown, D. D., and C. S. Weber. 1968. Gene linkage by RNA-DNA hybridization. II. Arrangement of the redundant gene sequences for 28s and 18s ribosomal RNA. J. Molec. Biol. **34**:681-697.

Brown, M. S. 1940. The relation between chiasma formation and disjunction. Univ. Texas Publ. No. 4032, p. 11-64.

Brown, M. S. 1947. A case of spontaneous reduction of chromosome number in somatic tissue of cotton. Amer. J. Bot. **34**:384-388.

Brown, M. S., and M. Y. Menzel. 1952. Polygenomic hybrids in *Gossypium*. I. Cytology of hexaploids, pentaploids, and hexaploid combinations. Genetics **37**:242-263.

Brown, S. W. 1949. Endomitosis in the tapetum of tomato. Amer. J. Bot. **36**:703-716.

Brown, S. W. 1954. Mitosis and meiosis in *Luzula campestris* D. C. Univ. Calif. Publ. Bot. **27**:231-278.

Brown, S. W. 1966. Heterochromatin. Science **151**:417-425.

Brown, S. W., and U. Nur. 1964. Heterochromatic chromosomes in the coccids. Science **145**:130-136.

Brown, S. W., and D. Zohary. 1955. The relationship of chiasmata and crossing over in *Lilium formosanum*. Genetics **40**:850-873.

Brown, W. V. 1946. Cytological studies in the Alismaceae. Bot. Gaz. **108**:262-267.

Brown, W. V. 1951. Apomixis in *Zephyranthes texana* herb. Amer. J. Bot. **38**:697-702.

Brown, W. V. 1960. Supernumerary chromosomes in a population of *Tradescantia edwardsiana*. Southwest Nat. **5**:49-60.

Brown, W. V., and E. M. Bertke. 1969. Textbook of Cytology. The C. V. Mosby Co., St. Louis.

Brown, W. V., and W. H. P. Emery. 1957. Persistent nucleoli and grass systematics. Amer. J. Bot. **44**:586-590.

Brown, W. V., and W. H. P. Emery. 1958. Apomixis in the Gramineae: Panicoideae. Amer. J. Bot. **45**:253-263.

Brown, W. V., and S. M. Stack. 1968. Somatic pairing as a regular preliminary to meiosis. Bull. Torrey Bot. Club **95**:369-378.

Buck, R. C. 1967. Mitosis and meiosis in *Rhodnius prolixus*: the fine structure of the spindle and diffuse kinetochore. J. Ultrastruct. Res. **18**:489-501.

Buffaloe, N. 1959. Some effects of colchicine on cells of *Chlamydomonas eugametos* Moewus. Exp. Cell Res. **16**:221-231.

Burdette, W. J. 1938. Tetraploid regions in the chromosomes of *Drosophila melanogaster*. Rec. Genet. Soc. Amer. **7**:66-67. (Abstr.).

Burnham, C. R. 1930. Genetical and cytological studies of semisterility and related phenomena in maize. Proc. Nat. Acad. Sci. U. S. A. **16**:269-277.

Burnham, C. R. 1932. An interchange in maize giving low sterility and chain configurations. Proc. Nat. Acad. Sci. U. S. A. **18**:434-440.

Burnham, C. R. 1956. Chromosomal interchanges in plants. Bot. Rev. **22**:419-552.

Burton, P. R., and D. G. Dusanic. 1968. Fine structure and replication of the kinetoplast of *Trypanosoma lewisi*. J. Cell Biol. **39**:318-331.

Butler, L. J., C. Chantler, N. E. France, and C. G. Keith. 1969. A liveborn infant with complete triploidy (69,XXX). J. Med. Genet. **6**:413-420.

Butterfass, T. 1967. Gepaarte chromosomen in zwei mitotischen metaphasen von diploiden zuckerrüben. Chromosoma **20**:442-444.

Buzzati-Traverso. 1950. Interspecific hybrids in the "obscura" group of *Drosophila*. Meeting of Soc. Study Evol., Columbus, Ohio.

Cairns, J. 1963. The bacterial chromosome and its manner of replication as seen by autoradiography. J. Molec. Biol. **6**:208-213.

Cairns, J. 1966. Autoradiography of HeLa cell DNA. J. Molec. Biol. **15**:372-373.

Cairns, J., and C. I. Davern. 1968. The mechanics of DNA replication in bacteria. In W. J. Peacock and R. D. Brock (editors). Replication and Recombination of Genetic Material. Australian Academy of Science, Canberra.

Callan, H. G. 1942. Heterochromatin in *Triton*. Proc. Roy. Soc. B **130**:324-335.

Callan, H. G. 1963. The nature of lampbrush chromosomes. Int. Rev. Cytol. **15**:1-34.

Callan, H. G. 1968. A radioautographic study of the time course of male meiosis in the newt *Triturus vulgaris*. J. Cell Sci. **3**:615-626.

Campbell, A. M. 1969: Episomes. Harper & Row Publishers, New York.

Cannon, H. G. 1922. A further account of the spermatogenesis of lice. Quart. J. Micr. Sci. **66**:657-667.

Cannon, W. A. 1902. A cytological basis for the mendelian laws. Bull. Torrey Bot. Club **29**:42-50.

Carniel, K. 1960. Beitrage zum Sterilitäts-und Befruchtungs problem von *Rhoeo discolor*. Chromosoma **11**:456-462.

Carothers, E. E. 1913. The mendelian ratio in relation to certain orthopteran chromosomes. J. Morph. **24**:487-511.

Carothers, E. E. 1917. The segregation and recombination of homologous chromosomes found in two genera of Acrididae (Orthoptera). J. Morph. **28**: 445-521.

Carothers, E. E. 1921. Genetical behavior of heteromorphic homologous chromosomes of *Circotettix* (Orthoptera). J. Morph. **35**:457-483.

Carr, D. H. 1965. Chromosome studies in spontaneous abortions. Obstet. Gynec. **26**:308-313.

Carson, H. L. 1962. Fixed heterozygosity in a parthenogenetic species of *Drosophila*. Univ. Texas Publ. No. 6205, p. 55-62.

Carson, H. L. 1967a. Permanent heterozygosity. Evolut. Biol. **1**:143-168.

Carson, H. L. 1967b. Selection for parthenogenesis in *Drosophila mercatorum*. Genetics **55**:157-171.

Carson, H. L. 1970. Chromosome tracers of the origin of species. Science **168**:1414-1418.

Caspersson, T., and J. Schultz. 1940. Nucleic acids in both nucleus and cytoplasm and the function of the nucleolus. Proc. Nat. Acad. Sci. U. S. A. **26**:507-511.

Caspersson, T., et al. 1968. Chemical differentiation along metaphase chromosomes. Expt. Cell Res. **49**:219-225.

Caspersson, T., et al. 1969. Chemical differentiation with fluorescent alkylating agents in *Vicia faba*. Metaphase chromosomes. Expt. Cell Res. **58**:128-140.

Caspersson, T., L. Zech, and E. J. Modest. 1970. Fluorescent labeling of chromosomal DNA: superiority of quinacrine mustard to quinacrine. Science **170**:762.

Cassagnau, P. 1968. Sur la structure des chromosomes salivaires de *Bilobella massoudi* Cass. (Collembola; Neanuridae). Chromosoma **24**:41-58.

Castle, W. E. 1909. A mendelian view of sex-heredity. Science **29**:395-400.

Castle, W. E. 1934. Possible cytoplasmic as well as chromosomal control of sex in haploid males. Proc. Nat. Acad. Sci. U. S. A. **20**:101-102.

Castro, D., M. Noronha-Wagner, and A. Camara. 1954. Two X-ray induced translocations in *Luzula purpurea*. Genet. Iberica **6**:3-18.

Cattanach, B. M. 1961. XXY mice. Genet. Res. **2**:156-158.

Cattanach, B. M. 1962. XO mice. Genet. Res. **3**:487-490.

Catcheside, D. G. 1937. Secondary pairing in *Brassica oleracea*. Cytologia, Fujii Jubil. Vol.: 336-343.

Catcheside, D. G. 1968. The control of genetic recombination in *Neurospora crassa*. In W. J. Peacock and R. D. Brock (editors). Recombination of Genetic Material. Australian Academy of Science, Canberra.

Cave, M. D. 1968. Chromosome replication and synthesis of nonhistone proteins in giant polytene chromosomes. Chromosoma **25**:392-401.

Chapman, G. B. 1959. Electron microscope observations on the behavior of the bacterial cytoplasmic membrane during cell division. J. Biophys. Biochem. Cytol. **6**:221-224.

Chase, S. S. 1949. Monoploid frequencies in a commercial double cross hybrid maize, and in its component single cross hybrids and inbred lines. Genetics **34**:328-332.

Chase, S. S. 1952. Production of homozygous diploids of maize from monoploids. Agron. J. **44**:263-267.

Chase, S. S. 1963. Androgenesis—its use for transfer of maize cytoplasm. J. Hered. **54**:152-158.

Chatton, E. 1920. Les Péridiniens parasites: morphologie, reproduction, ethologie. Arch. Zool. Exp. Genet. **59**:1-32.

Chauhan, K. P. S., and W. O. Abel. 1968. Evidence for the association of homologous chromosomes during premeiotic stages in *Impatiens* and *Salvia*. Chromosoma **25**:297-302.

Chen, A. T. L., and A. Falek. 1969. Centromeres in human meiotic chromosomes. Science **166**:1008-1010.

Chen, C. H., and J. G. Rose. 1963. Colchicine-induced somatic chromosome reduction in *Sorghum*. I. Induction of diploid plants from tetraploid seedlings. J. Hered. **54**:96-101.

Chen, P. L. 1966. Observation of the nuclear behavior in *Streptomyces cinnamonensis*. Amer. J. Bot. **53**:291-295.

Chiang, K., and N. Sueoka. 1967. Replication of chromosomal and cytoplasmic DNA during mitosis and meiosis in an eucaryote, *Chlamydomonas reinhardi*. Oak Ridge National Laboratory Symposium on Chromosome Mechanics at the Molecular Level, Tenn.

Chiarugi, A., and E. Francine. 1930. Apomissia in *Ochna serrulata*. Walp. Nuovo G. Bot. Ital. N. S. **37**:1-250.

Chicago Conference. 1966. Standardization in human cytogenetics. Birth defects: Original Article Series II:2. The National Foundation-March of Dimes **2**:2-9.

Chipman, R. H. 1925. A study of synizesis and synapsis in *Lilium superbum* L. Amer. J. Bot. **12**:1-18.

Chu, E. H. Y., and N. H. Giles. 1959. Human chromosome complements in normal somatic cells in culture. Amer. J. Hum. Genet. **11**:63-79.

Church, K., and D. Wimber. 1971. Meiosis in *Ornithogalum virens* (Liliaceae). II. Univalent production by preprophase cold treatment. Expt. Cell Res. **64**:119-124.

Clark, D. J., and O. Maaloe. 1967. DNA replication and the division cycle in *Escherichia coli*. J. Molec. Biol. **23**:99-112.

Clark, F. J. 1940. Cytogenetic studies of divergent meiotic spindle formation in *Zea mays*. Amer. J. Bot. **27**:547-559.

Clausen, J. 1954. Partial apomixis as an equilibrium system in evolution. Caryologia **1**:469-479.

Clausen, J. 1961. Introgression facilitated by apomixis in polyploid poas. Euphytica **10**:87-94.

Clausen, J., D. D. Keck, and W. M. Hiesey. 1945. Experimental studies on the nature of species. II.

Plant evolution through amphidiploidy and autoploidy with examples from the Madiinae. Carnegie Inst. Wash. Publ. No. 564.

Clausen, J., D. D. Keck, and W. M. Hiesey. 1947. Heredity of geographically and ecologically isolated races. Amer. Nat. **81**:114-133.

Clayberg, C. D. 1959. Cytogenetic studies of precocious meiotic centromere division in *Lycopersicon esculentum* Mill. Genetics **44**:1335-1346.

Cleland, R. E. 1936. Some aspects of the cytogenetics of *Oenothera*. Bot. Rev. **2**:316-348.

Cleland, R. E. 1951. Extra diminutive chromosomes in *Oenothera*. Evolution **5**:165-176.

Cleland, R. E. 1956. Chromosome structure in *Oenothera* and its effect on the evolution of the genus. Cytologia Symposia 1956:5-19.

Cleland, R. E. 1962. The cytogenetics of *Oenothera*. Advances Genet. **11**:147-237.

Cleland, R. E. 1968. Cytogenetic studies on *Oenothera*, subgenus Raimannia. Jap. J. Genet. **43**:329-334.

Cleland, R. E., et al. 1950. Studies in *Oenothera* cytogenetics and phylogeny. Indiana Univ. Publ. Sci. Ser. No. 16.

Cleveland, L. R. 1938. Longitudinal and transverse division in two closely related flagellates. Biol. Bull. **74**:1-24.

Cleveland, L. R. 1947. The origin and evolution of meiosis. Science **105**:287-289.

Cleveland, L. R. 1953. Studies on chromosomes and nuclear division. III. Pairing, segregation, and crossing-over. Trans. Amer. Phil. Soc. **43**:809-869.

Close, H. G., A. S. R. Goonetilleke, P. A. Jacobe, and W. H. Price. 1968. The incidence of sex chromosomal abnormalities in mentally subnormal males. Cytogenetics **7**:277-285.

Coe, E. 1959. A line of maize with high haploid frequency. Amer. Nat. **93**:381-382.

Coe, G. E. 1953. Cytology of reproduction in *Cooperia pedunculata*. Amer. J. Bot. **40**:335-343.

Coe, G. E. 1954. Chromosome numbers and morphology in *Habranthus* and *Zephyranthes*. Bull. Torrey Bot. Club **81**:141-148.

Cohen, J. A. 1967. Chemistry and structure of nucleic acids of bacteriophages. Science **158**:343-351.

Cohen, M. M., and R. G. Davidson. 1967. Down's syndrome associated with a familial (21q − ;22qt) translocation. Cytogenetics **6**:321-330.

Cohen, M. M., M. J. Marinello, and N. Back. 1967a. Chromosomal damage in human leukocytes induced by lysergic acid diethylamide. Science **155**:1417-1419.

Cohen, M. M., A. A. Sandberg, N. Takagi, and M. H. MacGileivray. 1967b. Autoradiographic investigations of centric fragments and rings in patients with stigmata of gonadal dysgenesis. Cytogenetics **6**:254-267.

Cohen, M. M., M. W. Shaw, and J. W. Maclure. 1966. Racial differences in the length of the human chromosome. Cytogenetics **5**:34-52.

Cohn, M. M., and H. F. Clark. 1967. The somatic chromosomes of five crocodilian species. Cytogenetics **6**:193-203.

Cole, R. M., and J. J. Hahn. 1962. Cell wall replication in *Streptococcus pyogenes*. Science **135**:722.

Colli, W., and M. Oishi. 1969. Ribosomal RNA genes in bacteria: evidence for the nature of the physical linkage between 16s and 23s RNA genes in *Bacillus subtilis*. Proc. Nat. Acad. Sci. U. S. A. **64**:642-649.

Collmann, R. D., and A. Stoller. 1962. A survey of mongoloid births in Victoria, Australia, 1942-1957. Amer. J. Public Health **52**:813-829.

Comings, D. E. 1966. Centromere: absence of DNA replication during chromatid separation in human fibroblasts. Science **154**:1463-1464.

Comings, D. E. 1967a. The duration of replication of the inactive X chromosome in humans based on the persistence of the heterochromatic sex chromatin body during DNA synthesis. Cytogenetics **6**:20-37.

Comings, D. E. 1967b. Histones of genetically active and inactive chromatin. J. Cell Biol. **35**:699-708.

Comings, D. E. 1968. The rationale for an ordered arrangement of chromatin in the interphase nucleus. Amer. J. Hum. Genet. **20**:440-460.

Comings, D. E. 1970. Half-chromatid aberrations and chromosome strandedness. Canad. J. Genet. Cytol. **12**:960-964.

Comings, D. E., and T. A. Okada. 1970a. Association of chromatin fibers with the annuli of the nuclear membrane. Exp. Cell Res. **62**:293-302.

Comings, D. E., and T. A. Okada. 1970b. Association of nuclear membrane fragments with metaphase and anaphase chromosomes as observed by whole mount electron microscopy. Exp. Cell Res. **63**:62-68.

Comings, D. E., and T. A. Okada. 1970c. Condensation of chromosomes onto the nuclear membrane during prophase. Exp. Cell Res. **63**:471-473.

Comings, D. E., and T. A. Okada. 1970d. Do half-chromatids exist? Cytogenetics **9**:450-459.

Comings, D. E., and T. A. Okada. 1970e. Whole-mount electron microscopy of the centromere region of metacentric and telocentric mammalian chromosomes. Cytogenetics **9**:436-449.

Comings, D. E., and T. A. Okada. 1971a. Whole mount electron microscopy of human meiotic chromosomes. Expt. Cell Res. **65**:99-103.

Comings, D. E., and T. A. Okada. 1971b. Fine structure of the synaptinemal complex. Expt. Cell Res. **65**:104-116.

Cooper, H. L., and K. Hirschhorn. 1962. Enlarged satellites as a familial chromosome marker. Amer. J. Hum. Genet. **14**:107-113.

Cooper, J. P., and H. W. Howard. 1952. The chromosome number of seedlings from the cross *Solanum demissum* backcrossed by *Solanum tuberosum*. J. Genet. **50**:511-521.

Cooper, K. W. 1938. Concerning the origin of the polytene chromosomes of Diptera. Proc. Nat. Acad. Sci. U. S. A. **24**:452-458.

Cooper, K. W. 1945. Normal segregation without

chiasmata in female *Drosophila melanogaster*. Genetics **30**:472-484.

Cooper, K. W. 1949. The cytogenetics of meiosis in *Drosophila*: mitotic and meiotic autosomal chiasmata without crossing over in the male. J. Morph. **84**:81-122.

Corey, R. R., and M. P. Starr. 1957. Genetic transformation of colony type in *Xanthomonas phaseoli*. J. Bact. **74**:141-145.

Correns, C. 1909. Zur Kenntniss der Rolle von Kern und Plasma bei der Vererbung. Z. Ind. Abst. Vererb. **2**:331-349.

Court-Brown, W. M. 1967. Human Population Cytogenetics. American Elsevier Publishing Co., Inc., New York.

Court-Brown, W. M., K. E. Buckton, P. A. Jacobs, I. M. Tough, E. V. Kuenssberg, and J. D. E. Knox. 1966. Chromosome studies on adults. Eugenics Lab. Mem. Ser. No. 42, Galton Laboratory. Cambridge University Press, London.

Court-Brown, W. M., P. A. Jacobs, and M. Brunton. 1965. Chromosome studies on randomly chosen men and women. Lancet **2**:561-562.

Craddock, E. 1970. Chromosome number variation in a stick insect *Didymuria violescens* (Leach). Science **167**:1380-1382.

Craig-Holmes, A. P., and M. W. Shaw. 1971. Polymorphism of human constitutive heterochromatin. Science **174**:702-704.

Creighton, H. B., and B. McClintock. 1931. A correlation of cytological and genetical crossing-over in *Zea mays*. Proc. Nat. Acad. Sci. U. S. A. **17**:492-497.

Crew, F. A. E., and R. Lamy. 1938. Mosaicism in *D. pseudoobscura*. J. Genet. **37**:211-228.

Crippa, M., E. H. Davidson, and A. E. Mirsky. 1967. Persistence in early amphibian embryos of informational RNA's from lampbrush chromosome stage of oogenesis. Proc. Nat. Acad. Sci. U. S. A. **57**:885-892.

Crosby, A. R. 1957. Nucleolar activity of lagging chromosomes in wheat. Amer. J. Bot. **44**:813-822.

Crozier, R. H. 1970. Karyotypes of twenty-one ant species (Hymenoptera; Formicidae), with reviews of the known ant karyotypes. Canad. J. Genet. Cytol. **12**:109-128.

Crumpacker, D. W., and V. M. Salceda. 1968. Uniform heterokaryotypic superiority for viability in a Colorado population of *Drosophila pseudoobscura*. Evolution **22**:256-261.

da Cunha, A. B., C. Pavan, J. S. Morgante, and M. C. Garrido. 1969. Studies on cytology and differentiation in Sciaridae. II. DNA redundancy in salivary gland cells of *Hybosciara fragilis* (Diptera, Sciaridae), Genetics (Suppl. 61) **1**, 1969:335-349.

Cutler, R. G., and J. E. Evans. 1967. Relative transcription activity of different segments of the genome throughout the cell division cycle of *Escherichia coli*. The mapping of ribosomal and transfer RNA and the determination of the direction of replication. J. Molec. Biol. **26**:91-105.

Cutter, E. G., and L. J. Feldman. 1970a. Trichoblasts in *Hydrocharis*. I. Origin, differentiation, dimensions, and growth. Amer. J. Bot. **57**:190-201.

Cutter, E. G., and L. J. Feldman. 1970b. Trichoblasts in *Hydrocharis*. II. Nucleic acids, proteins, and a consideration of cell growth in relation to endopolyploidy. Amer. J. Bot. **57**:202-211.

Daneholt, B., and J. E. Edstrom. 1967. The content of deoxyribonucleic acid in individual polytene chromosomes of *Chironomus tentans*. Cytogenetics **6**:350-356.

Danes, B. S., and A. G. Bearn. 1968. Localization of the cystic-fibrosis gene. Lancet **2**:1303.

Darevsky, I. S. 1966. Natural parthenogenesis in a polymorphic group of Caucasian rock lizards related to *Lacerta saxicola* Eversmann. J. Ohio Herpetol. Soc. **5**:115-152.

Darlington, C. D. 1929. Chromosome behavior and structural hybridity in the Tradescantiae. J. Genet. **21**:207-286.

Darlington, C. D. 1937. Recent Advances in Cytology. Ed. 2. Blakiston, McGraw-Hill Book Company, New York.

Darlington, C. D. 1939. The Evolution of Genetic Systems. Cambridge University Press, London.

Darlington, C. D. 1956a. Chromosome Botany. George Allen and Unwin, London.

Darlington, C. D. 1956b. Natural populations and the breakdown of classical genetics. Proc. Roy. Soc. (Biol.) **145**:350-364.

Darlington, C. D., and L. F. La Cour. 1950. Hybridity selection in *Campanula*. Heredity **4**:217-248.

Darlington, C. D., and G. W. Shaw. 1959. Parallel polymorphism in the heterochromatin of *Trillium* species. Heredity **13**:89-121.

Darlington, C. D., and P. T. Thomas. 1941. Morbid mitosis and the activity of inert chromosomes in *Sorghum*. Proc. Roy. Soc. (Biol.) **130**:127-150.

Darlington, C. D., and A. P. Wylie. 1956. Chromosome Atlas of Flowering Plants. The Macmillan Company, New York.

Davidson, E. H., M. Crippa, F. R. Kramer, and A. E. Mirsky. 1966. Genomic function during lampbrush chromosome stage of amphibian oogenesis. Proc. Nat. Acad. Sci. U. S. A. **56**:856-863.

Davidson, W. M., and D. R. Smith. 1954. A morphological sex difference in the polymorphonuclear neutrophil leucocytes. Brit. Med. J. **2**:6-9.

Davis, B. M. 1943. An amphidiploid in the F_1 generation from the cross *Oenothera franciscana* × *Oenothera biennis* and its progeny. Genetics **28**:275-285.

Dawson, M. H., and R. H. P. Sia. 1931. In vitro transformation of pneumococcal types. J. Exp. Med. **54**:681-699.

DeLamater, E. D. 1952. Preliminary observations on the occurrence of a typical mitotic process in micrococci. Bull. Torrey Bot. Club **79**:1-5.

DeLamater, E. D., and M. E. Hunter. 1951. Preliminary report of true mitosis in the vegetative cell of *Bacillus megatherium*. Amer. J. Bot. **38**:659-662.

Delbruck, M., and W. T. Bailey. 1946. Induced muta-

tions in bacterial viruses. Cold Spring Harbor Symp. Quant. Biol. **11**:33.
Delbruck, M., and G. Stent. 1957. On the mechanism of DNA replication. In W. D. McElroy and B. Glass (editors). The Chemical Basis of Heredity. The Johns Hopkins Press, Baltimore.
Demerec, M., and M. E. Hoover. 1936. Three related X-chromosome deficiencies in *Drosophila*. J. Hered. **27**:207-212.
Demerec, M., and M. G. Hoover. 1939. Hairy-wing — a duplication in *Drosophila melanogaster*. Genetics **24**:271-277.
Dempster, L. T., and G. L. Stebbins. 1968. A cytotaxonomic revision of the fleshy-fruited *Galium* species of the Californias and southern Oregon (Rubiaceae). Univ. Calif. Publ. Bot. **46**:1-52.
Denver Conference, 1960. Human chromosome study group. A proposed standard of nomenclature of human mitotic chromosomes. Suppl. Cerebral Palsy Bull. **2**:1-9, also AIBS Bull. Dec.:27-30.
Derman, H. 1947. Periclinal cytochimeras and histogenesis in cranberry. Amer. J. Bot. **34**:32-43.
Derman, H. 1953. Periclinal cytochimeras and origin of tissues in stem and leaf of peach. Amer. J. Bot. **40**:154-168.
Desai, S. 1962. Polyembryony in *Xanthoxylon* Mill. Phytomorphology **12**:184-190.
DeWet, J. M. J. 1965. Diploid races of tetraploid *Dichanthium* species. Amer. Nat. **99**:167-171.
DeWet, J. M. J. 1968. Diploid-tetraploid-haploid cycles and the origin of variability in *Dichanthium* agamospecies. Evolution **22**:394-397.
Dewey, D. R. 1965. Morphology, cytology, and fertility of synthetic hybrids of *Agropyron spicatum* × *Agropyron dasystachyum-riparium*. Bot. Gaz. **126**:269-275.
Dhillon, T. S., and E. D. Garber. 1960. The genus *Collinsia*. X. Aneuploidy in *C. heterophylla*. Bot. Gaz. **121**:125-133.
Diacumakos, E. G., L. Garnjobst, and E. L. Tatum. 1965. A cytoplasmic character in *Neurospora crassa*. The role of nuclei and mitochondria. J. Cell Biol. **26**:427-443.
Dice, L. R. 1940. Speciation in *Peromyscus*. Amer. Nat. **74**:287-298.
Dietrich, M. J. 1968. Organisation du fuseau de caryocinese en prometaphase dans les cellules-meres de microspores du Lis. C. R. Acad. Sci. **266**:579-581.
Dietz, R. 1966. The dispensability of the centrioles in the spermatocyte divisions of *Poles ferruginea* (Nematocera). In C. D. Darlington and K. R. Lewis (editors). Chromosomes Today, Vol. 1. Plenum Press, New York.
Digby, L. 1910. The somatic, premeiotic, and meiotic nuclear divisions of *Galtonia candicans*. Ann. Bot. **24**:727-758.
Djordjevic, B., and W. Szybalske. 1960. Genetics of human cell lines. J. Expt. Med. **112**:509-531.
Dobzhansky, T. 1929. Genetical and cytological proof of translocation involving the third and fourth chromosomes of *D. melanogaster*. Biol. Zentralbl. **49**:408-419.
Dobzhansky, T. 1930. Cytological maps of the second chromosome of *Drosophila melanogaster*. Biol. Zentralbl. **50**:671-685.
Dobzhansky, T. 1935. The Y chromosome of *Drosophila pseudoobscura*. Genetics **20**:366-376.
Dobzhansky, T. 1937. Further data on the variation of the Y chromosome in *Drosophila pseudoobscura*. Genetics **22**:340-345.
Dobzhansky, T. 1941. Genetics and the Origin of Species. Ed. 2. Columbia University Press, New York.
Dobzhansky, T. 1943. Genetics of natural populations. IX. Temporal changes in the composition of populations of *Drosophila pseudoobscura*. Genetics **28**:162-186.
Dobzhansky, T. 1947. A directional change in the genetic constitution of a natural population of *Drosophila pseudoobscura*. Heredity **1**:53-64.
Dobzhansky, T. 1948. Altitudinal and seasonal changes produced by natural selection in certain populations of *Drosophila pseudoobscura* and *Drosophila persimilis*. Genetics **33**:158-176.
Dobzhansky, T. 1949. Genetics and the Origin of Species, ed. 3. Columbia University Press, New York.
Dobzhansky, T. 1952. Changes induced by drought in *Drosophila pseudoobscura* and *Drosophila persimilis*. Evolution **6**:234-243.
Dobzhansky, T., and C. B. Bridges. 1928. The reproductive system of triploid intersexes in *Drosophila melanogaster*. Amer. Nat. **62**:425-434.
Dobzhansky, T., and H. Levene. 1948. Genetics of natural populations. XVII. Proof of operation of natural selection in wild populations of *Drosophila pseudoobscura*. Genetics **33**:537-547.
Dobzhansky, T., and O. Pavlovsky. 1955. An extreme case of heterosis in a Central American population of *Drosophila tropicalis*. Proc. Nat. Acad. Sci. U. S. A. **41**:289-295.
Dobzhansky, T., and J. Schultz. 1934. The distribution of sex-factors in the X-chromosome of *Drosophila melanogaster*. J. Genet. **28**:349-386.
Dobzhansky, T., and D. Socolov. 1939. Structure and variation of the chromosomes in *Drosophila azteca*. J. Hered. **30**:3-19.
Dobzhansky, T., and A. H. Sturtevant. 1938. Inversions in the chromosomes of *Drosophila pseudoobscura*. Genetics **23**:28-64.
Dodge, J. D. 1963a. Chromosome structure in the Dinophyceae. I. The spiral chromonema. Arch. f. Mikrobiol. **45**:46-57.
Dodge, J. D. 1963b. The nucleus and nuclear division in the Dinophyceae. Arch. Protistenk. **106**:442-452.
Dodge, J. D. 1964. Cytochemical staining of sections from plastic embedded flagellates. Stain Techn. **39**:381-385.
Donahue, R. P., W. B. Bias, J. H. Renwick, and V. A.

McKusick. 1968. Probable assignment of the Duffy Blood Group locus to chromosome 1 in man. Proc. Nat. Acad. Sci. U. S. A. **61**:949-955.

Doncaster, L., and H. G. Cannon. 1919. On the spermatogenesis of the louse, *Pediculus corporis* and *P. capitis*, with observations on the maturation of the egg. Quart. J. Microscop. Sci. **64**:303-328.

Doncaster, L., and G. H. Raynor. 1906. On breeding experiments with Lepidoptera. Proc. Zool. Soc. London **37**:125-133.

Dorn, G. L. 1967. A revised map of the eight linkage groups of *Aspergillus nidulans*. Genetics **56**:619-631.

Douglas, C. R., and M. S. Brown. 1971. A study of triploid and 3X-1 aneuploid plants in the genus *Gossypium*. Amer. J. Bot. **58**:65-71.

Driscoll, C. J., and N. L. Darvey. 1970. Chromosome pairing: effect of colchicine on an isochromosome. Science **169**:290-291.

Driscoll, C. J., N. L. Darvey, and H. N. Barber. 1967. Effect of colchicine on meiosis of hexaploid wheat. Nature **216**:687-688.

Duara, B. N., and G. L. Stebbins. 1952. A polyhaploid obtained from a hybrid derivative of *Sorghum halepense* × *S. vulgare* var. *sudanense*. Genetics **37**:369-374.

Dubinin, N. P., N. N. Sokolov, and G. G. Tiniakov. 1936. Occurrence and distribution of chromosome aberrations in nature. Nature **137**:1035-1036.

Dubuc, J. P., and R. C. McGinnis. 1970. Somatic association in *Avena sativa* L. Science **167**:999-1000.

Dunford, M. P. 1970. Interchange heterozygosity in diploid interspecific hybrids in *Grindelia* (Compositae). Amer. J. Bot. **57**:623-628.

Dunn, L. C. 1921. Unit character variation in rodents. J. Mammal. **2**:125-140.

DuPraw, E. J. 1965a. Macromolecular organization of nuclei and chromosomes. A folded fiber model based on whole mount electron microscopy. Nature **206**:338-343.

DuPraw, E. J. 1965b. The organization of nuclei and chromosomes in honey bee embryonic cells. Proc. Nat. Acad. Sci. U. S. A. **53**:161-168.

DuPraw, E. J. 1968. Cell and Molecular Biology. Academic Press Inc., New York.

East, E. M. 1933. The behavior of a triploid in *Nicotiana tabacum* L. Amer. J. Bot. **20**:269-287.

Edwards, J. H., C. Yuncken, D. I. Rushton, S. Richards, and U. Mittwock. 1967. Three cases of triploidy in man. Cytogenetics **6**:81-104.

Ehrendorfer, F., F. Krendl, E. Habeler, and W. Sauer. 1968. Chromosome numbers and evolution in primitive angiosperms. Taxon **17**:337-468.

Eigsti, O. J. 1942. A cytological investigation of *Polygonatum* using colchicine-pollen tube technique. Amer. J. Bot. **29**:626-636.

Einset, J. 1947a. Aneuploidy in relation to partial sterility in autotetraploid lettuce (*Lactuca sativa* L.). Amer. J. Bot. **34**:99-105.

Einset, J. 1947b. Chromosome studies in *Rubus*. Gentes Herb. **7**:181-192.

Eklund, M. W., F. T. Poysky, S. M. Reed, and C. A. Smith. 1971. Bacteriophage and the toxigenicity of *Clostridium botulinum* type C. Science **172**:480-482.

Ellar, D. J., D. G. Lundgren, and R. A. Slepecky. 1967. Fine structure of *Bacillus megaterium* during synchronous growth. J. Bact. **94**:1189-1205.

Elliott, F. C., and C. P. Wilsie. 1948. A fertile polyhaploid in *Bromus inermis*. J. Hered. **39**:363-367.

Ellis, J. R. 1962. *Fragaria-Potentilla* intergeneric hybridization and evolution in *Fragaria*. Proc. Linnaean Soc. London **173**:99-106.

Ellis, R. J. 1969. Chloroplast ribosomes: stereospecificity of inhibition by chloramphenicol. Science **163**:477-478.

Emerson, C. P., and T. Humphreys. 1971. Ribosomal RNA synthesis and the multiple, atypical nucleoli in cleaving embryos. Science **171**:898-901.

Emsweller, S. L., and H. A. Jones. 1934. The cytology of a backcross population derived from *(Allium cepa* × *fistulosum)* × *fistulosum*. Amer. Nat. **68**:467-468.

Endrizzi, J. E. 1958. The division of univalent chromosomes and a chromosome derivative in *Sorghum*. Caryologia **11**:208-216.

Endrizzi, J. E., and M. S. Brown. 1968. Cytological and genetical studies of the hybrid of *Gossypium herbaceum* L. and *triphyllum*. Genetica **39**:272-288.

Endrizzi, J. E., and R. J. Kohel. 1966. Use of telosomes in mapping three chromosomes in cotton. Genetics **54**:535-550.

Endrizzi, J. E., and L. L. Phillips. 1960. A hybrid between *Gossypium arboreum* L. and *G. raimondii* Ulb. Canad. J. Genet. Cytol. **2**:311-319.

Epling, C. 1947. Actual and potential gene flow in natural populations. Amer. Nat. **81**:104-113.

Epling, C., D. F. Mitchell, and R. H. T. Mattoni. 1953. On the role of inversions in wild populations of *Drosophila pseudoobscura*. Evolution **7**:342-365.

Esau, K., and R. H. Gill. 1965. Observations on cytokinesis. Planta **67**:168-181.

Evans, H. M., and O. Swezy. 1929. The chromosomes in man, sex and somatic. Mem. Univ. Calif. **9**:1-64.

Evans, W. L. 1954. Cytology of the grasshopper genus *Circotettix*, Amer. Nat. **88**:21-32.

Everett, G. M., and J. W. Borcherding. 1970. Somatic cell mating and segregation in chimeric frogs. Science **168**:850-852.

Fagerlind, F. 1940. Zytologic und Gametophytenbildung in der Gattung *Wikstroemia*. Hereditas **26**:23-50.

Falkow, S., and R. V. Citarella. 1965. Molecular homology of F-merogenote DNA. J. Molec. Biol. **12**:138-151.

Falkow, S., R. V. Citarella, J. A. Wohlhieter, and T. Watanabe. 1966. The molecular nature of R factors. J. Molec. Biol. **17**:102-116.

Fallieri, L. A., T. Camey, M. R. Schrieber, and G. Schreiber. 1969. DNA and nuclear size in early development of some invertebrates. Genetics (Suppl. 61) **1**, 1969:171-181.

Fan, H., and S. Penman. 1970. Mitochondral RNA synthesis during mitosis. Science **168**:135-138.

Fankhauser, G. 1937. The sex of a haploid, metamorphosed salamander *(Triton taeniatus* Laur.). Genetics **22**:192-193.

Fankhauser, G. 1938. Triploidy in the newt, *Triturus viridescens.* Proc. Amer. Phil. Soc. **79**:715-739.

Faussek, W. 1913. Zur Frage über den Bau des Zellkernes in den Speicheldrusen larve von *Chironomus.* Arkh. Mik. Anat. **82**:39-60.

Fawcett, D. W. 1956. The fine structure of chromosomes in the meiotic prophase of vertebrate spermatocytes. J. Biophys. Biochem. Cytol. **2**:403-406.

Feldman, M. 1965. Further evidence for natural hybridization between tetraploid species of *Aegilops* section *Pleionathera.* Evolution **19**:162-174.

Feldman, M. 1966. The effect of chromosomes 5B, 5D, and 5A on chromosomal pairing in *Triticum aestivum.* Proc. Nat. Acad. Sci. U. S. A. **55**:1447-1453.

Feldman, M., T. Mello-Sampayo, and E. R. Sears. 1966. Somatic association in *Triticum aestivum.* Proc. Nat. Acad. Sci. U. S. A. **56**:1192-1199.

Ferguson-Smith, M. A. 1961. Chromosomes and human disease. In A. G. Steinberg (editor). Progress in Medical Genetics. Vol. 1. Grune & Stratton, Inc., New York.

Ferguson-Smith, M. A. 1967. Clinical cytogenetics. In J. F. Crow and J. V. Neel (editors). Proceedings of Third International Congress Human Genetics. The Johns Hopkins Press, Baltimore.

Ferguson-Smith, M. A., and S. D. Handmaker. 1961. Observations on the satellited human chromosomes. Lancet **1**:638-640.

Feulgen, R., and H. Rossenbeck. 1924. Mikroskopische-chemischer Nachweis einer Nukleinsäure vom Typus der Thymonukleinsäure und die darauf berhende elektive Farbung von Zellkernen in mikroskopischen Präparaten. Zeits. Physiol. Chem. **135**:203-248.

Fialkow, P. J. 1967. Chromosomal breakage induced by extracts of human allogeneic lymphocytes. Science **155**:1676-1677.

Filner, P. 1965. Semi-conservative replication of DNA in a higher plant cell. Exp. Cell Res. **39**:33-39.

Flagg, R. O. 1958. A mutation and an inversion in *Rhoeo discolor.* J. Hered. **49**:185-188.

Ford, C. E., and J. L. Hamerton. 1956. The chromosomes of man. Nature **178**:1020.

Ford, C. E., J. L. Hamerton, and G. B. Sharman. 1957. Chromosome polymorphism in the common shrew. Nature **180**:392-393.

Ford, C. E., P. A. Jacobs, and L. Lajtha. 1958. Human somatic chromosomes. Nature **181**:1565.

Frankel, O. H., and A. Munday. 1962. The evolution of wheat. In The Evolution of Living Organisms. Symposium Royal Society, Victoria, Melbourne.

Franklin, N. C., W. F. Dove, and C. Yanofsky. 1965. A linear insertion of a prophage into the chromosome of *E. coli* shown by deletion mapping. Biochem. Biophys. Res. Commun. **18**:910-923.

Franz, V. 1928. Uber Bastard populatione in der Gattung *Paludina* (recte: *Viviparus*). Biol. Zentralbl. **48**:79-93.

Franzke, C. J., and J. C. Ross. 1957. A lineal series of mutants induced by colchicine treatment. J. Hered. **48**:47-50.

Fredga, K. 1965. A new sex determining mechanism in a mammal. Chromosomes of Indian mongoose *(Herpestes auropunctatus).* Hereditas **52**:409-420.

Freed, J. J. 1970. Somatic cell mating in frogs. Science **169**:1229-1230.

Frenster, J. H. 1965. Ultrastructural continuity between active and repressed chromatin. Nature **205**:1341-1342.

Freter, L. E., and W. V. Brown. 1955. A cytotaxonomic study of *Bouteloua curtipendula* and *B. uniflora.* Bull. Torrey Bot. Club **82**:121-130.

Friedländer, M., and J. Wahrman. 1970. The spindle as a basal body distributor. A study in the meiosis of the male silkworm moth, *Bombyx mori.* J. Cell Sci. **7**:65-89.

Frost, H. B. 1915. The inheritance of doubleness in *Matthiola* and *Petunia.* I. The hypothesis. Amer. Nat. **49**:623-636.

Frost, H. B. 1926. Polyembryony, heterozygosis, and chimeras in *Citrus.* Hilgardia **1**:365-402.

Frost, H. B. 1938a. Nucellar embryony and juvenile characters in clonal varieties of *Citrus.* J. Hered. **29**:423-432.

Frost, H. B. 1938b. The genetics of *Citrus.* Curr. Sci., special number on genetics, p. 24-27.

Fröst, S. 1959a. The cytological behaviour and mode of transmission of accessory chromosomes in *Plantago serraria.* Hereditas **45**:191-210.

Fröst, S. 1959b. Studies of the genetical effect of accessory chromosomes in *Centaurea scabiosa.* Hereditas **44**:75-111.

Fröst, S. 1960. A new mechanism for numerical increase of accessory chromosomes in *Crepis pannonica.* Hereditas **46**:497-503.

Fryxell, P. A. 1957. Mode of reproduction in higher plants. Bot. Rev. **23**:135-233.

Fujii, S. 1936. Salivary gland chromosomes of *Drosophila virilis.* Cytologia **7**:272-275.

Fujita, S., and K. Takamoto. 1963. Synthesis of messenger RNA on the polytene chromosomes of dipteran salivary gland. Nature **200**:494-495.

Furtado, J. S. 1968. Cytology of *Tremella rubromaculata.* Rev. Biol. **6**:411-419.

Furtado, J. S. 1969a. Ascal cytology of *Sordaria sclerogenia.* Protoplasma **67**:473-478.

Furtado, J. S. 1969b. Basidial cytology of *Exidia recisa.* Mycologia **61**:415-418.

Furtado, J. S. 1970. Ascal cytology of *Sordaria brevicollis.* Mycologia **62**:453-461.

Gabrusewycz-Garcia, N. 1964. Cytological and autoradiographic studies in *Sciara coprophila* salivary gland chromosomes. Chromosoma **15**:312-344.

Gagné, R., R. Tanguay, and C. Laberge. 1971. Differential staining pattern in man. Nature (New Biol.) **232**:29-30.

Gaines, E. F., and H. C. Aase. 1926. A haploid wheat plant. Amer. J. Bot. **13**:373-385.

Gairdner, A. E., and C. D. Darlington. 1931. Ring-formation in diploid and polyploid *Campanula percicifolia*. Genetica **13**:113-150.

Gairdner, A. E., and J. B. S. Haldane. 1933. A case of balanced lethal factors in *Antirrhinum majus*. J. Genet. **27**:287-291.

Gajewski, W. 1946. Cytogenetic investigations on *Anemone janczewskii*, a new amphidiploid species of hybrid origin. Acta Soc. Bot. Pol. **17**:129-194.

Gall, J. G. 1956. On the submicroscopic structure of chromosomes. Brookhaven Symp. Biol. **8**:17-29.

Gall, J. G. 1963. Chromosomes and cytodifferentiation. In M. Locke (editor). Cytodifferentiation and Macromolecular Synthesis. Academic Press Inc., New York.

Gall, J. G. 1966. Microtubule fine structure. J. Cell Biol. **31**:639-643.

Gall, J. G. 1968. Differential synthesis of the genes for ribosomal RNA during amphibian oogenesis. Proc. Nat. Acad. Sci. U. S. A. **60**:553-560.

Gall, J. G., and H. G. Callan. 1962. H^3 uridine incorporation in lampbrush chromosomes. Proc. Nat. Acad. Sci. U. S. A. **48**:562-570.

Gall, J. G., H. C. MacGregor, and M. E. Kidston. 1969. Gene amplification in the oocytes of Dytiscid water beetles. Chromosoma **26**:169-187.

Galton, M., K. Benerschke, and S. Ohno. 1965. Sex chromosomes of the chinchilla: allocycly and duplication sequence in somatic cells and behavior in meiosis. Chromosoma **16**:668-680.

Garen, A., and P. D. Skoar. 1958. Transfer of phosphorus-containing material associated with mating in *Escherichia coli*. Biochem. Biophys. Acta **27**:257-263.

Garnjobst, L., J. F. Wilson, and E. L. Tatum. 1965. Studies on a cytoplasmic character in *Neurospora crassa*. J. Cell Biol. **26**:413-425.

Gassner, G. 1967. Synaptinemal complexes: recent findings. J. Cell Biol. **35**:166A-167A. (Abst)

Gassner, G. 1969. Synaptinemal complexes in the achiasmatic spermatogenesis of *Bolbe nigra* Giglio-Tos (Mantoidea). Chromosoma **26**:22-34.

Gates, R. R. 1909. The stature and chromosomes of *Oenothera gigas* de Vries. Arch. Zellforsch. **3**:525-552.

Gay, H. 1956. Chromosome-nuclear membrane-cytoplasmic interrelations in *Drosophila*. J. Biophys. Biochem. Cytol. **2** (Suppl.):407-411.

Geitler, L. 1937. Die Analyse des Kernbaus und der Kernteilung der Wasserlaufer *Gerris lateralis* und *Gerris lacustris* und die Somadifferenzierung. Z. Zellforsch. **26**:641-698.

Geitler, L. 1938. Weitere cytogenitische Untersuchungen an natürlichen Populationen von *Paris quadrifolia*. Z. Ind. Abst. Vererb. **75**:161-190.

Gerald, P. S., S. Warner, J. Singer, P. Corcoran, and I. Umansky. 1967. A ring D chromosome and anomalous inheritance of haptoglobin type. J. Pediat. **70**:172-179.

German, J. 1964. Cytological evidence for crossing-over in vitro in human lymphoid cells. Science **144**:298-301.

German, J. 1967. Autoradiographic studies in human chromosomes. I. A review. In J. E. Crow and J. V. Neel (editors). Proceedings Third International Congress Human Genetics. The Johns Hopkins Press, Baltimore.

German, J. 1969. Chromosomal breakage syndromes. Birth defects: Original Articles Series **5**:117-131.

German, J. 1970. Studying human chromosomes today. Amer. Sci. **58**:182-201.

German, J., M. E. Walker, F. H. Stiefel, and H. Allen. 1968. MN blood-group locus: data concerning the possible chromosomal location. Science **162**:1014-1015.

Gersh, E. S. 1968. Mutants at the bobbed locus in *Drosophila melanogaster*: relation to ribosomal RNA synthesis. Science **162**:1139.

Gershenson, S. 1928. A new sex ratio abnormality in *Drosophila obscura*. Genetics **13**:488-507.

Gerstel, D. U. 1945. Inheritance in *Nicotiana tabacum*. XX. The addition of *Nicotiana glutinosa* chromosomes to tobacco. J. Hered. **36**:197-206.

Gerstel, D. U. 1953. Chromosomal translocations in interspecific hybrids of the genus *Gossypium*. Evolution **7**:234-244.

Giar elli, F., and R. M. Howlett. 1966. The identification of the chromosomes of the D group (13-15 Denver): an autoradiographic and measurement study. Cytogenetics **5**:186-205.

Gianelli, F., and R. M. Howlett. 1967. The identification of the chromosomes of the E group (16-18 Denver): an autoradiographic and measurement study. Cytogenetics **6**:420-435.

Gibor, A., and M. Izawa. 1963. The DNA content of the chloroplasts of *Acetabularia*. Proc. Nat. Acad. Sci. U. S. A. **50**:1164-1169.

Gilchrist, B. M., and J. B. S. Haldane. 1947. Sex linkage and sex determination in a mosquito, *Culex molestus*. Hereditas **33**:175-190.

Gilcrist, F. G. 1968. A Survey of Embryology. McGraw-Hill Book Company, New York.

Giles, N. H. 1942. Autopolyploidy and geographical distribution in *Cuthbertia graminea* Small. Amer. J. Bot. **29**:637-645.

Gilles, A., and L. F. Randolph. 1951. Reduction of quadrivalent frequency in autotetraploid maize during a period of 10 years. Amer. J. Bot. **38**:12-17.

Gillies, S., S. Bullivant, and A. R. Bellamy. 1971. Viral RNA polymerases: electron microscopy of Reovirus reaction cores. Science **174**:694-696.

Gimenez-Martin, G., J. F. Lopez-Saez, and A. Marcos-Moreno. 1965. Structure of the centromere in telocentric chromosomes. Experientia **21**:391-392.

Gläss, E. 1957. Das Problem der Genomsonderung in

den Mitosen unbehandelter Rattenlebern. Chromosoma **8**:458-492.

Godward, B. M. E. 1954. The "diffuse" centromere or polycentric chromosomes in *Spirogyra*. Ann. Bot. **18**:143-156.

Goldschmidt, R., and K. Katsuki. 1931. Vierte Mitteilung über erblichen Gynandromorphismus und somatische Mosaikbildung bei *Bombyx mori* L. Biol. Zentralbl. **51**:57-74.

Goldstein, L., and K. G. Lark. 1967. Mitosis: on the mechanism for invariable sister chromatid segregation. And rebuttal. Science **156**:1133-1134.

Gooch, P. E., and C. L. Fisher. 1969. High frequency of a specific chromosome abnormality in leukocytes of a normal female. Cytogenetics **8**:1-8.

Goodspeed, T. H. 1934. *Nicotiana* phylesis in the light of chromosome number, morphology and behavior. Univ. Calif. Publ. Bot. **17**:369-398.

Goodspeed, T. H. 1945. Cytotaxonomy of *Nicotiana*. Bot. Rev. **11**:533-592.

Gordon, M. 1947. Genetics of *Platypoecilus maculatus*. IV. The sex determining mechanism in two wild populations of the Mexican platyfish. Genetics **32**:8-17.

Gough, M., and M. Levine. 1968. The circularity of the phage P22 linkage map. Genetics **58**:161-169.

Gould, F. W., and Z. J. Kapadia. 1962. Biosystematic studies in the *Bouteloua curtipendula* complex. I. The aneuploid rhizomatous *B. curtipendula* of Texas. Amer. J. Bot. **49**:887-892.

Gould, F. W., and Z. J. Kapadia. 1964. Biosystematic studies in the *Bouteloua curtipendula* complex. II. Taxonomy. Brittonia **16**:182-207.

Granick, E. C. 1944. A karyosystematic study of the genus *Agave*. Amer. J. Bot. **31**:283-289.

Grant, V. 1958. The regulation of recombination in plants. Cold Spring Harbor Symp. Quant. Biol. **23**:337-367.

Grant, V. 1966a. Selection for vigor and fertility in the progeny of a highly sterile species hybrid in *Gilia*. Genetics **53**:757-775.

Grant, V. 1966b. The origin of a new species of *Gilia* in a hybridization experiment. Genetics **54**:1189-1199.

Grant, V. 1971. Plant Speciation. Columbia University Press, New York.

Grant, W. F. 1969. Decreased DNA content of birch *(Betula)* chromosomes at high ploidy as determined by cytophotometry. Chromosoma **26**:326-336.

Gratia, J. P. 1966. Studies on defective lysogeny due to chromosomal deletion in *Escherichia coli*. I. Single lysogens. Biken J. **9**:77-87.

Gray, A. P. 1954. Mammalian Hybrids, a Check List with Bibliography. Commonwealth Agr. Bur., Farnham Royal, Bucks, England.

Green, B. R., and H. Burton. 1970. *Acetabularia* chloroplast DNA: electron microscopic visualization. Science **168**:981-982.

Green, J. K., L. J. Krovetz, and W. J. Taylor. 1968. Two generations of 13-15 chromosomal mosaicism: possible evidence for a genetic defect in the control of chromosomal replication. Cytogenetics **7**:286-297.

Green, M. M. 1955. Pseudoallelism and the gene concept. Amer. Nat. **89**:64-71.

Green, P. 1955. Cytogenetic studies in *Poa*. III. Variation within *Poa nervosa*, an obligate apomict. Amer. J. Bot. **42**:778-784.

Gregory, R. P. 1914. On the genetics of tetraploid plants in *Primula sinensis*. Proc. Roy. Soc. (Biol.) **87**:484-492.

Grell, F. 1946. Cytological studies in *Culex*. I. Somatic reduction divisions. II. Diploid and meiotic divisions. Genetics **31**:60-94.

Grell, K. G. 1964. The protozoan nucleus. In J. Brachet and A. E. Mirsky (editors). The Cell. Academic Press Inc., New York.

Grell, K. G., and A. Ruthmann. 1964. Über die karyologie des Radiolars *Aulacantha scolymantha* und die feinstruktur seiner chromosomen. Chromosoma **15**:185-211.

Grell, R. F. 1964. Chromosome size at distributive pairing in *Drosophila melanogaster*. Genetics **50**:151-166.

Grell, R. F. 1967. Pairing at the chromosomal level. J. Cell. Physiol. **70**(Suppl. 1):119-146.

Grell, R. F., and A. C. Chandley. 1965. Evidence bearing on the coincidence of exchange and DNA replication in the oocyte of *Drosophila melanogaster*. Proc. Nat. Acad. Sci. U. S. A. **53**:1340-1346.

Griffith, F. 1928. The significance of pneumococcal types. J. Hyg. **27**:113-159.

Gross, J. D., and L. Caro. 1965. Genetic transfer in bacterial mating. Science **150**:1679-1684.

Grula, E. A., G. L. Smith, and M. M. Grula. 1968. Cell division in *Erwinia*: inhibition of nuclear body division in filaments grown in penicillin or mitomycin C. Science **161**:164.

Grun, P. 1955. Cytogenetic studies in *Poa*. III. Variation within *Poa nervosa*, an obligate apomict. Amer. J. Bot. **42**:778-784.

Grun, P. 1958. Plant lampbrush chromosomes. Exp. Cell Res. **14**:619-621.

Gustafsson, A. 1935. Studies on the mechanism of parthenogenesis. Hereditas **21**:1-112.

Gustafsson, A. 1939. The interrelation of meiosis and mitosis. I. The mechanism of agamospermy. Hereditas **25**:289-322.

Gustafsson, A. 1943. The genesis of the European blackberry flora. Lunds Univ. Arssk. **39**:1-200.

Gustafsson, A. 1944. The constitution of the *Rosa canina* complex. Hereditas **30**:405-428.

Gustafsson, A. 1946-1947. Apomixis in higher plants. 3 parts. Lunds Univ. Arssk. N. F. avd. 2, **42**(3):1-67; **43**(2):71-179; **43**(12):183-370.

Gustafsson, A., and A. Hakansson. 1942. Meiosis in some *Rosa* hybrids. Bot. Notiser 1942:331-343.

Guyer, M. F. 1902. Hybridism and the germ cell. Bull. Univ. Cincinnati No. 21.

Habert, E. W., and W. H. Beckert. 1971. Sex distinguishing chromatin in some invertebrates. Cytologia **36**:190-197.

Hadder, J. C., and G. B. Wilson. 1958. Cytological assay of C-mitosis and prophase poison reactions. Chromosoma **9**:99-104.

Haga, T. 1943. A reciprocal translocation in *Lilium hansonii* Leicht. Cytologia **13**:19-25.

Hagberg, A. 1954. Cytogenetic analysis of *erectoides* mutations in barley. Acta Agr. Scand. **4**:472-490.

Hair, J. B. 1956. Subsexual reproduction in *Agropyron*. Heredity **10**:129-160.

Hair, J. B., and E. J. Benzenberg. 1958. Chromosomal evolution in the Podocarpaceae. Nature **181**:1584-1586.

Hair, J. B., and E. J. Benzenberg. 1961. High polyploidy in a New Zealand *Poa*. Nature **189**:160.

Hakansson, A. 1946. Meiosis in hybrid nullisomics and certain other forms of *Godetia whitneyi*. Hereditas **32**:495-513.

Hakansson, A. 1951. Parthenogenesis in *Allium*. Bot. Notiser 1951:143-179.

Hakansson, A. 1954. Meiosis and pollen mitosis in X-rayed and untreated spikelets of *Heleocharis palustris*. Hereditas **40**:325-345.

Halkka, O. 1959. Chromosome studies on the Hemiptera-Homoptera *Auchenorrhyncha*. Ann. Acad. Sci. Fenn. AIV Biol. **43**:1-72.

Halkka, O. 1964. A photometric study of the *Luzula* problem. Hereditas **52**:81-97.

Hance, R. T. 1918a. Variations in the number of somatic chromosomes in *Oenothera scintillans* de Vries. Genetics **3**:225-275.

Hance, R. T. 1918b. Variations in somatic chromosomes. Biol. Bull. **35**:416-431.

Hansen-Melander, E. 1965. The relation sex chromosomes to chromocenters in somatic cells of *Microtus agrestis* (L). Hereditas **52**:357-366.

Hanson, G. P. 1962. Crossing-over in chromosome 3 as influenced by B-chromosomes. Maize Genet. Coop. News Letter **36**:34-35.

Harlan, J. R., and J. M. J. DeWet. 1963. Role of apomixis in the evolution of the *Bothriochloa-Dichanthium* complex. Crop Sci. **3**:314-316.

Harlan, J. R., J. M. J. DeWet, S. M. Nark, and R. J. Lambert. 1970. Chromosome pairing within genomes in maize-*Tripsacum* hybrids. Science **167**:1247-1248.

Harris, H., and J. F. Watkins. 1965. Hybrid cells derived from mouse and man: artificial heterokaryons of mammalian cells from different species. Nature **205**:640-646.

Harris, S. E., and H. S. Forrest. 1967. RNA and DNA synthesis in developing eggs of the milkweed bug, *Oncopeltus fasciatus* (Dallas). Science **156**:1613-1615.

Hastie, A. C. 1967. Mitotic recombination in conidiophores of *Verticillium albo-atrum*. Nature **214**:249-252.

Hauchteck, E., G. Murset, A. Prader, and E. Buhler. 1966. Siblings with different types of chromosomal aberrations due to D/E translocation of the mother. Cytogenetics **5**:281-294.

Havin, A. W., and H. D. Hill. 1966. B-chromosomes, their origin and relation to meiosis in interspecific *Lolium* hybrids. Amer. J. Bot. **53**:702-708.

Hawkes, J. G. 1956. A revision of the tuber-bearing solanums. Scottish Plant Breed. Sta. Ann. Rep. **33**:109.

Hay, E. D., and J. P. Revel. 1963. The fine structure of the DNP component of the nucleus. J. Cell Biol. **16**:20-51.

Hayes, W. 1952. Recombination in *bact. coli* K-12: unidirectional transfer of genetic material. Nature **169**:118-119.

Hayes, W. 1953. The mechanism of genetic recombination in *E. coli*. Cold Spring Harbor Symp. Quant. Biol. **18**:75-93.

Heddle, J. A., S. Wolff, D. Whissell, and J. Cleaver. 1967. Distribution of chromatids at mitosis. Science **158**:929-931.

Hegwood, M. P., and L. F. Hough. 1958. A mosaic pattern of chromosome numbers in the White Winter Pearmain apple and six of its seedlings. Amer. J. Bot. **45**:349-354.

Heilborn, O. 1936. The mechanics of so-called secondary association between chromosomes. Hereditas **22**:167-188.

Heitz, E. and H. Baur. 1933. Beweise fur die Chromosomennatur der Kernschiefen in den Knäuelkernen von *Bibio hortulanum*. Z. Zellforsch. **17**:67-82.

Helmstetter, C. H. 1967. The rate of DNA synthesis during the division cycle of *Escherichia coli* B/r. J. Molec. Biol. **24**:417-427.

Helmstetter, C. H., and S. Cooper. 1968. DNA synthesis during the division cycle of rapidly growing *Escherichia coli* B/r. J. Molec. Biol. **31**:507-518.

Helwig, E. R. 1942. Unusual integrations of the chromatin in *Machaerocera* and other genera of the Acrididae (Orthoptera). J. Morph. **71**:1-33.

Henderson, S. A. 1966. Time of chiasma formation in relation to the time of deoxyribonucleic acid synthesis. Nature **211**:1043-1047.

Henderson, S. A. 1967. The salivary gland chromosomes of *Dasyneura crataegi* (Diptera: Cecidomyiidae). Chromosoma **23**:38-58.

Heneen, W. K., and H. Runemark. 1962. Chromosome polymorphism and morphological diversity in *Elymus rechingeri*. Hereditas **48**:545-564.

Henigardner, R. 1968. Evolution of cellular DNA content in teleost fishes. Amer. Nat. **102**:517-523.

Herman, C. J., and L. W. Lapham. 1963. DNA content of neurons in the cat hippocampus. Science **160**:537.

Herreros, B., and F. Gianelli. 1967. Spatial distribution of old and new chromatid sub-units and frequency of chromatid exchanges in induced human lymphocyte endoreplication. Nature **216**:286-288.

Hershey, A. D. 1946. Spontaneous mutations in bacterial viruses. Cold Spring Harbor Symp. Quant. Biol. **11**:67-70.

Hershey, A. D., and E. Burgi. 1956. Genetic significance of the transfer of nucleic acid from parental to offspring phage. Cold Spring Harbor Symp. Quant. Biol. **21**:91-95.

Hershey, A. D., and M. Chase. 1951. Genetic recombination and heterozygosis in bacteriophage. Cold Spring Harbor Symp. Quant. Biol. **16**:47-50.

Hershey, A. D., and R. Rotman. 1948. Linkage among genes controlling inhibition of lysis in a bacterial virus. Proc. Nat. Acad. Sci. U. S. A. **34**:89-96.

Hershey, A. D., and B. Rotman. 1949. Genetic recombination between host-range and plaque-type mutants of bacteriophage in single bacterial cells. Genetics **34**:44-71.

Hess, O. 1963. Die mutation scute-8 von *Drosophila melanogaster* als ein Lethalfaktor, dessen Penetranz vom Heterochromatin abhangt. Verh. Deutsch. Zool. Ges. Zool. Anz. Suppl. **26**:87-92.

Hess, O. 1966. The function of the lampbrush loops formed by the Y chromosome of *Drosophila heydei* in spermatocyte nuclei. Molec. Gen. Genet. **103**:58-71.

Hess, O. 1967. Complementation of genetic activity in translocated fragments of the Y chromosome in *Drosophila heydei*. Genetics **56**:283-295.

Hess, O., and G. Meyer. 1968. Genetic activities of the Y chromosome in *Drosophila* during spermatogenesis. Advances Genet. **14**:171-223.

Highton, P. J. 1969. An electron microscopic study of cell growth and mesosomal structure of *Bacillus licheniformis*. J. Ultrastruct. Res. **26**:130-147.

Highton, R., and S. A. Henry. 1970. Evolutionary interactions between species of North American salamanders of the genus *Plethodon*. Evolut. Biol. **4**:211-256.

Hill, R. N., and J. J. Yunis. 1967. Mammalian X-chromosomes: change in patterns of DNA replication during embryogenesis. Science **155**:1120-1121.

Hindle, E., and G. Pontecorvo. 1942. Mitotic divisions following meiosis in *Pediculus corporis* males. Nature **149**:668.

Hinton, T., and K. C. Atwood. 1941. Terminal adhesions of salivary gland chromosomes in *Drosophila*. Proc. Nat. Acad. Sci. U. S. A. **27**:491-496.

Hiraoka, T. 1941. Studies of mitosis and meiosis in comparison. III. Behavior of chromonemata in the pre-leptotene stage of meiosis. Cytologia **11**:473-482.

Hiraoka, T. 1958. Somatic syndesis in *Daphne odora*. I. The chromosome behavior in mitosis. Jap. Acad. Proc. **34**:700-705.

Hirono, Y., and G. P. Rédei. 1965. Induced premeiotic exchange of linked markers in the angiosperm *Arabidopsis*. Genetics **51**:519-526.

Hirschhorn, K., H. L. Nadler, W. I. Waithe, B. I. Brown, and R. Hirschhorn. 1969. Pompe's disease: detection of heterozygotes by lymphocyte stimulation. Science **166**:1632-1633.

Hoar, C. S. 1931. Meiosis in *Hypericum punctatum*. Bot. Gaz. **92**:396-406.

Hoar, C. S., and E. J. Haertl. 1932. Meiosis in the genus *Hypericum*. Bot. Gaz. **93**:197-204.

Holliday, R. 1961. Induced mitotic crossing-over in *Ustilago maydis*. Genet. Res. **2**:231-248.

Holliday, R. 1964. A mechanism for gene conversion. Genet. Res. **5**:282-304.

Holliday, R. 1965. Induced mitotic crossing-over in relation to genetic replication in synchronously dividing cells of *Ustilago maydis*. Genet. Res. **6**:104-120.

Holt, C. M. 1917. Multiple complexes in the alimentary tract of *Culex pipiens*. J. Morph. **29**:607-627.

Hoskins, G. C. 1965. Electron microscopic observations of human chromosomes isolated by micrurgy. Nature **207**:1215-1216.

Hoskins, G. C. 1969. Electron micrographic observations of centromeres from unsectioned mammalian chromosomes isolated by micrurgy. Caryologia **22**:229-247.

Hotta, Y., M. Ito, and H. Stern. 1966. Synthesis of DNA during meiosis. Proc. Nat. Acad. Sci. U. S. A. **56**:1184-1191.

Howard, E. F., and W. Plaut. 1968. Chromosomal DNA synthesis in *Drosophila melanogaster*. J. Cell Biol. **39**:415-429.

Howard, H. W., and M. S. Swaminathan. 1952. Species differentiation in the genus *Solanum*, sect. tuberarium with particular reference to the use of interspecific hybridization in breeding. Euphytica **1**:20-28.

Hsu, T. C. 1948. The relations between heteropycnosis, spiralization and lampbrush formation of the chromosomes in the spermatogenesis of the Acrididae. J. Genet. **48**:311-315.

Hsu, T. C. 1952. Mammalian chromosomes in vitro. I. The karyotype of man. J. Hered. **43**:167-172.

Hsu, T. C. 1964. Mammalian chromosomes *in vitro*. XVIII. DNA replication sequence in the chinese hamster. J. Cell Biol. **23**:53-62.

Hsu, T. C. 1970. Distribution of constitutive heterochromatin in mammalian chromosomes. Abstract. Ninth American Somatic Cell Genetics Conference, Galveston, Texas.

Hsu, T. C., and F. E., Arrighi. 1966. Chromosomal evolution in the genus *Peromyscus* (Cricetidae, Rodentia). Cytogenetics **5**:355-359.

Hsu, T. C., F. E. Arrighi, R. R. Klevecz, and B. R. Brinkley. 1965. The nucleoli in mitotic divisions of mammalian cells in vitro. J. Cell Biol. **26**:539-553.

Hsu, T. C., and K. Benirschke (editors). 1967-1970. An Atlas of Mammalian Chromosomes. Volumes 1 to 4. Springer-Verlag New York Inc., New York.

Hsu, T. C., and T. T. Liu. 1948. Microgeographic analysis of chromosomal variation in a Chinese species of *Chironomus* (Diptera). Evolution **2**:49-57.

Huang, R. C., and J. Bonner. 1962. Histone, a suppressor of chromosomal RNA synthesis. Proc. Nat. Acad. Sci. U. S. A. **48**:1216-1222.

Hubbs, A. L., and C. L. Hubbs. 1932. Apparent parthenogenesis in nature, in a form of fish of hybrid origin. Science **76**:628-630.

Hubbs, Carl L. 1955. Hybridization between fish species in nature. Syst. Zool. **4**:1-20.

Hubbs, Clark. 1967. Analysis of phylogenetic relationships using hybridization techniques. Symp. Newer Trends in Taxonom. **34**:48-59.

Huettner, A. F. 1924. Maturation and fertilization in *Drosophila melanogaster*. J. Morph. Physiol. **39**:249-265.

Hughes, R. D. 1936. The morphology of the normal salivary chromosomes of *Drosophila virilis*. J. Hered. **27**:305-306.

Hughes, R. D. 1939. An analysis of the chromosomes of two subspecies, *Drosophila virilis virilis* and *Drosophila virilis americana*. Genetics **24**:811-834.

Hughes-Schrader, S. 1943a. Meiosis without chiasmata in diploid and tetraploid spermatocytes of the mantid *Callimantis antillarum* Saussare. J. Morph. **73**:111-141.

Hughes-Schrader, S. 1943b. Polarization, kinetochore movements, and bivalent structure in the meiosis of male mantids. Biol. Bull. **85**:265-300.

Hughes-Schrader, S. 1953. The nuclear content of desoxyribonucleic acid and interspecific relationships in the mantid genus *Liturgousa* (Orthoptera: mantoidea). Chromosoma **5**:544-554.

Hughes-Schrader, S. 1956. Polyteny as a factor in the chromosome evolution of the *Pentatomini* (Hemiptera). Chromosoma **8**:135-151.

Hungerford, D. A. 1971. Chromosome structure and function in man. I. Pachytene mapping in the male, improved methods and general discussion of initial results. Cytogenetics **10**:23-32, 33-37.

Huskins, C. L. 1948. Segregation and reduction in somatic tissues. J. Hered. **39**:311-325.

Huskins, C. L., and K. C. Cheng. 1950. Segregation and reduction in somatic tissues. IV. Reductional groupings induced in *Allium cepa* by low temperatures. J. Hered. **41**:13-18.

Huskins, C. L., and L. N. Steinitz. 1950. The nucleus in differentiation and development. II. Induced mitoses in differentiated tissues of *Rhoeo* roots. J. Hered. **39**:67-77.

Hutchinson, J. B. 1954. New evidence on the origin of the Old World cottons. Heredity **8**:225-241.

Huziwara, Y. 1959. Chromosomal evolution in the subtribe Asterinae. Evolution **13**:188-193.

Hyde, R. R., and H. M. Powell. 1916. Mosaics in *Drosophila ampelophila*. Genetics **1**:581-583.

Ichida, A. A., and M. S. Fuller. 1968. Ultrastructure of mitosis in the aquatic fungus *Catenaria anguillulae*. Mycologia **60**:141-155.

Inman, R. B. 1966. A denaturation map of the phage DNA molecule determined by electron microscopy. J. Molec. Biol. **18**:464-476.

Irwin, S., and J. Egozcue. 1967. Chromosomal abnormalities in leukocytes from LSD-25 users. Science **157**:313-314.

Ivanovskaya, E. V. 1939. A haploid plant of *Solanum tuberosum* L. C. R. Acad. Sci. U. R. S. S. **24**:517-520.

Jackson, R. C. 1962. Intraspecific hybridization in *Haplopappus* and its bearing on chromosome evolution in the Blepharodon section. Amer. J. Bot. **49**:119-132.

Jackson, R. C. 1964. Preferential segregation of chromosomes from a trivalent in *Haplopappus gracilis*. Science **145**:511-513.

Jackson, R. C. 1965. A cytogenetic study of a three-paired race of *Haplopappus gracilis*. Amer. J. Bot. **52**:946-953.

Jacob, F., and S. Brenner. 1963. Sur la régulation de la synthése du DNA chez les bactéries: l'hypothèse du rèplicon. C. R. Acad. Sci. Paris **256**:298-300.

Jacob, F., S. Brenner, and F. Cuzin. 1963. On the regulation of DNA replication in bacteria. Cold Spring Harbor Symp. Quant. Biol. **28**:329-348.

Jacob, F., and J. Monod. 1961. Genetic regulatory mechanisms in the synthesis of proteins. J. Molec. Biol. **3**:318-356.

Jacob, F., and E. L. Wollman. 1957. Analyse des groupes de laison génétique de differentes souches donatrices. C. R. Acad. Sci. Paris **245**:1840-1843.

Jacob, F., and E. L. Wollman. 1958. Genetic and physical determinations of chromosomal segments in *E. coli*. Symp. Soc. Exp. Biol. **12**:75-83.

Jacob, F., and E. L. Wollman. 1961. Sexuality and the Genetics of Bacteria. Academic Press Inc., New York.

Jacobs, P. A. 1969. Structural abnormalities of the sex chromosomes. Brit. Med. Bull. **25**:94-99.

Jacobson, C. B., and R. H. Barter. 1967. Intrauterine diagnosis and management of genetic defects. Amer. J. Obstet. Gynec. **99**:796-807.

Jagiello, G., J. Karnicki, and R. J. Ryan. 1968. Superovulation with pituitary gonadotropins: methods for obtaining meiotic metaphase figures in human ova. Lancet **1**:178-180.

Jain, H. K. 1966. Correlated synthetic activities of chromosomes. In C. D. Darlington and K. R. Lewis (editors). Chromosomes Today. Plenum Press, New York.

James, A. P., and B. Lee-Whiting. 1955. Radiation-induced genetic segregations in vegetative cells of diploid yeast. Genetics **40**:826-831.

James, S. H. 1965. Complex hybridity in *Isotoma petraea*. I. The occurrence of interchange heterozygosity, autogamy, and a balanced lethal system. Heredity **20**:341-353.

Janaki-Ammal, E. K. 1932. Chromosome studies in *Nicandra physaloides*. Cellule **41**:87-110.

Janaki-Ammal, E. K. 1940. Chromosome diminution in a plant. Nature **146**:839-840.

Janick, J., D. L. Mahoney, and D. L. Pfahler. 1959. The trisomics of *Spinacea oleracea*. J. Hered. **50**:47-50.

Janssens, F. A. 1909. La théorie de la chiasmatypie. Nouvelle interprétation des cinèses de maturation. Cellule **25**:389-411.

Janssens, F. A. 1924. La chiasmatypie dans les insectes. Spermatogenèse dans (1) *Stethophyma grossum*, (2) *Chorthippus parallelus*. Cellule **34**:135-359.

Jayan, P. K. 1970. Ecology of *Cenchrus ciliaris* complex. Thesis, Saurashtra University, Raykot, India.

Jenkins, J. A. 1929. Chromosome homologies in wheat and *Aegilops*. Amer. J. Bot. **16**:238-245.

Jessop, A. P., and D. G. Catcheside. 1965. Interallelic recombination at the *his*-1 locus in *Neurospora crassa* and its genetic control. Heredity **20**:237-256.

Jinks, J. L. 1964. Extrachromosomal Inheritance. Prentice-Hall, Inc., Englewood Cliffs, N. J.

John, B., and G. M. Hewitt. 1966. Karyotype stability and DNA variability in the Acrididae. Chromosoma **20**:155-172.

John, B., and K. R. Lewis. 1958. Studies on *Periplaneta americana*. III. Selection for heterozygosity. Heredity **12**:185-197.

Jokelainen, P. T. 1967. The ultrastructure and spatial organization of the metaphase kinetochore in mitotic rat cells. J. Ultrastruct. Res. **19**:19-44.

Jones, D. F. 1936. Segregation of color and growth-regulating genes in the somatic tissue of maize. Proc. Nat. Acad. Sci. U. S. A. **22**:163-166.

Jones, D. F. 1937. Somatic segregation and its relation to atypical growth. Genetics **22**:484-522.

Jones, K. 1970. Chromosome changes: reliable indicators of the direction of evolution? Taxon **19**: 172-179.

Jones, K., and C. Colden. 1968. The telocentric complement of *Tradescantia micrantha*. Chromosoma **24**:135-157.

Jorgenson, C. A. 1928. The experimental formation of heteroploid plants in the genus *Solanum*. J. Genet. **19**:133-211.

Käfer, E. 1961. The process of spontaneous recombination in vegetative nuclei of *Aspergillus nidulans*. Genetics **46**:1581-1609.

Käfer, E. 1963. Radiation effects and mitotic recombination in diploids of *Aspergillus nidulans*. Genetics **48**:27-45.

Kaltsikes, P. J., L. E. Evans, and E. N. Larter. 1969. Morphological and meiotic characteristics of the extracted AABB tetraploid component of three varieties of common wheat. Canad. J. Genet. Cytol. **11**:65-71.

Kapadia, Z. J., and F. W. Gould. 1964. Biosystematic studies in *Bouteloua curtipendula* complex. IV. Dynamics of variation in *B. curtipendula* var. *caespitosa*. Bull. Torrey Bot. Club. **91**:465-478.

Kaplan, W. D. 1953. The influence of Minutes upon somatic crossing-over in *Drosophila melanogaster*. Genetics **38**:632-651.

Karasaki, S. 1965. Electron microscopic examination of the sites of nuclear RNA synthesis during amphibian embryogenesis. J. Cell Biol. **26**:937-958.

Karasaki, S. 1968. The ultrastructure and RNA metabolism of nucleoli in early sea urchin embryos. Exp. Cell Res. **52**:13-26.

Kassanis, B. 1962. Properties and behaviour of a virus depending for its multiplication on another. J. Gen. Microbiol. **27**:477-488.

Kassanis, B. 1968. Satellitism and related phenomena in plant and animal viruses. Adv. Virus Res. **13**: 147-180.

Kato, K. 1930. Cytological studies of pollen mother cells of *Rhoeo discolor* Hance with special reference to the question of the mode of syndesis. Mem. Coll. Sci. Kyoto Imp. Univ., Ser. B. **5**:139-161.

Kaufmann, B. 1936. A terminal inversion of *Drosophila ananassae*. Proc. Nat. Acad. Sci. U. S. A. **22**:591-594.

Kaulenas, M. S., W. E. Foor, and D. Fairbairn. 1969. Ribosomal RNA synthesis during cleavage of *Ascaris lumbricoides* eggs. Science **163**:1201-1202.

Kawaguchi, E. 1936. Der einfluss der Eierbehandlung mit Zentrifugierung auf die Vererbung bei dem Seidenspinner. J. Fac. Agr. Hokkaido Imp. Univ. **38**:111-133.

Kayano, K. 1962. Cytological studies in *Lilium callosum*. IV. Transmission and multiplication of a small supernumerary B chromosome. Evolution **16**:86-89.

Keep, E. 1962. Interspecific hybridization in *Ribes*. Genetica **33**:1-23.

Kermicle, J. L. 1969. Androgenesis conditioned by a mutation in maize. Science **166**:1422-1424.

Kerr, W. E. 1969. Some aspects of the evolution of social bees (Apidae). Evolut. Biol. **3**:119-175.

Kessler, S., and R. H. Moos. 1969. XYY chromosome: premature conclusions. Science (letter) **165**:442.

Keyl, H. G. 1965. A demonstrable local and geometric increase in the chromosomal DNA of *Chironomus*. Experientia **21**:191-193.

Keyl, H. G., and C. Pelling. 1963. Differentielle DNA-replikation in den speicheldrüsen-chromosomen von *Chironomus thummi*. Chromosoma **14**:347-359.

Khoshoo, T. N. 1960. Cytogenetical evolution in the gymnosperms-karyotype. Proc. Summer School Bot. (Darjeeling) **1960**:119-135.

Khush, G. S. 1970. Aneuploidy. Academic Press Inc., New York.

Khush, G. S., and C. M. Rick. 1968. Cytogenetic analysis of the tomato genome by means of induced deficiencies. Chromosoma **23**:452-484.

Kihara, H. 1929. Conjugation of homologous chromosomes in the genus hybrids *Triticum* × *Aegilops* and species hybrids of *Aegilops*. Cytologia **1**:1-15.

Kihara, H. 1930. Genomanalyse bei *Triticum* und *Aegilops*. Cytologia **1**:263-270.

Kihara, H. 1936. Ein diplo-haploides Zwillingspaar bei *Triticum durum*. Agr. Hart. (Jap.) **11**:1425-1434.

Kihara, H., and Y. Katayama. 1933. Reifungsteilungen bei dem haploiden *Triticum monococcum*. Agr. Hart. (Jap.) **8**:1-17.

Kikkawa, H. 1934. Problems of crossing-over in triploid *Drosophila*, with criticism of Mather's interference. Amer. Nat. **68**:515-569.

Kiley, S., and J. F. Wohnus. 1968. Chromosomal analysis of *Rana pipiens*, *Bufo americanus* and their hybrid. Cytogenetics **7**:78-90.

Kimber, G., and R. Riley. 1963. Haploid angiosperms. Bot. Rev. **29**:480-531.

King, E. 1933. Chromosome behavior in a triploid *Tradescantia*. J. Hered. **24**:253-256.

Kirk, J. T. O. 1963. The deoxyribonucleic acid of broad bean chloroplasts. Biochem. Biophys. Acta **76**:417-424.

Kitani, Y. 1963. Orientation, arrangement, and association of somatic chromosomes. Jap. J. Hum. Genet. **38**:244-256.

Kitzmiller, J. B. 1963. Mosquito cytogenetics. A review of the literature, 1953-1962. Bull. W. H. O. **29**:345-355.

Kitzmiller, J. B., G. Frizzi, and R. H. Baker. 1967. Evolution and speciation within the Maculipennis complex of the genus *Anopheles*. In J. W. Wright and R. Pal (editors). Genetics of Insect Vectors of Disease. Elsevier Publishing Co., Amsterdam.

Kleinsmith, L. S., V. G. Allfrey, and A. E. Mirsky. 1966. Phosphorylation of nuclear protein early in the course of gene activation in lymphocytes. Science **154**:780-781.

Klevecz, R. R. 1969. Temporal coordination of DNA replication with enzyme synthesis in diploid and heteroploid cells. Science **166**:1536-1538.

Klingstadt, H. 1939a. Is the synaptic period of meiosis a modified mitosis? Soc. Sci. Fenn. Comment. Biol. **7**:1-9.

Klingstadt, H. 1939b. Morphology and spermatogenesis of *Chorthippus bicolor* Charp × *Ch. bigutulus* L. J. Genet. **37**:389-420.

Knipling, E. F. 1959. Sterile male method of population control. Science **130**:902-904.

Knott, D. R. 1956. A case of somatic reduction in a premeiotic cell in wheat. Canad. J. Bot. **34**:831-832.

Knott, D. R. 1959. The inheritance of rust resistance. Canad. J. Plant Sci. **39**:215-228.

Knott, D. R. 1964. The effect on wheat of an *Agropyron* chromosome carrying rust resistance. Canad. J. Genet. Cytol. **6**:500-507.

Koller, P. C. 1936. Structural hybridity in *Drosophila pseudoobscura*. J. Genet. **32**:79-102.

Koller, P. C. 1938. Asynapsis in *Pisum sativum*. J. Genet. **36**:275-306.

Kosswig, C., and A. Shengun. 1947. Intraindividual variability of chromosome IV of *Chironomus*. J. Hered. **38**:235-259.

Kostoff, D. 1930. Discoid structure of the spireme. J. Hered. **21**:373-324.

Kostoff, D. 1938a. The most probable place of location of the genes in the chromonemata. Nature **141**:749.

Kostoff, D. 1938b. Studies on polyploid plants. XXI. Cytogenetic behavior of the allopolyploid hybrids *Nicotiana glauca* Grah. × *Nicotiana langsdorffii* Weinm. and their evolutionary significance. J. Genet. **37**:129-209.

Koulischer, L., and S. Frechkop. 1966. Chromosome complement: a fertile hybrid between *Equus prjewalskii* and *Equus caballus*. Science **151**:94-95.

Krause, F. 1938. Examination of viruses by the electron microscope. Naturwissenschaften **26**:122-137.

Krishnaswamy, N. 1938. Cytological studies in a haploid plant of *Triticum vulgare*. Hereditas **25**:77-86.

Kroeger, H., and M. Lezzi. 1966. Regulation of gene action in insect development. Ann. Rev. Entomol. **11**:1-21.

Kubai, D. F., and H. Ris. 1969. Division in the dinoflagellate *Gyrodinium cohnii* (Schiller). A new type of nuclear reproduction. J. Cell Biol. **40**:508-528.

Kuempel, P. L. 1970. Bacterial chromosome replication. Advances Cell Biol. **1**:3-56.

Kunze, E. 1953. Untersuchungen über die paarungssaffinität bei Reisenchromosomen. Chromosoma **5**:501-510.

Kurabauaski, M. 1958. Evolution and variation in Japanese species of *Trillium*. Evolution **12**:286-310.

Kurnick, N. H., and I. H. Herskowitz. 1952. The estimation of polyteny in *Drosophila* salivary gland nuclei based on determination of desoxyribonucleic acid content. J. Cell. Comp. Physiol. **39**:281-299.

Kyhos, D. W. 1965. The independent aneuploid origin of two species of *Chaenactis* (Compositae) from a common ancestor. Evolution **19**:26-43.

La Chance, L. E., and M. Degrugillier. 1969. Chromosomal fragments transmitted through three generations in *Oncopeltus* (Hemiptera). Science **166**:235-236.

La Cour, L. T. 1952. Breakage and reunion in *Hyacinthus*. Heredity (Suppl.) **6**:163-179.

La Cour, L. F., and S. R. Pelc. 1958. Effect of colchicine on the utilization of labelled thymidine during chromosome reproduction. Nature **182**:506-508.

Lamb, B. C. 1969. Evidence from *Sordaria* that recombination and conversion frequencies are partly determined before meiosis, and for a general model of the control of recombination frequencies. Genetics **63**:807-820.

Lamm, R. 1941. Varying cytological behaviour in reciprocal *Solanum* crosses. Hereditas **27**:202-208.

Lamm, R. 1945. Cytogenetic studies in *Solanum* sect. *tuberosum*. Hereditas **31**:1-128.

Langan, T. A. 1968. Stimulation of histone phosphorylation by cyclic adenosine 3',5'-monophosphate: a potential mechanism for hormonal induction of RNA synthesis. J. Cell Biol. **39**:77-81.

Lange, R. J. de, D. M. Fambrough, E. L. Smith, and J. Bonner. 1968. Calf and pea histone IV. I. Amino acid composition and the identical COOH-terminal, 19-residue sequence. J. Biol. Chem. **243**:5906-5913.

Lark, K. G. 1966. Regulation of chromosome replication and segregation in bacteria. Bact. Rev. **30**:3-32.

Lark, K. G. 1967. Non-random segregation of sister chromatids in *Vicia faba* and *Triticum boeoticum*. Proc. Nat. Acad. Sci. U. S. A. **58**:352-359.

Lark, K. G., R. A. Consigli, and H. C. Minocha. 1966. Segregation of sister chromatids in mammalian cells. Science **154**:1202-1205.

Lark, K. G., T. Repko, and E. Hoffman. 1963. The effect of amino acid deprivation on subsequent deoxyribonucleic acid replication. Biochem. Biophys. Acta **73**:9-24.

Lawrence, C. W. 1961. The effect of the irradiation of different stages in microsporogenesis on chiasma frequency. Heredity 16:83-89.

Lawrence, C. W. 1963. The orientation of multiple associations resulting from interchange heterozygosity. Genetics 48:347-350.

Lawrence, C. W. 1965. Influence of non-lethal doses of radiation in *Chlamydomonas reinhardi*. Nature 206:789-791.

Lawrence, W. J. C. 1931. The secondary association of chromosomes. Cytologia 2:352-384.

Leadbetter, B., and J. D. Dodge. 1967. An electron microscope study of nuclear and cell division in a dinoflagellate. Arch. f. Mikrobiol. 57:239-254.

Leblond, C. P., and Y. Clermont. 1952. Spermiogenesis of rat, mouse, hamster and guinea pig as revealed by the "periodic acid-fuchsin sulfurous acid" technique. Amer. J. Anat. 90:167-216.

LeClerc, G. 1946. Occurrence of mitotic crossing over without meiotic crossing over. Science 102:553-554.

Lederberg, J. 1947. Gene recombination and linkage segregation in *E. coli*. Genetics 32:505-525.

Lederberg, J. 1949. Aberrant heterozygotes in *Escherichia coli*. Proc. Nat. Acad. Sci. U. S. A. 35:178-184.

Lederberg, J. 1957. Sibling recombinants in zygote pedigrees of *Escherichia coli*. Proc. Nat. Acad. Sci. U. S. A. 43:1060-1065.

Lederberg, J., L. L. Cavalli, and E. M. Lederberg. 1952. Sex compatibility in *E. coli*. Genetics 37:720-730.

Lederberg, J., and E. L. Tatum. 1946. Gene recombination in *E. coli*. Nature 158:558.

Leedale, G. F. 1958. Nuclear structure and mitosis in the Euglenineae. Arch. Mikrobiol. 32:32-64.

Lejeune, J., R. Turpin, and M. Gautier. 1959. Le mongolisme, premier example d'aberration autosomique humaine. Ann. Genet. 2:41-49.

Lejeune, J., B. Dutrillau, J. Lafourcade, et al., 1968. Endoreduplication selective du bras long du chromosome 2 chez une femme et sa fille. Compt. Rend. Acad. Sci. Paris, ser. D 266:24-26.

Lesley, J. W. 1928. A cytological and genetic study of progenies of triploid tomatoes. Genetics 13:1-43.

Lesley, M. M., and H. B. Frost. 1927. Mendelian inheritance of chromosome shape in *Matthiola*. Genetics 12:449-460.

Levan, A. 1932. Cytological studies in *Allium*. I. Chromosome morphological contributions. Hereditas 16:257-294.

Levan, A. 1935. Cytological studies in *Allium*. VI. The chromosome morphology of some diploid species of *Allium*. Hereditas 20:289-330.

Levan, A. 1937. Cytological studies in the *Allium paniculatum* group. Hereditas 23:317-370.

Levan, A. 1941. Syncyte formation in the pollen mother cells of haploid *Phleum pratense*. Hereditas 27:243-252.

Levan, A., K. Fredga, and A. A. Sandberg. 1964. Nomenclature for centromeric position on chromosomes. Hereditas 52:201-220.

Levan, A., and T. C. Hsu. 1959. The human idiogram. Hereditas 45:665-672.

Levenbook, L., E. C. Travaglini, and J. Schultz. 1958. Nucleic acids and their components as affected by the Y chromosome of *Drosophila melanogaster*. Exp. Cell Res. 15:43-79.

Levin, D. A. 1968. Multiple chromosome triplication in *Liatris*. Science 161:183-184.

Levitsky, G. A. 1931a. The morphology of chromosomes. Bull. Appl. Bot. Genet. Plant Breed. 27:19-174.

Levitsky, G. A. 1931b. The karyotype in systematics. Bull. Appl. Bot. Genet. Plant Breed. 27:220-240.

Levitsky, G. A., G. Araratian, G. Mardjaniskvil, and H. Shepeleva. 1931. Experimentally induced alterations of the morphology of chromosomes. Amer. Nat. 65:564-567.

Lewis, E. B. 1963. Genes and developmental pathways. Amer. Zool. 3:33-56.

Lewis, H. 1951. The origin of supernumerary chromosomes in natural populations of *Clarkia elegans*. Evolution 5:142-157.

Lewis, H., and C. Epling. 1946. Formation of a diploid species of *Delphinium* by hybridization. Amer. J. Bot. 33:21-S-22-S. (Abstr.)

Lewis, H., and C. Epling. 1959. *Delphinium gypsophilum*, a diploid species of hybrid origin. Evolution 13:511-525.

Lewis, H., and P. H. Raven. 1958. Rapid evolution in *Clarkia*. Evolution 12:319-336.

Lewis, K. R., and B. John. 1957. Studies on *Periplaneta americana*. II. Interchange heterozygosity in isolated populations. Heredity 11:11-22.

Lewis, K. R., and B. John. 1966. The meiotic consequences of spontaneous chromosome breakage. Chromosoma 18:287-304.

Lewis, W. H., R. L. Oliver, and T. J. Luikart. 1971. Multiple genotypes in individuals of *Claytonia virginica*. Science 172:564-565.

Lewis, W. H., R. L. Oliver, and Y. Suda. 1967. Cytogeography of *Claytonia virginica* and its allies. Ann. Missouri Bot. Garden 54:153-171.

Lewontin, R. C. 1967. An estimate of average heterozygosity in man. Am. J. Human Genet. 19:681-685.

Li, N., and R. C. Jackson. 1961. Cytology of supernumerary chromosomes in *Haplopappus spinulosus* ssp. *cotula*. Amer. J. Bot. 48:419-426.

Lima-de-Faria, A. 1950. The feulgen test applied to centromeric chromomeres. Hereditas 36:60-74.

Lima-de-Faria, A. 1952. Chromomere analysis of the chromosome complement of rye. Chromosoma 5:1-68.

Lima-de-Faria, A. 1959. Differential uptake of tritiated thymidine into hetero- and euchromatin in *Melanoplus* and *Secale*. J. Biophys. Biochem. Cytol. 6:457-466.

Lima-de-Faria, A. 1962. Genetic interiection in rye expressed at the chromosome phenotype. Genetics 47:1455-1462.

Lima-de-Faria, A., M. Birnstiel, and H. Jaworska. 1969. Amplification of ribosomal cistrons in the heterochromatin of *Acheta*. Genetics (Suppl. 61)**1,** 1969:145-159.

Lima-de-Faria, A., and H. Jaworska. 1964. Haplodiploid chimaeras in *Haplopappus gracilis*. Hereditas **52**:119-122.

Lima-de-Faria, A., and M. J. Moses. 1966. Ultrastructure and cytochemistry of metabolic DNA in *Tipula*. J. Cell Biol. **30**:177-192.

Lin, M. 1954. Chromosomal control of the nucleolar composition in maize. Chromosoma **7**:340-370.

Lindsley, D. L., and E. H. Grell. 1969. Spermiogenesis without chromosomes in *Drosophila melanogaster*. Genetics (Suppl. 61)**1,**1969:69-78.

Lindsley, D. L., L. Sandler, B. Nicoletti, and G. Trippe. 1968. Genetic control of recombination in *Drosophila*. In W. J. Peacock and R. D. Brock (editors). Replication and Recombination of Genetic Material. Australian Academy of Science, Canberra.

Lisco, H., and R. A. Conrad. 1967. Chromosome studies on Marshall Islanders exposed to fallout radiation. Science **157**:445-447.

Littau, V. C., C. J. Burdick, V. G. Alfrey, and A. E. Mirsky. 1965. The role of histones in the maintenance of chromatin structure. Proc. Nat. Acad. Sci. U. S. A. **54**:1204-1212.

Little, T. M. 1945. Gene segregation in autotetraploids. Bot. Rev. **11**:60-85.

Lockshin, R. A. 1966. Insect embryogenesis: macromolecular synthesis during early development. Science **154**:775-776.

London Conference. 1963. The London Conference on the normal human karyotype. Cytogenetics **2**: 264-268.

Longley, A. E. 1927. Supernumerary chromosomes in *Zea mays*. J. Agr. Res. **33**:769-784.

Longley, A. E. 1937. Morphological characters of teosinte chromosomes. J. Agr. Res. **54**:835-862.

Longley, A. E. 1938. Chromosomes of maize from North American Indians. J. Agr. Res. **56**:177-196.

Longley, A. E. 1945. Abnormal segregation during megasporogenesis in maize. Genetics **30**:100-113.

Longwell, A. C., and G. Svihla. 1960. Specific chromosomal control of the nucleolus and of the cytoplasm in wheat. Exp. Cell Res. **20**:294-312.

Lorkovic, Z. 1941. Die chromosomenzahlen in der spermatogenese der Tagfalter. Chromosoma **2**:155-191.

Lorz, A. P. 1947. Supernumerary chromosomal reproductions: polytene chromosomes, endomitosis, multiple chromosome complexes, polysomaty. Bot. Rev. **13**:597-624.

Loughman, W. D., et al. 1971. Acetylsalicyclic acid and chromosome damage. Science **171**:829-830.

Loughman, W. D., T. W. Sargent, and D. M. Israelstam. 1967. Leukocytes of humans exposed to lysergic acid diethylamide: lack of chromosomal damage. Science **158**:508-510.

Love, R. M., and C. A. Suneson. 1945. Cytogenetics of certain *Triticum-Agropyron* hybrids and their fertile derivatives. Amer. J. Bot. **32**:451-456.

Lowary, P. A., and C. J. Avers. 1965. Nucleolar variation during differentiation of *Phleum* root epidermis. Amer. J. Bot. **52**:199-203.

Lowe, C. H., and J. W. Wright. 1966. Chromosomes and karyotypes of cnemidophorine teiid lizards. Mammal Chrom. Newsletter **22**:199-200.

Lu, B. C. 1964. Polyploidy in the basidiomycete *Cyathus stercoreus*. Amer. J. Bot. **51**:343-347.

Lu, B. C. 1967. The course of meiosis and centriole behavior during the ascus development of the ascomycete *Gelasinospora calospora*. Chromosoma **22**:210-226.

Lu, B. C. 1969. Genetic recombination in *Corprinus*. I. Its precise timing as revealed by temperature treatment experiments. Canad. J. Genet. Cytol. **11**:834-847.

Lubs, H. A., and F. H. Ruddle. 1970. Chromosomal abnormalities in the human population: estimation of rates based on New Haven newborn study. Science **169**:495-497.

Lubs, H. A., and J. Samuelson. 1967. Chromosome abnormalities in lymphocytes from normal human subjects. Cytogenetics **6**:402-411.

Luck, D. J. L., and E. Reich. 1964. DNA in mitochondria of *Neurospora crassa*. Proc. Nat. Acad. Sci. U. S. A. **52**:931-938.

Luria, S. E., and R. Dulbecco. 1949. Genetic recombination leading to production of active bacteriophage from ultraviolet inactivated bacteriophage particles. Genetics **34**:93-125.

Lutz, A. M. 1917. Fifteen and sixteen chromosome mutants. Amer. J. Bot. **4**:53-111.

Luykx, P. 1970. Cellular mechanisms of chromosome distribution. Int. Rev. Cytol. Suppl. 2.

Lwoff, A., and A. Gutmann. 1950. Recherches sur un *Bacillus megatherium* lysogène. Ann. Pasteur **78**: 711-719.

Lyon, M. F. 1961. Gene action in the X-chromosome of the mouse *(Mus musculus* L.*)*. Nature **190**:372-374.

McCarthy, B. J., and E. T. Bolton. 1963. An approach to the measurement of genetic relatedness among organisms. Proc. Nat. Acad. Sci. U. S. A. **50**:156-164.

McClelland, G. A. H. 1962. Trans. Roy. Soc. Trop. Med. Hyg. **56**:4. Cited by McDonald, P. T., and K. S. Rai. 1970. *Aedes aegypti*: origin of a "new" chromosome from a double translocation heterozygote. Science **168**:1229-1230.

McClelland, G. A. H. 1966. Sex-linkage at two loci affecting eye pigment in the mosquito *Aedes aegypti* (Diptera: Culicidae). Canad. J. Genet. Cytol. **7**:192-198.

McClintock, B. 1929. A cytological and genetical study of triploid maize. Genetics **14**:180-222.

McClintock, B. 1931. Cytological observations of deficiencies involving known genes, translocations, and an inversion in *Zea mays*. Univ. Missouri Agr. Exp. Sta. Res. Bull. **163**:1-30.

McClintock, B. 1932. A correlation of ring-shaped chromosomes with variegation in *Zea mays*. Proc. Nat. Acad. Sci. U. S. A. **18**:677-681.

McClintock, B. 1933. The association of non-homologous parts of chromosomes in the mid-prophase of meiosis of *Zea mays*. Z. Zellforsch. **19**:191-237.

McClintock, B. 1934. The relation of a particular chromosomal element to the development of the nucleoli in *Zea mays*. Z. Zellforsch. **21**:294-328.

McClintock, B. 1939. The behavior in successive nuclear divisions of a chromosome broken at meiosis. Proc. Nat. Acad. Sci. U. S. A. **25**:405-416.

McClintock, B. 1941a. Spontaneous alternations in chromosome size and form in *Zea mays*. Cold Spring Harbor Symp. Quant. Biol. **9**:72-81.

McClintock, B. 1941b. The stability of broken ends of chromosomes in *Zea mays*. Genetics **26**:234-282.

McClintock, B. 1945. *Neurospora*. I. Preliminary observations of the chromosomes of *Neurospora crassa*. Amer. J. Bot. **32**:671-678.

McClintock, B., and H. E. Hill. 1931. The cytological identification of the chromosome associated with the R-G linkage groups in *Zea mays*. Genetics **16**:175-190.

McClung, C. E. 1901. Notes on the accessory chromosome. Anat. Anz. **20**:220-226.

McClung, C. E. 1902. The accessory chromosome-sex determinant? Biol. Bull. **3**:43-84.

McClung, C. E. 1927. Synapsis and related phenomena in *Mecostethus* and *Leptysma* (Orthoptera). J. Morph. Physiol. **43**:181-252.

McClure, H. M., K. H. Belden, W. A. Pieper, and C. B. Jacobson. 1969. Autosomal trisomy in a chimpanzee; resemblance to Down's syndrome. Science **165**:1010-1012.

McDonald, I. C. 1971. A male-producing strain of the house fly. Science **172**:489.

MacDonald, K. B., and W. R. Bruce. 1968. A mini-colony assay for the viability of mammalian cells in vitro. Exp. Cell Res. **50**:471-473.

McDonald, P. T., and K. S. Rai. 1970. *Aedes aegypti*: origin of a "new" chromosome from a double translocation heterozygote. Science **168**:1229-1230.

McFadden, E. S., and E. R. Sears. 1946. The origin of *Triticum spelta* and its free-threshing hexaploid relatives. Heredity **37**:81-89, 107-117.

Macfarlane, J. M. 1898. Observations on some hybrids between *Drosera filiformis* and *D. intermedia*. Publ. Univ. Penn. NS 5, Contrib. Bot. Lab. **5**:87-99.

McGinnis, R. C., and J. Unrau. 1952. A study of meiosis in a haploid of *Triticum vulgare* Vill. and its progenies. Canad. J. Bot. **30**:40-49.

MacGregor, H. C. 1965. The role of lampbrush chromosomes in the formation of nucleoli in amphibian oocytes. Quart. J. Microscop. Sci. **106**:215-228.

McGregor, J. F. 1970. The chromosomes of the maskinonge (*Esox masquinongy*). Canad. J. Genet. Cytol. **12**:224-229.

Mackensen, O. 1935. Locating genes on salivary chromosomes. Cytogenetic methods demonstrated in determining position of genes on the X-chromosome of *Drosophila melanogaster*. J. Hered. **26**:163-174.

McKusick, V. A. 1962. On the X chromosome of man. Quart. Rev. Biol. **37**:69-175.

McKusick, V. A. 1964. On the X Chromosome of Man. American Institute of Biological Sciences, Washington, D. C.

McKusick, V. A. 1968. Mendelian Inheritance in Man. Johns Hopkins Press, Baltimore.

McLaren, S., M. Woods, and R. Shea. 1966. Polyteny: a source of cryptic speciation among copepods. Science **153**:1641-1642.

McLeish, J. 1954. The consequence of localized chromosome breakage. Heredity **8**:385-409.

McMahon, R. M. 1956. Mitosis in polyploid somatic cells of *Lycopersicon esculentum* Mill. Caryologia **8**:250-256.

Magenis, R. E., F. Hecht, and E. W. Lavrien. 1970. Heritable fragile site on chromosome 16: probable localization of haptoglobin locus in man. Science **170**:85-87.

Magni, G. E., R. von Borstel, and S. Sora. 1964. Mutogenic action during meiosis and antimutogenic action during mitosis by 5-aminoacridine in yeast. Mutat. Res. **1**:227-230.

Magoon, M. L., R. W. Hougas, and D. C. Cooper. 1958. Cytogenetic studies of tetraploid hybrids in *Solanum*. J. Hered. **49**:171-178.

Maguire, M. P. 1965. The relationship of crossover frequency to synaptic extent at pachytene in maize. Genetics **51**:23-40.

Maguire, M. P. 1966. The relationship of crossing over to chromosome synapsis in a short paracentric inversion. Genetics **53**:1071-1077.

Maguire, M. P. 1967. Evidence for homologous pairing of chromosomes prior to meiotic prophase in maize. Chromosoma **21**:221-231.

Makino, S. 1951a. An Atlas of the Chromosome Numbers in Animals. Iowa State College Press, Ames, Iowa.

Makino, S. 1951b. Studies on the murine chromosomes. V. A study of the chromosomes of *Apodemus*, especially with reference to the sex chromosomes in meiosis. J. Morph. **88**:93-126.

Makino, S., M. S. Sasali, K. Yamada, and T. Kajii. 1963. A long Y chromosome in man. Chromosoma **14**:154-161.

Malheiros-Gardé, N., and A. Gardé. 1951. Agmatoploidia no genero *Luzula* D. C. Genet. Iberica **3**:155-176.

Manton, I. 1937. The problem of *Biscutella laevigata* L. II. The evidence from meiosis. Ann. Bot. N. S. **1**:439-462.

Manton, I. 1950. Problems of Cytology and Evolution in the Pteridophyta. Cambridge University Press, London.

Marchant, C. J. 1966. The cytology of *Spartina* and the origin of *S. x townsendii*. In C. D. Darlington and K. R. Lewis (editors). Chromosomes Today. Vol. 1. Plenum Press, New York.

Marchant, C. J. 1968. Chromosome patterns and nuclear phenomena in the cycad families Stangeriaceae and Zamiaceae. Chromosoma **24**:100-134.

Marden, P. M., D. W. Smith, and M. J. McDonald. 1964. Congenital abnormalities in the new-born infant. Paediatrics **64**:357-362.

Marks, G. E. 1957. Telocentric chromosomes. Amer. Nat. **91**:223-232.

Marquardt, H. 1952. Über die spontanen Aberrationen in der Anaphase der Meiosis von *Paeonia tenuifolia*. Chromosoma **5**:81-112.

Martens, R. 1960. The World of Amphibians and Reptiles. (Transl. by H. W. Parker) McGraw-Hill Book Company, New York.

Martin, G. M., and C. A. Sprague. 1969. Parasexual cycle in cultivated human somatic cells. Science **166**:761-763.

Martin, P. G. 1968. Differences in chromosome size between related plant species. In W. J. Peacock and D. Brock (editors). Replication and Recombination of Genetic Material. Australian Academy of Science, Canberra.

Martin, P. G., and D. L. Hayman. 1965. A quantitative method for comparing the karyotypes of related species. Evolution **19**:157-161.

Martin, P. G., and R. Shanks. 1966. Does *Vicia faba* have multistranded chromosomes? Nature **211**: 650-651.

Marton, L. 1934. The electronmicroscope: first attempt at its application to biology. Ann. Bull. Soc. Roy. Sci. Med. Nat. **92**:106-111.

Maslin, T. P. 1962. All female species of the lizard genus *Cnemidorphorus*, Teiidae. Science **135**:212-213.

Mather, K. 1933. The relation between chiasmata and crossing-over in diploid and triploid *Drosophila melanogaster*. J. Genet. **27**:243-259.

Mather, K. 1948. Significance of nuclear changes in differentiation. Nature **161**:872-874.

Matsuya, Y., and H. Green. 1969. Somatic cell hybrid between the established human line D 98 (presumptive HeLa) and 3T3. Science **163**:697-698.

Matthey, R. 1949. Les Chromosomes des Vertébrés. Librairie de l'Université, F. Rouge, Lausanne.

Matthey, R. 1951. The chromosomes of the vertebrates. Advances Genet. **4**:159-180.

Matthey, R. 1964. Evolution chromosomique du spéciation chex les *Mus* du sous-genre *Leggada* Gray 1837. Experimentia **20**:657-665.

Matthey, R. 1965. Un type nouveau de chromosomes sexuels multiples chez une souris africaine du groupe *Mus* (Leggada) *minutoides* (Mammalia-Rodentia). Male: $X_1 X_2 Y$. Female: $X_1 X_2 / X_1 X_2$. Chromosoma **16**:351-364.

Matthey, R. 1966. Cited by Mittwoch, U. 1967. Sex Chromosomes. Academic Press Inc., New York.

Matuszewski, B. 1965. Transition from polyteny to polyploidy in salivary glands of Cecidomyiidae. Chromosoma **16**:22-34.

Mavor, J. W. 1922. The production of non-disjunction by X-rays. Science **55**:295-297.

Mayr, E. 1969. Principles of Systematic Zoology. McGraw-Hill Book Company, New York.

Mazia, D. 1961. Mitosis and the physiology of cell division. In J. Brachet and A. E. Mirsky (editors). The Cell. Vol. 3. Academic Press Inc., New York.

Mechelke, F. 1953. Reversible strukturmodifikationen der Speicheldrüsenchromosomen von *Acricotopus lucidus*. Chromosoma **5**:511-543.

Melander, Y. 1950. Studies on the chromosomes of *Ulophysema öresundense*. Hereditas **36**:233-241.

Mella, B., and D. J. Lang. 1967. Leukocyte mitosis: suppression in vitro associated with acute infectious hepatitis. Science **155**:80-81.

Melnyk, J., H. Thompson, A. J. Rucci, F. Vanasek, and S. Hayes. 1969. Failure of transmission of the extra chromosome in subjects with 47,XYY karyotype. Lancet **2**:797-798.

Melnyk, J., and J. Unrau. 1959. Pairing between chromosomes of *Aegilops squarrosa* L. var. *typica* and *Secale cereale* L. var. *prolific*. Canad. J. Genet. Cytol. **1**:21-25.

Mendiola, L. R., C. A. Price, and R. R. L. Guillard. 1966. Isolation of nuclei from a marine dinoflagellate. Science **153**:1661-1663.

Menzel, M. Y. 1952. Polygenomic hybrids in *Gossypium*. III. Somatic reduction in a phenotypically-altered branch of a three-species hexaploid. Amer. J. Bot. **39**:625-633.

Menzel, M. Y., and M. S. Brown. 1952. Polygenomic hybrids in *Gossypium*. II. Mosaic formation and somatic reduction. Amer. J. Bot. **39**:59-69.

Menzel, M. Y., and M. S. Brown. 1954. The significance of multivalent formation in three-species *Gossypium* hybrids. Genetics **39**:546-557.

Menzel, M. Y., and D. W. Martin. 1971. Chromosome homology in some intercontinental hybrids in *Hibiscus* sect. *furcaria*. Amer. J. Bot. **58**:191-202.

Menzel, M. Y., and F. D. Wilson. 1963. An allododecaploid hybrid of *Hibiscus diversifolius*. J. Hered. **54**:55-60.

Mericle, L. W., and R. P. Mericle. 1967. Genetic nature of somatic mutations for flower color in *Tradescantia*, clone 02. Radiat. Bot. **7**:449-464.

Merril, C. R., M. R. Geier, and J. C. Petricciani. 1971. Bacterial virus gene expression in human cells. Nature **233**:398-400.

Meselson, M. 1964. On the mechanism of genetic recombination between DNA molecules. J. Molec. Biol. **9**:734-745.

Meselson, M. 1968. Reciprocal recombination in prophage Lambda. In Peacock, W. J., and Brock R. D. (editors). Replication and Recombination of Genetic Material. Austral. Acad. Sci., Canberra.

Meselson, M., and J. J. Weigle. 1961. Chromosome breakage accompanying genetic recombination in bacteriophage. Proc. Nat. Acad. Sci. U. S. A. **47**:857.

Metz, C. W. 1914. Chromosome studies in the Diptera. I. A preliminary survey of five different types

of chromosome groups in the genus *Drosophila*. J. Exp. Zool, **17**:45-59.
Metz, C. W. 1916a. Chromosome studies in the Diptera. II. The paired association of chromosomes in Diptera, and its significance. J. Exp. Zool. **21**:213-279.
Metz, C. W. 1916b. Chromosome studies in Diptera. III. Additional types of chromosome groups in the Drosophilidae. Amer. Nat. **50**:587-599,
Metz, C. W. 1922. Association of homologous chromosomes in tetraploid cells of Diptera. Biol. Bull. **43**:369-373.
Metz, C. W. 1925. Prophase chromosome behavior in triploid individuals of *Drosophila melanogaster*. Genetics **10**:345-350.
Metz, C. W. 1933. Monocentric mitosis with segregation of chromosomes in *Sciara* and its bearing on the mechanism of mitosis. Biol. Bull. Woods Hole **54**:333-347.
Metz, C. W. 1938. Chromosome behavior, inheritance and sex determination in *Sciara*. Amer. Nat. **72**:485-520.
Metz, C. W., and J. F. Nonidez. 1923. Spermatogenesis in *Asilus*. Arch. Zellforsch. **17**:438-449.
Meurman, O. 1933. Chromosome morphology, somatic doubling and secondary association in *Acer platanoides* L. Hereditas **18**:145-173.
Meyer, G. F. 1960. The fine structure of spermatocyte nuclei of *Drosophila melanogaster*. Proc. Europe Reg. Conf. Electron Microscop. **2**:951-954.
Meyer, G. F. 1964. A possible correlation between the submicroscopic structure of meiotic chromosomes and crossing-over. In M. Titlbach (editor). Electron Microscopy. Proceedings of Third European Reg. Conference on Electron Microscopy, Prague.
Michel, K. E., and C. R. Burnham. 1969. The behavior of nonhomologous univalents in double trisomics in maize. Genetics **63**:851-864.
Migeon, B. R., and C. S. Miller. 1968. Human-mouse somatic cell hybrids with single human chromosome (group E): link with thymidine kinase activity. Science **162**:1005-1006.
Miksche, J. P, 1967. Variation in DNA content of several gymnosperms. Canad. J. Genet. Cytol. **9**:717-722.
Miller, D. A., P. W. Allderdice, O. J. Miller, and W. R. Breg. 1971. Quinacrine fluorescence patterns of human D group chromosomes. Nature **232**:24-27.
Miller, D. D., and R. Roy. 1964. Further studies on variation of the Y chromosome of *D. affinis* subgroup species. Drosophila Information Service **39**:117.
Miller, D. D., and L. E. Stone. 1962. A reinvestigation of karyotype in *Drosophila affinis* and related species. J. Hered. **53**:12-24.
Miller, D. D., and R. A. Voelker. 1968. Salivary gland chromosome variation in the *Drosophila affinis* subgroup. I. The C chromosome of "western" and "eastern" *Drosophila athabasca*. J. Hered. **59**:86-98.
Miller, O. J. 1964. The sex chromosome anomalies. Amer. J. Obstet. Gynec. **90**:1078-1139.
Miller, O. J., and D. Warburton. 1968. The control of sex chromatin. Cytogenetics **7**:58-77.
Miller, O. L. 1963. Cytological studies in asynaptic maize. Genetics **48**:1445-1466.
Miller, O. L. 1964. Extrachromosomal nucleolar DNA in amphibian oocytes. J. Cell Biol, **23**:60A. (Abstr.)
Miller, O. L., and B. R. Beatty. 1969. Visualization of nucleolar genes. Science **164**:955-957.
Miller, O. L., B. R. Beatty, B. A. Hamkalo, and C. A. Thomas. 1970a. Electron microscope visualization of transcription. Cold Spring Harbor Symp. Quant. Biol. **35**:505-515.
Miller, O. L., B. A. Hamkalo, and C. A. Thomas. 1970b. Visualization of bacterial genes in action. Science **169**:392-395.
Miller, R. J., and D. E. Koeppe. 1971. Southern corn leaf blight: susceptible and resistant mitochondria. Science **173**:67-69.
Milner, G. R. 1969. Changes in chromatin structure during interphase in human normablasts. Nature **221**:71-72.
Milner, G. R., and F. G. J. Hayhoe. 1968. Ultrastructural localization of nucleic acid synthesis in human blood cells. Nature **218**:785-787.
Mirov, N. T. 1967. The Genus *Pinus*. The Ronald Press Company, New York.
Mirsky, A. E. 1943. In F. F. Nord and C. H. Werkman (editors). Advances in Enzymology and Related Subjects of Biochemistry. Interscience Publishers, Inc., New York.
Mirsky, A. E., and A. W. Pollister. 1946. Chromosin, a desoxyribose nucleoprotein complex of the cell nucleus. J. Genet. Physiol. **30**:117-126.
Mitra, S. 1958. Effects of X-rays on chromosomes of *Lilium longiflorum* during meiosis. Genetics **43**:771-789.
Mitra, J., M. O. Mapes, and F. C. Steward. 1960. Growth and organized development of cultured cells. IV. The behavior of the nucleus. Amer. J. Bot: **47**:357-368.
Mitra, J., and F. C. Steward. 1961. Growth induction in cultures of *Haplopappus gracilis*. II. The behavior of the nucleus. Amer. J. Bot. **48**:358-368.
Mittwoch, U. 1967. Sex Chromosomes. Academic Press Inc., New York.
Moens, P. B. 1964. A new interpretation of meiotic prophase in *Lycopersicon esculentum* (tomato). Chromosoma **15**:231-242.
Moens, P. B. 1968. Synaptinemal complexes of *Lilium tigrinum* (triploid) sporocytes. Canad. J. Genet. Cytol. **10**:799-807.
Moens, P. B. 1969. The fine structure of meiotic chromosome polarization and pairing in *Locusta migratoria* spermatocytes. Chromosoma **23**:1-25.
Moens, P. B. 1970. The fine structure of meiotic chromosome pairing in natural and artificial *Lilium* polyploids. J. Cell Sci. **7**:55-63.

Moffett, A. A. 1936. The origin and behaviour of chiasmata. XIII. Diploid and tetraploid *Culex pipiens*. Cytologia **7**:184-197.

Montgomery, T. H. 1901. A study of the chromosomes of the germ cells of Metazoa. Trans. Amer. Phil. Soc. **20**:154-230.

Montgomery, T. H. 1906. Chromosomes in the spermatogenesis of the Hemiptera-Homoptera. Trans. Amer. Phil. Soc. **21**:97-174.

Moore, K. L. 1966. The sex chromatin (26 articles by various authors). W. B. Saunders Co., Philadelphia.

Moore, R. J. 1956-1968. Index to plant chromosome numbers, yearly compilation. International Association of Plant Taxonomy, Utrecht.

Moorhead, P. S., and V. Defendi. 1963. Asynchrony of DNA synthesis in chromosomes of human diploid cells, J. Cell Biol. **16**:202-208.

Moorhead, P. S., P. C. Nowell, W. J. Mellman, D. M. Battips, and D. A. Hungerford. 1960. Chromosome preparations of leukocytes cultured from human peripheral blood. Exp. Cell Res. **20**:613-619.

Morgan, L. V. 1922. Non-criss-cross inheritance in *Drosophila melanogaster*. Biol. Bull. **42**:267-274.

Morgan, L. V. 1925. Polyploidy in *Drosophila melanogaster* with two attached X chromosomes. Genetics **10**:148-178.

Morgan, L. V. 1939. A spontaneous somatic exchange between nonhomologous chromosomes in *Drosophila melanogaster*. Genetics **24**:747-752.

Morgan, T. H. 1910. Sex-linked inheritance in *Drosophila*. Science **32**:120-122.

Morgan, T. H. 1911. An attempt to analyze the constitution of the chromosomes on the basis of sex-linked inheritance. J. Exp. Zool. **11**:365-414.

Morgan, T. H. 1916. The Engster gyandromorph bees. Amer. Nat. **50**:39-45.

Morgan, T. H., C. B. Bridges, and J. Schultz. 1935. Constitution of the germinal material in relation to heredity. Carnegie Inst. Wash. Yearbook **34**:284-291.

Morgan, T. H., A. H. Sturtevant, H. J. Muller, and C. B. Bridges. 1915. The Mechanism of Mendelian Heredity. Holt, Rinehart & Winston, Inc., New York.

Morris, R., and E. R. Sears. 1967. The cytogenetics of wheat and its relatives. In K. S. Quisenberry and L. P. Reitz (editors). Wheat and Wheat Improvement. American Society of Agronomy, Washington, D. C.

Morris, T. 1968. The XO and OY chromosome constitution in the mouse. Genet. Res. **12**:125-127.

Moses, M. J. 1956. Chromosomal structure in crayfish spermatocytes. J. Biophys. Biochem. Cytol. **2**:215-218.

Moses, M. J. 1968. Synaptinemal complex. Ann. Rev. Genet. **2**:363-412.

Mottier, D. M. 1905. The development of the heterotypic chromosomes in pollen mother cells. Bot. Gaz. **40**:171-177.

Mottier, D. M. 1914. Mitosis in pollen mother-cells of *Acer negundo*, L., and *Staphylea trifolia* L. Ann. Bot. **28**:115-133.

Mottier, D. M., and M. Nothnagel. 1913. The development and behavior of the chromosomes in the first or heterotypic mitosis of the pollen mother-cells of *Allium cernuum* Roth. Bull. Torrey Bot. Club **40**: 555-565.

Mounolou, J. C., H. Jacob, and P. P. Slonimski. 1966. Mitochondrial DNA from yeast "petite" mutants: specific changes of buoyant density corresponding to different cytoplasmic mutations. Biochem. Biophys. Res. Commun. **24**:218-224.

Moustacchi, E., and D. H. Williamson. 1966. Physiological variations in satellite components of yeast DNA detected by density gradient centrifugation. Biochem. Biophys. Res. Commun. **23**:56-61.

Mueller, G. C., and K. Kajiwara. 1966. Early- and late-replicating deoxyribonucleic acid complexes in HeLa nuclei. Biochem. Biophys. Acta **114**:108-115.

Mukherjee, A. S. 1961. Effect of selection on crossing over in the males of *Drosophila ananassae*. Am. Nat. **95**:57-59.

Muller, H. J. 1925. Why polyploidy is rarer in animals than in plants. Amer. Nat. **59**:346-353.

Muller, H. J. 1927. Artificial transmutation of the gene. Science **66**:84-87.

Muller, H. J. 1932. Some genetic aspects of sex. Amer. Nat. **66**:118-138.

Muller, H. J. 1939. Bibliography on the genus *Drosophila*. Imp. Bur. Animal Breed. Genet. Bibliogr. 5.

Muller, H. J., and E. Altenburg. 1928. Chromosome translocation produced by X-rays in *Drosophila*. Genetics Section at Convention in New York, December 29, 1928; Anat. Rec. **41**:100-103. (Abst.)

Muller, H. J., and T. S. Painter. 1929. The cytological expression of changes in gene alignment produced by X-rays in *Drosophilia*. Amer. Nat. **63**:193-200.

Muller, H. J., and T. S. Painter. 1932. The differentiation of the sex chromosomes of *Drosophila* into genetically active and inert regions. Z. Ind. Abst. Vererb. **63**:316-365.

Muntzing, A. 1931. Note on the cytology of some apomictic *Potentilla* species. Hereditas **15**:166-178.

Muntzing, A. 1938. Sterility and chromosome pairing in intraspecific *Galeopsis* hybrids. Hereditas **23**: 117-188.

Muntzing, A. 1939. Studies on the properties and ways of production of rye-wheat amphidiploids. Hereditas **25**:387-430.

Muntzing, A. 1940. Further studies on apomixis and sexuality in *Poa*. Hereditas **26**:115-190.

Muntzing, A. 1948. Accessory chromosomes in *Poa alpina*. Heredity **2**:49-61.

Muntzing, A. 1957. Frequency of accessory chromosomes in rye strains from Iran and Korea. Hereditas **43**:682-685.

Murdy, W. H., and H. L. Carson. 1959. Parthenoge-

sis in *Drosophila mangabeirai* (Malogolowkin). Amer. Nat. **93**:355-363.
Myers, W. M. 1943. Analysis of variance and covariance of chromosomal association and behavior during meiosis in clones of *Dactylis glomerata*. Bot. Gaz. **104**:541-552.
Myers, W. M. 1948. Studies on the origin of *Dactylis glomerata* L. Genetics **33**:117. (Abstr.)
Nagao, S., and K. Saki. 1939. Association of chromosomes in *Chelidonium majus*. L. Jap. J. Genet. **15**:23-28.
Nagata, T. 1963. The molecular synchrony and sequential replication of DNA in *Escherichia coli*. Proc. Nat. Acad. Sci. U. S. A. **49**:551-559.
Nagl, W. 1969. Banded polytene chromosomes in the legume *Phaseolus vulgaris*. Nature **221**:70-71.
Nakamura, K. 1928. On the chromosomes of a snake, *Natrix tigrina*. Mem. Coll. Sci. Kyoto Imp. Univ. Ser. B, **4**:1-18.
Nakamura, K. 1931. Studies on reptilian chromosomes. II. On the chromosomes of *Eumeces latiscutatus* (Hallowell), a lizard. Cytologia **2**:385-401.
Namboodiri, A. N., and R. J. Lowry. 1967. Vegetative nuclear division in *Neurospora*. Amer. J. Bot. **54**:735-748.
Nance, W. E., and E. Engel. 1967. Autosomal deletion mapping in man. Science **155**:692-694.
Nandi, H. K. 1936. The chromosome morphology, secondary association and origin of cultivated rice. J. Genet. **33**:315-336.
Nass, M. M. K. 1969. Mitochondrial DNA: advances, problems and goals. Science **165**:25-35.
Nass, M. M. K., and S. Nass. 1963. Intramitochondrial fibers with DNA characteristics. J. Cell Biol. **19**:593-596.
Nass, M. M. K., S. Nass, and B. A. Afgelius. 1965. The general occurrence of mitochondrial DNA. Exp. Cell Res. **37**:516-539.
Navashin, M. 1934. Chromosome alteration caused by hybridization and their bearing upon certain general genetic problems. Cytologia **5**:169-203.
Navashin, M. S. 1932. The dislocation hypothesis of evolution of chromosome numbers. Z. Ind. Abst. Vererb. **63**:224-231.
Nebel, B. R. 1936. Chromosome structure. X. An X-ray experiment. Genetics **21**:605-614.
Nebel, B. R., and E. M. Coulon. 1962. The fine structure of chromosomes in pigeon spermatocytes. Chromosoma **13**:272-291.
Nebel, B. R., and M. L. Ruttle. 1936. Chromosome structure. IX. *Tradescantia reflexa* and *Trillium erectum*. Amer. J. Bot. **23**:652-663.
Němec, B. 1910. Das Problem der Befruchtungsvorgange und andere zytologishe Fragen. Berlin.
Newell, W. 1915. Inheritance in the honey bee. Science **41**:218-219.
Newton, W. C. F. 1926. Chromosome studies in *Tulipa* and some related genera. J. Linnaean Soc. Bot. **47**:339-354.
Nitsch, J. P., and C. Nitsch. 1969. Haploid plants from pollen grains. Science **163**:86-87.

Noack, K. L. 1939. Uber *Hypericum*-Kreuzunger. Z. Ind. Abst. Vererb. **76**:569-601.
Nobs, M. A. 1963. Experimental studies on species relationships in *Ceanothus*. Carnegie Inst. Wash. Publ. No. 623.
Noda, S. 1966. Cytogenetics on the origin of triploid *Lilium tigrinum*. Bull. Osaka Gakuin Univ. No. 6, p. 85-140.
Noggle, G. R. 1946. The physiology of polyploid plants. I. Review of the literature. Lloydia **9**:153-173.
Nogusa, S. 1960. A comparative study of the chromosomes in fishes with particular consideration on taxonomy and evolution. Mem. Hyogo Univ. Agr. **3**:1-17.
Nordenskiöld, H. 1941. Cytological studies in triploid *Phleum*. Bot. Notiser 1941:12-32.
Nordenskiöld, H. 1951. Cyto-taxonomical studies in the genus *Luzula*. Hereditas **37**:325-355.
Nordenskiöld, H. 1956. Cyto-taxonomical studies in the genus *Luzula*. II. Hybridization experiments in the *Campestris-multiflora* complex. Hereditas **42**:7-73.
Novitski, E. 1946. Chromosome variation in *Drosophila athabasca*. Genetics **31**:508-524.
Novitski, E. 1951. Non-random disjunction in *Drosophila*. Genetics **36**:267-280.
Novitski, E., W. J. Peacock, and J. Engel. 1965. Cytological basis of "sex ratio" in *Drosophila pseudoobscura*. Science **148**:516-517.
Nowell, P. C. 1960. Phytohemagglutinin: an initiator of mitosis in cultures of normal human leukocytes. Cancer Res. **20**:462-466.
Nur, U. 1966. Nonreplication of heterochromatic chromosomes in a mealy bug, *Planococcus citri* (Coccoidea: Homoptera). Chromosoma **19**:439-448.
Nur, U. 1968. Synapsis and crossing over within a paracentric inversion in the grasshopper, *Camnula pellucida*. Chromosoma **25**:198-214.
Nuzzo, F., F. Caviezel, and L. De Carli. 1966. Y-chromosome and exclusion of paternity. Lancet **2**:260-266.
Nygren, A. 1946. The genesis of some Scandinavian species of *Calamagrostis*. Hereditas **32**:131-262.
Nygren, A. 1954a. Apomixis in angiosperms. Bot. Rev. **20**:577-649.
Nygren, A. 1954b. Investigations on North American *Calamagrostis*. Hereditas **40**:377-397.
Nygren, A., B. Nilsson, and M. Jahnke. 1968. Cytological studies in Atlantic salmon *(Salmo salar)*. Ann. Acad. Reg. Sci. Upsala **12**:21-52.
Odartchenko, N., and M. Pavillard. 1970. Late DNA replication in male mouse meiotic chromosomes. Science **167**:1133-1134.
Oehlkers, F., and P. Eberle. 1957. Spiralen und chromomeren in der meiosis von *Bellevalia romana*. Chromosoma **8**:351-363.
Ohmachi, F., and N. Ueshima. 1957. Variations of chromosome complements in *Euscirtus hemelytrus* De Haan (Orthoptera: Gryllodea). Bull. No. 15, Fac. Agr. Mie Univ., Tsu, Japan.

Ohno, S. 1965. Direct handling of germ cells. In J. J. Yunis (editor). Human Chromosome Methodology. Academic Press Inc., New York.

Ohno, S., J. Jainchill, and C. Stenius. 1963. The creeping vole (*Microtus oregoni*) as a gonosomic mosaic. I. The OY/XY constitution in the vole. Cytogenetics 2:232-239.

Ohno, S., W. Kaplan, and R. Kinosita. 1957. Heterochromatic regions and nucleolus organizers in chromosomes of the mouse, *Mus musculus*. Exp. Cell Res. 13:358-364.

Ohno, S., W. D. Kaplan, and R. Kinosita. 1961. Female germ cells in man. Exp. Cell Res. 24:106-110.

Ohno, S., H. P. Klinger, and N. B. Atkins. 1962. Human oogenesis. Cytogenetics 1:42-51.

Ohno, S., J. Muramoto, and L. Christian. 1967. Diploid-tetraploid relationship among old world members of the fish family Cyprinidae. Chromosoma 23:1-9.

Ohno, S., C. Stenius, E. Faisst, and M. T. Zenzes. 1965. Post-zygotic chromosomal rearrangements in rainbow trout (*Salmo irideus* Gibbons). Cytogenetics 4:117-129.

Oksala, T. 1944. Zytologische Studien an Odonaten. II. Die Entstehung der meiotischen Präkozität. Ann. Acad. Sci. Fenn. AIV 5:1-33.

Olby, R. C., and P. Gantrey. 1968. Eleven references to Mendel before 1900. Ann. Sci. 24:7-20.

Oliver, C. P. 1934. Fertile mosaics in *Drosophila*. Amer. Nat. 68:74-75.

O'Mara, J. G. 1939. Observations on the immediate effects of colchicine. J. Hered. 30:35-37.

O'Mara, J. G. 1951. Cytogenetic studies on *Triticale*. II. The kinds of intergeneric chromosome addition. Cytologia 16:225-232.

Omodeo, P. 1955. Cariologia dei Lumbricidae. II. Contributo. Caryologia 8:135-178.

Oppenheim, A. B., and M. Riley. 1966. Molecular recombination following conjugation in *Escherichia coli*. J. Molec. Biol. 20:331-357.

Ostergren, G. 1947a. Heterochromatic B-chromosomes in *Anthoxanthum*. Hereditas 33:261-296.

Ostergren, G. 1947b. Proximal heterochromatin, structure of the centromere, and the mechanism of its misdivision. Bot. Notiser 2:176-177.

Ostergren, G. 1949. Equilibria and movements of chromosomes. Proceedings of Eighth International Congress on Genetics, Hereditas Suppl. Vol.:688-689.

Ostergren, G. 1951. The mechanism of co-orientation in bivalents and multivalents. The theory of orientation by pulling. Hereditas 37:85-156.

Ostergren, G., and E. Vigfusson. 1953. On position correlations of univalents and quasi-bivalents formed by sticky chromosomes. Hereditas 34:33-50.

Overeem, C. van. 1921. Uber Formen mit abweichenden Chromosomenzahl bei *Oenothera*. Bot. Cent. 38:13-113.

Overton, J. B. 1922. The organization of the nuclei in the root tips of *Podophyllum peltatum*. Trans. Wisconsin Acad. Sci. Arts Letters 20:275-322.

Pace, L. 1913. Apogamy in *Atamosco*. Bot. Gaz. 56:371-394.

Painter, T. S. 1923. Studies in mammalian spermatogenesis. II. The spermatogenesis of man. J. Exp. Zool. 37:291-334.

Painter, T. S. 1924. The sex chromosomes of man. Amer. Nat. 58:506-524.

Painter, T. S. 1931. A cytological map of the X-chromosome of *Drosophila melanogaster*. Science 73:647-648.

Painter, T. S. 1933. A new method for the study of chromosome rearrangements and the plotting of chromosome maps. Science 78:585-586.

Painter, T. S. 1934a. A new method for the study of chromosome aberrations and the plotting of chromosome maps in *Drosophila melanogaster*. Genetics 19:175-188.

Painter, T. S. 1934b. The morphology of the X chromosome in the salivary glands of *Drosophila melanogaster* and a new type of chromosome map for this element. Genetics 19:448-469.

Painter, T. S. 1934c. Salivary chromosomes and the attack on the gene. J. Hered. 25:465-476.

Painter, T. S. 1935. The morphology of the third chromosome in the salivary gland of *Drosophila melanogaster* and a new cytological map of this element. Genetics 20:301-326.

Painter, T. S. 1941. An experimental study of salivary chromosomes. Cold Spring Harbor Symp. Quant. Biol. 9:47-53.

Painter, T. S., and A. B. Griffen. 1937. The structure and the development of the salivary gland chromosome of *Simulium*. Genetics 22:612-633.

Painter, T. S., and H. J. Muller. 1929. Parallel cytology and genetics of induced translocation and deletions in *Drosophila*. J. Hered. 20:287-298.

Pankratz, H. S., and C. C. Bowen. 1963. Cytology of blue-green algae. I. The cell of *Symploca muscorum*. Amer. J. Bot. 50:387-399.

Pardue, M. L., and J. G. Gall. 1970. Chromosomal localization of mouse satellite DNA. Science 168:1356-1358.

Pardue, M. L., S. A. Gerbi, R. A. Eckhardt, and J. G. Gall. 1970. Cytological localization of DNA complementary to ribosomal RNA in polytene chromosomes of Diptera. Chromosoma 29:268-290.

Pasternak, J. 1964. Chromosome polymorphism in the blackfly *Simulium vittatum* (Zett.). Canad. J. Zool. 42:135-158.

Patan, K. 1950. A correlation between separation of the two chromosome groups in somatic reduction and their degree. Genetics 35:128. (Abstr.)

Patil, S. R., S. Merrick, and H. A. Lubs. 1971. Identification of each human chromosome with a modified Giemsa stain. Science 173:821-822.

Patterson, J. T., and J. F. Crow. 1940. Hybridization in the mulleri group of *Drosophila*. Univ. Texas Publ. No. 4032, p. 251-256.

Patterson, J. T., and W. S. Stone. 1952. Evolution in the Genus *Drosophila*. The Macmillan Company, New York.

Patterson, J. T., W. S. Stone, and A. B. Griffen. 1940. The virilis complex in *Drosophila*. Genetics **25**:131-144.

Patterson, R. S., D. E. Wekdhaas, H. R. Ford, and C. S. Lofgren. 1970. Suppression and elimination of an island population of *Culex pipiens quinquefasciatus* with sterile males. Science **168**:1368-1370.

Pavan, C. 1965. Nucleic acid metabolism in polytene chromosomes and the problem of differentiation. Genetic control of differentiation. Brookhaven Symp. Biol. **18**:222-241.

Pavan, C., and M. E. Breuer. 1952. Polytene chromosomes in different tissues of *Rhynchosciara*. J. Hered. **43**:151-157.

Pavan, C., and A. B. da Cunha. 1969a. Chromosomal activities in *Rhynchosciara* and other Sciaridae. Ann. Rev. Genet. **3**:425-450.

Pavan, C., and A. B. da Cunha. 1969b. Gene amplification in ontogeny and phylogeny of animals, Genetics (Suppl. 61)**1**, 1969:289-304.

Pavan, C., T. Dobzhansky, and A. B. da Cunha. 1957. Heterosis and elimination of weak homozygotes in natural populations of three related species of *Drosophila*. Proc. Nat. Acad. Sci. U. S. A. **43**:226-234.

Pavan, C., and A. L. P. Perondini. 1967. Heterozygous puffs and bands in *Sciara ocellaris* Comstock (1882). Exp. Cell Res. **48**:202-206.

Pavan, C., A. L. P. Perondini, and T. Picard. 1969. Changes in chromosomes and in development of cells of *Sciara ocellaris* induced by microsporidian infections. Chromosoma **28**:328-345.

Pavlovsky, O., and T. Dobzhansky. 1966. Genetics of natural populations. XXXVII. The coadapted system of chromosomal variants in a population of *Drosophila pseudoobscura*. Genetics **53**:843-854.

Peacock, W. J. 1963. Chromosome duplication and structure as determined by autoradiography. Proc. Nat. Acad. Sci. U. S. A. **49**:793-801.

Peacock, W. J. 1965. Chromosome replication. Nat. Cancer Inst. Monogr. **18**:101-123.

Peacock, W. J. 1968. Chiasmata and crossing over. In W. J. Peacock and P. D. Brock (editors). Replication and Recombination of Genetic Materials. Australian Academy of Science, Canberra.

Peacock, W. J. 1971. Cytological aspects of the mechanism of recombination in higher organisms. In G. Kimber and G. P. Redei (editors). Stadler Genetics Symposium. Vols. 1 and 2. Univ. Missouri Agr. Exp. Sta., Columbia, Mo.

Pellew, C., and E. R. Sansome. 1931. Genetical and cytological studies on the relations between Asiatic and European varieties of *Pisum sativum*. J. Genet. **25**:25-54.

Peloquin, S. J., and R. W. Hougas. 1958. Fertility in two haploids of *Solanum tuberosum*. Science **128**:1340-1341.

Peloquin, S. J., and R. W. Hougas. 1960. Genetic variation among haploids of the common potato. Amer. Potato J. **37**:289-297.

Perez-Silva, J., and P. Alonso. 1966. Demonstration of polytene chromosomes in the macronuclear anlage of oxytrichous ciliates. Arch. Protistenk. **109**:65-70.

Perkowska, E., H. C. MacGregor, and M. L. Birnstiel. 1968. Gene amplification in the oocyte nucleus of mutant and wild-type *Xenopus laevis*. Nature **217**:649-650.

Perondini, A. L. P., and E. M. Dessen. 1969. Heterozygous puffs in *Sciara ocellaris*. Genetics (Suppl. 61)**1**, 1969:251-260.

Perrot, J. L. 1934. La spermatogenese et l'ovogénèse du Mallophage *Goniodes stylifer*. Quart. J. Microscop. Sci. **76**:353-377.

Person, C. 1955. An analytical study of chromosome behaviour in a wheat haploid. Canad. J. Bot. **33**:11-30.

Peto, F. H. 1935. Association of somatic chromosomes induced by heat and chloral hydrate treatments. Canad. J. Res. C **13**:301-314.

Philip, U. 1942. An analysis of chromosomal polymorphism in two species of *Chironomus*. J. Genet. **44**:129-142.

Phillips, L. L. 1966. The cytology and phylogenetics of the diploid species of *Gossypium*. Amer. J. Bot. **53**:328-335.

Phillips, R. L. 1969. Recombination in *Zea mays* L. I. Location of genes and interchanges in chromosomes 5, 6, and 7. Genetics **61**:107-116.

Pickett-Heaps, J. D., and D. H. Northcote. 1966. Cell division in the formation of the stomatal complex of the young leaves of wheat. J. Cell Sci. **1**:121-128.

Piko, L., A. Tyler, and J. Vinograd. 1967. Amount, location, priming capacity, circularity, and other properties of cytoplasmic DNA in sea urchin eggs. Biol. Bull. **132**:68-90.

Pipkin, S. B. 1940. Segregation and crossing over in a 2-3 translocation in *Drosophila melanogaster*. Univ. Texas Bull. No. 4032, p. 73-125.

Pipkin, S. B. 1960. Sex balance in *Drosophila melanogaster*: aneuploidy of long regions of chromosome 3, using the triploid method. Genetics **45**:1205-1216.

Pittenger, T. H., and M. B. Coyle. 1963. Somatic recombination in pseudo-wild-type cultures of *Neurospora crassa*. Proc. Nat. Acad. Sci. U. S. A. **49**:445-451.

Piza, S. de T. 1950. Observações chromossómicas en escorpiões Brasileiros. Cienc. Cutura **2**:202-206.

Plaut, W. 1968. On DNA replication in polytene chromosomes. In W. J. Peacock and R. D. Brock (editors). Replication and Recombination of Genetic Material. Australian Academy of Science, Canberra.

Pogo, B. G. T., V. G. Allfrey, and A. E. Mirsky. 1967. The effect of phytohemagglutinin on ribonucleic acid synthesis and histone acetylation in equine leukocytes. J. Cell Biol. **35**:477-482.

Pogo, B. G. T., A. O. Pogo, and V. G. Allfrey. 1969. Histone acetylation and RNA synthesis in rat liver regeneration. Genetics (Suppl. 61)**1**, 1969:373-379.

Policansky, D., and J. Ellison. 1970. "Sex ratio" in *Drosophila pseudoobscura:* spermiogenic failure. Science **169**:888-889.

Pontecorvo, G. 1953. The genetics of *Aspergillus nidulans*. Advances Genet. **5**:141-238.

Pontecorvo, G. 1958. Trends in Genetic Analysis. Columbia University Press, New York.

Pontefract, R. D., G. Bergeron, and F. S. Thatcher. 1969. Mesosomes in *Escherichia coli*. J. Bact. **97**:367-375.

Poulson, D. F., and C. W. Metz. 1938. Studies on the structure of the nucleolus-forming regions and related structures in the giant salivary gland chromosomes of Diptera. J. Morph. **63**:363-395.

Powers, L. 1945. Fertilization without reduction in quayule *(Parthenium argentatum* Gray) and a hypothesis as to the evolution of apomixis and polyploidy. Genetics **30**:323-346.

Prakken, R. 1943. Studies on asynapsis in rye. Hereditas **29**:475-495.

Prescott, D. M. 1966. The synthesis of total macronuclear protein, histone, and DNA during the cell cycle in *Euplotes eurystomus*. J. Cell Biol. **31**:1-9.

Price, S. 1955. Desynaptic pseudoassociations in *Secale montanum*. Science **122**:1190.

Price, S. 1959. Critique on apomixis in sugarcane. Econ. Bot. **13**:67-74.

Price, S. 1961. Maternal chromosome transmission in sugarcanes. Bot. Gaz. **122**:298-305.

Priest, J. H. 1969. Cytogenetics. Medical technology series. Lea & Febiger, Philadelphia.

Prokofyeva-Belgovskaya, A. A. 1935. The structure of the chromocenter. Cytologia **6**:438-443.

Pryor, L. D. 1959. Species distribution and association in *Eucalyptus*. Monogr. Biol. **8**:461-471.

Punyasingh, K. 1947. Chromosome numbers in crosses of diploid, triploid, and tetraploid maize. Genetics **32**:541-554.

Qasba, P. K., and H. V. Aposhian. 1971. DNA and gene therapy: transfer of mouse DNA to human and mouse embryonic cells by polyoma pseudovirions. Proc. Nat. Acad. Sci. U. S. **68**:2345-2349.

Rae, P. M. M. 1966. Whole mount electron microscopy of *Drosophila* salivary chromosomes. Nature **212**:139-142.

Ramirez, C., and J. J. Miller. 1962. Observations on vegetative nuclear division in *Saccharomyces cerevisiae*. Canad. J. Microbiol. **8**:603-608.

Randolph, L. F. 1928. Types of supernumerary chromosomes in maize. Anat. Rec. **41**:102-105.

Randolph, L. F. 1941a. An evaluation of induced polyploidy as a method of breeding crop plants. Amer. Nat. **75**:347-363.

Randolph, L. F. 1941b. Genetic characteristics of the B-chromosome in maize. Genetics **26**:608-631.

Randolph, L. F., and H. E. Fischer. 1939. The occurrence of parthenogenetic diploids in tetraploid maize. Proc. Nat. Acad. Sci. U. S. A. **25**:161-164.

Rasmussen, H. 1970. An adenyl cyclase control mechanism model. Science **170**:404-412.

Ratcliffe, S. G., M. M. Melville, A. L. Stewart, P. A. Jacobs, and A. J. Keay. 1970. Chromosome studies on 3500 newborn male infants. Lancet **1**:121-122.

Raven, P. H., and W. Kyhos. 1965. New evidence concerning the original basic chromosome number of angiosperms. Evolution **19**:244-248.

Raven, P. H., and H. J. Thompson. 1964. Haploidy and angiosperm evolution. Amer. Nat. **98**:251-252.

Ray-Chaudhuri, S. P., and G. K. Manna. 1952. A new type of segregation of the sex chromosomes in the meiotic divisions of the cotton stainer, *Dysdercus koenigii* (Fabr). J. Genet. **51**:191-197.

Redfield, H. 1930. Crossing-over in the third chromosome of triploids of *Drosophila melanogaster*. Genetics **15**:205-252.

Redfield, H. 1932. A comparison of triploid and diploid crossing-over for chromosome II of *Drosophila melanogaster*. Genetics **17**:137-152.

Rees, H. 1955. Genotypic control of chromosome behavior in rye, inbred lines. Heredity **9**:93-116.

Rees, H. 1957. Distribution of chiasmata in an asynaptic locust. Nature **180**:559.

Rees, H. 1961. Genotypic control of chromosome form and behavior. Bot. Rev. **27**:288-318.

Rees, H., F. M. Cameron, M. H. Hazarika, and G. H. Jones. 1966. Nuclear variation between diploid angiosperms. Nature **211**:828-830.

Reisman, L. E., A. Darnell, and J. W. Murphy. 1965. Abnormalities with ring chromosomes. Lancet **2**:445.

Renwick, J. H. 1969. Progress in mapping human autosomes. Brit. Med. Bull. **25**:65-73.

Revell, S. H. 1953. Chromosome breakage by X-rays and radiomimetic substances in *Vicia*. Heredity **6**:107-124.

Rhoades, M. M. 1933. A secondary trisome in maize. Proc. Nat. Acad. Sci. U. S. A. **19**:1031-1038.

Rhoades, M. M. 1936. Note on the origin of triploidy in maize. J. Genet. **33**:355-357.

Rhoades, M. M. 1940. Studies of a telocentric chromosome in maize, with special reference to the stability of its centromere. Genetics **25**:483-520.

Rhoades, M. M. 1942. Preferential segregation in maize. Genetics **27**:395-407.

Rhoades, M. M. 1952. Preferential segregation in maize. In J. W. Gowen (editor). Heterosis. Iowa State University Press, Ames, Iowa.

Rhoades, M. M. 1955. The cytogenetics of maize. In G. F. Sprague (editor). Corn and Corn Improvement. Academic Press Inc., New York.

Rhoades, M. M. 1961. Meiosis. In J. Brachet and A. E. Mirsky (editors). The Cell. Vol. 3. Academic Press Inc., New York.

Rhoades, M. M. 1968. Studies on the cytological basis of crossing over. In W. J. Peacock and R. D. Brock (editors). Replication and Recombination of Genetic Material. Australian Academy of Science, Canberra.

Rhoades, M. M., and E. Dempsey. 1953. Cytogenetic studies of deficient-duplicate chromosomes derived from inversion heterozygotes in maize. Amer. J. Bot. **40**:405-424.

Rhoades, M. M., and E. Dempsey. 1966. The effect of abnormal chromosome 10 on preferential segregation and crossing over in maize. Genetics **53**:989-1020.

Rhoades, M. M., and B. McClintock. 1935. The cytogenetics of maize. Bot. Rev. **1**:292-325.

Rhoades, M. M., and H. Vilkomerson. 1942. On anaphase movement of chromosomes. Proc. Nat. Acad. Sci. U. S. A. **28**:433-436.

Rick, C. M. 1971. Some cytogenetic features of the genome in diploid plant species. In G. Kimber and G. P. Redei (editors). Stadler Genetics Symposia. Vols. 1 and 2. Univ. Missouri Agr. Exp. Sta., Columbia, Mo.

Rick, C. M., and D. W. Barton. 1954. Cytological and genetical identification of the primary trisomics of the tomato. Genetics **39**:640-666.

Rick, C. M., W. H. Dempsey, and G. S. Khush. 1964. Further studies on the primary trisomics of the tomato. Canad. J. Genet. Cytol. **6**:93-108.

Rick, C. M., and N. K. Notani. 1961. The tolerance of extra chromosomes by primitive tomatoes. Genetics **46**:1231-1235.

Rieger, R. 1957. Inhomologenpaarung und meioseablauf bei haploiden formen von *Antirrhinum majus* L. Chromosoma **9**:1-38.

Riley, H. P. 1959. Chromosome behavior in *Tradescantia* triploid hybrids. Nucleus **2**:1-8.

Riley, R., and V. Chapman. 1958. Genetic control of the cytologically diploid behavior of hexaploid wheat. Nature **182**:713-715.

Riley, R., and V. Chapman. 1964. Cytological determination of the homeology of chromosomes of *Triticum aestivum*. Nature **203**:156-158.

Riley, R., J. Unrau, and V. Chapman. 1958. Evidence on the origin of the B genome of wheat. J. Hered. **49**:91-98.

Ris, H. 1956. A study of chromosomes with the electron microscope. J. Biophys. Biochem. Cytol. **2** (Suppl.):385-392.

Ris, H. 1966. Fine structure of chromosomes. Proc. Roy. Soc. Biol. **164**:246-257.

Ris, H., and B. L. Chandler. 1963. The ultrastructure of genetic systems in Prokaryotes and Eukaryotes. Cold Spring Harbor Symp. Quant. Biol. **28**:1-8.

Ris, H., and W. Plaut. 1962. Ultrastructure of DNA-containing areas in the chloroplast of *Chlamydomonas*. J. Cell Biol. **13**:383-391.

Ritossa, F. M. 1968. Unstable redundancy of genes for ribosomal RNA. Proc. Nat. Acad. Sci. U. S. A. **60**:509-516.

Ritossa, F. M., K. C. Atwood, D. L. Lindsley, and S. Spiegelman. 1966a. On the chromosomal distribution of DNA complementary to soluble RNA. Nat. Cancer Inst. Monogr. **23**:449-472.

Ritossa, F. M., K. C. Atwood, and S. Spiegelman. 1966b. A molecular explanation of the bobbed mutants of *Drosophila* as partial deficiencies of "ribosomal" DNA. Genetics **54**:819-834.

Ritossa, F. M., and G. Scala. 1969. Equilibrium variations in the redundancy of rDNA in *Drosophila melanogaster*. Genetics (Suppl. 61)**1**, 1969:305-317.

Ritossa, F. M., and S. Spiegelman. 1965. Localization of DNA complementary to ribosomal RNA in the nucleolus organizer region of *Drosophila melanogaster*. Proc. Nat. Acad. Sci. U. S. A. **53**:737-745.

Robbins, E., and T. W. Borun. 1967. The cytoplasmic synthesis of histones in HeLa cells and its temporal relationship to DNA replication. Proc. Nat. Acad. Sci. U. S. A. **57**:409-416.

Roberts, B., J. Whitten, and L. I. Gilbert. 1969. DNA granule synthesis by polytene chromosomes in the giant foot-pad nuclei of *Sarcophaga bullata* (Parker). Chromosoma **26**:215-244.

Roberts, M. R., and H. Lewis. 1955. Subspeciation in *Clarkia biloba*. Evolution **9**:445-454.

Roberts, P. A. 1966a. A tandem duplication that lowers recombination throughout a chromosome arm in *Drosophila melanogaster*. Genetics **54**:969-979.

Roberts, P, A. 1966b. Crossover suppressing translocations in *Drosophila melanogaster*. Genetics **52**: 469. (Abstr.)

Robertson, W. R. B. 1916. Chromosome studies. I. Taxonomic relationships shown in the chromosomes of Tettigidae and Acrididae. V-shaped chromosomes and their significance in Acrididae, Locustidae, and Gryllidae. J. Morph. **27**:179-331.

Robinow, C. F., and C. E. Caten. 1969. Mitosis in *Aspergillus nidulans*. J. Cell Sci. **5**:403-431.

Robinow, C. F., and J. Marak. 1966. A fiber apparatus in the nucleus of the yeast cell. J. Cell Biol. **29**:129-151.

Robson, E. B., P. E. Polani, S. J. Dart, P. A. Jacobs, and J. H. Renwick. 1969. Probable assignment of the alpha locus of haptoglobin to chromosome 16 in man. Nature **223**:1163-1165.

Roman, H. 1948. Directed fertilization in maize. Proc. Nat. Acad. Sci. U. S. A. **34**:36-42.

Roman, H. L. 1956. Studies of gene mutation in *Saccharomyces*. Cold Spring Harbor Symp. Quant. Biol. **21**:175-185.

Roodyn, D. B., and D. Wilkie. 1968. The Biogenesis of Mitochondria. Methuen, London.

Roper, J. A. 1966. The parasexual cycle. In G. C. Ainsworth and A. S. Sussman (editors). The Fungi, an Advanced Treatise. Vol. 2 of The Fungal Organism. Academic Press Inc., New York.

Rosenberg, O. 1903. Das Verhalten der chromosomen in einer hybriden Pflanze. Ber. Deutsch. Bot. Ges. **21**:110-119.

Rosenberg, O. 1904. Uber die tetradenteilung eines *Drosera*-Bastardes. Ber Deutsch. Bot. Ges. **22**:47-53.

Rosenberg, O. 1909. Cytologische und morphologische Studien an *Drosera longifolia* × *rotundifolia*. K. Svensk. Vet. Akad. Hamdl. **43**:1-64.

Rosenberg, O. 1917. Die Reduktionsteilung und ihre

Degeneration in *Hieracium*. Sv. Bot. Tidskr. **11**: 145-206.

Rosenberg, O. 1918. Chromosomenzahlen und Chromosomendimensionen in der Gattung *Crepis*. Arkh. Bot. **15**:1-16.

Rosenberger, R. F., and M. Kessel. 1968. Nonrandom sister chromatid segregation and nuclear migration in hyphae of *Aspergillus nidulans*. J. Bact. **96**: 1208-1213.

Rossen, J. M., and M. Westergaard. 1966. Studies on the mechanism of crossing over. II. Meiosis and the time of meiotic chromosome replication in the ascomycete *Neottiella rutilans* (Fr.) Dennis. C. R. Lab. Carlsberg **35**:233-260.

Roth, T. F., and M. Ito. 1967. DNA dependent formation of the synaptinemal complex at meiotic prophase. J. Cell Biol. **35**:247-255.

Roth, T. F., and L. G. Parchman. 1967. Diplotene achiasmatic chromosomes following normal synapsis at pachynema. Twenty-fifth Annual Meeting EMSA.

Rothwell, N. V. 1959. Aneuploidy in *Claytonia virginica*. Amer. J. Bot. **46**:353-360.

Rothwell, N. V., and J. G. Kump. 1965. Chromosome numbers in populations of *Claytonia virginica* from the New York metropolitan area. Amer. J. Bot. **52**:403-407.

Roy, G. P., and F. W. Gould. 1971. Biosystematic investigations of *Bouteloua hirsuta* and *B. pectinata*. I. Gross morphology. Southwest. Nat. **15**:377-387.

Rückert, J. 1895. Uber das selbststandigbleiben der väterlichen und mütterlichen Kernsubstanz während der ersten Einwicklung des befruchten *Cyclops*-Eies. Arch. Mikro. Anat. **45**:339-369.

Rudkin, G. T. 1965. Nonreplicating DNA in giant chromosomes. Genetics **52**:470. (Abstr.)

Rudkin, G. T. 1969. Non-replicating DNA in *Drosophila*. Genetics (Suppl. 61)**1**, 1969:227-238.

Rudkin, G. T., and S. L. Corlette. 1957. Disproportionate synthesis of DNA in a polytene chromosome region. Proc. Nat. Acad. U. S. A. **43**:964-968.

Rudkin, G. T., and J. Schultz. 1961. Disproportionate synthesis of DNA in polytene chromosome regions in *Drosophila melanogaster*. Genetics **46**:893-899.

Rudorf-Lauritzen, M. 1958. The trisomics of *Antirrhinum majus*. Proc. Tenth Int. Congr. Genet. **2**:243-244.

Ruess, A. L., S. Psuzausby, E. F. Lis, and K. Patan. 1962. The oral-facial-digital syndrome: a multiple congenital condition of females with associated chromosomal abnormalities. Pediatrics **29**:985-991.

Ruska, E., and M. Knoll. 1931. The magnetic concentrating electromagnet for fast electrons. Z. Tech. Phys. **12**:309-393.

Russell, L. B. 1964. Another look at the single-active-X hypothesis. New York Acad. Sci. **26**:726-730.

Rutishauser, A. 1956. Chromosome distribution and spontaneous chromosome breakage in *Trillium grandiflorum*. Heredity **10**:367-407.

Rutishauser, A., and E. Rothlisberger. 1966. Boosting mechanism of B-chromosome in *Crepis capillaris*. In C. D. Darlington and K. R. Lewis (editors). Chromosomes Today. Plenum Press, New York.

Ryter, A. 1968. Association of the nucleus and the membrane of bacteria: a morphological study. Bact. Rev. **32**:39-54.

Ryter, A., and F. Jacob. 1963. Étude au microscope electronique des relations entre mesosomes et noyaux chez *Bacillus subtilis*. C. R. Acad. Sci. **257**: 3060.

Ryter, A., and F. Jacob. 1964. Etude au microscope electronique de la liaison entre noyan et mesosome chez *Bacillus subtilis*. Ann. Inst. Pasteur **107**: 384-400.

Sadasivaiah, R. S., and K. J. Kasha. 1971. Meiosis in haploid barley—an interpretation of non-homologous chromosome associations, Chromosoma **35**: 247-263.

Saez, F. A. 1957. An extreme karyotype in an orthopteran insect. Amer. Nat. **91**:259-264.

Sagen, L. 1967. On the origin of mitosing cells. J. Theor. Biol. **14**:225-274.

Sakamura, T. 1918. Kurze Mitteilung über die Chromosomenzahlen und die Verandschaftsverhältnisse der *Triticum*-Arten. Bot. Mag. Tokyo **32**: 151-154.

Sandler, L. 1965. The meiotic mechanics of ring chromosomes in female *Drosophila melanogaster*. In J. I. Valencia and R. F. Grell (editors). International Symposium on Genes and Chromosomes, Structure and Function. Nat. Cancer Inst. Monogr. 18. U. S. Government Printing Office, Washington, D. C.

Sandler, L., D. L. Lindsley, B. Nicoletti, and G. Trippa. 1968. Mutants affecting meiosis in natural populations of *Drosophila melanogaster*. Genetics **60**:525-558.

Sandler, L., and E. Novitski. 1957. Meiotic drive as an evolutionary force. Amer. Nat. **91**:105-110.

Sarkar, K., and E. Coe. 1966. A genetic analysis of the origin of maternal haploids in maize. Genetics **54**:453-463.

Sarkar, P., and G. L. Stebbins. 1956. Morphological evidence concerning the origin of the B genome in wheat. Amer. J. Bot. **43**:297-304.

Sarkissian, I. V., and R. G. McDaniel. 1967. Mitochondrial polymorphism in maize. I. Putative evidence for de novo origin of hybrid-specific mitochondria. Proc. Nat. Acad. Sci. U. S. A. **57**:1262-1266.

Sarkissian, I. V., and H. K. Srivastava. 1967. Mitochondrial polymorphism in maize. II. Further evidence of correlation of mitochondrial complementation and heterosis. Genetics **57**:843-850.

Sarvella, P. 1958. Multivalent formation and genetic segregation in some allopolyploid *Gossypium* hybrids. Genetics **43**:601-619.

Sasaki, M. S., and A. Norman. 1966. DNA fibers from human lymphocyte nuclei. Expt. Cell. Res. **44**: 642-645.

Sax, H. J., and K. Sax. 1935. Chromosome structure and behavior in mitosis and meiosis. J. Arnold Arb. **16**:423-439.

Sax, K. 1922. Sterility in wheat hybrids II. Chromosome behavior in partially sterile hybrids. Genetics **7**:513-552.
Sax, K. 1931. Chromosome ring formation in *Rhoeo discolor*. Cytologia **3**:36-53.
Sax, K. 1935. The cytological analysis of species-hybrids. Bot. Rev. **1**:100-117.
Sax, K., and E. Anderson. 1933. Segmental interchange in chromosomes of *Tradescantia*. Genetics **18**:53-94.
Schmid, W. 1963. DNA replication patterns of human chromosomes. Cytogenetics **2**:175-193.
Schmid, W. 1967. Sex chromatin in hair roots. Cytogenetics **6**:342-349.
Schmid, W. 1969. Satellites on the long Y chromosome arm: a familial Y/autosome translocation in man. Cytogenetics **8**:415-426.
Schmid, W., and D. Vischer. 1967. A malformed boy with double aneuploidy and double-triploid mosaiciam 48,XXYY/77,XXXYY. Cytogenetics **6**:145-155.
Schneider, E. L., and N. P. Salzman. 1970. Isolation and zonal fractionation of metaphase chromosomes from human diploid cells. Science **167**:1141-1143.
Schneiderman, L. H., and C. A. B. Smith. 1962. Nonrandom distribution of certain homologous pairs of normal human chromosomes in metaphase. Nature **195**:1229-1230.
Schnell, L. O. 1948. A study of meiosis in the microsporocytes of interspecific hybrids of *Solanum demissum* × *S. tuberosum*, carried through four back crosses. J. Agr. Res. **76**:185-212.
Schrader, F. 1925. The cytology of pseudo-sexual eggs in a species of *Daphnia*. Z. Ind. Abst. Vererb. **40**:1-27.
Schrader, F. 1940. Touch-and-go pairing in chromosomes. Proc. Nat. Acad. Sci. U. S. A. **26**:634-636.
Schrader, F., and S. Hughes-Schrader. 1926. Haploidy in *Icerya purchasi*. Z. Wiss. Zool. **128**:182-200.
Schrader, F., and S. Hughes-Schrader. 1931. Haploidy in metazoa. Quart. Rev. Biol. **6**:411-438.
Schrader, F., and S. Hughes-Schrader. 1956. Polyploidy and fragmentation in the chromosomal evolution of various species of *Thyanta* (Hemiptera). Chromosoma **7**:469-496.
Schultz, J. 1934. The manifestation of dominants in the triploid. Drosophia Information Service **1**:55.
Schultz, J. 1941. Genes and chromosomes: structure and organization. Cold Spring Harbor Symp. Quant. Biol. **9**:50-58.
Schultz, J. 1947. The nature of heterochromatin. Cold Spring Harbor Symp. Quant. Biol. **12**:179-191.
Schultz, J. 1969. Hybridization, unisexuality, and polyploidy in the teleost *Poeciliopsis* (Poeciliidae) and other vertebrates. Amer. Nat. **103**:605-619.
Schultz, J., and T. Dobzhansky. 1933. Triploid hybrids between *Drosophila melanogaster* and *Drosophila simulans*. J. Exp. Zool. **65**:73-82.
Schultz, J., and D. A. Hungerford. 1953. Characteristics of pairing in the salivary gland chromosomes of *Drosophila melanogaster*. Genetics **38**:689 (abstract).
Schwarz, U., A. Asmus, and H. Frank. 1969. Autolytic enzymes and cell division of *Escherichia coli*. J. Molec. Biol. **41**:419-429.
Schwarzacher, H. G., and W. Schnedl. 1966. Position of labeled chromatids in diplochromosomes of endoreplicated cells after uptake of tritiated thymidine. Nature **209**:107-108.
Scragg, A. H., H. Morimoto, V. Villa, J. Nekhorocheff, and H. C. Halvorson. 1971. Cell-free protein synthesizing system from yeast mitochondria. Science **171**:908-910.
Sears, E. 1954. The aneuploids of common wheat. Univ. Missouri Agr. Exp. Sta. Res. Bull. No. 572.
Sears, E. R. 1944. Cytogenetic studies with polyploid species of wheat. II. Additional chromosomal aberrations in *Triticum vulgare*. Genetics **29**:232-246.
Sears, E. R. 1969. Wheat cytogenetics. Ann. Rev. Genet. **3**:451-468.
Sears, E. R., and A. Camara. 1952. A transmissible dicentric chromosome. Genetics **37**:125-135.
Sears, E. R., and M. Okamoto. 1958. Intergenomic chromosome relationships in hexaploid wheat. Proc. Tenth Int. Congr. Genet. **2**:258-259.
Sears, J. W. 1947. Relationships within the quinaria species group of *Drosophila*. Univ. Texas Publ. No. 4720, p. 137-156.
Sears, L. M. S., and S. Lee-Chen. 1970. Cytogenetic studies in *Arabidopsis thaliana*. Canad. J. Genet. Cytol. **12**:217-223.
Senanayake, Y. D. A., and R. S. Bringhurst. 1967. Origin of *Fragaria* polyploids. I. Cytological analysis. Amer. J. Bot. **54**:221-228.
Sharma, A. K., and D. Bhallachayie. 1953. Somatic reduction in untreated leguminous plants. Genetica **26**:410-414.
Sharman, B. C. 1943. A synthesis of a 42-chromosome wheat. Nature **152**:575-576.
Sharp, L. W. 1926. An Introduction to Cytology. Ed. 2. McGraw-Hill Book Co., New York.
Shatkin, A. J. 1971. Viruses with segmented ribonucleic acid genomes: multiplication of influenza virus reovirus. Bact. Rev. **35**:250-266.
Shaw, M. W. 1968. In Lozzio, C. B., and A. I. Chernoff (editors). Symposium on Human Cytogenetics. Defects Evaluation Center, University of Tennessee.
Shechter, Y., and B. L. Johnson. 1968. The probable origin of *Oryzopsis contracta*. Amer. J. Bot. **55**:611-618.
Sheridan, W. F., and H. Stern. 1967. Histones of meiosis. Exp. Cell Res. **45**:323-330.
Sherman, M. 1946. Karyotype evolution: a cytogenetic study of seven species and six interspecific hybrids of *Crepis*. Univ. Calif. Publ. Bot. **18**:369-408.
Shinke, N. 1934. Spiral structure of chromosomes in meiosis in *Sagittaria aginashi*. Mem. Coll. Sci. Kyoto Imp. Univ. Ser. B **9**:367-392.

Shull, G. H. 1910. Inheritance of sex in *Lychnis*. Bot. Gaz. **49**:110-125.

Siddiqi, O. H. 1963. Incorporation of parental DNA into genetic recombinants of *E. coli*. Proc. Nat. Acad. Sci. U. S. A. **49**:589-592.

Signer, E. R. 1966. Interactions of prophages at the *att*80 site with the chromosome of *Escherichia coli*. J. Molec. Biol. **15**:243-255.

Silow, R. A. 1944. The genetics of species development in Old World cottons. J. Genet. **46**:62-77.

Simantel, G. M., and J. G. Rose. 1964. Colchicine-induced somatic chromosome reduction in sorghum. IV. An induced haploid mutant. J. Hered. **55**:3-5.

Simon, E. 1965. Recombination in bacteriophage T-4: a mechanism. Science **150**:760-763.

Simpson, C. E., and E. C. Bashaw. 1969. Cytology and reproductive characteristics in *Pennisetum setaceum*. Amer. J. Bot. **56**:31-37.

Singleton, J. R. 1953. Chromosome morphology and the chromosome cycle in the ascus of *Neurospora crassa*. Amer. J. Bot. **40**:124-144.

Sinha, A. K. 1967. Spontaneous occurrence of tetraploidy and near-haploidy in mammalian peripheral blood. Exp. Cell Res. **47**:443-448.

Sinha, A. K. 1968. Presumptive trisomy for human chromosome number 3. Acta Genet. **18**:584-592.

Sinha, A. K., and R. L. Bejar. 1968. Long Y-chromosome associated with enlarged satellites on a D-chromosome. Southern Med. J. **61**:5-9.

Sinha, A. K., and J. J. Nora. 1969. Evidence for X/X chromosome translocation in humans. Ann. Hum. Genet. **33**:117-124.

Sinha, U., and J. M. Ashworth. 1969. Evidence for the existence of elements of a para-sexual cycle in the cellular slime mold, *Dictyostelium discoideum*. Proc. Roy. Soc. B **173**:531-540.

Sinclair, J. H., and B. J. Stevens. 1966. Circular DNA filaments from mouse mitochondria. Proc. Nat. Acad. Sci. U. S. A. **56**:508-514.

Sinclair, J. H., B. J. Stevens, P. Sanghair, and M. Rabinowitz. 1967. Mitchondrial-satellite and circular DNA filaments in yeast. Science **156**:1234-1237.

Six, E. 1966. Specificity of P-2 for prophage site I on the chromosome of *Escherichia coli* strain C. Virology **29**:106-125.

Skovsted, A. 1933. Cytological studies in cotton. I. The mitosis and the meiosis in diploid and triploid Asiatic cotton. Ann. Bot. **47**:227-251.

Skovsted, A. 1934. Cytological studies in cotton. II. Two interspecific hybrids between Asiatic and New World cottons. J. Genet. **28**:407-424.

Skovsted, A. 1937. Cytological studies in cotton. IV. Chromosome conjugation in interspecific hybrids. J. Genet. **34**:97-134.

Slayton, H., T. Kiho, C. E. Hall, and A. Rech. 1968. An electron microscopic study of large bacterial polyribosomes. J. Cell. Biol. **37**:583-590.

Smillie, R. M., D. Graham, M. R. Dwyer, A. Grieve, and N. F. Tobin. 1967. Evidence for the synthesis in vivo of the photosynthetic electron-transfer pathway on chloroplast ribosomes. Biochem. Biophys. Res. Commun. **28**:604-610.

Smith, D. W., J. M. Docter, R. E. Ferrier, J. L. Frias, and A. Spock. 1968. Possible localization of the gene for cystic fibrosis of the pancreas to the short arm of chromosome 5. Lancet **2**:309-312.

Smith, E. B. 1968. Supernumerary chromosomes in *Haplopappus validus* (Rydb.) Cory. Evolution **22**: 748-750.

Smith, F. H. 1932. The structure of the somatic and meiotic chromosomes of *Galtonia candicans*. Cellule **41**:243-263.

Smith, H. O., and M. Levine. 1965. Gene order in prophage P-22. Virology **27**:229-231.

Smith, L. 1936. Cytogenetic studies in *Triticum monococcum* L. and *T. aegilopoides* Bal. Missouri Agr. Exp. Sta. Res. Bull. No. 248.

Smith, L. 1942. Cytogenetics of a factor for multiploid sporocytes in barley. Amer. J. Bot. **29**:451-456.

Smith, Luther. 1951. Cytology and genetics of barley. Bot. Rev. **17**:1-51, 285-355.

Smith, S. G. 1941. A new form of spruce sawfly identified by means of its cytology and parthenogenesis. Sci. Agr. **21**:245-305.

Smith, S. G. 1942. Polarization and progression and pairing. II. Premeiotic orientation and the initiation of pairing. Canad. J. Res. **20D**:221-229.

Smith, S. G. 1953. Chromosome number of Coleoptera. Heredity **7**:31-48.

Smith, S. G. 1956. Spermatogenesis in an elaterid beetle. J. Hered. **47**:2-10.

Smith, S. G. 1960. Chromosome numbers of Coleoptera. Canad. J. Genet. Cytol. **2**:66-88.

Smith, S. G. 1965. Heterochromatin, colchicine, and karyotype. Chromosoma **16**:162-165.

Smith-White, S. 1948. Polarised segregation in the pollen mother cells of a stable triploid. Heredity **2**:119-129.

Smith-White, S. 1955. The life history and genetic system of *Leucopogon juniperinus*. Heredity **9**:79-91.

Smith-White, S., and C. R. Carter. 1970. The cytology of *Brachycome linearioba*. II. The chromosome species and their relationships. Chromosoma **30**:129-153.

Smith-White, S., C. R. Carter, and H. M. Stace. 1970. The cytology of *Brachycome*. I. The subgenus *Eubrachycome*: a general survey. Aust. J. Bot. **18**:99-125.

Snow, R. 1960. Chromosomal differentiation in *Clarkia dudleyana*. Amer. J. Bot. **47**:302-309.

Snyder, L. A. 1951. Cytology of inter-strain hybrids and the probable origin of variability in *Elymus glaucus*. Amer. J. Bot. **38**:195-202.

Snyder, R. W., and F. E. Young. 1969. Association between the chromosome and the cytoplasmic membrane in *Bacillus subtilis*. Biochem. Biophys. Res. Commun. **35**:354-362.

Solari, A. J. 1965. Structure of the chromatin in sea urchin sperm. Proc. Nat. Acad. Sci. U. S. A. **53**:503.

Solari, A. J., and L. L. Tres. 1967. The ultrastructure of the human sex vesicle. Chromosoma **22**:16-31.

Solbrig, O. T., L. C. Anderson, D. W. Kyhos, P. H. Raven, and L. Rüdenberg. 1964. Chromosome numbers in Compositae V Astereae. Amer. J. Bot. **51**:513-519.

Somers, C. E., T. P. Wagner, and T. C. Hsu. 1960. Mitosis in vegetative nuclei of *Neurospora crassa*. Genetics **45**:801-810.

Sonnenblick, B. P. 1950. The early embryology of *Drosophila melanogaster*. In M. Demerec (editor). The Biology of *Drosophila*. Wiley, London.

Soost, R. K. 1951. Comparative cytology and genetics of asynaptic mutants in *Lycopersicon esculentum* Mill. Genetics **36**:410-434.

Soudek, D., R. Lorova, and K. Adamek. 1968. Pericentric inversion in a family with a 21/22 translocation. Cytogenetics **7**:108-117.

Southern, D. I. 1969. Stable telocentric chromosomes produced following centric misdivision in *Myrmeleotettix maculatus* (Thumb). Chromosoma **26**:140-147.

Sparks, R. S., and D. T, Arakaki. 1966. Intrasubspecific and intersubspecific chromosomal polymorphism in *Peromyscus maniculatus* (deer mouse). Cytogenetics **5**:411-418.

Sparvoli, E., H. Gay, and B. P. Kaufmann. 1965. Number and pattern of association of chromonemata in the chromosomes of *Tradescantia*. Chromosoma **16**:415-435.

Spizizen, J. 1958. Transformation of biochemically deficient strains of *Bacillus subtilis* by deoxyribonucleate. Proc. Nat. Acad. Sci. U. S. A. **44**:1072-1078.

Stack, S. M. 1969. Somatic and premeiotic homologous chromosome pairing and the development of the synaptinemal complex in certain low chromosome number plants. Ph.D. Dissertation. University of Texas, Austin.

Stack, S. M. 1971. Premeiotic changes in *Ornithogalum virens*. Bull. Torrey Bot. Club **98**:207-214.

Stack, S. M., and W. V. Brown. 1969a. Somatic and premeiotic pairing of homologues in *Plantago ovata*. Bull. Torrey Bot. Club. **96**:143-149.

Stack, S. M., and W. V. Brown. 1969b. Somatic pairing, reduction and recombination: an evolutionary hypothesis of meiosis. Nature **222**:1275-1276.

Stadler, L. J. 1928. Genetic effects of X-rays in maize. Proc. Nat. Acad. Sci. U. S. A. **14**:69-75; also Science **68**:186.

Staiger, H. 1956. Der chromosomendimorphismus bein prosobranchier *Purpura lapillus* in Bezehung zur Ökologie der art. Chromosoma **6**:419-478.

Stalker, H. D. 1954. Parthenogenesis in *Drosophila*. Genetics **39**:4-34.

Stalker, H. D. 1961. The genetic systems and modifying meiotic drive in *Drosophila paramelanica*. Genetics **46**:177-202.

Star, A. E. 1970. Spontaneous and induced chromosome breakage in *Claytonia virginica*. Amer. J. Bot. **57**:1145-1149.

Stebbins, G. L. 1932. Cytology of *Antennaria*, II. Parthenogenetic species. Bot. Gaz. **94**:322-345.

Stebbins, G. L. 1938a. Cytological characteristics associated with the different growth habits in the dicotyledons. Amer. J. Bot. **25**:189-198.

Stebbins, G. L. 1938b. Cytogenetic studies in *Paeonia*. II. The cytology of the diploid species and hybrids. Genetics **23**:83-110.

Stebbins, G. L. 1941. Apomixis in angiosperms. Bot. Rev. **7**:507-542.

Stebbins, G. L. 1950. Variation and Evolution in Plants. Columbia University Press, New York.

Stebbins, G. L. 1970. Variation and evolution in plants: progress during the past twenty years. In M. K. Hecht and W. C. Steere (editors). Essays in Evolution and Genetics in Honor of Theodosius Dobzhansky. Supplement to Evolutionary Biology. Appleton-Century-Crofts, New York.

Stebbins, G. L., and A. Day. 1967. Cytogenetic evidence for long continued stability in the genus *Plantago*. Evolution **21**:409-428.

Stebbins, G. L., and S. Ellerton. 1939. Structural hybridity in *Paeonia californica* and *P. brownii*. J. Genet. **38**:1-36.

Stebbins, G. L., and J. A. Jenkins. 1939. Apospopric development in North American species of *Crepis*. Genetics **21**:1-34.

Stebbins, G. L., and S. S. Shah. 1960. Developmental studies of cell differentiation in the epidermis of monocotyledons. II. Cytological features of stomatal development in the Gramineae. Develop. Biol. **2**:477-500.

Steele, M. W., W. R. Breg, A. I. Eidelman, D. T. Leon, and T. A. Terzakis. 1966. A B-group ring chromosome with mosaicism in a newborn child with cri du chat syndrome. Cytogenetics **5**:419-429.

Steffenson, D. M. 1963. Evidence for the apparent absence of DNA in the interbands of *Drosophila* salivary chromosomes. Genetics **48**:1289-1301.

Steil, W. N., 1951. Apogamy, apospory, and parthenogenesis in the pteridophytes. Bot. Rev. **17**:90-104.

Steinitz-Sears, L. M. 1963. Chromosome studies in *Arabidopsis thaliana*. Genetics **48**:483-390.

Stent, G. S. (editor). 1960. Papers on Bacterial Viruses. Little, Brown, and Company, Boston.

Stent, G. S. 1963. Molecular Biology of Bacterial Viruses. W. H. Freeman & Co., Publishers, San Francisco.

Stephens, S. G. 1947. Cytogenetics of *Gossypium* and the problem of the New World cottons. Advances Genet. **1**:431-442.

Stephens, S. G. 1950. The internal mechanism of speciation in *Gossypium*. Bot. Rev. **16**:115-149.

Stern, C. 1931. Zytologisch-genetische Untersuchungen als Beweise fur die Morgansche Theorie des Factorenaustauschs. Biol. Zentralbl. **51**:547-587.

Stern, C. 1936. Somatic crossing over and segregation in *Drosophila melanogaster*. Genetics 21:625-630.

Stern, C. 1939. Somatic crossing-over and somatic translocations. Amer. Nat. 73:95-96.

Stern, C. 1969. Gene expression in genetic mosaics. Genetics (Suppl. 61)1, 1969:199-211.

Stevens, N. M. 1908. A study of the germ cells of certain Diptera with reference to the heterochromosomes and the problem of synapsis. J. Exp. Zool. 5:359-374.

Stewart, R. N., and H. Derman. 1970. Determination of number and mitotic activity of shoot apical initial cells by analysis of mericlinal chimeras. Amer. J. Bot. 57:816-826.

Stoltz, D. R., K. S. Khera, R. Bendall, and S. W. Gunner. 1970. Cytogenetic studies with cyclamate and related compounds. Science 167:1501-1502.

Stomps, T. J. 1910. Kernteilung und synapsis bei *Spinacia oleracea* L. Dissertation (published). Amsterdam.

Stone, D., E. Lamson, Y. S. Chang, and K. W. Pickering. 1969. Cytogenetic effects of cyclamates on human cells in vitro. Science 164:568-569.

Storey, W. B. 1968. Somatic reduction in cycads. Science 159:648-650.

Strasburger, E. 1904. Über Reduktionsteilung. Sitz. Konig. Preuss. Akad. Wiss. 18:1-28.

Strasburger, E. 1905. Typische und allotypische Kernteilung. Jahresber. Wiss. Bot. 42:1-71.

Strasburger, E. 1909. Zeitpunkt der Betstimmung des Geschlechts, Apogamie, Parthenesis, und Reduktionsteilung. Jena.

Strauch, G., E. Engel, P. D. Taft, L. Atkins, and A. P. Forbes. 1965. Syndrome de Klinefelter à caryotype 46,XX aux niveau cutane sanguin et testiculaire. Ann. Endocr. 26:727-738.

Straus, J. 1954. Maize endosperm tissue grown in vitro. II. Morphology and cytology. Amer. J. Bot. 41:833-839.

Streisinger, G., R. H. Edgar, and G. H. Denhardt. 1964. Chromosome structure of the genome of phage T-4. I. The circularity of the linkage map. Proc. Nat. Acad. Sci. U. S. A. 51:775-779.

Strickberger, M. W., and C. J. Wills. 1966. Monthly frequency changes of *Drosophila pseudoobscura* third chromosome gene arrangements in a California locality. Evolution 20:592-602.

Strid, A. 1968. Stable telocentric chromosomes formed by spontaneous misdivision in *Nigella doerfleri* (Ranunculaceae). Bot. Not. 121:153-164.

Stubblefield, E., and G. C. Mueller. 1962. Molecular events in the reproduction of animal cells. II. The focalized synthesis of DNA in the chromosomes of HeLa cells. Cancer Res. 22:1091-1099.

Sturtevant, A. H. 1913. The linear arrangement of six sex-linked factors in *Drosophila* as shown by their mode of association. J. Exp. Zool. 14:43-59.

Sturtevant, A. H. 1931. Known and probable inverted sections of the autosomes of *Drosophila melanogaster*. Carnegie Inst. Wash. Publ. No. 42, p. 1-27.

Sturtevant, A. H. 1951. The relation of genes and chromosomes. In L. C. Dunn (editor). Genetics in the 20th Century. The Macmillan Company, New York.

Sturtevant, A. H., and G. W. Beadle. 1936. The relations of inversions in the X-chromosome of *Drosophila melanogaster* to crossing over and disjunction. Genetics 21:554-604.

Sturtevant, A. H., and T. Dobzhansky. 1930. Reciprocal translocations in *Drosophila* and their bearing on *Oenothera* cytology and genetics. Proc. Nat. Acad. Sci. U. S. A. 16:533-536.

Sturtevant, A. H., and Th. Dobzhansky. 1936. Geographical distribution and cytology of "sex-ratio" in *Drosophila pseudoobscura* and related species. Genetics 21:473-490.

Stutz, E., and H. Noll. 1967. Characterization of cytoplasmic and chloroplast polysomes in plants: evidence for three classes of ribosomal RNA in nature. Proc. Nat. Acad. Sci. U. S. A. 57:774-781.

Sueoka, N. 1968. A model of separation of daughter chromosomes by periodic condensation in bacteria. In W. J. Peacock and R. D. Brock (editors). Replication and Recombination of Genetic Material. Australian Academy of Science, Canberra.

Sueoka, N., K. S. Chiang, and J. R. Kates. 1967. Deoxyribonucleic acid replication in meiosis of *Chlamydomonas reinhardi*. J. Molec. Biol. 25:47-66.

Sumner, A. T., H. J. Evans, and R. A. Buckland. 1971. A new technique for distinguishing between human chromosomes. Nature (New Biol.) 232:31-32.

Sunderland, N., and J. McLeish. 1961. Nucleic acid content and concentration in root cells of higher plants. Exp. Cell Res. 24:541-554.

Suomalainen, E. 1940. Polyploidy in parthenogenetic Curculionidae. Hereditas 26:51-64.

Suomalainen, E. 1969. Evolution in parthenogenetic Curculionidae. Evolut. Biol. 3:261-296.

Sutton, W. S. 1902. On the morphology of the chromosome group in *Brachystola magna*. Biol. Bull. 4:39-47.

Sutton, W. S. 1903. The chromosomes in heredity. Biol. Bull. 4:231-251.

Svardson, G. 1941. Somatic pairing in *Salmo* and *Coregonus* and its hypothetical explanation. Arkh. Zool. 33B:1-6.

Swanson, C. P. 1940. The distribution of inversions in *Tradescantia*. Genetics 25:438-465.

Swanson, C. P. 1957. Cytology and Cytogenetics. Prentice-Hall, Inc., Englewood Cliffs, N. J.

Swanson, C. P., T. Merz, and W. J. Young. 1967. Cytogenetics. Prentice-Hall, Inc., Englewood Cliffs, N. J.

Sweadner, W. R. 1937. Hybridization and the phylogeny of the genus *Platysamia*. Ann. Carneg. Mus. 25:163-242.

Swift, H. 1969. Nuclear physiology and differentiation: a general summary. Genetics (Suppl. 61)1, 1969:439-461.

Swift, H., and E. M. Rasch. 1954. Nucleoproteins in *Drosophila* polytene chromosomes. J. Histochem. Cytochem. **2**:456-458. (Abstr.)

Sykes, M. G. 1908. Nuclear division in *Funkia*. Arch. Zellforsch. **1**:380-398.

Tai, W. 1970. Multipolar meiosis in diploid crested wheatgrass, *Agropyron cristatum*. Amer. J. Bot. **57**:1160-1169.

Takagi, N., and A. A. Sandberg. 1968. Chronology and pattern of human chromosome replication. Cytogenetics **7**:135-143.

Takenouchi, T. 1970. Three further studies of the chromosomes of Japanese weevils (Coleoptera: Curculionidae). Canad. J. Genet. Cytol. **12**:273-277.

Tan, C. C. 1936. Genetic maps of autosomes in *Drosophila pseudoobscura*. Genetics **21**:796-807.

Tartof, K. D. 1971. Increasing the multiplicity of ribosomal RNA genes in *Drosophila melanogaster*. Science **171**:294-297.

Tartof, K. D., and R. P. Perry. 1970. The 5s RNA genes of *Drosophila melanogaster*. J. Molec. Biol. **51**:171-184.

Taylor, J. H. 1958a. Sister chromatid exchanges in tritium-labeled chromosomes. Genetics **43**:515-529.

Taylor, J. H. 1958b. The mode of chromosome duplication in *Crepis capillaris*. Exp. Cell Res. **15**:350-357.

Taylor, J. H. 1959. Autoradiographic studies of nucleic acids and proteins during meiosis in *Lilium longiflorum*. Amer. J. Bot. **46**:477-484.

Taylor, J. H. 1965. Distribution of tritium-labeled DNA among chromosomes during meiosis. J. Cell Biol. **25**(No. 2, Pt. 2):57-67.

Taylor, J. H. 1968. Rates of chain growth and units of replication in DNA of mammalian chromosomes. J. Molec. Biol. **31**:579-594.

Taylor, J. H., and R. D. McMaster. 1954. Autoradiographic and microphotometric studies of desoxyribose nucleic acid during microsporogenesis in *Lilium longiflorum*. Chromosoma **6**:489-521.

Telfer, M. A., D. Baker, G. R. Clark, and C. E. Richardson. 1968. Incidence of gross chromosomal errors among tall criminal American males. Science **159**:1249-1250.

Teplitz, R. L., P. E. Gustafson, and O. L. Pellett. 1968. Chromosomal distribution in interspecific in vitro hybrid cells. Exp. Cell Res. **52**:379-391.

Tewari, K. K., W. Votsch, M. Mahler, and H. R. Mackler. 1966. Biochemical correlates of respiratory deficiency. VI. Mitochondrial DNA. J. Molec. Biol. **20**:453-481.

Tewari, K. K., and S. G. Wildman. 1966. Chloroplast DNA from tobacco leaves. Science **153**:1269-1271.

Therman, E. 1951. Somatic and secondary pairing in *Ornithogalum*. Heredity **5**:253-269.

Thomas, C. A., M. Rhoades, and L. A. MacHattie. 1968. The molecular genetics of viral DNA. In W. J. Peacock and P. D. Brock (editors). Replication and Recombination of Genetic Material. Australian Academy of Science, Canberra.

Thomas, R. 1965. The smaller teiid lizards (*Gymnophthalamus* and *Bachia*) of the southeastern Caribbean. Proc. Biol. Soc. Wash. **78**:141-154.

Thompson, H., J. Melnyk, and F. Hecht. 1967. Reproduction and meiosis in XYY. Lancet **2**:831.

Thompson, J. B. 1956. Genotypic control of chromosome behavior in rye. II. Disjunction at meiosis in interchange heterozygotes. Heredity **10**:99-108.

Thorneycroft, H. B. 1966. Chromosomal polymorphism in the white-throated sparrow, *Zonotrichia albicollis* (Gmelin). Science **154**:1571-1572.

Tischler, G. 1918. Untersuchungen über den Reisenwuchs von *Phragmites communis* var. *pseudodonex*. Ber. Den. Bot. Ges. **36**:549-558.

Tjio, J. H., and A. Levan. 1956. The chromosome number of man. Hereditas **42**:1-6.

Tjio, J. H., and T. T. Puck. 1958. The somatic chromosomes of man. Proc. Nat. Acad. Sci. U. S. A. **44**:1229-1237.

Tjio, J. H., and J. Wang. 1962. Chromosome preparations of bone marrow cells without prior in vito culture or in vivo colchicine administration. Stain Techn. **37**:17-19.

Tobgy, H. A. 1943. A cytological study of *Crepis fuliginosa, C. neglecta*, and their F_1 hybrid, and its bearing on the mechanism of phylogenetic reduction in chromosome number. J. Genet. **45**:67-111.

Tobias, P. V. 1956. Chromosomes, Sex-Cells, and Evolution in a Mammal. Percy, Lund, Humphries, and Co., London.

Tokuyasu, K., S. C. Madden, and L. J. Zeldis. 1968. Finestructural alterations of interphase nuclei of lymphocytes stimulated to growth activity in vitro. J. Cell Biol. **39**:630-660.

Tometorp. G. 1939. Cytological studies on haploid *Hordeum distichum*. Hereditas **25**:241-254.

Tomizawa, J. 1960. Genetic structure of recombinant chromosomes formed after mating in *Escherichia coli* K-12. Proc. Nat. Acad. Sci. U. S. A. **46**:91-101.

Tomizawa, J., and N. Anraku. 1964. Molecular mechanisms of genetic recombination in Bacterephage. J. Molec. Biol. **8**:508-515, 516-540.

Trosko, J. E., and D. S. Wolff. 1965. Strandedness of *Vicia faba* chromosomes as revealed by enzyme digestion studies. J. Cell Biol. **26**:130-135.

Trujillo, J. M., C. Stenius, L. Christian, and S. Ohno. 1962. Chromosomes of the horse, the donkey, and the mule. Chromosoma **13**:243-248.

Tsuchiya, T. 1953. Fertility autotetraploids and their hybrids in barley. I. Meiosis and fertility in some autotetraploids. Rep. Kihara Inst. Biol. Kyoto **6**:46-52.

Tsuchiya, T. 1959. Genetic studies in trisomic barley. I. Relationships between trisomics and genetic linkage groups in barley. Jap. J. Bot. **17**:14-28.

Tupper, W. W., and H. H. Bartlett. 1916. A comparison of the wood structure of *Oenothera stenomeres* and its tetraploid mutation *gigas*. Science **43**:292.

Turner, B. L. 1967. Chromosome survey of *Podolepis* (Compositae-Inuleae). Aust. J. Bot. **15**:445-449.

Turner, B. L., W. L. Ellison, and R. M. King. 1961.

Chromosome numbers in the Compositae. IV. North American species, with phyletic interpretations. Amer. J. Bot. **48**:216-223.

Turner, B. L., and D. Flyr. 1966. Chromosome numbers in the Compositae. X. North American species. Amer. J. Bot. **53**:24-33.

Turner, B. L., and D. Horne. 1964. Taxonomy of *Machaeranthera* sect. Psilactis (Compositae-Astereae). Brittonia **16**:316-331.

Turner, B. L., and W. H. Lewis. 1965. Chromosome numbers in the Compositae. IX. African species. J. S. Afr. Bot. **31**:207-217.

Turner, B. L., M. Powell, and R. M. King. 1962. Chromosome numbers in the Compositae. VI. Additional Mexican and Guatemalan species. Rhodora **64**:251-271.

Turpin, R., and J. Lejeune. 1965. Les Chromosomes Humains. Gauthier-Villars, Paris.

Ugent, D. 1970. The potato. Science **170**:1161-1166.

Ullerich, F. H. 1966. Karyotyp und DNS-Gehalt von *Bufo bufo, B. viridis, B. bufo × B. viridis*, und *B. calamita*. Chromosoma **18**:316-324.

Unnerus, V., J. Fellman, and A. de la Chapelle. 1967. The length of the human Y chromosome. Cytogenetics **6**:213-227.

Unrau, J. 1958. Cytogenetics and wheat-breeding. Proc. Tenth Int. Cong. Genet. **1**:129-141.

Upcott, M. 1937. The external mechanism of the chromosomes. VI. The behavior of the centromere at meiosis. Proc. Roy. Soc. (Biol.) **124**:336-361.

Uzzell, T. M. 1964. Relations of the diploid and triploid species of the *Ambystoma jeffersonianum* complex (Amphibia, Caudata). Capeia 1964:257-300.

Uzzell, T. M., and S. M. Goldblatt. 1967. Serum proteins of salamanders of the *Ambystoma jeffersonianum* complex, and the origin of triploid species of this group. Evolution **21**:345-354.

Vanderlyn, L. 1948. Somatic mitosis in the root tip of *Allium Cepa* — a review and a reorientation. Bot. Rev. **14**:270-318.

Van't Hoff, J. 1966. Experimental control of DNA of synthesizing and dividing cells in excised root tips of *Pisum*. Amer. J. Bot. **53**:970-976.

Vasek, F. C. 1956. Induced aneuploidy in *Clarkia unguiculata* (Onagraceae). Amer. J. Bot. **43**:366-371.

Vasek, F. C. 1962. "Multiple spindle" — a meiotic irregularity in *Clarkia exilis*. Amer. J. Bot. **49**:536-539.

Vendrely, R. 1955. The deoxyribonucleic acid content of the nucleus. In E. Chargaff and J. Davidson (editors). The Nucleic Acids. Vol. 2. Academic Press Inc., New York.

Vereyskaya, V. N. 1965. Meiosis in males of bisexually reproducing allotetraploid in the silkworm *Bombyx mori*. (Transl.) Genetics (USSR)**1965**(4):69-88.

Vernon, G. M., and E. R. Witkus. 1970. Behavior of supernumerary chromosomes in *Puschkinia libanotica*. Bull. Torrey Bot. Club **97**:289-295.

Vig, B. G. 1968. Spontaneous chromosome abnormalities in root and pollen mother cells of *Aloe vera* L. Bull. Torrey Bot. Club **95**:254-261.

Vincent, W. S., and O. L. Miller. 1966. International symposium on the nucleolus, its structure and function. Nat. Cancer Inst. Monogr. No. 23.

Voelker, R. A. 1967. Further studies on the genetics of *Drosophila affinis*. Genetics **56**:593. (Abstr.)

Voelker, R. A. 1970. Relative fitnesses of XO and XY males in *Drosophila affinis*. Ph.D. Dissertation. University of Texas, Austin.

Volpe, E. P., D. Duplantur, and E. M. Earley. 1970. Clarification of alleged "cytologically verified" hybrid between a toad and a frog. Cytogenetics **9**:161-172.

Volpe, E. P., and E. M. Earley. 1970. Somatic cell mating and segregation in chimeric frogs. Science **168**:850-852.

Volpe, E. P., and B. M. Gebhardt. 1966. Evidence from cultured leucocytes of blood cell chimerism in exparabiotic frogs. Science **154**:1197-1199.

Wagenaar, E. B. 1969. End-to-end chromosome attachments in mitotic interphase and their possible significance to meiotic chromosome pairing. Chromosoma **26**:410-426.

Wagenaar, E. B., and R. S. Sadasivaiah. 1969. End-to-end, chainlike associations of paired pachytene chromosomes of *Crepis capillaris*. Canad. J. Genet. Cytol. **11**:403-408.

Wagner, R. P. 1969. Genetics and phenogenetics of mitochondria. Science **163**:1026-1031.

Wahrman, J. 1966. A carabid beetle with only eight chromosomes. Heredity **21**:154-159.

Wahrman, J., and K. Fried. 1970. The Jerusalem prospective newborn survey of mongolism. Ann. N. Y. Acad. Sci. **171**:341-360.

Wahrman, J., and M. Friedländer. 1966. Nucleoli persisting throughout the meiosis of *Dielocroce baudii* (Neuroptera: Nemopteridae). Proc. Fifteenth Meeting Genet. Soc. Israel **2**:123.

Wahrman, J., and R. Goitein. Hybridization in nature between two chromosome forms of spiny mice (Rodentia: Murinae). Chromosomes Today. Vol. 3.

Wahrman, J., R. Goitein, and E. Nevo. 1969. Mole rat *Spalax*: evolutionary significance of chromosome variation. Science **164**:82-84.

Wahrman, J., R. Voss, T. Shapiro, and A. Ashkenazi. 1967. The Philadelphia 1 chromosome in two children with chronic myeloid leukemia. Israel J. Med. Sci. **3**:380-391.

Wahrman, J., and A. Zahavi. 1955. Cytological contributions to the phylogeny and classification of the rodent genus *Gerbillus*. Nature **175**:600-602.

Wahrman, J., and A. Zahavi. 1958. Cytogenetic analysis of mammalian sibling species by means of hybridization. Proc. Tenth Int. Congr. Genet. **2**:304-305.

Wald, H. 1936. Cytologic studies on the abnormal development of the eggs of the *claret* mutant type of *Drosophila simulans*. Genetics **21**:264-281.

Walen, K. H. 1965. Spatial relationships in the repli-

cation of chromosomal DNA. Genetics **51**:915-929.
Walker, P. M. B. 1968. How different are the DNAs from related animals? Nature **219**:228-232.
Walker, R. I. 1959. Chromosome behaviour in F_1 hybrids between *Solanum demissum* and three diploid species. Bull. Torrey Bot. Club **86**:31-40.
Walker, T. G. 1962. Cytology and evolution in the fern genus *Pteris* L. Evolution **16**:27-43.
Wallace, H., and M. L. Birnstiel. 1966. Ribosomal cistrons and the nucleolar organizer. Biochem. Biophys. Acta **114**-296-310.
Walters, J. L. 1941. The distribution of segmental interchange types in *Paeonia californica* and *P. brownii*. Amer. J. Bot. **28**:726-727. (Abstr.)
Walters, J. L. 1942. Distribution of structural hybrids in *Paeonia californica*. Amer. J. Bot. **29**:270-275.
Walters, J. L. 1956. Spontaneous meiotic chromosome breakage in natural populations of *Paeonia californica*. Amer. J. Bot. **43**:342-354.
Walters, M. S. 1954. A study of pseudobivalents in meiosis of two interspecific hybrids of *Bromus*. Amer. J. Bot. **41**:160-171.
Walters, M. S. 1970. Evidence on the time of chromosome pairing from the preleptotene spiral stage in *Lilium longiflorum* "Croft." Chromosoma **29**:375.
Walters, M. S., and D. U. Gerstel. 1948. A cytological investigation of a tetraploid *Rhoeo discolor*. Amer. J. Bot. **35**:141-150.
Warburton, D., A. D. Miller, O. J. Miller, P. W. Allerdice, and A. de Capoa. 1969. Detection of minute deletions in human karyotypes. Cytogenetics **8**:97-108.
Warmke, H. E. 1946. Sex determination in sex balance in *Melandrium*. Amer. J. Bot. **33**:648-660.
Warmke, H. E., and A. F. Blakeslee. 1940. The establishment of a 4n dioecious race in *Melandrium*. Amer. J. Bot. **27**:751-762.
Warters, M., and A. B. Griffen. 1950. The telomeres of *Drosophila*. J. Hered. **41**:183-190.
Wasserman, A. O. 1970. Polyploidy in the common tree toad *Hyla versicolor* Le Conte. Science **167**:385-386.
Watkins, G. M. 1935. A study of chromosome pairing in *Yucca rupicola*. Bull. Torrey Bot. Club **62**:133-150.
Watkins, J. F., and D. M. Grace. 1967. Studies on the surface and antigens of interspecific mammalian cell heterokaryons. J. Cell Sci. **2**:193-204.
Watson, J. D., and F. H. C. Crick. 1953. Molecular structure of nucleic acids. A structure for deoxyribose nucleic acid. Nature **171**:737-738.
Weaver, E. C. 1960. Somatic crossing over and its genetic control in *Drosophila*. Genetics **45**:345-357.
Webster, P. L., and J. Van't Hoff. 1969. Dependence on energy and aerobic metabolism of initiation of DNA synthesis and mitosis by G_1 and G_2 cells. Exp. Cell Res. **55**:88-94.
Webster, P. L., and J. Van't Hoff. 1970. DNA synthesis and mitosis in meristems: requirements for RNA and protein synthesis. Amer. J. Bot. **57**:130.

Wedberg, H. L., H. Lewis, and C. S. Venkatesh. 1968. Translocation heterozygotes and supernumerary chromosomes in wild populations of *Clarkia williamsonii*. Evolution **22**:93-107.
Weinberg, R. A., U. Loening, M. Williams, and S. Penman. 1967. Acrylamide gel electrophoresis of HeLa cell nucleolar RNA. Proc. Nat. Acad. Sci. U. S. A. **58**:1088-1095.
Weiss, M. C., and H. Green. 1967. Human-mouse hybrid cell lines containing partial complements of human chromosomes and functioning human genes. Proc. Nat. Acad. Sci. U. S. A. **58**:1104-1111.
Weitkamp, L. 1969. Chromosome location of MN blood group locus. Science **164**:1187-1188.
Welshons, W. J. 1965. Analysis of a gene in *Drosophila*. Science **150**:1122-1129.
Wernham, H. F. 1912. Floral evolution IX. New Phytol. **11**:373-377.
Westergaard, M. 1958. The mechanism of sex determination in dioecious flowering plants. Advances Genet. **9**:217-281.
Westergaard, M. 1964. Studies on the mechanism of crossing over. I. Theoretical consideration. C. R. Lab. Carlsberg **34**:359-405.
Westergaard, M., and D. von Wettstein. 1970. Studies on the mechanism of crossing over. IV. The molecular organization of the synaptinemal complex in *Neottiella* (Cooke) Saccardo (ascomycetes). C. R. Lab. Carlsberg **37**:239-268.
Westmoreland, B. C., W. Szybalski, and H. Ris. 1969. Mapping of deletions and substitutions in heteroduplex DNA molecules of bacteriophage lambda by electron microscopy. Science **163**:1343-1348.
Wettstein, R., and J. R. Sotelo. 1967. Electron microscope serial reconstruction of the spermatocyte I nuclei at pachytene. J. Microscop. **6**:557-576.
Whitaker, T. W. 1934. Chromosome constitution in certain monocotyledons. J. Arnold Arb. **15**:135-143.
Whitaker, T. W. 1936. Fragmentation in *Tradescantia*. Amer. J. Bot. **23**:517-519.
White, M. J. D. 1948a. Animal Cytology and Evolution. Cambridge University Press, London.
White, M. J. D. 1948b. The cytology of Cecidomyiidae. IV. J. Morph. **82**:53-80.
White, M. J. D. 1951. Cytogenetics of orthopteroid insects. Advances Genet. **4**:267-341.
White, M. J. D. 1954. Animal Cytology and Evolution. Cambridge University Press, London.
White, M. J. D. 1957. Cytogenetics of the grasshopper *Moraba scurra*. I. Meiosis of interracial and interpopulation hybrids. Aust. J. Zool. **5**:285-304.
White, M. J. D. 1968. Models of speciation. Science **159**:1065-1070.
White, M. J. D. 1970. Heterozygosity and genetic polymorphism in parthenogenetic animals. In M. K. Hecht and W. C. Steere (editors). Essays in Evolution and Genetics in Honor of Theodosius Dobzhansky. Supplement to Evolutionary Biology. Appleton-Century-Crofts, New York.

White, M. J. D., J. Cheney, and K. H. L. Key. 1963. A parthenogenetic species of grasshopper with complex structural heterozygosity (Orthoptera: Acridoidea). Aust. J. Zool. **11**:1-19.

Whitfield, P. R., and D. Spencer. 1968. The biochemical and genetic autonomy of chloroplasts. In W. J. Peacock and P. D. Brock (editors). Replication and Recombination of Genetic Material. Australian Academy of Science, Canberra.

Whiting, P. W. 1935. Sex determination in bees and wasps. J. Hered. **26**:263-278.

Whiting, P. W. 1939. Multiple alleles in sex determination in *Habrobracon*. J. Morph. **66**:323-352.

Whiting, P. W. 1943. Multiple alleles in complementary sex determination of *Habrobracon*. Genetics **28**:365-382.

Whiting, P. W. 1945. The evolution of male haploidy. Quart. Rev. Biol. **20**:231-260.

Whiting, P. W., and A. R. Whiting. 1925. Diploid males from fertilized eggs in Hymenoptera. Science **62**:437-438.

Wieman, H. L. 1917. The chromosomes of human spermatocytes. Amer. J. Anat. **21**:1-22.

Wilson, E. B. 1896. The Cell in Development and Inheritance. The Macmillan Company, New York.

Wilson, E. B. 1902. Mendel's principles of heredity and the maturation of the germ cells. Science **16**:416.

Wilson, E. B. 1905. The chromosomes in relation to the determination of sex in insects. Science **20**:500-502.

Wilson, E. B. 1906a. The Cell in Development and Inheritance. Ed. 2. The Macmillan Company, New York.

Wilson, E. B. 1906b. Studies on chromosomes. J. Exp. Zool. **3**:1-40.

Wilson, E. B. 1909. Studies on chromosomes. J. Exp. Zool. **6**:69-99.

Wilson, E. B. 1910. Studies on chromosomes. VI. A new type of chromosome combination in *Metapodius*. J. Exp. Zool. **9**:53-78.

Wilson, E. B. 1912. Studies on chromosomes. VIII. Observations on the maturation-phenomena in certain Hemiptera and other forms, with considerations on synapsis and reduction. J. Exp. Zool. **13**:345-349.

Wilson, E. B. 1913. A chromatoid body simulating an accessory chromosome in *Pentatoma*. Biol. Bull. Woods Hole **24**:392-411.

Wilson, E. B. 1925. The Cell in Development and Heredity. The Macmillan Company, New York.

Wilson, F. D., and P. A. Fryxell. 1970. Meiotic chromosomes of *Cienfuegosia* species and hybrids and *Hampea* species (Malvaceae). Bull. Torrey Bot. Club **97**:367-376.

Wilson, G. B., and K. C. Cheng. 1949. Segregation and reduction in somatic tissues. II. The separation of homologous chromosomes in *Trillium* species. J. Hered. **40**:3-6.

Wilson, H. J. 1968. The fine structure of the kinetochore in meiotic cells of *Tradescantia*. Planta **78**:379-385.

Wimber, D. E. 1968. The nuclear cytology of bivalent and ring-forming rhoeos and their hybrids. Amer. J. Bot. **55**:572-574.

Wimber, D. E., and D. M. Steffensen. 1970. Localization of 5S RNA genes on *Drosophila* chromosomes by RNA-DNA hybridization. Science **170**:639-641.

Winge, O. 1940. Taxonomic and evolutionary studies in *Erophila* based on cytogenetic investigations. C. R. Lab. Carlsberg **23**:41-74.

Winiwarter, H. von. 1901. Recherches sur l'ovogenèse et l'organogenèse de l'ovaire des mammifères (Lapin et Homme). Arch. Biol. **17**:33-200.

Wohnus, J. F., and K. Benirschke. 1966. Chromosome analysis of four species of marmosets (*Callithrix jacchus, Tamarinus nigricollis, Cebuella pygmaea*). Cytogenetics **5**:94-105.

Wolf, B., A. Newman, and D. A. Glaser. 1968. On the origin and direction of replication of the *Escherichia coli* K12 chromosone. J. Molec. Biol. **32**:611.

Wolfe, S. L. 1968. The effect of prefixing on the diameter of chromosome fibers isolated by the Langmuir trough—critical point method. J. Cell Biol. **37**:610-620.

Wolfe, S. L., and B. John. 1965. The organization and ultrastructure of male meiotic chromosomes in *Oncopeltus fasciatus*. Chromosoma **17**:85-103.

Wolff, S., and J. A. Heedle. 1968. Some chromosome studies with tritiated thymidine, In W. J. Peacock and P. D. Brock (editors). Replication and Recombination of Genetic Material. Australian Academy of Science, Canberra.

Wollman, E. L., and F. Jacob. 1955. Sur le mécanisme du transfert de matériel génétique au cours de la recombinaison chez *E. coli* K-12. C. R. Acad. Sci. **240**:2449-2451.

Wolstenholme, D. R., and I. B. Dawid. 1968. A size difference between mitochondria DNA molecules of urodele and anuran amphibia. J. Cell Biol. **39**:222-228.

Wood, C. E. 1955. Evidence for the hybrid origin of *Drosera anglica*. Rhodora **57**:105-130.

Wood, T. H. 1967. Genetic recombination in *Escherichia coli*: clone heterogeniety and the kinetics of segregation. Science **157**:319-321.

Woollam, D. H. M., and E. H. R. Ford. 1964. The fine structure of the mammalian chromosome in meiotic prophase with special reference to the synaptinemal complexes. J. Anat. **98**:163-173.

Woollam, D. H. M., E. H. R. Ford, and J. W. Millen. 1966. The attachment of pachytene chromosomes to the nuclear membrane in mammalian spermatocytes. Exp. Cell Res. **42**:657-661.

Wright, J. W., and C. H. Lowe. 1967. Hybridization in nature between parthenogenetic and bisexual species of whiptail lizards (genus *Cnemidophorus*). Amer. Mus. Nat. Hist. Novit. No. 2286.

Wurster, D. H., and K. Benirschke. 1970. Indian muntjac, *Muntiacus muntjak:* a deer with a low diploid chromosome number. Science **168**:1364-1366.

Yamashita, K. 1950. A synthetic complex heterozygote in einkorn wheats: *aegilopoides monococcum.* Proc. Jap. Acad. **26**:66-71.

Yarnell, S. H. 1930. Genetic and cytological studies in *Fragaria.* Genetics **16**:422-454.

Yarnell, S. H. 1931. A study of certain polyploid and aneuploid forms in *Fragaria.* Genetics **16**:455-489.

Yerganian, G. 1957. Cytologic maps of some isolated human pachytene chromosomes. Amer. J. Hum. Genet. **9**:42-54.

Yoshikawa, H., and N. Sueoka. 1963. Sequential replication of *Bacillus subtilis* chromosome. I. Comparison of marker frequencies in exponential and stationary growth phases. Proc. Nat. Acad. Sci. U. S. A. **49**:559-566.

Yotsuyangi, Y. 1962. Etudes sur le chondriome de la levure. J. Ultrastruct. Res. **7**:121-140.

Yunis, J. J., and W. G. Yasmineh. 1970. Satellite DNA in constitutive heterochromatin of the guinea pig. Science **168**:263-265.

Yunis, J. J., and W. G. Yasmineh. 1971. Heterochromatin, satellite DNA, and cell function. Science **174**:1200-1205.

Zang, K. D., and E. Back. 1968. Quantitative studies on the arrangements of human metaphase chromosomes. I. Individual features in the association pattern of the acrocentric chromosomes of normal males and females. Cytogenetics **7**:455-470.

Zimmering, S. 1955. A genetic study of segregation in a translocation heterozygote in *Drosophila.* Genetics **40**:808-825.

Zinder, N. D. 1953. Infective heredity in bacteria. Cold Spring Harbor Symp. Quant. Biol. **18**:261-269.

Zinder, N. D., and J. Lederberg. 1952. Genetic exchange in *Salmonella.* J. Bact. **64**:679-699.

Zingmark, R. G. 1970. Ultrastructural studies on two kinds of mesocaryotic dinoflagellate nuclei. Amer. J. Bot. **57**:586-592.

Zohary, D., and M. Feldman. 1962. Hybridization between amphidiploids and the evolution of polyploids in the wheat *(Aegilops-Triticum)* group. Evolution **16**:44-61.

Zweidler, A. 1964. Die Replikation der chromosomen von *Allium cepa.* Arch. Klaus-Stift. Vererb. **39**:54-64.

Artemia
  parthenogenesis, 163
  polyploidy, 164
Ascaris, 44
Aspergillus, 53
  nonrandom segregation, 68
  somatic recombination, 60
Aster
  basic chromosome numbers, 145, 146
  chromosome evolution, 146
Asymmetrical cell division, 48
Asynapsis, 75, 101-102
Aulacantha, chromosomes, 56, 157
Autopolyploidy, 164-166

B

Bacteria
  conjugation, 286
  cotransduction, 291
  cytogenetics, 276-279, 283-291
  genes in action, 284, 285
  lysogeny, 290
  meromixis, 287
  merozygote, 287
  prophage, 290
  sexduction, 287
  sex factors, 287-289
    F^+ and Hfr, 287
  transduction, 289
  transformation, 286
Balanced lethals, 183-188
  inversion, 234
Balbiani rings, 26, 36, 37
Band heterozygosity, 29, 37, 239
Barr body
  general occurrence, 106
  Lepidoptera females, 106
  in XO and XY, 106, 172
Basic chromosome numbers, 144
B-chromosomes; see Supernumerary chromosomes
Biscutella, 167
Blood chimeras, 16
Bobbed locus, 11, 18
Bolbe nigra, 77, 101
Bombyx, 162
Bothriochloa-Dichanthium, 72, 170, 246, 248
Bouquet, 75, 76, 77
Bouteloua
  agamic complex, 248
  aneuploids, 154
Brachycome, 210
Bridge-fragment anaphase, 201-204
Bufo
  DNA amounts, 15
  x *Rana,* "hybrid", 219

C

C amounts of DNA, 13, 16
Calamagrostis, apomict, 243, 246
Cell hybrids
  human × mouse, 275
  somatic reduction, 68
Centric fission, 140
Centric fusion, 139
  beetles, 140

Centric fusion—cont'd
  grasshoppers, 140
  true, 140
Centrioles, non-function in mitosis, 43
Centromere, 43, 114, 121-125
  diffuse, 44
  kinetochore, 43
  non-replicative in mitosis, 38
Chaenactis
  diploid decrease, 191
  dysploid speciation, 191
Chiasmata, 85, 87-89
  vs. crossovers, 84
  genetic control, 99, 101
  localization, 99
Chironomus
  nucleolus organizers, 8
  salivaries in development, 36
Chromatin, 12-20, 119
  lengths, 15
Chromomeres, 75, 76
Chromosome
  arm ratio, 211
  centromeric index, 212
  knobs, 213
  relative length, 211
  structure, 113-127
Chromosome numbers, 135-147
  basic (x), 144
    Angiospermae, 147
    Astereae, 147
    determination, 146
    examples, 145-147
  centric fusion, 139
  constancy, 142
  Drosophila, 142
  evolution of, 144-147
  grasshoppers, 139, 140
  intraindividual variation, 140, 141, 143, 144
  Poa, 142
  polymorphism, 142, 212
  Robertsonian changes, 137; see also Robertsonian changes
  sizes, 136
  spectacular changes, 141
Chromosome polymorphism; see also Chromosome numbers; Robertsonian changes
  Claytonia, 142
  Euscirtus, 143
  fish, 140, 143
  Peromyscus, 145
  Salmo, 141
  snail, 141
  Sorex, 137, 140
Chromosomes
  acentric fragments, 92
  AI forms, 92
  differential mutability, 126
  diffuse centromere, 44, 114, 162, 170-171, 210
  end associations, 64
  holokinetic, 44, 162, 170-171, 210
  interphase, structure, 13
  isochromosomes, 153, 155, 268
  nonrandom segregation, 51
  quadriradial, 269

Chromosomes—cont'd
  replicating
    condensed, 14
    diffuse, 14
  silent regions, 126
  somatic pairing, 50
  somatic reduction, 18, 68, 69, 98, 172
  strandedness, 15, 73, 119, 120
  supernumerary, 128-134
    in animals, 132
    boosting mechanism, 130, 131
    Clarkia, 133
    Crepis, 130, 131
    Drosophila, 133
    effects, 132
    gene content, 128
    Haplopappus, 128, 129, 132, 134
    heterochromatin, 131
    increase mechanisms, 129, 130, 131
    Lolium, 133
    nondisjunction, 129
    occurrence, 132
    origins, 133
    parasites, 128, 132, 134
    preferential fertilization, 130
    Tradescantia, 129, 130, 131, 134
    translocation, 133
    types, 130
    Zea, 128-134
  telocentric, 115, 116, 140
    centric fission, 140
    in centric fusion, 139
    in evolution, 140
    in Robertsonian change, 137-141
  telophase decondensation, 47
  unicentric, 44
Citrus, apomict, 243
Clarkia
  C. dudleyana, 217
  interspecific hybrids, 193, 224
  translocations, 181
Claytonia, 119, 142, 143
Cnemidophorus, 164
Colchicine
  autopolyploidy, 165
  effects, 43, 45, 253
  polyploid *Chlamydomonas*, 67
  somatic pairing, 63
  spindle inhibition, 45
  in technique, 253
  unilateral chromatid replication, 70
Collinsia, aneuploids, 153
Complex heterozygotes, 183-187
  Isotoma, 183
  Oenothera, 184
  Raimannia, 236
  Rhoeo, 186
Complexes, 183-187
Crepis
  apomixis, 243, 246, 247
  basic number, 145
  chromosomal evolution, 145
  chromosome end associations, 65
  chromosome numbers, 142, 145
  dysploid decrease, 156, 188

Crepis—cont'd
  karyotypes, 209
  nucleolus inhibition, 12
Crossovers, theoretical number, 84
Cuthbertia, 165
Cycad root cells, chromosome reduction, 18, 68, 69, 98
Cytogenetics, history of, 1-6
Cytogenetics of hybrids, 215-232
Cytology, history of, 2

D

Dactylis, 167
Dalton, 279
Daphne, somatic pairing, 60
Datura
  haploidy, 149
  translocation, 180, 182
  trisomy, 153
Deletions
  Drosophila, 32
  human, 267
  viral, 281-283
  Xenopus, 11
Delphinium, 167
Didymuria, 144
Differentiation in G_1, 40
Dinoflagellate mitosis, 55
Diploidization, 166
  Anemone, 166
  Gilia, 136
  wheat, 166
  Zea, 166
Diplopolyploidy, 237-239
  dog roses, 237
  Leucopogon, 237
  Miastor, 239
Diplospory, 242, 244
Directed segregation, 179
Distributive pairing, 64, 66
Dosage compensation, 106
DNA
  amounts, 136
  amplification, 9-12, 16-18, 26, 27, 36, 272
    gene, 17
    whole chromosome, 16
  cytoplasmic, 17
  diminution
    chromosomal, 18
    partial chromosomal, 18
    whole nuclei, 18
  hybridization, uses, 13
  inconstancy, 15
  informational (I-DNA), 18
  puffs, 17
  redundant, 20
  replication, 20
    rates, 15
  ribosomal, 8
  ribosomal redundancy, 8, 9
  specific variations, 136
  variable amounts, 15
    in fish, 19
Dosage compensation, 261
Down's syndrome, 265
Draba, 192

Drosera, 225
Drosophila
  asynapsis, 107
  attached X, 110
  basic chromosome number, 144
  bobbed locus, 11, 18
  chromocenter, 17
  chromosome numbers, 142
  disjunction, 180
  dysploid decrease, 144, 188, 213
  dysploid increase, 193
  gene magnification, 18
  haploids, 149
  inversions, 32-36, 196-198, 204-206
  meiotic genes, 99, 101
  minutes and pairing, 59, 60
  monosomic, 155
  nucleolus organizers, 8, 28
  nucleolus organizer redundancy, 8, 9, 11
  obligate heterozygosity, 234
  parthenogenesis, 164
    inversions, 234
  polytene chromosomes, 24-36
  primary nondisjunction, 109
  sex determination, 104-110
  sex ratio, 111
  somatic recombination, 51, 59, 61
  spermateliosis, genes repressed, 179
  spermiogenic failure, 111
  supernumerary chromosome, 133
  triploidy, 159
  XO males, 11
Dysploid increase
  Allium, 195
  Clarkia, 193
  Draba, 192
  Drosophila, 193
  evolutionary, 192-195
  Gilia, 192
  Nicotina, 192
  translocations, 193-195
  without translocations, 192
Dysploidy, 156, 188-195; *see also* Translocations, dysploidy

E

Elymus
  glaucus, 216, 217
  rechingeri, 212
Endopolyploidy, 171-173
Epigenesis, 2
Episome, 28, 276, 287, 289
Erophila (Draba), 192
Escherichia coli, 278, 280, 283-291
  ribosomal DNA cistrons, 11
Euchromatin, 19
Exchange; *see* Translocations
Exchange pairing, 64, 66

F

Folded fiber model of chromosome, 119
Founder species, 144
Fragaria
  evolution, 231
  interspecific hybrids, 231

Fragaria — cont'd
  triploidy, 160
  unreduced gametes, 232
Fragile site, 272
  endoreplication, 272
  gene amplification, 17
Fragment, acentric, 45
Fundamental number, 139
Fungi
  mitotic model, 52-55
  parasexual cycle, 60, 61
  premeiotic pairing, 61
  somatic recombination, 60, 67

G

G_1 period, 13
G_2 period, 13
Galax, 165
Gene activity visualized, 10, 90, 284, 285
Gene amplification, 9-12, 16-18, 26, 27
  in humans, 17, 273
Gene magnification, 18
Genes
  meiotic, 74, 98-102
  regulatory, 13
  repression, 19
  silent, 126
Genetic drift, 212
Genetic inactivity in spermateliosis, 179
Genetics, history of, 1
Genome segregation, 67
Genophores, 277
  bacteria, 277, 278, 280, 288
    "origin", 287
  lengths, 278, 280
  mitochondria and plastids, 280, 291, 292
  multiple, 278
  permuted, 281
  prokaryon, 277
  recombination
    bacterial, 289
    viral, 281
  replication, 277
  separation, 278
  supercoiling, 290
  unwinding rate, 278
  virus, 279
    incomplete (satellite), 299
    multiple, 278
Gerbillus, Robertsonian change, 137
Gerris
  Barr body in XO male, 172
  endopolyploidy, 172
Giemsa heterochromatin technique, 20, 121, 274
Gilia
  allodiploid, 136
  dysploid increase, 192
  interspecific hybrids, 136
  recombinational speciation, 136
Gonomery, 50
Gossypium
  evolution, 221-223, 227
  genomes, 188, 221-223, 227
  hybrids, 221-223, 227
  translocations, 188

Grasshoppers
 chromosome changes, 139
 disjunction, 180
 intraspecific hybrids, 220
Grindelia, 225
Gynandromorphs, 144

H

Haploidy, 148-151
 abnormal, 149
 Datura, 149
 gametophytes, 149
 meiosis in, 151
 Nicotina, 150
 nonhomologous pairing, 65, 76, 151, 160
 normal, 148
 polyhaploidy, 66, 72, 148, 170, 246
 synaptinemal complex, 151
 tomato, 149, 151
 values, 150
 wheat, 149, 151
 Zea, 150, 151
Haplopappus gracilis
 chromosome numbers, 147
 dysploid decrease, 189
 mosaics, 144
 somatic pairing, 62, 63
 supernumerary chromosomes, 128, 129, 132, 134
Heterochromatin, 19
 and histones, 19
 constitutive, 20, 121, 274
 facultative, 20
 genetically inactive, 20
 H-arms, 140
 late replicating, 19
 redundancy, 20
 staining methods, 20, 274
Heteroduplex visualized, 282
Heterozygosity obligate, 233
 amphiploid, 236
 asexuality, 237
 band heterozygosity, 29, 37, 239
 diplopolyploidy, 237-239
  dog roses, 237
  Leucopogon, 237
  Miastor, 239
 inversions, 234
  Drosophila, 234
  white-throated sparrows, 234
 permanent, 233-240
 types, 233
 translocation, 174-195, 235, 236
  Isotoma, 183
  Oenothera, 184-186, 235, 236
  Rhoeo, 186, 187, 236
Heterozygous puffs, 29, 30
Histone replication, 13
Holokinetic chromosomes, 44, 114, 162, 170-171, 210
Human cytogenetics, 252-275
 aging, effects of, 265, 273
 chromosome data, 258
 chromosome numbers, 252
 cytogenetic anomalies, 262
  autosomal, 264-275
  deletions, 267

Human cytogenetics—cont'd
 cytogenetic anomalies—cont'd
  deletions—cont'd
   cri du chat, 267
   Philadelphia, 267
  sex chromosome, 262-264
 dosage compensation, 261
 Down's syndrome, 265
 fragile site, 272
 genetic engineering, 271
 heterochromatin
  constitutive, 255, 274
  facultative, 261
 isochromosomes, 268
 karyotype, 254
  Denver, London, Chicago, 254
  non-Denver, 256-258
  satellites, 255, 256
  technique
   autoradiography, 255
   fluorescence, 274
   Giemsa, 274
   Moorhead, et al., 253
  variations, normal, 256
 Klinefelter's syndrome, 262
 mapping genes, 273
 meiosis, 258-260
  oogenesis, 258
  sex chromosomes, 260
  spermatogenesis, 259
  synaptinemal complex, 259
 monosomics, 267
 multiple X females, 264
 oogenesis, 258, 259
 quadriradial chromosomes, 269-271
 replication, 255
 ring chromosomes, 271
 sex chromatin (Barr) body, 260
  dosage compensation, 261
  drumstick, 261
  frequency, 260
  nature, 261
  sexing, 262
 spermatogenesis, 259
 synaptinemal complex, 259
 transduction, 275
 translocation Down's syndrome, 266
 translocations, 265-267
 triploidy, 269
 trisomies
  21, 265
  D-group, 265
  E-group, 265
  Turner's syndrome, 264
  X chromosome, 255, 256, 259
  XX males, 264
  XY females, 264
  XYY syndrome, 264
  Y chromosome, 255, 256, 259
Hybridization of nucleic acids, 8, 9, 13, 281, 282
Hybrids of animals, 161-164, 218-221
 Ambystoma, 162
 Bombyx, synthetic, 162
Hybrids, interspecific
 animals, 218-221

Hybrids—cont'd
  animals—cont'd
    Bufo × *Rana*, 219
    Chironomus, 37
    Drosophila, 32, 34, 220
    mosquitoes, 32, 34
  diploid × diploid, 218-225
    Chaenactis, 191
    Clarkia, 193, 224
    Crepis, 188
    Datura, 180
    Delphinium, 167
    Gilia, 136, 192
    Gossypium, 221-223
    Grindelia, 225
    Haplopappus, 189
    Tradescantia, 204
    wheat, 180
  diploid × tetraploid, 225-227
    Drosera, 225
    Nicotina, 192
    Oryzopsis, 227
    wheat, 226
  hexaploid × diploid
    potato, 229
  hexaploid × hexaploid
    Spartina townsendii, 231
    wheat × *Agropyron*
  hexaploid × tetraploid
    wheat, 230
  octoploid × diploid
    Fragaria, 231
  pentaploid × tetraploid
    potato, 229
  tetraploid × tetraploid
    Agropyron, 278
    wheat, 228
Hybrids, intraspecific, 216-232; *see also* Parthenogenesis
  Clarkia, 193, 216, 217
  Drosophila, 32, 35
  Elymus glaucus, 216, 217
  Galeopsis, 217
  grasshoppers, 180, 220
  Hyla, 163
  Moraba, 180, 220
  Rhoeo, 186
Hymenoptera, haploid males, 148

I

Idiogram, 207
I-DNA, 18
Interkinesis, 72
Interphase, 7-22
  DNA amounts, 13, 16
  heterochromatin, 19, 58, 63
  vs. postmitotic, 7
  premeiotic, 51, 58, 73, 95, 96
  subdivisions of, 13
  without DNA replication, 18, 68, 69, 172
Introgressive hybridization, 215
Inversions, 32-36, 196-206
  anaphase bridges, frequencies, 201, 204
  anaphase results, 203
  crossover types, 201-204

Inversions—cont'd
  Drosophila, 196-198, 204-206
  evolution study, 204-206
  formation, 196
  loops, 34, 35, 197, 199-204
  meiosis, 199-204
  multiple, 34, 35
  occurrence, 32-36, 205
  Paeonia, 205
  pairing and synapsis, 198-202
    anaphase bridges, 201-204
    crossing-over, 199-204
    pairing failure, 200, 201
    reverse synapsis, 200, 201
    rod pairing, 200, 201
  paracentric, 197
  Paris, 205
  pericentric, 197
  salivary gland chromosomes, 32-36, 197, 204-206
  Zea, 198-204
Isochromosomes, 153, 155, 268
I-somes, 18
Isolabeling, 22
Isotoma, 183

K

Karyotypes, 207-214
  asymmetrical, 210
  chromosome data, 211
    number, 208
  conservative, 209
  Crepis, 209
  evolutionary changes, 213
    dysploidy, 192-195, 213
    inversions, 196-206, 213
    knobs, 213
    translocations, 174-195
  Sagittaria, 207
  satellite, 212
    genome repression, 12
    on long arm, 212
    numbers, 212
    on short arm, 212
  symmetrical, 210
  Yucca-Agave, 210
Kinetochore, 43
Klinefelter's syndrome, 262

L

L chromosomes, 17
Late replicating chromatin, 19
Lethals, balanced, 183-188; *see also* Balanced lethals
Leucopogon
  diplopolyploidy, 237-239
  unipolar spindle, 46
Lilium
  mostly adjacent disjunction, 180
  theoretical number of crossovers, 84
  triploidy, 159
Luzula, 44, 75
  agmatoploidy, 171
  autopolyploidy, 165

M

Maize; *see Zea*

Meiocytes, 73
Meiosis, 72-102
  anaphase I, 92
    bridges without fragments, 100
    bridges without inversion, 100
    lagging univalents, 93
  anaphase II, 93
  apomeiosis in apomicts, 245
  bouquet, 75, 76, 77
  bridge-fragments, 92
  classical scheme, 73-94
  diakinesis, 90
  diplotene, 87-90
    diffuse stage, 88
    lampbrush chromosomes, 88-90
    repulsion, 87, 88
  disjunction, 92
    adjacent-alternate, 93
  DNA replication, 73, 75, 85
  environmental effects, 98
  evolution of, 96-98
  failure with fertilization, 72
  without fertilization, 72
  haploids, 151
  human, 258-260
  in hybrids; see Hybrids
  interkinesis, 93
  interlocking, 95
  leptotene, 75
  metaphase I, 92
  metaphase II, 93
  multiple spindles, 100
  neoclassical scheme, 94-98
  nondisjunction, 92
  pachytene, 76
    time of recombination, 84
    time of the SC, 77-85
  postreduction, 112
  precocity theory, 60, 97
  preleptotene, 75
  premeiotic interphase, 73, 95, 96
  premeiotic pairing, 51, 57-64, 94
  prereduction, 112
  Prometaphase, 90
  prophase I, 75
    not like mitotic prophase, 95-96
  prophase II, 93
  pseudoassociations, 102
  quasibivalents, 102
  recombination similar to somatic, 84, 98
  repair DNA, 85
  synizesis, 76
  telophase I, 93
  telophase II, 93
  unipolar spindles, 46
  zygotene, 75
Meiotic drive, 92, 111
  Drosophila, 112
  maize, 111
  Musca, 112
Meiotic genes, 74, 98-102
  Claret of *Drosophila*, 101
  divergent spindle, 100
  maize, 101
  Neurospora, 99

Melandrium, sex determination, 105, 111
Mesosomes, 49, 50
Metakinesis, 43
Micronucleoli, 17
Mitochondria, cytogenetics, 291-294
  corn leaf blight, 293
  petite, 292, 294
  pokey, 292
Mitochondria and plastids
  genophores, 280, 291
  mitochondrial genetics, 292, 293
  as prokaryotes, 291
Mitosis, 38-56
  anaphase, 45
  asymmetrical division, 48
  centriolar reproduction, 47
  centromeres, 44
  C (colchicine), 45, 70, 253, 272
  cytokinesis, 47
  dinoflagellate, 55
  errors, 40
  fungal, a model, 52-55
  interphase, 38
    preparation for, 40
  in maize, 39
  metaphase, 45
  nondisjunction, 45
  nonrandom segregation, 51
  nuclear envelope formation, 46
  nucleolar formation, 46
  persistent nucleoli, 42
  postmitotic, 38
  premeiotic pairing, 51, 57-64, 94
  prokaryotic, 48-50
  prometaphase, 43
  prophase, 41-43
    centrioles, 42
    chromosomal arrangement, 41, 64, 65
    chromosomal condensation, 41
    microtubules, 42
    nuclear envelope, 41, 42
    nucleolus, 42
  restitution nuclei, 48, 244
  somatic pairing, 50, 51, 57-64
  spindle convergence, 45
  telophase, 46
  "track" mitosis of fungi, 54
Mitotic chromosomes, 113-127
  acentric, 114
  acrocentric, 114, 115
  centromeres, 114, 121-125
  characterization, 114, 256-258
    arm ratio, 114, 258
    centromeric index, 114, 258
    relative length, 114
  cytogenetic uses, 115
  dicentric, 114
  fine structure, 119-121
  folded fiber model, 119
  gross structure, 113-119
  holokinetic, 44, 114, 162, 170, 171, 210
  isochromosomes, 153, 155, 268
  metacentric, 114, 115
  monokinetic, 114
  nucleolus organizers, 7-12, 116, 255, 256

Mitotic chromosomes—cont'd
  polycentric, 44, 114, 162, 170, 171, 210
  primary constriction, 114
  satellites, 116, 117, 255, 256
  secondary constrictions, 114, 117, 255
  subacrocentric, 115
  submetacentric, 114, 115
  telocentric, 115, 116, 140
  types
    autosomes, 125
      A, 128
      B, 128
    heterosomes, 125
    sex, 103-112, 125
    supernumerary, 125, 128-134
  unicentric, 114
  variations, 117-119
Mitotic cycle, 13, 40
Mitotic metaphase chromosomes, 113-127
Monosomics
  Drosophia, 155
  human, 155, 267
  uses, 155, 156
  wheat, 155
Moraba, 237
  intraspecific hybrids, 180, 220
  translocations, 180
Mosaics, 144
  human, 261-264
Mosquitos
  achiasmatic X, *golden* gene, 101
  chromosome number, 209
  endopolyploidy, 172
  gut lining cells, 69, 98, 172
  salivaries, 32-35
  sex determination, 104
  translocations, 176-179
Muntiacus, 141, 142, 192

N

Neocentromeres, 44, 92, 100
Neottiella, 73, 77, 80, 82, 83, 85
Neurospora
  meiotic genes, 99
  mitochondrial genetics, 292, 293
  pokey mutant, 292
  postmeiotic mitosis, 94
  somatic recombination, 61
Nicotiana, interspecific hybrids, 224
Nondisjunction
  meiotic, 92
  mitotic, 45, 46
  supernumerary chromosomes, 129
Nucleolar genes
  in action, 10, 11
  redundancy, 9
Nucleoloids, 17
Nucleoli, 7-12, 28
  amphibian oocyte, 9-11
  cells without, 46
  extrachromosomal, 8-11
  organization, 8
Nucleolus organizer, 7-12, 116, 255, 256
  of *Chironomus*, 8
  in embryos, 9

Nucleolus organizer—cont'd
  function, 8
  genetic control of, 11, 12, 18
  locations of, 7
  in micronuclei, 12
  numbers, 116
  positions, 116
  redundancy, 8-11, 18
    ontogenetic change, 18
  related to nucleoli, 12
  of *Rhynchosciara*, 28
  sizes, 116
Nucleus, 7-22
  poles and antipoles, 41

O

Odontophrynus, 162, 163
Oenothera
  balanced lethals, 184
  complex heterozygosity, 184-186
  evolution of complexes, 184, 185
  Raimannia, species pairs, 236
  translocations, 182-186
    disjunction, 180, 184
Oncopelteus
  chromosome fine structure, 121
  holokinetic chromosomes, 171
Ophioglossum, 157
Ornithogalum
  asynapsis by cold, 98
  somatic pairing, 51

P

Paeonia
  inversions, 205
  translocation rings, 183
Pairing; *see also* Premeiotic pairing; Somatic pairing
  exchange and distributive, 64, 66
  factors affecting, 76
  in haploids, 65, 76, 151, 160
  nonhomologous, 65, 76, 151, 160
  premeiotic, 51, 57-64, 94
  somatic, 50, 51, 57-64
  vs. synapsis, 94
Panorpa, 77, 86, 87
Parasexuality, 61, 98
  human cells, 274
Paris, inversions, 205
Parthenogenesis
  amphibians, 162
  Artemia, 164
  Drosophila, 164
  fish, 162
  lizards, 162, 164
  plants, 245
  weevils, 164
  Zea, 165
Partial chiasmatype theory, 85
Pediculus, premeiotic pairing, 57
Pentaploidy, 161
Peromyscus, chromosome forms, 139, 143
Petunia, aneuploidy, 152
Plantago
  chromosomal evolution, 145
  premeiotic pairing, 58, 63

Poa
  agamic complex, 247
  apomicts
    nervosa, 243
    pratensis, 247
  chromosome numbers, 142
Podocarpaceae, 139
Poeciliopsis, 150, 162
Polyhaploidy, 66, 72, 148, 170, 246
Polymorphism, balanced, 233
Polyploidy, 157-173
  agmatoploidy, 170
  allopolyploidy, 167
  amphibians, 162-164
  amphidiploid, 166
  in animals, 159, 161-164
    parthenogenesis, 162-164
    vs. sex chromosomes, 161
    sexual, 162
  Artemia, 163, 164
  autopolyploidy, 164-166
    meiosis, 165
    physiology, 165
    production, 165
  autoallopolyploidy, 168
  diploidization, 166
  endopolyploidy, 171
  in evolution, 173
  genomic allopolyploidy, 168
  irreversible, 170
  lizards, 162, 164
  parthenogenesis, 163
  pentaploidy, 161
  Poeciliopsis, 162
  polyhaploidy, 170
  Saccharum, 169
  segmental allopolyploidy, 167
  triploidy, 157-164
    meiosis in, 160
  very high, 157-159, 169
  weevils, 164
Polymitotic gene of maize, 18
Polyteny, 16, 17
Polytene chromosomes, 23-37
  asynapsis, 34
  Balbiani rings, 26, 36, 37
  bands and interbands, 29, 30
  bands as phenotypes, 34
  chromocenter, 27
  deletions, 32
  in developmental studies, 36, 37
  discovery of cytogenetic value, 31
  DNA puffs, 17, 26, 27
  evolution studies, 32, 35, 36, 204, 205
  gene amplification, 17
  Hawaiian species, 36
  in hybrids, 24, 34, 35, 37, 197
  inversion loops, types, 34, 35
  inversions, 32, 197
    D. azteca, 197
    D. pseudoobscura, 32
  maps, 31, 33
  mosquito, 32-35
  in non-Diptera, 23
  nucleic acid hybridization, 8

Polytene chromosomes—cont'd
  nucleolus organizer, 28, 29
  origin and development, 24
  pairing, 26
  puffs, 26, 34, 36
    DNA, 17, 26, 27
    heterozygous puffs, 29, 30
  replication in, 27
  sizes, 23, 26
  structure, 26
  super giants, 25, 26
  telomeres, 27
  translocations, 32
  variable pairing, 24, 26
Postmitotic cells, 15, 16
Potato
  evolution, 167
  interspecific hybrids
    hexaploid × diploid, 229
    pentaploid × tetraploid, 229
  polyhaploid, 170
  triploidy, 159
Precocity theory of meiosis, 4, 60, 97
Preferential distribution, 129, 130
Preferential fertilization, supernumerary chromosomes, 130
Premeiotic interphase, 95, 96
  DNA replication, 73
  prochromosomes, 58, 59
Premeiotic pairing, 51, 57-64, 94
  in animals, 57-59
  in fungi, 61
Premeiotic replication
  Chylamydomonas, 73
  Gelasinospora, 73
  Neottiella, 73
Prochromosomes, 19
  pairing, 58, 59
Prometaphase, 43
Prophage, 281, 290
Pseudogamy, 243
Puffs, 26, 29, 30, 34, 36
  DNA, 17, 26, 27
  heterozygous, 29, 30
  RNA, 26, 29, 34, 36

Q

Quadriradial chromosomes, 269

R

Rana, 219
Recombination
  index, 135
  meiotic, 84-87
  meiotic, like somatic, 95
  somatic, 51, 60, 61
  time of meiotic, 64
Recombinational speciation, 136
Redundancy, genetic, 9, 13, 20
Replication
  of DNA, 15, 21
  rates of, 15
  semiconservative, 21
Restitution nuclei, 48, 244

Rhoeo
    bivalent pairing form, 187
    circle of 12 chromosomes, 186, 187, 236
    complex heterozygote, 186-187
    contrasted to *Oenothera*, 186
    meiosis
        in autotetraploid, 187
        in bivalent pairing form, 187
    species pairs, 236
Rhynchosciara, 26-29, 37
    cryptopolyteny, 17
    in development, 37
    polytene chromosomes, 26-29
    super giant salivaries, 25, 26
RNA, 8
    5s, 8
    nucleolar, 8, 9
    puffs, 17, 26, 29, 34, 36
Robertsonian changes, 137, 139
    crocodilians, 140
    grasshoppers, 139
    Gerbillus, 137
    mice, 139
    Podocarpaceae, 139
    snail, 141
    Sorex, 137, 138
    Thais, 137
Rosa canina
    diplopolyploidy, 237-239
    unipolar spindle, 46
Rubus, 245, 248
Rumex, sex chromosomes, 111

S

Saccharum, 169
Sagittaria, 119
    karyotype, 207, 209
Salivary gland chromosomes; *see also* Polytene chromosomes
    band heterozygosity, 29, 37, 239
    first cytogenetic use, 4, 30
Salmo, 141
Sacrophaga, 17
Satellites, chromosomal, 116, 117, 256
Schizaea, 157
Sciara
    echromosome elimination, 18
    gene amplification, 17
    heterozygous puff, 29, 30
    unipolar spindle, 46
Secale, supernumerary chromosomes, 130, 131, 134
Secondary association, 59, 146
Segregation, nonrandom, 51, 68
    hypotheses, 69
Selective autogamy, 234
Semiconservative replication, 21
Semigamy, 245
Semisterility, 179
Sex chromatin body, 106, 172, 260; *see also* Barr body
Sex chromosomes, 103-112
    complex systems, 108, 109
    Drosophila Y, 105, 107
    evolution, 106-110
    historical, 103
    mongoose, 109

Sex chromosomes — cont'd
    plants, 106, 110, 111
    polymorphisms, 109
    XY to XO, 105, 107
Sex determination, 103-112
    human, 105
    Hymenoptera, 106
    Melandrium, 105, 177
    in mosquitos, 104
    Musca, 107
    reversal in fish, 106
    vs. sex production, 104
Sex ratio, 93, 111
Silent genes, 126
Sister chromatid exchange, 21
Solanum; *see* Potato
Somatic pairing, 50, 51, 57-64
    in animals, 57-60
    in fungi, 60
    in plants, 60-64
Somatic pairing and meiosis, 64, 98
Somatic recombination, 51, 97, 98
    in fungi, 60, 61
Somatic reduction, 66-68, 98, 172
Sorex, Robertsonian change, 119, 137, 140
Sorghum
    genome segregation, 67
    somatic reduction, 67
Spalax, Robertsonian change, 119, 137
Spartina townsendii, 231
Speciation, recombinational, 136
S-period, 13
Spindle, 39, 43, 45, 46, 53
    multipolar, 100
    plaques, 53
    unipolar, 46
Substitution line, 156
Supernumerary chromosomes; *see* Chromosomes, supernumerary
Synapsis, 76-87, 94-98
    classical, 76
    in haploids, 76, 151
    neoclassical, 94, 96
    vs. pairing, 94
    somatic, 57-71, 94
Synaptinemal complex, 77-87, 94
    in achiasmatic meiocytes, 101
    crossover frequencies, 77, 84
    function, 77, 84, 94
    in haploids, 151
Synizesis, 76

T

Telocentric chromosomes, 115, 116, 140
Thais, 137
Thyantor, 163
Tomato
    aneuploidy, 152
    chromosomal silent regions, 126
    differential mutability, 126
    haploid, 149
    trisomics, 154, 155
Townsendia, 246
Tradescantia
    anaphase bridges, frequency, 204

Tradescantia—cont'd
  supernumerary chromosomes, 129, 130, 131, 134
  triploidy, 160
Transcription to translation, summary, 13
Transduction, 275, 281, 289
Transformation, 286
Translocations, 174-195
  adjacent disjunctions, 179
  alternate disjunction, 179
    chiasmata, 182
    mechanisms, 181, 182
    requirements for, 180
    ring sizes, 182
  balanced lethals, 183-188
  cockroaches, 183
  complex heterozygosity, 183-188
  directed segregation, 179
  disjunction, 179-182
    adjacent, 179
    alternate, 179
    genetic control, 180
  Down's syndrome, 266
  Drosophila, 180, 182, 188, 193
  dysploid decrease, 144-147, 188-192
    Chaenactis, 191
    Crepis, 188
    Drosophila, 144
    Haplopappus, 189
    Muntiacus, 141, 192
  dysploid increase, 192-195
    Clarkia, 193
    Drosophila, 193
  end arrangements, 183
  formations, 174, 175
  Gossypium, 188
  grasshoppers, disjunction, 180
  Grindelia, 225
  human, 265-267
  Isotoma, 183
  mitotic pairing, 178
  mosquitoes, 176-179
  Oenothera, 184-186
  Paeonia, 183
  reduced crossing-over near, 176
  Rhoeo, 186
  rings, large, 182
    disjunction, 180-182
    semisterility, 179
    species, 182
  wheat
    disjunction, 180
    rings of 4 to 14, 180, 182
Trillium
  allocycly, 121
  chromosomes, 211
  chromosome size, 117
  and SC, 102
Triploidy, 157
  animals, 161, 195
  human, 269
  meiosis in, 160
  origins, 159
  parthenogenetic animals, 161-164
Trisomic ratio, 154
Trisomy, 153-155

Trisomy—cont'd
  meiosis, 154
Triticum; see Wheat
Turner's syndrome, 264

U

Unilateral chromatid replication, 70
Unique genes, 13, 30
Ustilago
  mitosis, 42
  somatic recombination, 61

V

Verticillium, somatic recombination, 60, 67
Vicia
  chromosome fine structure, 120
  DNA
    amounts, 16
    variation, 136
  mitotic metaphase chromosomes, 210
Virus
  cytogenetics, 279-283
  deletions, 283
  genophore, 277, 281-283
    incomplete (satellite), 279
    multiple, 278
  genophore substitution, 282, 283
  heteroduplex, 281
  prophage, 281
  reaction cores, 283
    transcribing, 11
  recombination, 281
  satellite, 279
  temperate phage, 290
Vivipary, plant, 242

W

Wheat (including *Aegilops*)
  Agropyron hybrids, 231
  disjunction, 180
  evolution, 223, 226, 228, 230
  5B locus, 63, 155, 224
  genomes, 223, 226, 228, 230
  haploid, 149
  hybrids, 223, 226, 228, 230
  interspecific hybrids
    diploid × diploid, 223
    diploid × tetraploid, 226
    tetraploid × tetraploid, 228
  isochromosomes, 155
  meiotic genes, 99
  monosomics, 155
    uses, 156
  nucleolus organizer variation, 12
  nullisomics, 155
  pairing genes, 63
  polyhaploid, 170
  substitution lines, 156
  telocentric chromosomes, 115
  translocation rings, 180, 182
White-throated sparrows, 144, 234

X

Xenopus
  nu deletion, 11

Xenopus—cont'd
   nucleolus organizer redundancy, 8, 9, 11
XY females, 264
XYY syndrome, 264

Y

Yeast mitochondrial genetics, 292
Yucca, 51
Yucca-Agave, 210

Z

Zea
   autopolyploidy, 165, 166
   chromosomal silent regions, 126
   diploidization, 166

Zea—cont'd
   haploid, 150
   inversions, 198-204
   monosomy, 155
   neocentromere, 100
   nonhomologous pairing, 65
   number of translocations, 176
   pachytene cytogenetics begun, 4
   parthenogenesis, 165
   polymitotic gene, 101
   supernumerary chromosomes, 128-134
   telocentric chromosomes, 116
   translocation alternate disjunction, 100
   trisomics, 154, 155